# The Portable
# Pediatrician

{SECOND EDITION}

Laura Walther Nathanson, M.D., FAAP

Quill
A HarperResource Book
*An Imprint of* HarperCollins*Publishers*

THE PORTABLE PEDIATRICIAN. Copyright © 2002 by Laura Walther Nathanson, M.D., FAAP. All rights reserved. Printed in the United States. No part of this book may be used or reproduced in any manner whatsoever without written permission except in the case of brief quotations embodied in critical articles and reviews. For information address HarperCollins Publishers Inc., 10 East 53rd Street, New York, New York 10022.

HarperCollins books may be purchased for educational, business, or sales promotional use. For information, please write to: Special Markets Department, HarperCollins Publishers Inc., 10 East 53rd Street, New York, New York 10022.

A paperback edition was published in 1994 by HarperCollins under the title *The Portable Pediatrician for Parents*.

SECOND EDITION

Library of Congress Cataloging-in-Publication Data has been applied for.

ISBN 0-06-093847-1

04  05  06  WB/RRD  10  9  8  7  6  5  4  3

*Book design by Jennifer Ann Daddio*
*Illustrations by Kay Life*

To Chuck and Sara, with love.

# Acknowledgments

In the First Edition, I said "Thank You" to the gigantic village that raised this particular pediatrician. This time, I'd like to recognize some of the people who made working on this revision a pleasure.

I have been extremely lucky to be a part of a practice that has always put patients' best interest above the contingencies of Managed Care. My special thanks to our Managing Partner, Fred Frumin, M.D., who has made this his guiding principle: I could not write about Pediatrics with a clear conscience and a happy heart without that assurance. And to my other partners, Rosalind Dockweiler, M.D., Gary Gross, M.D., Nick Levy, M.D., Christine Ito Wood, M.D., Sangita Bhasin, M.D., Julie Snyder-Blok, M.D., Ron Park, M.D., and Melissa Reinhardt, M.D.: What good doctors!

Along the same lines, I want to thank our office staff. You can't write (or practice medicine) wholeheartedly if you're not part of a team you trust. I particularly want to thank our office manager, Rita Adams, for her steady hand on the wheel; Referral Office director Diane Molina for protecting us from so many Managed Care snafus; and Medical Assistant Lourdes Carr, for her ability to solve problems before they happen.

My friends and extended family have been more than supportive. In particular, thank you, Sam Popkin, for your unqualified enthusiasm: Folks, if you were given a copy of the First Edition as a present, there's a good chance it came from Sam.

Thank you to my always on-target agent, Sandra Dijkstra, and to Joe Sweeney, fitness trainer extraordinaire, for endowing both Sandra and me with all this energy and musclepower.

Writing a monthly column for *Parents* magazine has made this revision easier. Thanks especially to Kate Lawler, Betty Wong, Catherine Winter, and Janet Gold.

My editors at HarperCollins, Matthew Benjamin and Megan Newman, have been exceptionally efficient and helpful, and my hat is off to the copyeditor! I want to thank

Kay Life, whose drawings capture so well the fun of being a parent and/or pediatrician.

Most of all, I salute my husband, Chuck Nathanson. Despite his own demanding schedule, he finds the time to untangle my prose, demystify my software, bring me coffee, and in general stand in for The Muse. Finally, a big thanks to our daughter Sara, who has made parenthood so much fun—even now that she's grown up—that I just had to go and write about it.

# Contents

## Part One: The Well Child

# Part Two: Illness and Injury

# Part Three: Pediatric Concerns and Controversies

# Part Four: Glossary of Medical Terms with Pronunciations

# Preface to Second Edition

It is my fondest hope that this Second Edition of *The Portable Pediatrician* winds up tattered and torn, splattered with coffee stains, down on the floor beside (or under; I'm not fussy) your bed. That's where the First Edition often wound up—as many readers tell me—and I'm honored. After all, if you're a pediatrician who's portable, that's where you belong: where the action is, and *when* the action is, which is often at night.

Parents tend to reach for a child care book in crises, medical or emotional. Crises require clear advice *and* a response that lets you know somebody's really listening. I hope you sense that the person on the other end of these pages has been on the other end of the phone, live, tens of thousands of times. I've intended both editions of *The Portable Pediatrician* to be a source not just of medical or developmental information, but of companionship.

But I hope you reach for it at calmer, happier times, too. Childhood is an amazing drama and it's really fun to have the program notes at hand. That way, you can be primed to cherish the moment. You can also be prepared to deal with it. That's important—not just for parents, but for children. Children feel safe and behave better when parents give off an aura of competence.

In the years since the First Edition, I've continued to practice pediatrics, all the while receiving feedback from readers. As a result, I am very much aware that some areas need revision. Hence, the Second Edition.

## Setting Limits

The first time round, I concentrated on helping parents understand the reasons for children's obnoxious behaviors, but I was not very clear on how to prevent them, nip them in the bud, or discourage them. By obnoxious behaviors, I include:

*Saying No, again and again. *Hitting, pinching, kicking. *Digging in the heels

*and refusing to budge. *Tearing around
wildly and making annoying noises just to
get your goat. *Interrupting, procrastinating, whining. *Mouthing off: talking
back, shrieking x-rated words. *Getting
into things, making a mess on purpose.
*Refusing to use the potty despite knowing
how. *Sibling battles—pushing, nagging,
yelling, grabbing, tattling, turning the
backseat of the car into a war zone. . . .*

Let's see. Have I left anything out? Oh
yes:

*Not listening. *Refusing to stay in one's
own bedroom. *Worst of all: visiting this
obnoxiousness upon one's beloved parent
but behaving perfectly (perfectly!) for every
other adult.*

There are many pejorative terms for children who act this way. Pediatricians, who do
not have to put up with them, call them
*Oppositional.*

Such behaviors tend to begin around the
age of a year and peak at the age of eighteen
months to Two. Then, if all goes well on the
parenting technique front, they should begin
to fade away. By around the age of Three to
Four, Cherub ought to have achieved a deep
understanding that parents are ultimately in
charge, and such oppositional behavior
ought to be pretty much a thing of the past.
At that point—but not before then—
Cherub is developmentally ripe to benefit
from the techniques of Authoritative Parenting: Giving an explanation when you set a
limit; presenting Cherub with the challenge

of making choices, when that's appropriate;
and even allowing the privilege of negotiating, on occasion.

It's *getting to that point* that's tricky. So in
the Second Edition, I've provided stage-by-
stage (which means Chapter by Chapter)
guidance in discipline skills so that these
parental skills build as Cherub matures.
There's also a set of essays in Part III on
*Oppositional Behavior, Potty Refusal,* and *Sibling Battles.*

About spanking. Rather than go into
moral diatribes against spanking, I have
learned to assume that parents resort to it if
their children become sufficiently oppositional, simply because they can't think of
anything else to do at that point, and who
can really blame them? The trick is to prevent, or nip in the bud, or squelch effectively, oppositional behavior before you get
that desperate.

## The Epidemic of Overweight Children

In the First Edition, I knew very well that
our culture had already come down with an
overwhelming case of The Chubbies, and
tried to provide guidance on the subject.

Today, we are faced with a true epidemic
of chubbiness. In fact, we have become so
accustomed to seeing chubby children in real
life and the media that chubby looks normal
and normal looks thin.

Extreme chubbiness in childhood can
trigger all kinds of problems, from deep dis-

tress of the soul to sleep apnea (trouble breathing at night, due to extra tissues in the upper airway), from exacerbation of asthma to the risk of early puberty—girls starting breast development in second or third grade. To say nothing of disorders previously the province of overweight adults, such as high blood pressure and Type 2 diabetes.

Even so, many experts feel that pediatricians should not intervene until a child has already attained a worrisome weight. They fear that doing so may trigger an overreaction in parents—that they might "starve" a child or become punitive and overcontrolling. They worry that restricting a child's diet (even to the extent of eliminating sodas, french fries, and heavy desserts) may lead to eating disorders later on.

I don't agree. All the parents I know want the best for their children, and want the real scoop from their pediatrician. They can be trusted to be caring and to use common sense. So listen up! If you don't fit this description, skip the Chubbiness sections. In fact, go buy a different book.

The truth is, once a child is overweight, dealing with the problem is very difficult. Chubbiness is best prevented. Next-best is nipping it in the bud. That means that parents need to know how to monitor Cherub's weight, and need to be alert for an upward trend in the "weight for height" ratio. They need the tools to figure out what's causing the trend, and what to do about it.

In this Second Edition, every Well-Visit chapter includes a *Chubbiness Watch*. This section helps you figure out whether your particular Cherub is gaining weight appropriately, and flags the main lifestyle habits that may steer a child that age into Chubbiness. The essay in Part III, called "*Chubby or Not, Here We Come,*" reviews the subject and tries to demystify and simplify those exasperating growth charts.

## New Concerns and Developments in Medicine

I've added a short essay on *Autism/Pervasive Developmental Delay/Aspberger Syndrome*, in Part III. I've updated the section on *Managed Care* in Chapter One—the *Pre-Baby Visit* chapter: skip it at your own peril. There's an update on new controversies in the *Allergy* essay in Part III, and a bit on drug-resistant bacteria in the essay *Trouble in the Middle Ear*, in Part III. And little bits and pieces of updated information (and useful websites) throughout.

In the *Preface* to the First Edition, I said that I cherish the Well-Child Visits—the basis for this book. Much of my workday is filled officiating at them. I still am aware that as Rites of Passage, these encounters leave something to be desired. Nobody gets dressed up, there may be some screaming, and the refreshments tend to lack sophistication. But think of how nearly unique these visits are. They cut across religions and economic class, cultural attitudes, and degree of education. At their best, these visits unite parents and pediatrician as a team, focusing on a particular child in a particular family with specific values and culture, stresses and strengths.

The real problem is that the Well Visits are too short to get everything done.

Children and parents whiz by in the whirl of time. I grab them out of the current, hold them still for a second, and say: Hey, look, here we are! Here's where we've come from, here's where we are now, and here's what you might want to pay attention to between this visit and the next time we—hey, wait, don't leave, bye now!

And out the door they go.

Not enough time, not enough time. That's why I write these books.

# Introduction to the Second Edition

Here's how the book is organized.

## Part One: Well-Child Visits

Each chapter is a self-contained description of the age it covers, and each is designed to be read as a unit. If you didn't read previous chapters, or have forgotten completely what they said, don't worry; the chapter at hand will refresh your memory or refer you back to a previous chapter.

Therefore, each chapter focuses on the same issues (unless they are inappropriate for that age or duplicate a discussion of a previous chapter). These include:

- **Portrait of the Age:** What life with a child this age tends to feature. I've tried to re-create typical interactions of the age. I hope that parents will see a bit of their own child in these portraits, though temperaments and behaviors vary widely. Mostly, these interactions are supposed to show that behavior which can strike a loving parent as worrying or upsetting or downright obnoxious is often developmentally inevitable. I try to show parents and caretakers in the act of struggling not to take such behavior personally. When I witness this heroism in the office setting, I am filled with a joyful respect. Parents should honor themselves with the same.

- **Separation Issues:** The inevitable ones, in which the baby discovers he or she is a separate individual, and the social ones, like bedtime and daycare. It isn't just the child who has separation issues; it's parents, too. These are different for each developmental age.

- **Limits:** Separation issues and limit-setting are the flip sides of the same task: establishing boundaries between loving adults and beloved children. The more you're bonded to a child, the more important, and sometimes the more difficult, it is to set limits. The first Three

Years are crucial in this regard. Parents need to learn how to be loving and still be in charge. They need to become comfortable doing so, and babies need to sense that comfort. The essays on *Oppositional Behavior, Potty Resistance,* and *Sibling Rivalry* set forth an overview of that process, and each Well-Visit chapter tries to help parents a little bit further down that important path.

- **Day-to-Day Issues:**

  Milestones of development: *These aren't checklists, but discussions of how to make sure the child is doing what is expected for age.*

  Sleep: *Expectations and problems.*

  Growth: *In this edition, I have paid special attention to helping children escape our current epidemic of childhood obesity. This is treated more fully in the essay in Part III:* "Chubby or Not, Here We Come!"

  Teeth: *How to prevent problems.*

  Bathing: *Reminders about safety and hygiene.*

  Diapering, dressing, and clothing: *Reminders about safety (nobody needs a zipped-up penis catastrophe), hints about self-care.*

  Activities, Toys, and Equipment: *Appropriate ones, expensive ones that never get played with, and dangerous ones.*

  Safety Issues and Medicine Chests: *We start out with a baby-safe home and a medicine chest appropriate for a newborn's welfare,* **and then we add on as the child grows.**

- **Health and Illness:** What pattern to expect for this age range, and why. Specific problems that are most likely to appear now, and ways to deal with them.

- **Common Scary but Usually Innocent Problems:** Starting at about Six Months, these can appear. Fever convulsions, night terrors, vanishing vaginas (labial adhesions), and so on. Even if you only glance at this section ahead of time, you'll be prepared a bit if something weird and frightening occurs—to your child or to a playmate.

- **Window of Opportunity:** Habits, skills, and information that is easiest to impart (or to squelch) during this period.

- **What If:** This section discusses certain events that could occur, focusing on how a child this age can be prepared for them. These include starting daycare, extended separation from the parents, traveling with the baby, conceiving another child, the arrival of a second sibling, the maturing of a younger brother or sister, divorce and custody issues, surgery and hospitalization, and moving.

- **The Well-Child Visit:** This gives you an idea of what's likely to occur in the visit to the pediatrician that's coming up, and how to prepare you and your child for the encounter.

- **Looking Ahead:** These sections point you toward to the next stage of development.

## Part Two: Illness and Injury

Since very young babies are **in a category of their own**, illness and injury for babies from birth to two or three months of age are covered completely in the first Well-Child chapters of Part I: *Birth to Two Weeks* and *Two Weeks to Two Months*, and partly in the chapter *Two Months to Four Months*.

In this part, I've tried to look at illness and injury from the point of view of the parents I've seen and talked with over the years.

> One: Frightening Behaviors: *The urgent problems covered are Trouble Breathing, Fever, Seizures (convulsions, fits, breath-holding spells), Anaphylayis, Not Acting Right, Looking Right or Smelling Right, Dehydration, Waking Up Crying at Night and Night Terrors. For each symptom, we cover the basics, how to assess the situation, and how to act on the problem.*
>
> Two: First Aid: *The common problems covered here include Head Bonks, Injuries to Necks, Eyes, Noses (including foreign bodies and nosebleeds), Mouths and Teeth, Arms and Hands, Legs and Feet. I've also included here (instead of in Major Crises): Cuts and Bleeding, Scrapes and Heat Burns, Human and Animal Bites, Insect Stings, and Poisoning. These may indeed be major crises, but usually they aren't serious.*
>
> Three: Body Parts, Body Functions and What Ails Them: *Starting from the top of the body and working down, this section covers the body. Headache and Stiff Neck,*

> *Eyes, Ears, Nose, Mouth, Throat, Voice, Airway and Lungs, Coughs, Abdominal Pain, Vomiting, Diarrhea, Trouble Pooping/Constipation, Genitals, Trouble Peeing, Skin, Hips, Legs, and Feet.*
>
> Four: Illnesses, Both Common and Uncommon: *There is also a section on common illnesses with unfamiliar names and uncommon illnesses with familiar names.*

## Part Three: Essays

Some issues that cause recurrent communication problems between parents and pediatricians. These essays lay out the pediatric point of view so that parents can see what is going on in the mysterious mind of their child's doctor. The topics are: • Growth Patterns • Viruses, Bacteria, and Antibiotics • Immunizations • Allergies • Otitis Media (middle ear infection) • The Child Who Is "Sick All the Time" • Severe Behavior Concerns (Autism, ADD), Oppositional Behavior, Potty Resistance, and Sibling Battles.

## Part Four: Glossary with Pronunciations of Medical Terms

This is an informal glossary in the sense that the pronunciations are in my own (midwestern/Boston/California) accent, and the definitions are somewhat informal but nonetheless accurate.

## Sources

All discussions reflect standard of care in up-to-date pediatric literature and the American Academy of Pediatrics. Discussion of developmental milestones is based on the work of Martin Stein and Burton L. White, as well as on the coverage in such pediatric textbooks as Nelson (Behrman) and Hoekelman. Sources for the moral development of the child include the work of William Damon, Ph.D. Dental recommendations are those of the American Academy of Pediatric Dentistry. The source for speech advice is the American Speech, Language, and Hearing Association, and the Speech Foundation of America. Nutritional advice is based on recommendations of the Academy of Pediatrics. Recommendations for handling infectious disease in the daycare setting are based on the Report of the Committee on Infectious Diseases (AAP 2000).

However, all these excellent sources have been filtered through my brain and are colored by own training and experience, and if there are any errors, they are mine alone.

Of course, no book can cover all symptoms and conditions. Always call your pediatrician if you have any doubts or questions about whether your child's illness is one that can be handled "by the book"—this book or any other.

## A Note About Prose Style and Other Matters

In every such book, the writer has to confront the matter of the personal pronouns he and she. I have spent a great deal of energy trying to avoid referring to the pediatrician as a person of one sex or the other.

Having exhausted myself in this endeavor, I refer to the babies and children in this book as he or she without forethought, simply as the mood strikes me. Actually, as I described a characteristic or activity, one of my young patients would usually come to mind, and I would write with the vision of that unnamed child—male or female—in my head. I have gone over these descriptions to make sure that I am not being sexist. When I refer to parents, it is usually as "you" rather than as the mother or the father. This "you" also, of course, refers to anyone who considers himself or herself as being in the position of the parent.

None of the babies, children, or family members is described physically, but they all (who can talk) speak standard English or that which I in my innocence regard as standard English. This is true even if the child I have in mind as I write is of a different ethnic, linguistic, or racial background. It's very hard to get accents and dialects right on the printed page.

I do not assume that babies are born to parents rather than adopted by them, that everyone lives in a house, that mothers stay home to take care of children, that fathers do not stay home to take care of children, or that everyone eats meat or with Western utensils.

Finally, nobody mentioned in the book is fictional, but all descriptions are composites of more than one real child. When an anecdote about a specific child is used (and all are true) the names have been changed. There is one exception: that of our daughter Sara, who has given permission that her identity be revealed.

Part One

---

# The Well Child

# The Pre-Baby Visit

## As if You Didn't Have Anything Else to Do

## To-Do List

- Figure out your insurance plan!
- Make a prenatal or pre-adoption visit with your chosen pediatrician
- If you're planning to nurse, get the best guide:

    *The Ultimate Breastfeeding Book of Answers* (Prima), by Jack Newman, M.D. and T. Pitman
- Buy and figure out how to install a car seat for the baby
- Turn the hot water heater down to 120 degrees

Back when dinosaurs roamed the earth, before Managed Medical Care, parents interviewed pediatricians to decide which one to choose before the baby arrived. These days, parents often have the choice made for them by their insurance plan. So the focus of the parental visit has changed for most people, and maybe for the better. Parents aren't fixed in a consumer mode, and pediatricians aren't locked into their best behavior. The visit is a social one, and the whole experience can be more relaxed and more fun.

But why make a parental visit at all, if you already know who the pediatrician, or group of pediatricians, is going to be? If you're already stuck with each other?

I think the most important reason is to bring home the realization that the big climax of pregnancy is not the act of delivery, but the arrival of the child: the delivery is but a means to an end. This may seem obvious intellectually. But there is something about obstetrical visits that induces a kind of fog over post-delivery events. The ob-gyn waiting room is quiet. Nearly everyone is an adult. You look at the size of the bellies, and compare them with the one housing your own fetus. You overhear the nurses and receptionists reciting the litany of pregnancy and labor. Most adopting parents these days participate to some degree in the pregnancy of the birth mother, so they get fogged-in also.

Walk into the pediatric waiting room, and savor the contrast: the pungent odors, the wild spectrum of pitch and decibel. Be careful you don't step on somebody who comes up to your ankles. Try to figure out what that obviously beloved scrap of material, coiled around that sucked thumb, used to be. Hear the office staff rejoicing that Jessica finally has pee'd in the cup, extolling Thomas's juicy pinchable thighs, offering stickers to Enrique for being so brave after his many pokes. You put your hand on the pregnant belly and feel your fetus kick, or you think about the readied nursery at home, and suddenly you realize: This is a baby we're talking about here.

Before this appointment, it's useful to have considered some basic questions. That way, you know what you want to discuss with your Pediatrician To Be. Here's my list, with my answers. Think about what questions *you* want to ask.

- How important is it to nurse when you really don't want to? How important is it to try to nurse if you are pretty sure you won't be able to do so? If we use formula, how can we discuss intelligently with the pediatrician which one to use?
- What are the most frequent medical interventions babies need right at birth, how scary are they, and what do they mean?
- What medical routines can be expected in the hospital nursery? Are they really necessary, especially the ones that are uncomfortable for the baby, or involve separating the baby from the parents?

- What are the pros and cons of circumcision?
- How far ahead of time, and on what basis, should we choose a daycare provider? What if we have to put the baby in daycare really soon after birth—how can we keep the baby from being endangered physically or emotionally? How can we cope ourselves?
- What should we try to learn about our pediatrician, even if we don't have a choice about which one to see?
- What's most important to understand about the insurance plan for the baby?

Before we turn to these questions, however, a word about the most important item you'll need to purchase before the baby arrives: the car seat.

## CAR SEAT FOR THE NEWBORN

Don't even think of bringing the baby home from the hospital in someone's lap. Many hospitals won't discharge a baby if you don't have a car seat. They'll try humor and tact first; but if you persist, they will call The Law: Car seats for babies are required in all states and the District of Columbia. And for good reason. If you have an accident, the possibility of death or injury is reduced by 80% when a car seat is used. If you are riding in a cab, take the car seat along.

Infant car seats must face backwards. They should be installed in the center of the backseat. The car seat should be installed so that it is not in a seat with an air bag in front or next to it: the detonation of the air bag

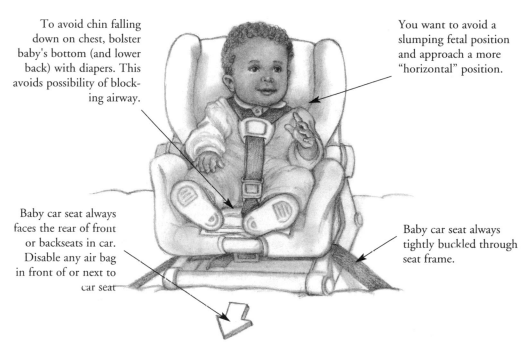

To avoid chin falling down on chest, bolster baby's bottom (and lower back) with diapers. This avoids possibility of blocking airway.

You want to avoid a slumping fetal position and approach a more "horizontal" position.

Baby car seat always faces the rear of front or backseats in car. Disable any air bag in front of or next to car seat

Baby car seat always tightly buckled through seat frame.

could prove fatal. Consult your car manual for indications for disconnecting the air bag.

Babies should sit facing backward as long as they can fit into the car seat: the minimum age to face forward is One Year, the minimum weight 20 pounds. (Both milestones should be achieved.) If your baby is under about six pounds at the time of hospital discharge, make sure that you have positioned the baby in the reclining, rather than the sitting, position: you don't want Cherub's chin down on his chest. Such a curled-up, fetal position can close off the airway.

You will need a seat with a "five-point" harness—shoulder straps, lap belt, crotch strap. Many good brands are available, but you must be sure that the one you choose is compatible with your vehicle. Check the manual.

Car seats with shields can overwhelm an infant; *Consumer Reports* suggests these be reserved for toddlers. A padded armrest may help a baby be comfortable, but is no protection in a crash.

Once you've chosen your seat, check it out through the Auto Safety Hotline: 800-424-9393 to make sure there hasn't been a recall. Have the manufacturer, model name and number, and date of manufacture ready.

## How Important Is It, Really, to Nurse?

Most of the parents I see know full well that breast milk is optimal for babies.

But does that really outweigh other factors? • What if the mother is returning to work at four, eight, twelve weeks? • What if she has inverted nipples or has had breast surgery, and suspects or has been told nursing will be difficult? • What if she doesn't, um, have "that kind" of relationship with her breasts? • What if the baby's father isn't very enthusiastic? • What if the mother wants to be sure the father will bond early, and strongly, with the baby, and thinks nursing will get in the way? • What if the baby is born prematurely, and you're faced with pumping for weeks until he or she is big enough to nurse? • What if the baby is adopted, or is spending her fetal life in the womb of a surrogate, and you know that it is possible to nurse without having been pregnant, but that it takes a great deal of effort and commitment?

Parents need to know the trade-off: is it worth a small, or a large, struggle to nurse? There are two aspects to consider.

- **Medical problems.** Will nursing prevent allergies, ear infections, diarrhea, obesity? Is there a particular reason that your particular baby would benefit from nursing more than the average baby?
- **Bonding.** If a mother doesn't nurse, can she and the baby bond well? If she does nurse, will the baby bond well with the father, even though feeding is such a strong part of nurturing?

When I guide individual parents, here are the points I stress:

- If your baby is premature, breastfeeding really is important, important enough to make considerable sacrifices. The more premature the baby, the more important to try to nurse.
- Nursing helps to protect babies from crib death, or sudden infant death syndrome (SIDS).
- Nursing for the first four months, avoiding formula and foods, helps to protect a baby from middle ear infections for the whole first year of life.
- If you are going to be visiting or living where hygiene is poor or highly unreliable, nursing is a powerful, unique protection against infant diarrhea.
- Nursing for even three weeks or so is worthwhile.
- Even deeply engrained attitudes about breasts, and their erotic versus maternal functions, can change.

## ABOUT BONDING

Babies bond in a multitude of ways, mostly touch and smell. *Both parents can bond, no matter what the feeding method.* Besides, by two weeks of age, a full-term thriving nursing baby usually can start having an occasional (up to once a day) bottle containing, ideally, expressed breast milk.

The most important factor in bonding between father and baby is that no one act as "gatekeeper," discouraging the father in subtle ways. Gatekeepers are people who allow or forbid access to the baby and who monitor how others handle the newborn. Leaving the father and the baby alone, in a spirit of

# FISH SAFETY: WHAT KINDS TO AVOID OR LIMIT

Fish is a great source of protein and healthy fats. However, some large fish (both freshwater and ocean) may be contaminated with mercury. Excess of mercury is toxic to the central nervous system, especially in the case of the fetus, baby, and young child.

The FDA (Food and Drug Administration) monitors the safety of commercial, "store-bought" fish and ocean and coastal fish. The FDA recommends that children and women of childbearing age *avoid* eating shark, mackerel, swordfish, and tilefish. The FDA also advises that pregnant women, and those who may become pregnant, to *limit* their consumption of other commercial fish to an average of 12 ounces per week.

The FDA may have issued different or supplemental advisories by the time you read this. Check the phone number or website below.

The EPA (Environmental Protection Agency) monitors the safety of non-commercial, freshwater fish—fish caught by family and friends in lakes, streams, ponds. The EPA recommends that pregnant, possibly pregnant, and nursing mothers, as well as "young children" (no age range given), limit such fish to one meal per week. For adults, that's 6 ounces cooked fish or 8 ounces uncooked fish; for children, 2 ounces of cooked fish.

According to the CDC (Committee on Disease Control) young children should not eat raw fish or shellfish at all, due to the risk of bacterial food poisoning.

To help keep mercury out of our waters, safely get rid of any mercury thermometers you have on hand, and don't get any new ones.

For more information:
EPA: *http//www.epa.gov/ost/fish*
FDA: 888–SAFEFOOD or *www.cfsan.fda.gov*

---

tranquility and confidence, is usually all it takes to forge a strong father-baby bond.

Whichever mode of feeding you decide upon, the main points are to enjoy the baby and to keep her well-nourished. Babies grow and change so rapidly. If you cling to strong feelings, of guilt, or even of vanity, about the method of feeding, you won't be prepared when feeding becomes routine, taken for granted, and the baby's playing a whole new game. Whether you nurse or bottle-feed is not going to determine your style of parenting, the joyfulness of your relationship, the character traits of your child, or your methods of discipline.

## QUESTIONS ABOUT NURSING

*Is human milk superior to formula?*
**Yes, in several ways:**

Immunologically: *Human milk, especially during the first three weeks after delivery, is rich in immune substances. For example; if a nursing mother catches a virus, the mother makes antibodies to the virus, and these then go through the milk to help the baby. Nursing babies are better protected from a whole menu of illnesses, from Rotovirus diarrhea to RSV bronchiolitus.*

Nutritionally: *The tricky ratio of calcium to phosphorus, crucial for normal growth, is automatically correct. The protein and fat in human milk is healthier and more readily absorbed than that in formulas. Electrolytes, like sodium and chloride, and trace elements like zinc and copper are automatically present in correct amounts, not vulnerable to the rare manufacturing error. Vitamins and iron are built-in and present in absorbable form. Extra water is automatically added in times of dryness and heat.*

Hygienically: *Human milk is clean. It can't give a baby an infectious disease, unless contaminated by giving it in a dirty bottle or expressing it with a dirty pump.*

Aesthetically: *There is some evidence that babies become bored when milk tastes the same day in and day out, and may even stop eating well. Formula always tastes the same. Human milk tastes different from day to day depending on the mother's diet, and some research shows that this variety*

*pleases babies. Even strong flavors like garlic, rather than causing colic, may enhance the appeal of the breast milk for many babies. From a parent's point of view, the poop from nursing babies smells more pleasant than formula stools, and even what we politely call "Spit Up" isn't very offensive.*

*Is sucking at the breast better than sucking at the bottle, no matter what is in the bottle?*
At the breast, babies latch on, form a suction pressure with cheeks, and glug down the milk. Their lips and tongue don't work. At the bottle, the baby has to work with the lips and tongue.

Some studies show that babies who suck from a bottle can force the fluid plus their own secretions back through the eustachian tube into the middle ear, thus giving themselves a set-up for a middle ear infection. (See the essay on *Otitis Media* in Part III.)

*Is nursing psychologically superior to bottle feeding?*
**Not necessarily.** It is absolutely possible to give a bottle in a close, warm way, and with skin-to-skin contact. Whichever makes parents most comfortable and happiest with themselves and their roles as parents and mates is the psychologically superior form of feeding.

*If the baby nurses, is it important that he or she never drinks from a bottle or touches formula for months and months?*
I am personally certain that every nursing baby ought to be educated to take a

"reminder" bottle, ideally containing expressed breast milk. I suggest starting at about two weeks of age, given that nursing is well established and the baby's weight gain is good. If you wait much longer, some babies become very stubborn in their refusal to do so. I still remember the distress caused by a certain enchanting four-month-old whose mother had to abruptly interrupt nursing. Jane refused bottle, cup, spoon, syringe for thirty-six hours. We all sprouted a few gray hairs, except Jane.

### Is formula ever definitely better for a baby?

**Very, very rarely**. It is rare that a mother cannot nurse because of a disease. However, some dangerous viruses can be passed from mother to baby through breast milk. A mother with Herpes lesions on her breast must not nurse. Nor should a mother with any of the HIV or HTLV (AIDS) viruses. If a mother is a carrier of hepatitis B, she can nurse if her baby receives a gamma globulin shot plus an immunization against the virus.

Very rarely, a baby may develop sensitivity to an ingredient in breast milk. Most often this is to dairy or soy protein. In this case, it's often advised to give a high-tech, non-allergic formula until the mother has been on a diet that excludes these proteins long enough for the baby to tolerate the breast milk again.

Even more rarely, a baby (even with no family history of problems) will have an inborn metabolic disorder that makes it dangerous to take human milk because of its sugar or protein content. These metabolic disorders are screened for in a blood sample before the baby is discharged home from the nursery.

### Can every woman who wants to, nurse? How can I best prepare to nurse?

**Most women who wish to nurse can do so**. There are a very few situations in which nursing just doesn't work. Previous breast surgery, especially if the nipple was involved, can sometimes prevent the central nervous system/hormone reflex that generates breast milk. A few women lack the hormone, prolactin, which is vital to nursing. A tiny, tiny number of women may have inadequate breast tissue. These may all make nursing truly impossible.

Women who want to nurse do well to read at least one good text and attend a breastfeeding class, if one is available. See the "To-Do List" at the top of this chapter for the title of the book I think is the very best one.

If you have or suspect a special problem, such as any of the conditions above or something like flat or inverted nipples, a visit with a lactation specialist or to the La Leche league can be most helpful—before, as well as after, the baby arrives.

### Can a mother nurse her adopted baby?

**Yes, in many cases**, if she starts far enough ahead of the adoption, has access to a supportive lactation counselor, and is energetic and committed. Nursing is a neurological activity in which the baby's sucking stimulates the hypothalamic area of the brain, which then releases milk-encouraging hormones. You needn't be pregnant to nurse. But be sure to discuss this carefully with the pediatrician and with a lactation consultant experienced in this enterprise.

*Can I continue to nurse even though I'm going to have to interrupt it for a time? What if I need to travel, say, for a week without the baby?*

**If nursing is well established**, if you stay rested and relaxed, and if you are consistent in expressing the milk frequently, you will probably do fine.

If this is a possible scenario, you want to be very sure the baby becomes adept at going back and forth from breast to bottle, long before you get on the plane. This usually means starting the trip after both nursing and expressing milk with a pump have become very well established, certainly not before about a month of age at the very earliest.

*What about wet nurses, or a breast milk bank?*

The problem with wet nurses is twofold. The first main concern is that viruses can be passed from the nursing woman to the baby, or vice versa. The main worries are hepatitis B and AIDS.

Second, wet nurses on medications or drugs can pass these to the baby through the milk. You'd need to know, trust, and perhaps monitor the wet nurse thoroughly.

A third, very rare concern is that a nursing woman's white blood cells are present in the breast milk, and exceedingly rarely can harm a baby with an immature or deficient immune system, because they recognize the baby's tissue as foreign and attack the baby, who lacks effective defenses. This is called a graph versus host reaction.

Sophisticated medical advice is necessary if such an alternative is considered. Even if it's your sister (or even your mother) who'd be the wet nurse.

## Questions about Formula

*Are soy formulas better than milk formulas? Or vice versa?*

A baby who has cramping, crying, loose stools, rash, or any other sign of not tolerating his formula needs to see his pediatrician. The problem may indeed be due to the formula or to something completely unrelated to the formula.

Babies who glug down just about anything that comes their way can do just fine on either milk or soy formulas. Very adaptable babies can even go back and forth between the two. All commercial formulas—milk-based, meat-based, or soy, brand-name or generic—must conform to the FDA's standards for adequate safety and nutrition.

Milk formulas have one big advantage over soy formulas. Modifying cow's milk for human babies is less prone to human errors of omission and commission. Soy milk is made "from scratch." Moreover, cow's milk sugar is the same as that of human milk, and it assists the absorption of calcium. Finally,

The American Academy of Pediatrics urges that all formula fed babies be given high-iron formulas. Iron deficiency can cause developmental impairment, and a baby can be iron deficient without being pale or showing signs of anemia on screening blood tests, such as hemoglobin or hematocrit tests.

cow's milk–based formulas are generally less expensive than those based on other proteins.

Concern about catching the dreadful illness mad cow disease (BSE, or *bovine spongioform encephalopathy*) from milk products is unwarranted. The infectious agent for mad cow disease is found in a cow's nervous system, and cannot find its way into milk.

The worry about hormones in cow's milk formula is also misplaced. The growth hormone given to cows (BST) to increase milk production is the same hormone that is made by the cows themselves, and its presence in milk is carefully monitored. (So is the presence of antibiotics. Any lot with detectable antibiotics is discarded.)

However, some babies are intolerant to the protein in cow's milk, with symptoms of excess crying, cramping, loose stools, often with obvious or microscopic blood in the stool. Rarely, a baby can have a life-threatening allergic reaction, with hives and shock. Many of these babies will do fine on a soy formula, but about 20% of babies allergic to cow's milk will also be allergic to soy, as well. For those babies, there are truly hypo-allergenic formulas such as Nutramigen, Alimentum, or Pregestimil.

Rarely, a baby may be unable to digest the sugar, lactose, in cow's milk. I say rarely because this sugar is identical to the sugar in breast milk. For such a baby, there are two alternatives: Lactose-free cow's milk formula, such as Enfamil's Lactofree, or a soy formula that contains sucrose or glucose.

*Why are there so many kinds of formulas?*

Formula companies have attempted to analyze breast milk down to the last molecule and tried to re-create it in the factory. But no formula perfectly duplicates breast milk. Their competing claims are based on sophisticated analyses of the protein, fat, sugar, vitamins, minerals, and trace elements in human milk, and on their claims to have outdone their competitors in achieving duplicate profiles of human milk.

*Do I need to buy one of the brand names, when the generic store brand is so much less expensive?*

Both brand names and generic, as I say, are monitored for safety and nutrition by the FDA. The generic formulas are cheaper for two reasons.

Some cow's milk formulas are advertised as "low allergy" or "hypoallergenic." Beware! This only means that 90% of babies known to be allergic to cow's milk will tolerate these other formulas. There is no way of telling whether any *particular* allergic baby will tolerate them, and a few babies have had severe, life-threatening reactions to these "low allergy" formulas. Such babies need very special, prescription formulas.

1. They are making use of research performed by the brand-name companies, so they have no research and development expenses to make up. Since they are not involved in continuing research, they are not on the cutting edge when it comes to refinements such as the perfect lipids for brain growth, the perfect protein for easy digestibility, and so on.
2. The generic formulas do not sell to the Federal WIC program, which makes formula available to low-income families. The brand-name formula companies do sell to WIC, and make no profits on these sales.

So there are medical, financial, and political issues here. Some people feel that the brand-name formula companies should be reimbursed for their pioneering research. They feel that the higher prices of the brand names are justified, and that the "generics" get a "free ride." The brand names are the companies that have not only developed regular formulas that are very close to breast milk, nutritionally; they have also designed the special, life-saving formulas for babies who have rare metabolic disorders and cannot tolerate either breast milk or regular formulas. Moreover, these companies continue to research subtle improvements.

Others feel that anything that helps poor families feed babies healthfully—such as inexpensive generic formula—is a good thing.

*When is it necessary to use a formula that is neither cow's milk nor soy based?*

Rarely, a baby may need a very special, high-tech formula such as Pregestimil, Alimentum, or Nutramigen. That happens if a baby is allergic to both cow's milk protein and soy. Even more rarely, a baby may be born with a metabolic condition such as PKU (phenylketonuria) in which even these formulas are dangerous. Such a baby needs a "designer" formula actually prescribed for him or her.

Before you choose any high-tech formulas, the pediatrician should be consulted. They are very expensive and don't smell terrific, but have saved lives and sanity (the formulas, not the pediatrician, though that may be true also).

Homemade formulas based on condensed, evaporated, or whole milk are dangerous. Such concoctions are miles and miles from human breast milk. They contain too much of the wrong kind of fat; the wrong calcium:phosphorus ratio; not anywhere near enough iron. Babies need special formulas to thrive in all ways. If the problem is expense, parents should apply to the WIC program for help.

## What Are the Most Frequent Unexpected Special Interventions Full-Term Babies Need Right at Birth?

Many parents will have taken childbirth classes, read books, and toured their hospital before delivery, and have a pretty good idea about the mechanics of vaginal delivery and hospital policies (such as having a Significant Other at the delivery, flexible schedules, and rooming in). They also usually have seen a film, or have been given a description, of cesarean section deliveries.

It isn't unusual for such classes to glance over, or not even mention, medical problems that teachers feel may unnecessarily alarm parents. However, it's my opinion that parents like feeling in control, and appreciate knowing ahead of time the most common untoward events that can occur right at birth. Realizing that these events are not uncommon, and that most babies do just fine, can be very soothing.

Most deliveries, no matter how lengthy, intense, or chaotic, don't bother the baby much. Not infrequently, however, a per-

fectly normal full-term baby with a perfectly normal delivery will turn up at birth with a problem. The most common are:

- Meconium (newborn stool) in the amniotic fluid: pooping before being born;
- Not taking an effective breath soon after birth; and
- Retaining normal fluid in the lungs longer than usual.

### MECONIUM IN THE AMNIOTIC FLUID: POOPING BEFORE BIRTH

It's amazing this doesn't happen more often. Some babies poop because they are at or beyond their due dates, and just can't wait for the next gas station. Others have transient distress, perhaps when the umbilical cord gets compressed and briefly cuts off their oxygen supply, and stool by reflex. There is nothing the mother did or didn't do that causes this, and nothing that she could have done to prevent it.

The newborn stool, meconium, is thick and tarry, and very irritating to the lungs if this thick tarry substance is inhaled. If the amount of meconium is small, and very diluted with amniotic fluid, and the baby is very vigorous, inhaling this lightly stained fluid probably isn't going to hurt. But if the baby is already showing signs of distress, the staff will try hard to make sure the baby inhales as little meconium as possible.

And these efforts may make the delivery less peaceful than you had in mind.

Usually, in this situation, baby's mouth and nose are suctioned very thoroughly as soon as the head appears, before the rest of the body is delivered and the baby starts to breathe. This can take a few minutes, and is usually accompanied by everyone in the delivery room shouting to the mother, "Don't push, don't push!"

If the baby shows signs of having been in distress, it is necessary for someone (a pediatrician, anesthesiologist, or specially trained nurse) to look down the airway with a special light, to see if there is any meconium there. If it looks as if the baby might already have inhaled meconium, perhaps even before being born, the designated person can place a tube into the airway and suction it out. This may have to be done several times.

Often, this maneuver prevents "meconium aspiration pneumonia," and nothing else is needed. (Once in awhile, despite all efforts, some meconium has gotten into the lungs before the baby is born, when he takes "fetal breaths." In this case, the baby may need more specialized help.)

Suctioning out the airway before the baby takes a breath delays, of course, the baby's programmed first big breath. So it's not unusual for a baby thus suctioned to require a "jump start" in order to start up breathing: see below.

In any event, when "poop happens," the final part of the delivery is not calm and smooth, and the attention to the baby can seem urgent and frightening. Knowing what's going on, and that most babies do just fine, can be of great help.

## THE JUMP START: HELP WITH THE FIRST BREATH

It's also surprising that more babies don't need some help starting to breathe. After all, during fetal life the lungs are filled with fluid. The pressure needed to fill them with air the first time or two is considerable. (What a relief a baby must feel when he discovers that the first breaths are the hardest; it's not going to stay that difficult.)

For any one of a number of reasons, a baby may not breathe effectively right at birth. If there is a warning on the fetal monitor that this might happen, a second physician, usually a pediatrician, or a specially trained baby nurse, is called in. But delivery room and nursery nurses, as well as obstetricians and anesthesiologists, are trained to help in this relatively common event.

Usually all that is needed is stimulation and oxygen blown from a mask on the baby's face. Sometimes it is necessary to "bag" the baby, forcing in oxygen through the mask by squeezing the attached bag. And once in awhile, more measures are needed: a tube directly into the baby's airway, or medicine given into one of the vessels of the umbilical cord.

Once again, most babies do just fine.

## RETAINED LUNG FLUID (TRANSIENT TACHYPNEA, OR TEMPORARY RAPID BREATHING OF THE NEWBORN)

The lungs of the fetus are filled with fluid. During the first minutes to hours of life, as the baby breathes, that fluid is forced out of the lung tissue and into the circulation where the baby gets rid of it over the next few days. Sometimes, especially if the baby was born by caesarean section and the lungs weren't massaged by labor, the fluid doesn't exit as rapidly as usual. This makes the baby breathe rapidly and sometimes turn a bit blue.

The most important duty of the pediatrician is to make sure (usually with blood tests and X rays) that this, not something else (such as pneumonia), is the cause of the symptoms; and, of course, to provide oxygen and support for the baby. Often the baby will breathe extra oxygen from a "hood" or clear plastic cylinder that fits over her head. And frequently the baby will be breathing too rapidly to nurse or take a bottle right away, and an intravenous line may need to be started.

The things to remember are that usually this all resolves in about forty-eight hours and complications are uncommon, and that the baby isn't really in any pain. It's very important not to think of the baby as "sick" or "fragile" or "malformed" in any way, and not to feel guilty or responsible for this quite understandable adjustment problem.

## A PERSPECTIVE ON SUCH BIRTH PROCEDURES

Because these situations aren't infrequent, there are lots of very bright, happy, normal children around who had low Apgar scores, especially for the first minute of life. The Apgar score is an assigned score that totals up signs of how well the baby is adapting to life outside the womb. We are all grateful that Dr. Virginia Apgar was the one who developed the scoring, because her name gives the mnemonic:

**A** Appearance: *Blue, pale, or pink? If blue, just the hands and feet, or the whole baby?*
**P** Pulse rate: *How rapid is the heartbeat?*
**G** Grimace: *Does the baby make a face when you suction her nose?*
**A** Activity: *Does the baby move spontaneously? How vigorously?*
**R** Respirations: *How effectively and frequently is the baby breathing?*

Each parameter can be judged a 0, 1, or 2, with a "perfect" total score of 10. The test is performed at one, five, and (if problems) ten minutes of age.

Notice that there is room here for observer input. One doctor's Apgar 8 can be another doctor's 6 or 10. Some cynics swear that only the newborns of highly placed medical personnel ever get a 10.

The Apgar score is used to monitor— not to decide whether to start—special measures to help the baby. No one is going to wait around doing nothing for one minute, and then note that the Apgar score is only, say, 2.

Finally, it is clear that even very low scores (3 and under) at one minute and even five minutes do not automatically mean that a child is in for trouble later on. Your pediatrician will want to talk with you carefully if your baby had any kind of problem at or after birth, but in most cases it is quite possible to be very optimistic.

Some experts hint that in the future Apgar scores may not be obtained routinely at all. It has served an important function, however, in focusing attention on how babies adjust to life outside the womb. And its name is a tribute to a woman who made enormous contributions to the care of all newborns.

### THE TRAUMA OF DELIVERY: A PERSONAL POINT OF VIEW

I do not agree with the idea that birth is a terrible shock and trauma to the baby. Rather, it is the culmination of a long period of preparation. We have no real reason to think that birth is painful.

While you and I may wince at the idea of the compression of the skull, a baby's skull is designed to be molded. Maybe it feels good. The brain itself has no nerves and feels no pain. We may imagine that the birth canal feels "tight" and the baby feels squeezed, but we are undoubtedly projecting our own claustrophobia on the situation, heightened by our imagining how awful it would be not to be able to breathe, a consideration clearly not entertained by the emerging baby.

And we have no reason to think that the

cry that babies give with the first breath is shocked or terrified. The fact is, that first breath has got to be a big one, to expand the lungs. And when it is exhaled, a noise will emerge. That noise is a cry. What else would one expect the baby to do? Yodel?

Moreover, I don't think that babies are terribly offended by a noisy delivery room. We have good evidence that noise is transmitted very nicely into the womb. Any mother who experiences music, shouting, traffic, construction noises, animal sounds, and so on has a newborn who has heard it all too.

Bright lights and flash bulbs don't harm a baby; often they don't even seem to notice. Even in the brightest of delivery rooms, babies will go into the normal alert state after birth, opening their eyes and gazing about with that steel-blue, newborn stare.

Gentle handling, talking to the baby in tones of welcome, admiration, and reassurance, and avoiding unnecessary procedures are all part of good medical care, and I take them for granted. But I don't think babies need or "want" to be born in specially darkened, specially hushed rooms, or underwater, or that they need to be immediately immersed in a warm bath.

Finally, the parent-newborn bond is not one of epoxy, where the timing is crucial and it either takes or it doesn't. Bonding is not one feeling, nor is it all positive; it has many ingredients and takes place over time. It encompasses awe and exhilaration, but also boredom (the baby sleeps all the time) and disappointment (she got Grandpa Frank's nose) and worry (what's that funny noise he makes?) and even anger (we so wanted a girl!).

The one thing that can, no question, interfere with bonding is being convinced that if it doesn't happen right after delivery, something is irretrievably lost. This is simply, and absolutely, not so.

## What Medical Routines Can Be Expected in the Nursery, and Why?

### TREATMENT TO PREVENT EYE INFECTIONS

All babies receive medication to try to prevent eye infections they could have acquired during delivery. We have Helen Keller to thank for this instituted routine; she felt strongly about preventing blindness whenever possible. The erythromycin eye ointment prevents both gonorrheal and chlamydial infections.

### VITAMIN K TO ASSURE NORMAL CLOTTING OF THE BLOOD

Vitamin K enables blood to clot. If a baby does not have enough vitamin K, she may have bleeding that is very hard to stop.

It's a vitamin that has to be activated by bacteria in the intestine. Some babies do not make vitamin K themselves—especially babies who are exclusively breastfed, or who have liver disease. Such babies can develop life-threatening bleeding into the brain, as late as seven months of age.

For that reason, all babies are given an injection of vitamin K at birth. Some pediatricians like to give a second or even third dose of vitamin K to breastfed babies just to be sure.

As of this update, oral vitamin K has not been proven to be as effective as injectable vitamin K in preventing clotting problems.

### DETERMINATION OF THE BABY'S BLOOD GROUP AND TYPE

A baby's blood group and type are not determined routinely. If the mother's blood type is O, many physicians order the tests on the baby's blood, collected from the umbilical cord at birth. This is because the baby could have inherited from the father blood that is type A or B. Since this situation (Mom type O; Baby type A, B, or AB) may predispose to jaundice of a degree that needs treatment, it's helpful to know ahead of time, especially if the baby is going to be discharged at an early age. Babies whose mother are Rh negative always have their blood typed to see if they are Rh positive. If they are, their mothers need to have a RhoGAM shot to prevent Rh sensitization in later pregnancies. (I am not going to go into the Rh aspects further. If you are Rh negative, you know all about it from your obstetrician.)

### NEWBORN BLOOD SAMPLE SCREENING TESTS

The purpose of the screening tests is to identify a condition before the baby actually has symptoms. That way, any treatment can be started early, and the benefits can be maximal. Most of these conditions are rare, and inherited in such a way that a family history is often not suspected. Many inherited diseases do not as yet have a reliable screening test. These include cystic fibrosis and the mucopolysaccharidoses. There are three categories of conditions tested for at this time:

1. *Metabolic diseases:* The baby lacks the chemistry necessary for normal growth and development, and needs a special diet or dietary supplement.
2. *Hormonal diseases:* The baby needs hormone supplements for normal growth and development.
3. *Disorders of blood formation:* The baby's blood cells contain abnormal hemoglobin.

Every state has different regulations, determining which conditions are to be screened for. You can find out what your state tests for by contacting either your hospital of delivery or your state's department of public health.

Some parents feel that they would like to have the most complete screening available. It is possible for parents to order and purchase a kit to do so. This kit allows a tiny blood sample (usually taken from a heel stick), obtained at birth or later, to be screened for more than thirty disorders. The kit costs $25, and this fee covers follow-up guidance on the test results. You order the kit and take it with you to the hospital. Your pediatrician or the hospital staff draws the blood, which you then mail immediately to the testing center.

Here are two organizations offering such a kit:

The Institute of Metabolic Disease
Baylor University Medical Center
Dallas, TX
Phone: 800–4Baylor

NeoGen Screening
Pittsburgh, PA
Phone: 412–220–2300
Order on the Web: *www.neogenscreening.com*

No screening test is 100% accurate. There will be some normal babies whose tests come back abnormal, and parents will worry unnecessarily. And there will be a tiny, tiny number whose tests are normal, but who turn out to have one of the underlying conditions.

## Newborn hearing test

Between one and three of every thousand children (about 24,000 in the United States each year) is born with moderate to severe hearing impairment. It's the most common congenital disability, and impossible to diagnose from the baby's physical examination or behavior until the baby is much older—the diagnosis often isn't made until the child is two and a half to three years old.

But if a baby is diagnosed with a hearing problem and is successfully treated before six months of age, her language development is likely to be in the normal range. If treatment is delayed after that, normal language skills are much harder to attain.

As with other screening tests, the newborn hearing test is not perfect, and some babies will be flagged as having problems when their hearing is actually normal. So

## RISK FACTORS THAT DESERVE A
## NEWBORN HEARING TEST

- Family history (sibling, parents, grandparents) of deafness possibly genetic in nature
- Baby is less than 35 weeks' gestation
- History of medications that could affect hearing (very rare, these days)
- Complicated neonatal course—in Neonatal ICU, or on a ventilator for more than five days
- Physical features of the baby that would suggest possible hearing problems (ask your pediatrician)—for instance, very small or oddly shaped ears, skin tags, or pits

don't panic if your baby is asked to repeat the test. Also, about 30% of hearing problems develop *after* the newborn period. Don't take a normal newborn test as lifelong reassurance of normal hearing.

A baby with risk factors for hearing problems should definitely be screened.

### IMMUNIZATION AND/OR TREATMENT FOR HEPATITIS B

Hepatitis B is a virus disease that infects the liver. It is possible for a woman to have had the disease and not know it but be a carrier. Most obstetricians test a woman during the pregnancy. If she is a carrier, the baby needs to have a special gamma globulin shot (HBIG) and a vaccine given immediately after birth. All babies are immunized for Hepatitis B during the first year of life, but babies of carriers need this special early care. If the mother's carrier status is not known, pediatricians will give the vaccine and/or the gamma globulin unless test results can be obtained right away.

### TESTING THE BABY'S BLOOD SUGAR

It's not unusual to need to perform a test on a drop of the baby's blood to make sure the blood sugar level is high enough. If a baby is larger or smaller than average, or has had a difficult womb-to-room transition (a "jump start," for instance), or is jittery or a little listless, the blood sugar may be lower than is desirable. Usually, a drop of blood is taken from the baby's warmed heel, and results are ready in just a minute or two. If the blood sugar is low, sometimes a bit of glucose-rich water is given by mouth: the importance of this outweighs the desire to avoid bottles in a breastfed baby. Rarely, a sugar solution must be given into a vein.

## CARE OF THE UMBILICAL CORD

Umbilical cord care varies considerably. In some places, the cord is clamped; in others, it's tied. Some pediatricians like to have parents clean the cord with alcohol two or three times a day; others paint the cord with a vivid blue solution (it's called Triple Dye). Both of these methods counter infection. Both of these also tend to encourage the cord to stay attached longer. The umbilical cord has no sensation, no nerves at all, so whatever is done doesn't hurt.

# What are the Pros and Cons of Circumcision?

Circumcision is the surgical removal of the foreskin from the glans of the penis. The foreskin is a cuff that protects the glans when the penis is not erect, and retracts when it is. When the foreskin is not removed, a cheesy secretion called *smegma* develops underneath.

At birth, the foreskin usually is so tightly adhered to the glans that it can't be retracted except with great force. In an uncircumcised boy, the foreskin gradually becomes retractable. This may happen in the first two years, but occasionally not until puberty.

Medical, as opposed to ritual or religious, circumcision usually is performed sometime in the first two weeks of life in full-term, healthy babies. After that, the baby shows more focused and long-lasting distress at the procedure. If a baby requires circumcision

after the newborn period, it usually must be done under general anesthesia.

Here are the questions parents ask me:

*What are the medical pros and cons of newborn circumcision?*

*Cons:*

1. It's got to really, really hurt. Even if the doctor gives local anesthesia, that stings too; and it's anybody's guess how well it works.

2. It's performed without the baby's consent.

3. After the procedure, which is performed under sterile conditions, the open wound is then exposed to the baby's stool, which is not sterile.

4. As documented below, most of the medical benefits of circumcision—prevention of cancer of the penis, and cancer of the cervix in sexual partners; and a decreased risk of sexually transmitted disease—can be duplicated in uncircumcised men by using careful hygiene.

5. Urinary tract infections may be more common in uncircumcised boys, but they are still rare, and even if they occur, can be diagnosed, treated, and followed up so that permanent damage is very rare. Yes, a baby with a fever of unknown origin does need a urinalysis and culture performed. And it's true that you need a very clean urine sample, and that this is hard to obtain in an uncircumcised boy. But unless the baby is quite ill you may not have to catheterize the baby or stick a needle

into the bladder. You can clean the penis and then wait alertly with the baby undiapered until he pees, and then catch a midstream specimen in a well-aimed cup. If the specimen shows signs of infection, then the more distressing procedure will have to be done. But if this specimen is sterile, you can forget about a bladder infection as the cause for the fever.

6. Finally, there is some evidence that removing the foreskin can blunt sexual sensation. If Nature put the foreskin there, why should human beings so brutally remove it?

*Pros:*

1. Studies show that applying a prescription numbing cream called EMLA, or giving local anesthesia, or even having the baby suck on a sugar-dipped pacifier decreases the crying. Even if there is still some pain, babies don't seem to be bothered after the procedure itself is over.

2. Yes, it's done without the child's consent. But then if the child has to have an emergency circumcision later on, because of infection of the uncircumcised penis, that is done without consent also. We don't know how frequently this happens, but experience from England suggests about 1% of the time.

3. When an older child has to have a circumcision, it is done under general anesthesia, which carries risks. To say nothing of the trauma to a three- or five-year-old's *Amour Propre.*

4. Even though the wound is then exposed to a dirty field, infections are rare.

5. Studies show that cancer of the penis, a very rare cancer, can be prevented by circumcision. They also show that cancer of the cervix in women is much more uncommon in the mates of circumcised males. Moreover, many sexually transmitted diseases, including AIDS, appear to be less easily acquired by circumcised males. Even though these may be prevented by good hygiene, who's going to promise that the boy will always be in a position, or a mood, to practice cleanliness?

6. Urinary tract infections are ten times more common in uncircumcised male infants. It is true, even so, that such infections are very uncommon in boys. However, the mere statistic means that any baby boy with a fever with no obvious cause may have to have a urinalysis and culture performed, because an undiagnosed and untreated urinary tract infection can damage the kidneys.

   Not only can this be expensive, but difficult. Because the foreskin is still tightly attached, the area cannot be cleaned well, and the resulting collection of urine can be contaminated. This means that to obtain a very accurate urine collection, the urine may have to be obtained by an invasive procedure: either a catheter through the urethra of the penis into the bladder, or a direct needle tap of the bladder.

7. The reports of blunted sexual sensation are generally studies of men who were

circumcised after puberty. Perhaps boys who were circumcised at birth have developed compensatory sexual sensation in other genital areas. As for Nature, she still goes on making a number of items we no longer have any use for, such as appendixes and wisdom teeth.

*If the father is circumcised or not circumcised, won't the little boy feel bad if he doesn't follow the same route?*

I have no idea, but I suspect that the real question is, won't the father—or mother—feel bad? I also suspect that little boys are more impressed by size difference than by the presence or absence of foreskin, but who knows.

*What about locker rooms?*

I can't find any studies on this, and no wonder; you'd have to plant cameras and bugs to find out. Some fathers tell me that this is a big issue, with rude nicknames for both conditions. Others say it is size that counts, not the presence or absence of circumcision. Still others say that merely looking at somebody else's equipment is enough to get your kneecaps broken. The boys themselves don't seem to want to discuss the issue.

# What if we have to put the baby in daycare soon after birth?

## LET'S LOOK AT IT THIS WAY

The first two weeks after the birth are not only exciting and joyous but very harrowing and demanding.

From two weeks to two months, parents feel increasingly as if they're getting the hang of it; but there may be a lot of crying and perhaps a bout of colic to contend with. From two months on, having a new baby becomes more and more fun.

By age three months, most babies should be sleeping through the night. They coo and smile a lot, and focus on things and people with glee. Colic has run its course.

So it doesn't seem fair that three months is considered by many employers to be a *generous* length for maternity leave. (Paternity leave is so rare, still, that I can only include it by implication.) It can be tough to leave a baby you've just started to understand and enjoy.

It may be even harder to return to work earlier than three months. Leaving a baby who seems amazingly vulnerable and fragile in the care of someone else can be very difficult. Yet six to eight weeks is the *usual* (as opposed to generous) maternity leave.

Choosing a daycare situation always takes thought, no matter what the age of the child. However, for mothers who must return soon after the birth, especially those who must return sooner than three months, I very strongly urge that the daycare decision be made before the baby is born.

From contacts with so many parents, I have a sense of how heavily this decision can weigh if it is postponed. Perhaps the problem is that after birth, mothers are so exhausted and in the throes of hormonal tides. Perhaps the demands of caring for the new baby make time seem to pass incredibly swiftly. Perhaps choosing a daycare situation for a real, rather than fantasized, baby feels like abandonment.

And perhaps being able to visualize the daycare person and situation even before you yourself hold the baby makes the transition easier. In any event, I strongly suggest making the decision beforehand.

The ideal daycare situation for a very young baby is that which is as close to one baby/one adult as possible. If, as is usually the case, there is more than one baby per adult, the most ideal arrangement is when all the babies are very young, or when the other children are quite a lot older, beyond the toddler and preschool years.

There are several reasons for this:

- The caretaker can have a routine for feeding, diapering, cuddling, soothing that is not interrupted by other highly demanding young children.
- Young children can be both curious about and very jealous of small babies and require constant supervision not to hurt them.
- One of the "jobs" of toddlers and preschoolers is to catch infectious diseases in order to build up lifetime immunity to them. They're equipped to do this job because their own immune sys-

tems are relatively mature. But young babies don't have mature immune systems, and are very vulnerable to serious consequences of "ordinary childhood diseases." They are most vulnerable for the first three months of life, though the immune system doesn't approach real adult levels until about two years of age.

Solutions to the daycare problem I've heard from parents include:

- Seeking out other parents from prenatal classes who might be interested in watching one or two other age-matched babies.
- Banding together with two or three other parents and hiring a mutually satisfactory daycare person for all the babies.
- Investigating grandparents, not only of their own family, but those of their friends and prenatal classmates.
- Inveigling their prenatal class teacher into starting infant babycare, in which she takes several babies for the first six months only.

In making a decision about a daycare person for babies this age, the most important factors are:

- Personal hygiene and health, with no smoking permitted in the household, even when the babies aren't there, and no pets that could endanger the baby in any way.
- An environment that permits health and hygiene. The area for diapering should

be separate from the area devoted to preparing bottles. There should be an area for handwashing, with a soap dispenser and paper towels.

- Warmth and a physically nurturing disposition, along with patience and the ability to tolerate crying. It's important that she knows that holding a very young baby or picking up the baby as soon as he cries is appropriate and won't "spoil" him.
- Competence and experience enough to recognize a potentially serious problem and the linguistic ability to be able to get help.
- Ability to accept parental suggestions about safety, schedules and hygiene.

How do you find a daycare person or nanny? Word of mouth, ads, and agencies. It takes time, energy, shrewdness, and luck to find a good one. I have been urged by a number of parents to add a caveat: *Even though an agency may say its nannies are bonded and/or prescreened, check the references yourself.* Don't be satisfied with a cautious or waffling recommendation from a previous employer who may be reluctant to "ruin her chances for a good job."

# What should we try to find out about our Pediatrician at the Prenatal or Pre-adoption Visit?

Pediatrics is a very strange medical specialty.

The pediatrician is invited to participate fully in the mundane details of care such as diaper choice and how to burp and in the emotional crises of everyday life such as potty training and whining. And yet the pediatrician also has to be ready and able to deal with truly life-threatening emergencies.

Pediatricians see babies from birth for sick visits and injuries and for routine well-child care. The well visit includes a developmental history and/or exam, a complete physical exam with special attention to the particular foibles of that age (congenital hip problems in infancy, abdominal masses in toddlers, developmental problems all the way along), and "anticipatory guidance" about health and developmental questions.

So it's a good idea to get to know this person, or group of people, well.

The hard-core questions are:

- Are the physician and the partners in the group *Board-Certified* (completed all training and taken the exams for the American Board of Pediatrics) or *Board-Eligible* (has completed training but has not been in practice long enough to take the exams)?
- What are the office hours? Many pediatricians offer evening and weekend hours for sick visits, and some even for well-baby visits.
- What kind of after-hours coverage is available? If your physician is in solo practice, does he or she sign out to the local emergency room, and if so, who covers that emergency room? Does the emergency room doctor at least consult with the pediatrician by phone? If your physician is in a group practice, who and

how well trained are the other members of the group? If a child must be seen at night or on the weekend, what does that involve?

• Is there a separate emergency line, and what is the protocol if there is a true emergency? Is the office equipped to deal with emergencies? Has your doctor (and/or his or her partners) taken the American Heart Association's Newborn Resuscitation and Pediatric Advanced Life Support courses? Has your local emergency room trained its doctors and nurses for pediatric emergencies, and stocked the necessary child-sized equipment and medications?

• What happens if your child needs hospitalization? What hospitals have children's wards? Does your physician attend to the hospitalized child in person, or confer with doctors at the children's hospital?

• Who answers phone calls about medical problems?

Many pediatric offices include nurse practitioners or physician assistants (often referred to as P.A.s) on their staffs, and they are usually excellent sources of care.

Then there are questions of style:

• Does your doctor genuinely enjoy children, and try to make them comfortable during the examination and treatment? As one mother complained about her child's previous doctor, "There wasn't any foreplay!" Does your pediatrician treat children as real people?

• Are there areas of shared values? Does your doctor disapprove of mothers working, of babies attending daycare, or of fathers staying home with the baby?

• Does your doctor think of mothers and fathers as being equally capable of nurturance and kindness, of intelligence and competence? Does your pediatrician assume that fathers will feed, diaper, play with, and medicate their children? That mothers can and will discipline, stay calm in a crisis, and want and understand explanations that are as complex as necessary? If yours is an unconventional family arrangement, will this in any way prejudice your child's care?

• Does your doctor feel comfortable in explaining the rationale for medical decisions, outlining options, and discussing opposing viewpoints?

It's easiest to discover the answers to these questions if both parents can come to the prenatal visit, and if you can observe the physician under field conditions, so to speak. Arrive early and chat with the people in the waiting room. Keep the exam door cracked so you can hear the interactions in the hallway and other rooms.

If your pediatrician seems a bit rushed, or wooden, or abrupt, or otherwise not quite the wonderful guru of childhood you'd hoped for, be charitable and optimistic. Pediatrics is a contact sport: perhaps your middle-aged pediatrician just had to chase a toddler down a corridor, or crawl under an exam table, or remove an unidentifiable item from somebody's nose. Perhaps your youthful pediatrician is pregnant. Perhaps your pediatrician of any age is preoccupied with a very sick or worrisome child, or has to go to the bathroom and hasn't had a moment to all afternoon.

Be charitable. Remember that this is someone who goes to sleep at night never knowing if the next two A.M. phone call will be a diaper rash or the delivery room requesting urgent attendance at a delivery of premature twins.

# What's the most important thing to understand about the Baby's Insurance?

**Don't even think of skipping this part.**

Everybody hates confronting the realities of insurance, but it is crucial these days to know the answers to the questions below. Some parents just tackle one question a day before the baby is born. Write down the answers.

- How and when do you enroll the new baby on your plan? If you wait too long after birth, you may miss a crucial deadline.
- Does your insurance cover the delivery and normal newborn care? Circumcision, if you want one? If so, in the hospital or in the office?
- Does your insurance cover the hospital at which the delivery will take place? Does your pediatrician have privileges at that hospital, or will you need to have an "on call" pediatrician attend the birth, if a

pediatrician is needed? After the baby is born, which emergency room is covered on your plan? It may not be the hospital of birth, or your closest hospital, or a hospital with a Children's Ward or Children's Emergency room.

- Are you on an HMO? An HMO requires that you see participating physicians, and use participating services (hospitals, emergency rooms, labs, X rays, etc.) only.
- Who is your baby's Primary Care Provider? Make sure that the person you have chosen for your baby's pediatrician is the baby's official Primary Care Provider. You need to have a card with the name and phone number of your baby's Primary Care Provider on it. The Primary Care Provider is responsible for either staying within the health plan guidelines for services, or for obtaining authorizations for procedures not listed in the guidelines.
- Is your baby's Primary Care Provider/Pediatrician a member of an IPA (Independent Physicians Association)? The IPA has its own guidelines, which may be more restrictive than your Health Plan's guidelines. If a service is denied, find out if it was the IPA or the Health Plan that denied it. If it was the IPA that denied it, you appeal to the IPA first; if the denial is repeated, then appeal to the Health Plan.
- If you are on a PPO, POS, EPO, or any other variation of managed care, check what physicians and services are listed as preferred providers. These will be the

ones the plan covers most completely. When you go out of the plan's preferred provider list, there is always a penalty for doing so. What is the penalty for going out of *your* plan?

- Does your insurance pay for well visits? (Beware the temptation to ask the pediatrician to call a well visit a sick visit if the insurance doesn't cover well ones. The diagnosis may turn into a liability later on, when you reapply for insurance.)
- Does your insurance cover immunizations? If not, is there a "free clinic" available, and what is it like? Are you eligible?
- Does your insurance pay for prescriptions? For dental care? For mental health professionals?
- What is the deductible and/or copayment for office visits? Are there any waivers for preexisting conditions?
- Will your pediatric office do the billing for you, or will you have to pay and then be reimbursed? (This has very little to do with the kindness or willingness of the office. Insurance companies usually make their own rules.)

Now, don't you feel better? You know more about your health care plan than anyone else in your peer group.

## Looking Ahead

Congratulations! You are now fully prepared to get the most out of the Pre-Baby visit. Just one last caution: Don't ask your pediatrician any of those Managed Care questions. Most of us don't have a clue. Instead, make a buddy of the person who runs the referral office. Bring him or her some chocolate. That person may turn out to be a very important person in your life.

**Now, go turn down that hot water heater to 120 degrees, and I'll see you after the baby arrives.**

## { T W O }

# Birth to Two Weeks:

## Wonder and Worry

### To-Do List

- Make sure the baby's enrolled on your insurance.
- Schedule a follow-up visit with the pediatrician no later than five days after hospital discharge (three days for first-time nursing babies). At the same time, make an appointment for a well visit at two weeks of age.
- Feed on demand—but make sure the baby demands to be fed *frequently* enough. Count the pees and poops to make sure Cherub is getting enough milk. When in doubt, weigh Cherub at the pediatrician's.
- Put the baby on her back to sleep, day and night; keep her on her back to play until the umbilical cord has fallen off and the area is healed.

### Portrait of the Age: Already a Person

"Is this the way they're supposed to look?" Angie's father, Sam, stares at the form in the nursery crib.

Angie, tongue lolling, gazes up at him. Sam has never before seen anyone able to keep one eye fully open and one fully shut with no appearance of strain. Perhaps, he thinks, the lid of one or the other got broken, somehow, during the delivery. He is trying hard to focus on the baby's face; the shape of her head disturbs him profoundly. As he watches, she yawns hugely. Are some of her facial muscles missing? Never has a mouth opened so wide. He can see the little thing hanging down in back. At least she has now scrunched both eyes tightly shut. This must mean that the eye that was stuck open, at any rate, is capable of working fully.

Sam feels slightly queasy. How can he possibly take on this responsibility, making sure his daughter opens her mouth to eat, shuts her eyes, shapes her head—breathes, even! He shuts his own eyes, the better to think, maybe pray. And hears a sound. At first he thinks it means something terrible. The noise is as loud as a firecracker.

But as he listens, it seems to beat to him in a gentle, reassuring code. A message, a gift, a statement that Angie can perhaps be depended on to find her own way, that he can trust her, just a little at first, to be her own person. He opens his eyes and watches his daughter, calmly sucking her hand, as her chest bounces intermittently.

"The healthier the baby, the louder the hiccups," the nurse says cheerfully.

Whether a baby arrives by birth or by adoption, it's a little like meeting a blind date with whom you know you are destined to fall in love. The difference is that the baby is such a total enigma. When confronted with an enigma, one natural response is excitement and joy; another is anxiety. Most people feel both, and the balance changes from minute to minute and from hour to hour.

When our daughter was born, I was completely ignorant about healthy babies. I was more comfortable starting an IV in a preemie than burping Sara. After all, I'd been trained about IVs and preemies.

While I wouldn't have missed either the thrill or the worry for the world, I could have used more information. It would not have calmed either the excitement or the anxiety very much, but I would have been able to place what I was feeling and observing in some kind of framework. Now that I have

accompanied so many new mothers and fathers through the first weeks after birth, I have an idea of what that framework is.

Here, then, is what I wish I'd known when Sara was a newborn.

- The first two weeks are unbelievable, whether for single parents or parents with partners. Everything is new and thrilling and awkward, and the idea of life ever being routine again seems a fantasy. Sort of a combination of a new job, a new lover, and a very demanding sport.

- From two weeks to two months, the baby cries a great deal but you do feel more in command of the situation. Besides, during this time the baby starts to smile, and what a great reward that is.

- From six weeks to three months, Cherub becomes more and more fun. By at least three months of age, most full-term babies are sleeping through the night, gleeful and chortling, and nearly all colic has resolved. Coparents are now communicating normally, but there is only one topic of conversation.

- You simply can't predict a baby's long-term temperament from the experience of the first two weeks. The components of temperament are: how active the baby is spontaneously, without stimulation; how intensely she reacts to external events; how easily she can soothe herself; and how readily someone else can soothe her. Birth and the adjustments to living outside the uterus are heavily demanding. A baby who seems very jumpy, or very quiet, or very unhappy, may just be reacting physically. After all, what if your baby were judging your temperament? Would this two weeks be a fair sample?

## Separation Issues: Bonding

When Sam has a glimmer of an idea that he can trust Angie to be her own self, it's a special initiation into parenthood.

One of the secrets of getting to know the new baby is to allow her her mystery. There is only so much one can do to control the new baby's behavior. Let her take the lead, and don't impose things: schedule, entertainment, social engagements. Some babies love to be held and cuddled; some prefer to have only one or the other part of their bodies held and stroked. Some can sleep through all kinds of havoc; others become agitated and alarmed at the sound of their diaper tabs being torn. Some like rhythmic motion; others like to be moved as little as possible.

All, however, seem to be sensitive to body language in adults. Inexperience and awkwardness are normal, and babies do not resent less than expert handling. But anxiety and anger seem to be communicated all too easily. Then the baby becomes upset, the adult's negative emotions are heightened, and a vicious cycle can ensue. If this seems to be happening, there is really only one thing to do: put the baby down in her crib, get help, take a break, feel better, and start again.

There are so many things you *don't* have to worry about yet: giving her a bath, brushing her teeth, teaching her manners or morals. Hold the thought.

There has been so much emphasis on bonding (in many hospitals, the nurses are asked to chart how well parents appear to be bonding to the baby) that alas, it may feel like yet another test to pass, a burden to carry.

Try to relax and take your time. Bonding, as I've said elsewhere, isn't like Superglue—either it takes right away, or you've made a mess. Bonding is slow and complex and made up of many emotions, both pleasurable and uncomfortable. All parents feel some worry, self-doubt, ambivalence. Some may have initial sorrow or even anger because they wanted a baby of different sex, appearance, or temperament.

Most importantly, bonding doesn't require immediate time with the baby just after birth. It's great if you can have that time, but if the baby or the mother or the father must be otherwise engaged, don't imagine that bonding has been disrupted. You will just start where you left off.

A new parent has to adjust not only to the arrival of a specific, mysterious baby, but to a brand-new culturally and personally defined role. It is easy for these two kinds of adjustment to become confused, and indeed they do overlap. However, it is useful to separate getting to know the baby from getting to know the role of Parent.

If each parent gets too enmeshed in his or her own role, and can't function as a helpful audience for the other, the first two weeks can be more difficult than necessary. Mothers can help fathers by trusting them to care for the baby and not instructing or hovering. Fathers can help mothers by taking the initiative to do so. Grandparents can help most by taking their cues from the parents, and, if help is asked for, sticking primarily to housework and errands.

Finally, some parents may be haunted by what the wonderful child psychologist Selma Fraiberg calls "ghosts in the nursery." A new baby recalls for all of us our own childhoods, our feelings for our parents, our early feelings about ourselves and the world. Sometimes these feelings can be overwhelming. If they are, the best thing is to recognize that you are in the grip of an important, universal, and potentially helpful experience, and that you are entitled to find the counselor and confidante (your pediatrician can help) who will guide you towards insight.

## Setting Limits: Not on the baby's behavior, but on the adults'!

The baby doesn't need limits set on behavior at this age; you couldn't, even if you wanted to. But sometimes adults, no matter how loving, try to do so. For instance, it's not unusual for fellow-caretakers to have wildly varying notions about baby care, notions that can range from the inappropriate to the downright dangerous. Here are some that often need arbitration:

*Back is best.*
The safest sleep position for babies—day and night—is on their backs. The back posi-

tion reduces the chance of sudden infant death syndrome (SIDS) significantly. Rarely, an exception is made for a baby with a special medical condition. Your physician will tell you if this is necessary, and may also prescribe an "apnea monitor" so that any too-long pauses in breathing trigger an alarm. Also, go easy on letting Cherub engage in tummy-down play until the umbilical cord is off and healed. You don't want her scrunching it or getting it wet—that can lead to infection.

*Jiggling the baby to soothe or entertain him is potentially dangerous.*
Any jiggling that makes a baby's head wobble is a very bad idea. Babies shouldn't be tossed around, even by adults with excellent reflexes. It isn't being dropped that's the main danger; it's bleeding into the brain. The neck muscles are very weak and the blood vessels lining the brain are fragile. Bouncing the baby, even with the head level, is also a problem.

*"Spoiling" the baby is not a possibility.*
Babies this age cannot be spoiled.

A new baby cries from need. It's not as if she has a lot of other options. What the baby needs to learn at this age is trust—that her cries are heeded promptly. In fact, studies show that babies under three months of age who are picked up immediately, within thirty seconds, of starting to cry, cry much less thereafter. *This is true even if the baby doesn't stop crying after being picked up.*

*Visitors are a mixed blessing.*
This is a very individual matter. Some people require peace and quiet to regroup; others feel abandoned and need cheerful voices to recharge their batteries.

One solid rule, however, is that no visitor should endanger the baby's health. Thus, no smoking in the house; no sick visitors, even if they say "it's just an allergy"; and—this is not going to win me friends—no toddlers and preschoolers. Any sociable young child has a high likelihood of carrying a virus, even though she is apparently healthy. Viruses are most contagious just before symptoms appear.

Having visitors should be relaxing and energizing, not exhausting. You know exactly what I'm talking about here.

**So if you want to restrict visitors, it's easy: Tell people the pediatrician has told you to do so. That is true—I just did.**

*Pacifiers*
When it comes to pacifiers, three considerations are important:

- **Make sure that the pacifier is really what the baby wants and needs.** New babies under two weeks of age always should be offered milk whenever they're willing to suck; they have a lot of weight gain to accomplish. Older babies may want cuddling, or a change of scene, or your nose to play with, and the pacifier is a poor substitute for such attention.
- **Make sure the pacifier is safe.** Pull on it to check that the nipple can't become detached from the base; it could be

choked on if it came loose. The molded one-piece pacifiers are safest. The base of the pacifier must be larger than Cherub's mouth: discard the small infant versions when they are outgrown. Never attach a pacifier to something that encircles Cherub's neck! Strangling is a real danger.

- **Make sure it's given up by six months of age.** Beyond six months, pacifier use is associated with an increased risk of middle ear infections. Moreover, Cherub is more and more likely to form a dependency on the pacifier which will become harder to deal with. Pacifier use over a year of age is associated with dental problems and with speech and language difficulties. There are no studies that indicate a pacifier prevents thumb sucking later in life.

*Schedules are for later on in life.*
No matter what you've read, this is not the time to establish or enforce a schedule. A baby should be fed whenever hungry. The only real danger with demand feeding is that the baby may *not* demand to be fed often *enough*. (See the later section on "happy to starve" babies.) So please, no time charts to schedule feedings and no using a pacifier to stretch the time between the feedings.

# Day to Day

## MILESTONES

How can a newborn baby be said to have Milestones? Easy. These are the Milestones accomplished by no longer being a fetus.

Unlike the older baby, these do not take the form of achievements as we normally think of achievements. Rather, they take the form of observable phenomena: changes in physical appearance, behaviors, biological abilities.

## THE NEWBORN FROM HEAD TO TOE

As with Sam, when parents first see their newborn, it takes awhile to process all that brand new information. You see parts at first, not a whole baby. I suspect this is why parents focus on counting the fingers and toes: familiar, reliable body parts.

Every now and then an obviously-upset-but-trying-not-to-appear-so parent or relative will say, jokingly, "He looks just like E.T." So the first thing to keep firmly in mind is: Spielberg designed E.T. on purpose to have some of the most appealing features of a fetus. It is E.T. that looks like the baby, not the other way round.

Now, taking it body part by body part, here are the features that cause parents unnecessary concern, the features unique to newborns:

## PHYSICAL FEATURES

### Heads
- Head shapes range from the molded skull of a vaginal delivery (in its extreme form fondly referred to as a Banana Head) to the elf-point or flattened face of a breech. The longer the time the head has been molded, the longer it takes to resume a rounded shape.
- There is usually only one "soft spot" or fontanel. This is the open space where the plates of the skull haven't come together. It is located in the front, and it is shaped like an elongated diamond. Usually it isn't any bigger than two finger-breadths in each direction. If it is much larger, or if there is another fontanel at the back of the head, the pediatrician may want to check for thyroid deficiency or other problems.
- Ridges and protuberances of the skull bones are common where the plates of the skull are coming together after the molding that occurs during pregnancy, labor, and delivery.
- Facial lack of symmetry follows the same rule. Most often a baby's lopsided smile or cry will straighten out over a three to-six month period. No permanent nerve damage was done; just some pressure. Check with your pediatrician, just to be sure. Sometimes such asymmetry means that the baby has a wry neck, or *torticollis*, from the position in the womb. If that's the case, you may be asked to help stretch the neck muscles with positioning and exercises.

  Very very rarely, a facial asymmetry can occur because a suture—the place where the plates of the skull come together—has fused prematurely. This is called *craniosynostosis*. If this is suspected, your pediatrician will refer you to a pediatric neurosurgeon to correct the situation.
- Bumps on the head are common. A spongy swelling right where the baby

emerged is usually a *caput*, a little retained fluid at the point of pressure. It goes away in one or two days. Another type of swelling feels like a water-filled lump, and the skull surrounding it feels a bit like a cracked eggshell. This has the intimidating name of *cephalohematoma* (it translates as "head bruise") and occurs when there is bleeding between the outside of the skull and the membrane covering the outside of the skull. It doesn't cause brain damage. It does take a long time, sometimes up to three months, to go away.

### Skin

- Bluish hands and feet are normal. Babies are born with extra red blood cells, and the circulation to the extremities is a bit sluggish.
- Most babies have birthmarks, most commonly patches on the nape of the neck ("stork bites") or on the upper lip and eyelids ("angel kisses"). These affectionate names arise form the frequency of the marks and their tendency to disappear within the first few years. Many babies also have bluish spots with the unfortunate name of Mongolian Spots. Sometimes these are mistaken for bruises. They can appear anywhere on the body and are perfectly normal; they are frequent in any family of African, Asian, or Mediterranean descent. These usually fade by mid to late childhood.
- Peeling is normal. No creams or lotions are necessary, and indeed may themselves cause problems.

- Rashes are common in newborns and usually innocent, but any rash needs to be checked by (or at least described over the phone to) the pediatrician. The most common normal rashes are:
  1. A flamboyant newborn rash with the scary name of *erythema toxicum* often appears in the first three or four days. Each spot consists of a tiny white bump surrounded by a red patch. It is so common your doctor may forget to mention it, even if the baby is covered with it. It will go away by itself.
  2. *Milia* are tiny dots like a sprinkling of salt on the nose and often the rest of the face. Each tiny dot is a little cyst. Milia disappear over the first few weeks to months.

### Eyes

- Babies' eyes don't focus well at first. However, you ought to be able to assure yourself that they do move in all directions—that one isn't just stuck looking inward at the nose, for instance. Babies are born farsighted, so they certainly can see across the room; but they prefer to look at things close-up, at about a 12- to 18-inch distance.
- Puffy eyelids are common and normal. (If you'd stood on your head for three months, the position of most newborns during the last trimester, your eyelids would be puffy too.) Red spots in the white of the eye are too; they are tiny broken blood vessels from the exertions of delivery. (Red and swollen, rather

than puffy, eyelids are not normal and require immediate attention.)

- Babies like to open one eye at a time, and often prefer to open one eye more than the other. Rarely, a baby's eyelid is incapable of opening spontaneously. Such a drooping lid is called *ptosis* (TOE-sis). If the drooping lid covers the pupil of the eye, that eye doesn't see anything. Therefore, nerve impulses from it do not go to the brain and stimulate the part of the brain in charge of seeing. If the brain is deprived of nerve impulses for longer than a few weeks after birth, it can "forget" how to see, even after the eyelid is fixed and the eye is open. So if you suspect ptosis, see your pediatrician promptly, in case you need an urgent referral to a pediatric eyelid surgeon.

- White, yellow, or even greenish goop in an otherwise normal-appearing eye usually is not abnormal. It's the residue of tears (white blood cells left when the water evaporates, plus normal tear mucus) that are backed up due to a clogged tear duct. The tear duct carries tears from the eye to the inside of the nose, and when it becomes clogged, the tears gather in the eye or run down the cheek. However, if the white of the eye is red or if the lids are red or swollen, this is an infection that needs immediate attention.

### BLOCKED TEAR DUCTS

A baby under three months of age with a teary eye or eyes often has a blocked tear duct, but your pediatrician needs to see the baby to be sure—you don't want to miss infection or the rare case of glaucoma.

Sometimes parents think that the tears mean that the baby "has finally started making tears" and are glad to see them. *Au contraire.* All babies make tears by about two weeks of age: if they didn't, the eyes would be seriously dry. But normally, the tears are carried by a tiny tube from the inside corner of the eye into a little sac called the tear sac, and then into the nose, where they're absorbed by the nasal lining.

If it's a blocked tear duct, what's happened is that the little tube that drains the tears from the inner part of the eye into the inside of the nose has not fully opened by the time the baby is born. It is partially closed.

When this happens, it's a good idea to try to keep the tears flowing, so that they don't build up and cause the tear sac to get swollen and maybe infected.

Your pediatrician will want you to massage the tube to squish the mucus down and out into the nose, unblocking the tube. If there is any sign of infection, you'll be prescribed antibiotic drops to use just before you massage. Some ophthalmologists suggest massaging in the down direction only, to push the tears out into the nose, if possible. Others suggest massaging in the up direction only, to push the tears up out of the duct system so they don't get trapped in the tear sac. Still others suggest massaging in both directions. Probably, they all work.

So do what your own expert tells you to do. What I can help with is getting any prescribed drops into the eyes:

- Have the baby vertical on somebody's shoulder, and give the baby something to suck. This helps the eyelids relax: it's sort of a reflex. Look at the baby's eye. (If the white of the eye (*conjunctiva*) is looking red, or the pupil cloudy, or if there is swelling and redness of the skin in the area, call your pediatrician.)
- Wipe away tears, crust and goop.
- With the baby's eye open, put a little traction just under the lower lid. It should flip out at you a bit, making a pocket. Place a couple of drops in the flipped-out lining of the lower lid.

Remember, you're unblocking the tube, not curing the underlying narrowing. So don't be surprised if the tube gets blocked again after you've fixed it the first time. Most babies' tear ducts grow so that the problem disappears in half of them by about six months, and nearly all by a year of age.

Once in a great while, despite care, the tear sac does become distended with backed-up tears. When it does, it looks like a blueberry right next to the inner part of the eye, between the eye and the nose. You don't want it to get infected! Call your pediatrician. Sometimes very aggressive massage will help, but some pediatric ophthalmologists feel that draining it surgically, before infection can occur, is the safest course. An infected tear sac looks like a big pink-red marble, in the same location. A baby with this condition usually needs hospitalization.

*Nose*

- Most flattened noses reshape themselves just fine. One exception: if the septum of the nose is very slanted so that one nostril looks a lot bigger than the other, this is a deviated septum and treatment in the first couple of days, done easily and with no need for anesthesia, can prevent problems later. Ask your pediatrician.

*Mouth*

- Tiny hard white beads on the gums and roof of the mouth are normal. They are called Epstein Pearls. They'll go away in time.
- Sucking blisters on the lips mean that the baby has been sucking on something, hand, thumb, arm, or her own lip, for quite a while in the uterus.

*Breasts*

- Swollen breasts are common in both boys and girls, from the mother's hormones. Rarely, a drop of milk, "witch's milk," may appear. That's normal. Don't let anyone squeeze the breasts; that can cause infection. A red, painful breast needs prompt medical attention.

*Umbilicus*

The cut cord makes some people queasy at first. Knowing that the cord has no nerves, and that manipulating it gives no pain, doesn't help; perhaps it is a metaphysical queasiness. My suggestion is to recognize the feeling and just live with it.

Some babies have a cuff of skin that protrudes from the abdomen, sheathing the

Do NOT put castor oil, clay, or any other home remedy on the cord. Oil can seal in bacteria and give rise to a serious infection. Clay has on occasion been reported to contain tetanus bacteria—a life-threatening disease can ensue. Do not use any iodine preparations such as Betadine on the cord, as the iodine can be absorbed into the bloodstream and make the baby ill.

lower part of the cord. This is normal, and when the cord falls off that cuff usually tucks itself right back into the belly button. The ones that don't may become "outies."

As to the cord, many physicians either paint it with a rather Sci-Fi blue dye that prevents infection or tell parents to clean the area with alcohol or with dry gauze pads several times a day. If you have been asked to do the latter, try to get into the little pocket between cord and skin. Slight oozing of blood is normal. Redness of the skin, blood that trickles or flows, pus, or a foul odor is not. Keeping the baby on his back, asleep and awake, will keep the umbilicus dry and uncrushed. So will folding the diaper down below the area.

After the cord falls off, there may be a little glistening dirty-gray nubbin left called an *umbilical granuloma*. It's healing tissue that got a bit enthusiastic. It should be treated by the pediatrician so as not to be a focus of infection, but it isn't an urgent, scary problem.

A tiny amount of dried or a drop of fresh blood may appear now and then. After the cord is off, wipe the blood off with a swab of baby oil rather than with alcohol; alcohol may promote more bleeding. Anything more than this tiny bit of oozing needs prompt pediatric attention.

Many African-American babies and preemies, as well as some full-term babies of other racial backgrounds, will have an umbilical hernia. This is a spreading of the abdominal muscles that allows a bit of intestinal contents to protrude. Most often, this is a normal variant that closes in time—usually by about age four. Rarely, it can indicate a problem, such as low thyroid hormone, and it is always worth checking with your pediatrician.

*Girl's Genitals*
The vulva is swollen from maternal hormones, and the amount of discharge is often copious. Sometimes it is mucousy and thick, sometimes clear. There is often white cheesy vermix trapped in the folds, as well. It isn't necessary or good to scrub all this; just try to sponge away any stool that gets involved.

*Boy's Genitals*
- The penis should be straight, without a curve. A curve is called a *chordee* (kor-DEE) and should have surgical correction. The opening for the urine should be right at the tip-end of the glans. If it is on the lower edge of the glans, or on the shaft of the penis, that's called

---

## SURPRISE!

Many babies have a little vaginal bleeding in the first week or so as the uterus reacts to the withdrawal of the mother's hormones. *Be sure everyone who changes her diaper knows that this can happen and is normal.* I know of at least one grandfather who responded with chest pain.

---

*hypospadias* (high-poh-SPAY-dee-us) and needs to be corrected. A baby boy with either of these conditions should not be circumcised, as the foreskin is used in surgical repair.

- **Circumcision:** The site looks very angry and red at first. It's pretty upsetting to look at, but fortunately the babies themselves don't appear bothered by it after the procedure is over, perhaps because they can't see it. As the circumcision heals, a yellow-green crust forms. This isn't infection; it's healing tissue. Infection is rare, and heralded by a red shaft of the penis and the presence of white or yellow pus, not crust. For a discussion of circumcision, see the previous chapter. Once the penis is circumcised, the exposed glans is not pink but a kind of orchid-lavender.

- In an uncircumcised baby, the foreskin sticks tightly to the glans of the penis and can't be easily retracted. Leave it alone! Don't let anyone forcibly pull it back.

- Sometimes one or both testicles are not down fully or even partially into the scrotum. Your physician will discuss this with you. Much of the time the testicles descend fully by the time the baby is six months old and do not need correction.

- You may discern that there appear to be two testicles on one side, or that one testicle feels much larger than the other. Almost always, what you are feeling is a normal collection of fluid in the scrotal sac. This is called a *hydrocoele* (HIGH-dro-seel), and most of the time it will disappear by the age of a year. Do bring this condition to the attention of your pediatrician.

- There is an enormous range of normal size. Sometimes the penis looks remarkably long; sometimes it is only about two centimeters, even if you stretch it all the way from where it starts at the pubic bone to its tip. Sometimes the penis is buried in the pad of fat, and is nearly invisible. It is your physician's job to reassure you that all is normal, and then it is your and other observers' task to avoid rude comments, teasing, and auguries for the future. The scrotum, too, can vary enormously. Sometimes the testicles are encased in a tight little smooth sac; sometimes the sac is big, heavy and wrinkled.

If a baby is in the breech position for any amount of time before birth, the chance of *hip dysplasia*—the hip socket forming incorrectly—is higher. Many pediatricians check all breech babies with a hip ultrasound (see Glossary) within a few weeks of birth. If there is a family history of hip dysplasia—a sibling, parent, grandparent, aunt, uncle, or first cousin with the problem—such an ultrasound may also be suggested.

*Hands, legs, and feet*

- It's normal for hands and feet to look bluish and to feel cold, while the rest of the baby is rosy and warm. This is called *acrocyanosis* (AK-roh-sigh-an-OH-sis) and is the consequence of newborn circulation.

- For months, the baby's legs "want" to be folded into the intra-uterine position. For a baby with the usual presentation, the lower legs look quite bowed. These straighten out in time.

- It is crucial that your physician check the baby's hips carefully for potential displacement. Most will be just fine. If a hip is found to be dislocated or dislocatable, early treatment to keep the legs positioned in abduction—knees pointing out—generally corrects the problem. This usually means simply using double diapers at first, and then a contraption of straps with velcro fastening called affectionately by some parents a Rhino Kicker. This contraption in no way impedes a baby's motor development, even when you kind of hope it might slow it down a bit.

- Feet look curved in. If this is something that needs attention, your doctor will talk with you. Most of the time, nothing is needed. Sometimes stretching the foot, special shoes (before the baby starts to walk), or in severe cases brief casting is recommended.

- Toenails always look tiny, funny, and ingrown. They seem not to grow in length at all for the longest time, even a whole year. Rarely are they truly ingrown, but if the skin is red and shiny, ask your physician.

## BEHAVIOR AND ABILITIES

*Facial expressions*
Newborns make really strange faces. They cross their eyes, stick out their tongues, stare at you with one eye open and one shut.

*Breathing*
Babies breathe irregularly. They normally breathe rapidly for half a minute or so, then take a lull of ten to fifteen seconds, then resume with a big breath. If you count for a whole minute while the baby is asleep, you'll find the baby takes forty to sixty breaths a minute. If the rate is *consistently* sixty or

A rule of thumb for judging whether a new baby is getting enough milk: She should have *at least* one wet diaper and one poopy diaper for every day she is old. That is, in the first twenty-four hours after birth, one of each; in the second twenty-four hours, two of each; and so on. (Don't worry, the numbers stop going up at between six and eight days.) The poops should be real poops, not just stains, and by Day Five should be bright yellow, like mustard out of the squeeze bottle.

higher when the baby is asleep call your pediatrician.

A frequent normal baby noise has the sophisticated medical name of "Schnurgles." This is a snort, chuffle, and gurgle that results from milk in the back of the throat and nose. It is a gentle sound, often just at night, in a baby who is otherwise entirely well. When in doubt, ask your doctor.

*Oddities*

Besides eating, drinking, peeing, and pooping, babies perform a number of other feats that often confound or worry parents. Hiccuping, even a lot of hiccuping, is normal.

Sneezing is normal. (Repeated coughing is not.) Trembling the chin is normal, and doesn't mean the baby is cold, frightened, or having a convulsion. Gas, or wind, the nicer English term, is normal. Many babies emit from both ends with perfect comfort and lack of inhibition. Along with this are loud gastric noises; as with the Duchess in the

limerick,* the Rumblings Abdominal can be Simply Phenomenal. If the appetite and other behaviors are normal, don't worry.

Occasional jittering of arms and legs is almost always normal. If you put your hand on the jittering part and it stops jittering, or it stops when the baby sucks, everything is all right. If it doesn't, or if Cherub is jittery most of the time she's awake, call the pediatrician promptly. Sometimes this can be a sign of low blood sugar or calcium, or other problems.

*Abilities*

Every month something new is discovered about what a newborn can do. This is all fascinating, but I don't think this is necessarily the time to dwell on the subject in much detail. Full-term newborns love to look at faces, the color red, and invisible entertainment to one side of the adult staring at them. If you work at it, you can make them follow an object with their eyes, but not past the midline. And why would you want to?

*I sat next to the Duchess at tea.
It was just as I thought it would be.
Her rumblings abdominal
Were simply Phenomenal
And everyone thought it was me.

It's important that a baby not lose much more than 7% of her birth weight, and that by four or five days weight loss has tapered off. By day five or six, the baby should start to gain.

This is why nursing babies need a visit to the pediatrician, or from the home health nurse with a scale, no later than three days after hospital discharge; bottle-fed babies, by five days after discharge.

They can hear quite acutely, and talking, singing, music, and funny noises are probably welcome.

They can smell even more acutely, and can recognize their own mother's breast milk, and probably both parents. This is no time to change your aftershave, or for a nursing mother to wear scent of any kind.

They are, of course, very responsive to touch. Some love to be cuddled, some like freedom and stroking better.

They can soothe themselves to a degree, girls as a rule more effectively than boys, but the variation among individuals is greater than the variation between the sexes.

They can discriminate between tastes, and love sugar, which seems to act as an analgesic in this age group (and perhaps for some of us much, much older as well).

## Sleep

Newborns sleep anywhere from sixteen to twenty hours in a twenty-four hour period. The rest of the time they eat, cry, stare into space. Many have the same pattern of sleep at night as they do in the daytime. Some seem to have days and nights "confused," and sleep a lot in the day, then stay up all night. In the first two weeks, it is fruitless to try to reverse this schedule. Go with it. After two weeks of age, many babies naturally fall into a more reasonable schedule; the rest usually can be coaxed into one.

## Growth

Nearly all nursing babies, and many formula-fed ones, lose weight after birth. Babies are born with extra fluid that is lost over the first few days. During this time, they often don't seem to want to feed very often. However, it's important to keep after a sleepy baby, waking her up every two to three hours in the daytime, and at least every three hours at night and coaxing her to suck.

When they are down to their lean (fighting) weight, and are beginning to get the hang of feeling hunger and then eating, they are supposed to start to demand feedings much more frequently. Nursing babies may request a feeding every one and a half to three hours; bottle-fed babies every two to four.

Both nursing and formula-fed babies

should be back up to, or over, their birth weight by about ten days of age.

## TEETH

Rarely, a baby shows up at birth with a tooth. Sometimes such a "natal tooth" can be left in place, but usually it's pulled: such teeth are often poorly formed and rooted, and you don't want a tooth to come out and be choked on. Nor do you want a nursing mother to be chewed on.

### Fluoride

The buds of all the primary teeth are in place, and the permanent teeth start to form at birth. By the time a child is ready for second grade, all those permanent teeth will be calcified and ready to go. So from birth on, these "invisible teeth" need protection from cavities. Fluoride is incorporated into the lattice of the tooth and strengthens it.

In too high a dose, it can cause little white specks in the enamel of the permanent teeth, so it's important to give just the right amount. The right amount depends upon Cherub's age, and upon the amount of fluoride Cherub ingests via milk, water, foods, and swallowed toothpaste. This is a tricky estimation. Some bottled water contains fluoride even though that's not on the label. Some "unfluoridated" city water contains significant amounts of natural fluoride. Foods prepared and then shipped for sale may contain fluoride from the water supply of origin. And on and on. The following are guidelines that seem to be firm and reliable for brand-new babies:

- Breastfed babies get no fluoride from mother's milk, even when Mom is drinking fluoridated water. All nursing babies can benefit from supplemental fluoride drops.
- If your baby is getting fluoride from formula, either in the formula itself or from the water you mix it with, you don't need a fluoride supplement. Your pediatrician will know the fluoride concentration of your community water supply, and whether the baby is getting the right amount of fluoride from this water.

If you are using soy formula (Isomil, Prosobee, Soyalac, Nursoy, etc.), dilute it with NON-fluoridated water, because soy formulas already contain the amount of fluoride needed by newborns and infants. (If you can't, or haven't been, don't panic. It's not a toxic overdose by any means. But try not to.)

## NUTRITION AND FEEDING

Never give honey or karo syrup to a baby under a year of age, as they may contain spores of the bacteria *C. botulinus* and make the baby very sick. Also, don't change formula or give juice without checking with the pediatrician.

For guidelines on eating fish, see page 7.

### Supplements for nursing babies

- **Vitamin K:** Some pediatricians like to give nursing babies a second, or even a third, injection of vitamin K to prevent late onset bleeding problems. See section in Chapter 1.

BEST NURSING GUIDE

*The Ultimate Breastfeeding Book of Answers*, by Jack Newman, M.D. and T. Pitman,
Prima Publishing Co. Published in 2000.
Nurse as often as the baby shows an interest and will suck.
If you need to use a pump, get a good one; cheap ones can do damage.
Monitor your success by counting wet and poopy diapers.

- **Vitamin D:** Nursing babies who don't absorb about ten minutes a day of sunshine need vitamin D to prevent the bone-softening condition rickets. Nursing babies who don't get outside, or who live where there is little sun, or who have very dark skin are all candidates for vitamin D supplementation. (Formula contains the daily requirement of vitamin D.)

## NURSING

Right after birth, many babies have an eager, alert period of nursing. Then, for the next day or so, they seem to lose interest, and only nurse every four or five hours, and not ravenously at that. This is a normal pattern (given that the baby is otherwise fine). However, it is still important to try to wake Cherub at least every three hours and see if he will suck.

This little "resting phase" occurs because babies are designed to lose a little weight after birth. Nursing gives Cherub colostrum (the yellowish-white, thick premilk, rich in immunity) and helps stimulate the real milk.

Giving formula "because she isn't nurs-ing" during these first few days isn't a good idea unless there is a specific reason to do so: for instance, if the baby has a low blood sugar or is very large and ravenous. A baby given formula, especially by bottle, may decide it's just too difficult to suckle at the breast.

Then, after she's lost about 7% of her birth weight, just as the baby gets hungry, the milk comes into the breasts. This usually occurs at about three or four days of age. Often, the timing is just right, and the milk is there in abundance exactly when the baby is ready for it. Sometimes the baby has some hungry hours before the milk comes in. Sometimes, too, the milk comes in before the baby is ready, and the breasts can become heavily engorged.

*The most important keys to successful nursing:*

### CHERUB NEEDS TO GET A GOOD HOLD ON YOU

During the first few days of colostrum, the breasts are soft. Because they are pliable, this gives you a chance to make sure the baby

### Breastfeeding Cuddle Hold

The goal is to have baby "milk" the dark circle around your nipple (areola)—sucking only on the nipple will not provide any milk and will get you sore.

Sit in an upright open position (to allow free arm movement) with your back supported vertically.

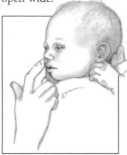

Hold baby's head in one hand.

Use pillow to support baby's weight (or raise one knee).

The Cross-Cuddle or "Football" Hold is highly recommended for the JAM.

1. Stroke cheek or touch baby's mouth so it is open wide.

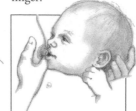

Support head firmly with one hand.

2. Squeeze the nipple-areola with thumb and finger.

Gently bring baby's head up close to your breast.

3. Baby's nose buried in breast.

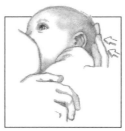

The LATCH-ON: JAM the baby all the way onto the breast.

4. Use finger of free hand to make a space for the baby to breathe.

This halts most prolonged latch-on problems.

1. Start with Cuddle Hold—head higher than rest of baby's body.

Put free hand on your ribcage under your breast and push gently upwards. This lifts breast upwards so the nipple protrudes and points downward. Gently bring baby's head towards nipple.

2. Switch holding arms and hold baby with the opposite arm. The added leverage of the arm length effectuates a strong JAM motion.

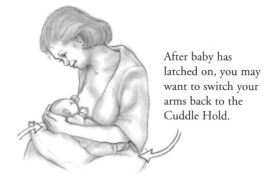

After baby has latched on, you may want to switch your arms back to the Cuddle Hold.

JAM head into breast and bury nose.

is getting a good hold on the nipple and positioning his tongue correctly. If he is on the tip of the nipple or his tongue is on top rather than below, you'll get sore. To fully appreciate how he's nursing, try holding him like a football while someone observes: this gives you a better view.

The baby should get a large part of the nipple and areola (dark ring) into his mouth. If it hurts to nurse, the positioning is wrong. Break the suction and reposition him.

- Bring the baby to the breast, not the breast to the baby. While it's cozy and classic to have the baby nestle with your hand on his bottom, it may not be the best way to get him latched on well, at least at first. You may do better holding his head in your hand, in the cross-the-body or in the football position.
- Any way you hold him, try to get the whole nipple and as much as the areola in as possible. Wait until he opens his mouth or cries—you can tickle his upper lip to get him to do so—and he's actively rooting around. Then JAM his head on the breast, nearly burying his nose in the breast. After he starts to suck, you may need to provide breathing space for that nose, by pressing against the breast with your finger.
- Use lots of pillows to rest your arms and the baby. Don't waste your strength on weight lifting.

## NURSE WHENEVER CHERUB WANTS TO, AND WAKE CHERUB UP IF THAT'S NOT OFTEN ENOUGH

*Don't try to impose a schedule on a baby this young.* After the first few days, the baby should want to nurse about every two to three hours round the clock, eight to twelve times in a twenty-four-hour period.

It is obvious that this doesn't leave time for anything besides catnaps, basic hygiene, nutrition, and refreshing helpful visitors. Discourage unwanted visits and phone calls with the time-honored excuse: "Doctor's orders." You are being truthful; I just ordered it. One of the most frequent causes of nursing problems is the mother not getting enough rest.

It's a good idea to start out nursing a brand-new baby about five minutes on each side, and gradually increase the time, perhaps by one minute each session.

Each baby has a unique nursing style, ranging in intensity from the Gourmet who licks and tastes, to the Satisfied Customer who nurses without fuss until full, and then stops, to the Barracuda, who grabs onto that nipple and sucks the daylights out of it. The Gourmet may nurse for a total of twenty to thirty minutes. The Satisfied Customer may take an efficient twenty, and the Barracuda may have guzzled herself full in only ten.

A baby who consistently takes more than thirty minutes to nurse is probably getting hardly any milk from the breasts, even though she doesn't seem hungry or upset. Record her wet and poopy diapers; then call your pediatrician to ask about having a weight check to make sure she's getting enough milk.

### DEAL PROMPTLY WITH SORE NIPPLES, ENGORGEMENT, AND MASTITIS (INFECTION OF THE BREAST FROM A BLOCKED MILK DUCT)

- Sore nipples mean that the baby isn't latching on properly. Take a peek at the latching-on advice on page 44. You may need a consultation with a lactation counselor to help, but first you need first aid. Apply warm wet compresses or warm teabags (already softened by being immersed in boiling water, then cooled) as often as needed for pain.

- Engorgement means breasts like boulders. It's not cute and it's not funny, and you need to take care of it right away. If the breasts are too swollen, the nipple won't stick out and the baby can't latch on. If they hurt a lot, nursing is painful. Naturally you'll tense up and so will the baby, and you'll understandably want to avoid the whole scene. Finally, the heavy pressure of the milk may, ironically, prevent the breasts from producing enough milk, and this may create quite a long-lasting diminished milk supply.

- When your breasts get so swollen and heavy, they have to be emptied promptly. Hot packs may soften them enough so that the baby, especially if she's a large Barracuda, can empty them.

- But, you say, if I nurse so vigorously, won't I just make more milk and get more engorged? No. Much of the engorgement is swelling that has nothing to do with milk supply. You need that frequent nursing now; the swelling will go down in a few days.

- If Cherub can't empty the breasts effectively, you need a really good pump, preferably a well-designed electric one. Many lactation experts recommend the MEDELA line of pumps.

- An inefficient pump, even a battery-operated one, can make matters worse by irritating the nipples while not emptying effectively, and a hand pump is likely to be useless. Empty enough milk to soften the breasts, and apply cold packs afterward unless the baby is eager to nurse. If so, nurse the baby and then apply the cold packs.

- Mastitis means that the breast tissue looks red and feels hot and hard because a duct has gotten blocked off and the tissue is infected. You may also have a fever or flu-like symptoms. If you have a red-hot hard area, apply moist heat and massage the lump. Nurse the baby on that side as much as you can. Cherub will not catch the infection. You may need antibiotics (that are safe for Cherub) as well, so call your obstetrician or your pediatrician.

**DON'T GIVE A BOTTLE UNLESS YOU ARE ADVISED TO DO SO BY YOUR PEDIATRICIAN OR LACTATION CONSULTANT**

When a baby sucks on the breast, she uses her cheek muscles to create suction. On a bottle, she uses her lips and tongue. Getting in the habit of drinking from the bottle may cause her "nipple confusion"—when she's put back on the breast, she won't know how to suck.

When the milk is late coming in, or when you have a "happy to starve" baby, you may need to give a jump start of formula.

There are two ways to give formula without using the bottle. You can fill a syringe with formula (or expressed breast milk, if your milk is in), and tape it to your chest with a soft flexible feeding tube attached that brings the milk into the baby's mouth as she's on the breast. That way, she can suck on the nipple and obtain milk, even if there's not much in the breast yet. The second method is called "finger-feeding." In this method, you let the baby suck on your clean finger while you instill formula or expressed milk into the corner of her mouth with a dropper or a syringe.

Once in a great while, a baby who is reluctant to suck on anything perks up and gets the idea when given a bottle, and may

then transfer that skill to the breast. But that's unusual, and I'd defer to the wisdom of your pediatrician or lactation counselor. When Cherub is two weeks old, your pediatrician will advise you about when to start a "reminder" bottle.

## SPECIAL NURSING PROBLEMS

### Too little milk

Too little milk in the breasts can stem from the after-effects of breast surgery, from exhaustion, from the mother's being on the older side and thus, hormonally speaking, out of practice.

But the most common cause of too little milk getting into the baby is too little *effective* nursing. This is a vicious cycle that occurs when a baby doesn't get enough milk, becomes sleepy, nurses less, stimulating the breasts less, so there isn't enough milk. The treatment is to increase nursing, perhaps after a feeding of formula or expressed breast milk to give the baby energy and break the vicious cycle. This situation is a very good reason to make sure the baby's weight is checked within a few days of birth.

The least frequent cause of too little milk is dehydration. It makes sense for a nursing mother to drink enough that she has to pee at least every three hours, and the urine should look pretty clear, not deep yellow (and vitamins make no difference—they don't turn urine yellow). But overhydration is not going to make more milk.

Once in awhile, maybe in five percent of cases of inadequate milk supply, the problem is hormonal, but unrelated to age. There are two prescription medications, Reglan and Thorazine, that can help increase the milk supply. They both can have side effects, however; discuss carefully with your pediatrician.

*Too much milk, too fast*

We should all be so lucky. But too much can be a problem too. Sometimes the milk comes roiling out in a manner suggesting Roman fountains. The baby gulps and glubs and cries, gets so upset he can't finish, and then is hungry again two hours later. The frequent nursing simply produces more milk!

Here are some suggestions:

- Express half an ounce or so before the baby gets to the breast.
- Nurse with both you and the baby lying on your sides, parallel to each other, so the milk can dribble down her face without her having to glug and cope with the overflow.
- Nurse only on one side each feeding, so you get only one let-down response to deal with. Apply cold packs after nursing to reduce the amount of milk also.
- If this doesn't work, a lactation counselor can teach you ways to diminish the flow.

## DIET FOR NURSING MOTHERS

For guidelines on eating fish, see page 7.

You'll be using an extra 500 calories a day when nursing is well established. Clearly, these should be highly nutritious calories, not junk food. Each day's diet should include three servings of protein. This can be dairy, meat, fish, fowl, or beans. For vitamin A and fiber, at least three servings of green and yellow vegetables are recommended. Then there should be two of something rich in vitamin C (citrus fruit, broccoli, cabbage, tomatoes). Whole-grain cereals and breads and fruits can make up the rest of the meal.

As far as "gassy" foods go, not to worry. Beans or cabbage may upset the stomach of the mother who eats them, but there is no evidence that the very large molecules that produce gas travel into the bloodstream. If they don't get into the bloodstream of the mother, they don't get into the milk and into the baby.

Garlic does cross into the breast milk (it even crosses the placenta: I've smelt the breath of several just-been-borns who I swear had been ordering take-out pizza in there). But most babies seem to like it (yes, this has been studied) and it doesn't seem to cause any problems.

You'll need 1200 mg of calcium daily to replace what goes into the milk and what you need for your own body. If you don't get it, the baby won't suffer: the calcium stored in your bones will go into the breast milk. Of course, you can imagine what weeks and months of this will do to your bones. Dairy products, calcium-fortified juice or breads, tofu (but check the label), broccoli, collard greens, and sardines are all options.

Most obstetricians recommend nursing mothers take a multivitamin supplement and drink at least eight glasses of water daily.

## "HAPPY TO STARVE" BABIES: BEWARE!

Once in awhile, a very placid baby will fool everybody by seeming contented even though he isn't getting enough milk. A baby more than two days old who nurses only briefly and goes longer than three hours between feedings may have this potentially serious problem. Such a baby is sometimes described as "happy to starve." It's as if just a little snack makes him contented enough to sleep, and when he goes to feed again, the lack of calories makes him nurse ineffectively. Such a baby will urinate and stool infrequently, and the stool will be green. But the baby's behavior may give you no other clue, and it's easy to be fooled into thinking of him as a "good" or "easy" baby.

If you have any doubts, Cherub needs to be weighed—promptly—on your pediatrician's scale.

### FORMULA

See Chapter 1 for discussion of various formulas.

In the first few days, most formula fed babies don't eat very frequently. Then, when the "water weight" is gone and they are more used to the work of living outside the uterus, they really wake up.

During the first two weeks, many formula-fed babies want to take only one to three ounces every two to three hours. During this period, when the baby is just discovering what hunger is and how to fix it, it's best to let Cherub take the lead. At about two weeks of age (see next chapter) most can be coaxed to a more grown-up schedule.

The key thing is not to make the baby unhappy.

A good rule of thumb for a full-term baby approaching the age of two weeks is to offer one ounce of formula for every two pounds she weighs at each feeding. (An eight pounder would get four ounces.) If the baby can take this amount, feedings start to get spaced at three- to four-hour intervals.

### Preparing formula

In most areas of the country, it isn't necessary to sterilize bottles and nipples, but just to wash them in hot water. Ask your pediatrician; you could save yourself a great deal of work, to say nothing of the pungent

If you're using well water, check to be sure it does not contain too much nitrates, fluoride, lead, mercury, or other contaminants.

## FORMULA AND SAFETY

When mixing formula, it is crucial to get the proportions of formula and water exactly right. If you are instructing a helper, make sure you do so clearly, and do a quality-control check every now and then. Too much or too little water can make a baby quite sick.

Once the baby has sucked from the bottle, the bottle can't be set aside (even in a refrigerator) for refeeding. Refeeding the same bottle can give a baby thrush (a yeast infection in the mouth) and, rarely, real illnesses. Toss the bottle after forty-five minutes to an hour.

Beware of the microwave. Hot spots can develop. Let the bottle rest for at least a minute, swirl it to mix hot and cold, and test the milk carefully. Don't microwave the nipple! It can stay really hot, and you don't have a good way to test it.

aroma of rubber nipples that have been forgotten on the stove and boiled dry.

Check the fluoride content of the water you're using (see *Fluoride,* above).

### Giving the bottle

It probably goes without saying that both adult and baby should be comfortable, with the baby's head firmly supported, also that the bottle should always be held, not propped, in the mouth.

It is surprisingly easy to forget, however, that an eagerly sucking baby—especially a Barracuda—can suck so hard she blocks off the nipple. If the baby sucks eagerly, then starts to fret or scream, check by removing the bottle. A collapsed nipple will then expand with a little gasping hiss. Fixing the problem may mean simply maneuvering the bottle as the baby sucks, removing it every so often to get air back in the nipple. Or it may mean

experimenting with nipples of different design.

### BURPING

Burping should be an intransitive verb: it is not something you can force on someone else. The best you can do is get the baby into a receptive position for a bubble of swallowed air to rise and emerge from the mouth. If nothing happens after a minute or so, resume feeding or, if the baby is done, put her on her side. So she burps herself, or so she spits up a bit.

## BATHING

The bath Cherub was given after birth is all that is needed in these first two weeks.

After the umbilical cord separates, the circumcision is healed, and the adults are able to communicate in sentences rather than inarticulate grunts, the first bath can be undertaken. But even then it's more of a photo op than a necessity.

In the meantime, wash the face with water if the baby spits up and pat dry. For diapering, see below.

## DIAPERING, GENITALS, PEE AND POOP, DRESSING

### Diapering

Wash the genitals and bottom with plain water, pat dry. A layer of Vaseline makes a nice Teflon-type surface so that the next stool-removal is easy.

Little girls' vulvar areas are very folded and mysterious, and parents often wonder how thorough cleansing has to be. Try gently to remove any trapped stool. The whitish, clear, mucousy discharge need not and should not be scrubbed away. Don't forget that a little bit of blood in the discharge, even with a small clot, is normal.

Circumcised baby boys need gentle care. Thanks to nature's anatomical foresight, it is unusual for stool to get into the circumcision site. If it does, wash very gently with a mild solution of soap and water, rinse well, pat dry, blow dry. Any time you change the diaper, apply Vaseline or KY jelly to the diaper in front so that the penis doesn't rub against the diaper.

With uncircumcised baby boys, don't try to pull back the foreskin at all. Don't use powders; the powder can become trapped under the foreskin.

In fact, don't use powder at all. Several reasons:

- It's a terrible irritant to the lungs if you or the baby breathes it in. Spills and accidents can occur even at this age.

- It can irritate mucous membranes as well. An uncircumcised boy can get powder trapped under the foreskin where it can cause adhesions. Circumcised boys can also get adhesions around the line where the glans meets the shaft of the penis.
- It tends to make adults sneeze on the baby.

If you use very absorbent disposable diapers, you may not be able to tell whether or not the baby has urinated. Try placing a little wad of absorbent cotton fluff in the diaper. If it gets wet, you know the baby urinated. If the baby was recently circumcised, put Vaseline on the raw end of the penis so the fluff doesn't stick to it.

*Pee and poop*

If you think you are expending a ridiculous amount of energy monitoring Output, you're right. And it's going to stay that way for awhile. What comes out of the baby is a good indication of the baby's general health, and of how much has gone in. Remember that urine is sterile, unless there's an infection. Comforting for when you sustain a direct hit.

### PEE

**As I mentioned above, you may see a coral-colored stain on the diaper.** It isn't blood; it is uric acid crystals, which result from normal breakdown of DNA. It is normal and will go away as the baby eats more. You can tell it is not blood because it is pink, not red, and because the color doesn't change to brown on exposure to air.

When your baby boy pees with his diaper off, after you shriek, observe the stream. It should cover a good arc, not a dribble. He should be peeing with no effort at all, no grunt or scrunched-up face. (If he shows effort, or dribbles, notify your pediatrician: the baby may have an obstruction to the outflow of urine as it travels through the penis. Girls, of course, aren't at risk for this.) If he (or she; some girls have remarkable power) gets you in the face, you get a year of good luck. A direct hit in mouth or eyes gets you a decade. I'm somewhere into my second century of good luck.

### POOP

When it comes to stool, or poop, or caca, or bowel movements, you are entering a new world of experience which requires a new vocabulary. Food analogies are most accurate for describing this important phenomenon. Stools change as the baby matures and as they reflect both what the baby is ingesting and how he deals with it. Watch the color change, and the texture.

The first stool, as every new parent knows, is meconium. Meconium is the fetal stool, and is composed of sterile debris in the amniotic fluid. It is green-black and sticky like licorice.

Why is it green-black? you may ask. What has the baby been eating in there? Did the mother ingest too much spinach or licorice?

It is green-black because it contains bilirubin, a yellow-green breakdown product of red blood cells. For more on bilirubin, see the section on *Newborn Jaundice*.

As the milk begins to be digested and the meconium is expelled, the stools change. For a breastfed baby, this transitional period can

## THE MANY COLORS OF POOP

*Birth to Age Two or Three Days*: Greenish-black, sticky meconium
*Three to Five Days*: gravy, pesto sauce, mashed celery
*Six Days and Onward*: Mustard out of the squeeze bottle: yellow, watery, and loud

give a colorful array of any of the following types of stool: brown guacamole, mashed celery, spinach puree, or broth. For a bottle-fed baby, the transition is usually to brown or green cottage cheese.

Once all the meconium is flushed out, a nursing baby's stools are the brilliant dark yellow of mustard, and in fact sound rather like mustard being ejected from the squeeze bottle: *gurgle gurgle rumble squish splat*. The term "stool velocity" begins to have meaning here: most babies, undiapered, can hit a wall several feet away.

If a nursing baby has a rare green stool, it is probably just a fluke, reflecting the presence of bile in the stool. If the stools are consistently green, it's important to check with the pediatrician. Most often, this means the baby isn't getting enough milk.

Formula-fed babies ought to have stools the consistency of cottage cheese. Sometimes watery or hard stools may occur. The watery stool may reflect lack of absorption, allergy, or infection. The hard stool doesn't mean any of these, but is still a problem. Anything harder than soft peanut butter needs attention. Don't change formulas, or give juice, without calling your pediatrician, but don't let hard stools persist, either.

*Clothing*

Cotton rather than wool or polyester is most likely to keep the baby warm and rash free. The best way to keep the baby from losing heat to the environment is to put a hat on the baby, and it need not be a fancy one. Dressing the baby in layers—diaper, shirt, "sacque" with a purse-string tie, then blankets—gives you simplicity plus flexibility.

To make sure the baby is warm enough, compare what the baby is wearing to what you are wearing to keep comfortable at the temperature of your environment. Dress the baby with the same layers that keep you comfortable; then add one more layer. This goes unless the temperature is very warm, in which case you may need to subtract a layer.

Protect the baby from winds and drafts and direct sun.

## ACTIVITIES

Don't expect to engage in any activities other than feeding the baby, changing the diaper, and trying to sleep when the baby does.

## TOYS AND EQUIPMENT

Newborns need only their own bodies and those of loving adults for "toys."

## MEDICINE CHEST

You'll want to have a rectal thermometer (underarm temperatures aren't accurate enough for babies this age), but please get an electronic one. Glass thermometers contain mercury. When the thermometers are thrown away, crushed, and burned, the vaporized mercury gets into the air and then the water, poisoning fish and other water life.

You'll also need a bulb syringe to suction out mucus; rubbing alcohol and cotton pads for the care of the umbilical cord; petroleum jelly for the diaper area and circumcision. You also might want to have salt water nose drops (buffered saline, available over the counter at the pharmacy).

Not in the medicine chest, but handy to have around, is a can of oral electrolyte solution such as Pedialyte, just in case your pediatrician recommends it for a feeding or two if the baby gets an upset stomach. You can buy the fluid your pediatrician prefers without prescription at food and drug stores.

Don't have cotton swabs on sticks, as adults will be tempted to clean inside the baby's ears with them; it's all too easy to do damage that way. You don't need acetaminophen or other fever drops, as you never want to treat or mask a fever in this age group: even a slight rise in temperature can herald serious illness, and needs prompt medical attention.

## OTHER EQUIPMENT

A car seat, safely installed and used exactly according to directions, is the *sine qua non* of safety equipment. Nursery monitors are nice, but can wait until later, as the baby is going to be pretty close to you all the time, even at night. You probably won't be using a crib yet; but if you are, see the section that corresponds to this one in the Two Month Old chapter.

## SAFETY ISSUES

- Remember that the baby needs to sleep on her back, not her side or her tummy, day and night.
- No secondhand smoke! Be firm. It's associated with SIDS, childhood leukemia, and respiratory problems, among other awful or nasty entities.
- Turn down the temperature on the hot water heater to about 120° to avoid accidental burns.
- Make sure the baby always rides in the car seat, facing backwards, in the backseat. Never on a lap.
- Never place the baby's car seat or infant seat with the baby in it on anything it could fall off of.
- Don't place a baby on a soft surface that could smother: water beds, pillows, mattresses. Sleeping with the baby in bed with you is not a good idea; though very rare, rolling onto and smothering the baby has occurred. Also a danger: the baby gets wedged between mattress and wall, or head of bed.

# Health and Illness

*Part II of this book discusses symptoms and illnesses of babies and children older than Four Months of age. Newborns, from birth*

## NO FETAL POSITIONING

Make sure that when you put the baby in a carrier, infant seat, swing, or car seat, he keeps head and body in the same plane, more or less. *You don't want him to curl up into the fetal position with chin on chest.* This can cut off the airway and deprive the baby of oxygen.

Make sure that the baby's car seat doesn't put the baby into a sitting position where the chin falls down on the chest. Most full-term babies are perfectly fine in the car seat, but 30% of premature babies will slump into a position that can compromise the airway.

If your baby is under six pounds at hospital discharge, bring the car seat up to the room and try him out in it. Have an experienced nurse check. If he's curled up, with his chin on his chest, try putting some folded receiving blankets or a few diapers under the baby's bottom to get him more horizontal.

*up to Two or Three Months old, are different from even slightly older babies: more fragile, prone to different conditions, and still adjusting to life outside the womb. Symptoms and illnesses for this age group, Birth to Two Weeks, are contained entirely in the section in this chapter. If your baby has any worrisome condition not covered here, or if you do not understand or are uncomfortable with the discussion in this chapter, get in touch with your pediatrician.*

### EXPECTATIONS

One of the toughest parts of having a brand-new baby is that you are uncertain of what is normal, so you aren't sure what means trouble. Bear in mind that most babies stay perfectly healthy these first two weeks, and that this is the hardest time of adjustment. Also, you should feel free to call your pediatrician,

at any hour, if you are concerned that the baby is sick. The guidelines below will help. But if you think your baby is just "not acting right," or "not looking right," call—even if, at this point, you can't be more specific.

### COMMON MINOR SYMPTOMS, ILLNESSES, AND CONCERNS

Many of the characteristics parents notice in new babies aren't really problems; they're normal—rashes like erythema toxicum and normal bleeding from the vagina, for instance.

#### Jaundice

Jaundice is a yellow coloring to the whites of the eyes and the skin. It is usually a normal and innocent finding in babies from three to ten days old. Nearly all babies will have some degree of jaundice in that time. The under-

## JAUNDICE NEEDS PROMPT MEDICAL ATTENTION IF

- It begins in the first 36 hours of life.
- The baby is acting sick or even just "not right."
- The baby isn't taking feedings well or seems to be not gaining weight.
- You think the baby's skin looks yellow all the way down to his knees.
- It begins after the baby turns five days old.
- It hasn't gone away by two weeks of age.

lying cause of most newborn jaundice is usually just the consequences of normal adjustments to being born.

However, there are two circumstances in which jaundice is a medical problem needing prompt attention. The first is when the level of jaundice gets too high. What "too high" is depends on the age, weight, and health of the baby and the cause of the jaundice. The second is when the cause of the jaundice is an illness or malformation.

The substance that causes jaundice is called *bilirubin*.

### Normal jaundice

Normal jaundice appears at around the third day of life, and deepens over the next two or three days. First the baby's eyes look yellow, then the face, then the chest, then sometimes the belly. Then the jaundice fades, retreating up the body in the reverse of its appearance. The baby acts fine, eating well, sleeping comfortably, with no fever or other signs of illness. At around a week of age, the jaundice begins to fade away in the

reverse order that it came: the eyes are the last to clear.

Why do normal babies turn yellow with such regularity? A variety of reasons:

- **There's more bilirubin made by newborns.** Jaundice is caused by a yellow molecule called bilirubin. Bilirubin is what becomes of red blood cells when they disintegrate. And they always are disintegrating, and we are always making new ones. Newborns have more red blood cells than adults do, and the red cells are designed to disintegrate more rapidly, so of course there's more bilirubin for the baby to get rid of.

- **Newborns aren't as good at getting rid of it.** After the newborn period, we get rid of bilirubin by sending it via the blood to the liver, which repackages it and puts it out into the intestine. Newborn livers often haven't mastered this as efficiently as adult ones, so there's another reason for increased blood levels of bilirubin.

- **In newborns, the bilirubin keeps getting recycled.** Beyond the newborn period, bilirubin exits in the stool and is never heard from again. Of course, a fetus can't get rid of the bilirubin: a fetus does not poop. So, while the baby is in the womb, the bilirubin exits into the stool. And the stool just sits there in the intestine, getting fuller and fuller of bilirubin. That's why meconium is the dark yellow-green-black it is: it's full of bilirubin.

Nature has designed fetuses in such a way that some of the bilirubin in the intestine gets reabsorbed into the fetal bloodstream. This bilirubin goes through the placenta to the mother's blood, and she gets rid of it for fetus.

• After birth, babies' intestines keep on recycling the bilirubin. But they no longer have the placenta and mother to get rid of it, so it stays in their own blood. If there's enough of it, the bilirubin leaks from blood to skin and the baby turns yellow.

All of these factors play a role in determining that most babies have at least a little jaundice. There is a school of thought that believes this normal jaundice, so carefully designed to happen, has a role in newborn biochemistry that has not yet been discovered.

## SERIOUS OR POTENTIALLY SERIOUS SYMPTOMS, ILLNESSES, AND CONCERNS

When a brand-new baby displays a serious problem, or even one that might just possibly be serious, you need to get help right

away. If you can't reach your pediatric group right away, zip to the emergency room and do not allow the personnel to make you wait in line. Be charming (but insistent) if you can; be obnoxious, if necessary.

Of course, most of the time this will all be a false alarm. If you have been extraordinarily rude in your anxiety, just send a note and a box of cookies. Everyone understands about being a brand-new parent.

These are the three most prominent categories of major problems in brand-new babies:

• A normal process that becomes exaggerated and endangers the baby, such as too much jaundice from normal causes.
• A malfunction of an organ that seemed to be normal at birth but which is revealed as abnormal as the baby adjusts to life outside the womb, such as a defect in the heart, a metabolic problem, or a liver problem causing jaundice.
• Serious infection. This could be pneumonia, *sepsis* (infection in the blood stream), *pyelonephritis* (infection in the kidneys), *meningitis* (infection in the spinal fluid), or *omphalitis* (infection of the umbilical cord).

In nearly every case, if diagnosis is made early in the course of the problem and treatment begun promptly, the baby can do very well.

### Worrisome jaundice

Worrisome jaundice occurs when the level of bilirubin gets too high, or when the underlying cause of the jaundice is a malformation of the liver, a metabolic abnormality, or a serious infection (see box on page 56).

## THE LEVEL OF BILIRUBIN GOES UP TOO HIGH OR TOO RAPIDLY

At the usual levels of bilirubin, there is no danger to healthy, full-term babies. However, under some circumstances, the bilirubin can leech out of the skin and into the brain and other organs, doing damage. The bilirubin level at which this can happen varies considerably, depending on the baby's health, degree of maturity, age, weight, and the cause of the jaundice.

There are several risk factors that can predispose to a too-high bilirubin:

- The baby is premature, 36 weeks or younger; or small, five pounds eight ounces or smaller. This makes it easier for lower levels of bilirubin to leach into the brain and other organs.
- The extra bilirubin is due not to the normal changes above, but to a process called *hemolysis*, which means the breaking up, or lysis, of red blood cells. This can occur especially if the mother's blood is type O and the baby's blood type is other than O. That could be A, B, or AB. (It can also occur if the mother is Rh negative and the baby is Rh positive, though this has become rare since the development of RhoGAM shots.)*

Hemolysis produces byproducts that make it easier for lower levels of bilirubin to leech into the brain and other organs.

- The normal changes causing increase in bilirubin are very much exaggerated so that the barrier against leeching into the tissue is overwhelmed. Most pediatricians don't like to see the bilirubin level go above 20 without starting treatment, even though most healthy babies tolerate such levels just fine.

## THE JAUNDICE IS DUE TO AN UNDERLYING ABNORMALITY THAT COULD BE SERIOUS

- Once in a blue moon, a baby has jaundice not because of any of these reasons but because his liver isn't functioning properly. This may be because of a liver infection, an infection in the bloodstream, a metabolic abnormality, or an anatomic problem with the drainage from the liver. Such a baby usually acts sick—lethargic, not feeding well, poor weight gain, poor color. He may have very dark urine; he may have trouble clotting his blood, also. A baby with this problem does need very special prompt help.

Most pediatricians keep a special eye on babies who could be specially endangered by jaundice—those who are sick, premature, or whose blood type potentially clashes with the mother's blood type.

---

*In both these blood group situations, the mechanism is the same. If a little of the baby's blood gets into the mother's circulation, and her body recognizes it as foreign, she'll make antibodies to those red cells. These antibodies return through the placenta to the fetus, and tear down the red blood cells. Hence: hemolysis.

## WHAT TO DO IF CHERUB LOOKS YELLOW

If Cherub is less than 36 hours old, call your pediatrician at once. If your baby looks yellow, but is over 36 hours old and is acting fine, call your pediatrician the same day, but not as an emergency. Many pediatricians will take a quick peek in the office without requiring a formal visit to see if the baby needs a blood test and/or a formal appointment. Or your pediatrician may simply direct you to take Cherub to the lab for a blood test.

If your baby who is yellow is also acting sick, he needs to be seen right away, even in the middle of the night, even if the jaundice looks mild to you.

### HOW CAN YOU TELL IF JAUNDICE IS WORRISOME IF THE BABY IS ACTING FINE?

But once you're home with a baby who has no risk factors but who starts to look yellow, how can you tell whether the jaundice is too severe?

This is a tricky question, as babies look more or less yellow under different lighting conditions and babies have a whole rainbow of natural skin color. Many skin pigments have a natural yellow tone mixed in. It's very hard for parents to judge their own baby's skin color. Best solution: have the pediatrician take a look.

### TREATMENT OF JAUNDICE

**If the jaundice is due to an underlying problem, such as infection or liver disease, both the underlying problem and the jaundice need to be treated.**

In most cases, when jaundice needs treatment, there is no underlying problem. It's merely the intensity of the jaundice that requires attention. The level of bilirubin in the blood needs to be reduced so that it doesn't leech out into the brain and other organs.

When that's the situation, the treatment usually consists of blue light from special phototherapy units. The light frequency breaks down the bilirubin in the skin so that the kidneys, as well as the liver, can get rid of it. Extra breast milk or formula (not water or sugar water) can help flush it out, and also help the baby poop out the bilirubin-rich meconium.

Sometimes the special phototherapy units can be set up at home; sometimes the baby needs to be readmitted, briefly, to a hospital unit. These units tend to look like a nude beach, with the babies basking underneath blue lights, their eyes chicly shaded.

Sometimes indirect sunlight is advised. This isn't a very efficient way of giving blue light for jaundice, but done properly it won't do any harm. Done improperly, however, a baby can be roasted. Never put the baby in a small space (like the backseat or ledge of a car) and let the sun blare in. Monitor the baby closely for flushed skin and sweating. It isn't worth endangering the baby from heat (or cold, if you have a wintry or air-conditioned home) to get a little edge on the jaundice. Make sure you have clear directions from your pediatrician.

### Fever of any degree; temperature below normal

Even a tiny fever in a newborn can be an early sign of a serious problem. The normal rectal temperature in a baby under two months of age is between 98° and 100.4°.

#### TOO HOT

A rectal temperature of **100.4°** or higher is considered real fever in a newborn. However, sometimes such a temperature is due to overheating. Given that the baby is acting fine and eating well, and if she has been dressed very warmly or has been cuddled by a hot adult, you can undress her and take her temperature every fifteen minutes for an hour. If it goes up during that time, call the pediatrician at once. If it stays at 100.4° after an hour of cooling, call your pediatrician at once.

If her temperature returns to normal in an hour, monitor it once an hour for four hours just to be sure. As babies grow older, fever all by itself becomes less important as an indicator of possible serious sickness. By three months, Cherub is more able to show just how sick she is.

#### TOO COOL

If the temperature is under 98° but the baby is acting fine, it may be that she is underdressed for the environmental temperature. Put on her hat, add a blanket, and take the temperature again in half an hour. If it is still below 98° or (of course) if she is acting sick, call the pediatrician.

Once in awhile, a temperature that is below normal can be a sign of serious illness in a newborn. And a newborn can have a normal temperature and still be sick. If the baby is acting sick, especially in any of the ways described in this section, don't be falsely reassured by a temperature that is normal or less than normal.

### Not acting or looking right

A new baby who shows a change in behavior needs close observation. Infection or the malfunctioning of heart, lungs, kidneys, liver, or other organs don't announce themselves with the clarity that they do in older children and adults.

Signs that a serious problem could be brewing include:

- Not waking spontaneously to feed within two hours of her usual time. Refusing to nurse or feed, showing no interest when offered a feeding at least one hour after the "regular" time. (This should sort out which baby is merely lengthening the feeding schedule.)
- Crying inconsolably for more than half an hour straight, despite feeding. Make sure that the baby really is able to get the milk, however; ineffective nursing is a common source of such crying. If in doubt, give a nursing baby some *expressed* breast milk.

Never give acetaminophen (Tylenol, Tempra, etc.) or ibuprofen (Motrin, Advil) to a baby this young. You need to know if Cherub has a fever, and the medication will mask that information.

- Very jittery movements that don't stop when you place a hand on the arm or leg; jittery movements that last for as long as a whole minute.
- Acting very quiet and limp; not moving arms and legs as normal; floppy when held or handled.
- Sweating while feeding; taking more than half an hour with feedings.
- Working hard to breathe (see *Chest*, below); or consistently breathing 60 times a minute or more even at rest. Grunting with each breath.
- Blueness around the lips.
- Skin that is pale, gray, or mottled like marble (unless it always looks marbled and your pediatrician has assured you this is normal for your particular baby).

## SIGNS OF SPECIFIC ILLNESSES

### Skin

A rash that looks like blisters, with a red area around the blisters. Remember that the normal rash of *e. toxicum* consists of little white dots on a pink patch. Blisters contain fluid; they're greyish or yellowish, not white. Such blisters may be caused by the **herpes** virus, transmitted during the end of pregnancy, labor, or delivery. A baby can show up with herpes despite a negative culture from the mother, or an absence of a history of herpes, or even a cesarean section delivery.

Any sores with pus, crusting, or a shiny appearance can mean bacterial infection. So can redness and swelling of the skin in any area: eyelids, breasts, fingers and toes. A red and shiny area around the area of the umbilical cord may indicate a serious infection.

A rash that looks like bruises, or like tiny little red pin pricks which don't blanche when you rub them, may be *petechiae*—a sign that the blood isn't clotting properly. **All these conditions need urgent attention.**

### Eyes

If the *white* of an eye is inflamed, or the baby seems to squint when presented with a bright light, this can mean infection or other problems, and should prompt an immediate visit, whether or not there is goop or discharge from the eye. (A *spot* of red in the white of the eye or a little red ring encircling the pupil is almost always due to a normal little bleed occurring after the stress of labor. This is apparent right after birth, and your pediatrician should check it and assure you of its nature. These red spots take weeks to disappear.)

Swelling of the eye or the eyelids also calls

Herpes infection in the newborn is an emergency. Such a rash needs immediate evaluation, even if Cherub is acting fine and has no fever. Herpes acquired from the mother during delivery can show up in babies any time in the first four weeks of life. The sooner the diagnosis and treatment (with intravenous acyclovir, or other antiviral agent), the more likely Cherub is to recover without complications.

for a rapid assessment. (Puffy eyelids from birth are normal for about the first week. After that, they need to be evaluated.)

See also the discussion of droopy eyelid, *ptosis,* on page 35.

### Nose

A persistently stuffy nose in the first two weeks or so can mean that the lining is puffy from the mother's hormones; or that the septum (wall) down the center of the nose got pushed out of shape during labor or delivery; or that there is a bony obstruction partly blocking air getting through the back of the nose. All of these may need treatment, so consult your pediatrician.

A runny nose in the first two months is usually due to a cold, but rarely may indicate a more serious infection. If the nose is really dripping, or if the mucous is very thick or blood tinged, an appointment with the pediatrician is necessary.

### Chest

Before you get worried about abnormal breathing, remember that normal newborn babies breathe irregularly. They often go *"breathe-breathe-breathe-breathe, long pause, big sigh."* They may breathe so quietly you

keep wanting to check to make sure everything is OK. They may make soft schnurgly noises or intermittent soft whistles. They sneeze and hiccup. When they are asleep, they usually breathe under sixty breaths a minute. When they are awake, the rate of breathing depends on their activity.

All new babies' tummies move a bit with each breath. But their nostrils don't flare, and they don't grunt with each breath. The muscles between the ribs don't suck in and out, nor does the muscle in the notch of the collarbone. They don't cough or wheeze.

Abnormal breathing in a baby can reflect a problem with, or an infection in, the heart or lungs. It can also reflect fever, a generalized infection, a neurological problem, or a problem with metabolism.

If the baby seems to be breathing rapidly, time her for one minute while she is asleep. If the rate is more than sixty a minute for three different counts, it's important to talk with the doctor or have her seen quickly.

Breathing with effort is of equal or more concern. If the baby is using extra muscles, you will see his abdomen pulling in and out (not just gently moving) with each breath, and perhaps the muscles between the ribs sucked in or out too. Cherub may give a

## SIGNS OF AN URGENT PROBLEM

If Cherub cries in pain before vomiting; if the vomitus is yellow, green, or brown; if her stomach is hard and distended even after vomiting; or if she is pale and listless between vomiting episodes, get help at once. She could have an acute bowel obstruction or an overwhelming infection.

grunt with each and every breath, minute after minute. This, too, requires a rapid assessment by a physician.

### Umbilical cord

The umbilical cord dries up and falls off by two or three weeks of age. While it's still attached, a sign of infection is redness of the skin around the cord (the skin, not the red lining of the belly button). This may not look bad but actually can be an urgent problem, because the umbilical cord leads deep into the body so can carry infection inside rapidly. Get attention at once.

A tiny bit of blood oozing from the umbilicus and making a dried spot on skin or diaper isn't worrisome, but if you see blood oozing and continuing to ooze, call right away as this can mean trouble clotting the blood in general.

### Vomiting and spit up

Spitting up and vomiting are fairly common in the first two weeks. Usually, it means the baby's eyes are too big for his stomach, as your grandmother used to say, and he drank too much milk. If the baby gently burbles a little milk out, even with each feeding, but is

obviously comfortable, gaining weight, and eating eagerly, he is just eating a lot and showing his appreciation.

If Cherub starts with occasional vomiting but then the vomiting becomes more and more frequent, it is important to call your pediatrician. The condition called *pyloric stenosis,* in which the muscle that empties the stomach becomes tight and obstructing, comes on in this gradual way and is not rare.

If Cherub seems otherwise healthy and comfortable but is a "spitter," she probably has normal *gastroesophageal reflux* (GER). All babies regurgitate to a degree. Signs of GER need to be evaluated and treated if there are complications:

- Failing to gain weight normally;
- Aspirating regurgitated milk into the lungs, causing pneumonia or recurrent wheezing episodes;
- Frequent pain from acid reflux: crying and arching the back, crying during or just after feedings.

Most often, blood in vomitus means that the nursing mother's nipples are bleeding, and the baby is swallowing the blood. If the baby isn't nursing, or the mother's nipples aren't at all cracked, or if the baby is acting sick, an immediate call and visit is warranted.

### Poop

More than a few drops of blood in the stool, whether or not Cherub is in pain, require an immediate examination. There may be a lesion in the intestine that is bleeding. Take the poopy diapers with you! If there are just a few drops or streaks, the most likely cause is a little tear in the lining of the anus. Make a same-day appointment, and don't forget to take the diaper.

Stools that change after the baby has established normal milk-feeding stools may mean infection, allergy, or underfeeding. Watery, green, or very foul-smelling stools most likely stem from an intolerance to something in the milk, or from a viral infection, and need a same-day or next-day appointment, depending on how Cherub is feeling. Hard stools in an otherwise well, formula-fed baby warrant a call, but unless the baby is in pain or otherwise acting sick, the call can wait until the morning.

Frequent watery stools can rapidly dehydrate a new baby, and need a same-day appointment. If there is delay in reaching your pediatrician, switch the baby to an electrolyte solution such as Pedialyte until you get further advice—whether Cherub is a nursing baby or formula-fed.

## GENITALS

Most potentially serious genital problems occur in boys.

### Penis woes

See the discussion of the healing circumcision, above.

- If the shaft of the penis is red or swollen, whether Cherub is circumcised or not, call or see the pediatrician right away as this may indicate infection.
- If Cherub had the style of circumcision that leaves the little plastic ring left on, and the shaft looks swollen, again call right away because the ring can "choke" the penis.
- If there is bleeding from the circumcision after the baby comes home, call right away.

### Scrotum

Usually a full-looking scrotum, or enlarged testicles, only means fluid in the sac (hydrocoeles) and is innocent. But this is usually noted at birth. If it appears after birth make a non-urgent appointment to show the pediatrician. It could be either a hydrocoele or a hernia, or both.

If the area is red or painful to the touch, or the baby is fussy, call immediately, as this could mean a hernia that has become trapped—that is, a bit of his "innards" has emerged from the abdomen through a gap in the muscle wall, and then has swollen so that it can't get back in again. If this persists, circulation to the "innards" can be cut off: a real emergency situation. Another possible cause is a testicle that has twisted on its stalk. Both these conditions need emergency treatment.

## Minor injuries

The most common and heartrending minor injury in this age group occurs when the baby's finger is cut when the parents try to trim the nail. The cut is almost never of any real medical concern, particularly since babies heal so beautifully without even a scar most of the time. But the pediatrician ought to be called, mostly to prevent the other parent from attacking the one who did the cutting.

## Major injuries

These are rare, and in general involve burns, car accidents with the baby on someone's lap, accidentally being dropped or stepped on by a sibling, attacked by a pet, or entrapped in a pillow or other soft surface such as a waterbed.

Isn't it great that they are all preventable!

## Window of Opportunity

Mostly, this is a window of opportunity for parents to enter into their new roles with gusto. This may mean learning ways of dealing with well-meant but unwelcome advice: just because Aunt Melba says you should, you don't need to rub castor oil into the umbilical cord. Trust your own better judgment. A good all-purpose reply, that you can use many times in a row to the same person is, "That's an interesting suggestion. I'll be sure to ask the baby's doctor about it." Trusting your coparent with the baby, with-

out anxious looks, sighs, hastily stifled exclamations, and constant monitoring, is another skill learned now.

All the baby needs to learn is that someone always comes in response to the very first cry.

## What If?

### The Baby has to Stay in the Hospital

The hardest part here is to keep reminding yourself that you are the parent even though the baby is receiving most of the care from others. Not just high-tech care, but feeding, bathing, and burping. I strongly urge you to take an active role as soon and as thoroughly as possible in participating in that care. Offer, with vigor, to feed, bathe, and change your baby. All hospitals should be well equipped and very supportive in assisting nursing mothers, as well.

Even if you did not plan to nurse, I would suggest that you reconsider, even just for the period the baby is hospitalized. Not only is the breast milk medically far superior to formula for sick or premature babies, but the nursing is an automatic entrée into getting and keeping the attention of an often-busy staff.

Both fathers and mothers can endear themselves to the staff, and by doing so be invited to participate in more and more aspects of care, by being sensitive to the staff's duties and needs. Warm interest, compliments, nonhostile requests for information, and, of course, food treats are all welcome. A note to a nursing supervisor or administrator about a special act of expertise or kindness is a great way to win nurses' hearts.

It is terribly hard to think of such things when recuperating from childbirth and with the worry and sadness of a sick or premature baby, but I suggest them anyway. They will give you something concrete to do, will win you praise and esteem, and will probably increase your comfort level when participating in your baby's or babies' care.

## POSTPARTUM BLUES AND DEPRESSION: NOT JUST A MOTHER'S PREROGATIVE

Most new mothers weep easily. "I cry all the time," Molly says calmly, tears rolling down her cheeks. "Sometimes something triggers it, like a story on the news or a really cute pair of baby socks. Sometimes it just happens." When asked how she feels, Molly snuffles and says, "Fine. Why do you ask?"

As many as 80% of new first-time mothers feel some degree of sadness in the first weeks after giving birth, from many different causes. Most often, it's not a constant feeling; it comes and goes. It feels like a combination of worry (about baby, job, family, the world), stress, and fatigue and sometimes genuine grief (missing being pregnant; wanting a baby of the sex opposite than the one you got). If a mother feels pretty good much of the time, and occasionally really happy, the blues will usually run their course over the first few months. Getting as much rest,

exercise, and support from family and friends as possible all help. Getting time for oneself, for pleasure, is a must.

It isn't just mothers who get the blues. New fathers do, too, they just don't get any publicity. For the first weeks after birth, fathers are expected to nurture, not to be nurtured. They are supposed to squelch all negative feelings, from anger to worry, so that they don't upset the mother. On top of which there's fatigue, a new (and ever more stringently) culturally defined role, plus the surprising emotional upheaval of participating in the birth. And if new mothers have trouble organizing their thoughts into coherent expression, fathers have a double duty: organizing the thoughts, and pretending it's easy. Every new father is entitled to a few tears, kind treatment, and praise in abundance.

Perhaps most important for both parents, though, is a sense of perspective. Bonding takes time. I can't say it too strongly. And by bonding I mean feelings that the baby is one's own. Pleasurable feelings. It isn't at all unusual to spend the first two weeks in a state of bewilderment or neutrality or mild hostility towards the new baby. If you can relax into it, and not expect any more of yourself than responsible civilized *behavior,* the feelings will take care of themselves. By the time the baby is about eight weeks old, you will be so entranced you won't even remember how you felt at first.

**Postpartum depression,** however, is another kettle of fish and, as far as it is defined medically, is restricted to mothers. When it is severe enough to be depression rather than

> ## BEYOND THE BLUES:
>
> If you think you are depressed, or if you know a mother with postpartum depression, it is of the highest priority that you get help. Even when a depressed mother is able to tend to a baby's basic needs, the baby is affected in mood and development.

blues, women become pretty much incapacitated. They don't want to handle the baby. They don't like the baby's smell. The baby's crying acts on them like a fingernail down a blackboard. They can't think what to do next, and can't get anything done.

What makes things worse is that they often don't act sad. They don't weep. Instead, their behavior speaks for them—but only if someone can understand and interpret it. Once in awhile a mother with postpartum depression will recognize that she's in deep trouble—that she is tempted to hurt herself or the baby, or that she simply cannot care for the baby's basic needs. More often, she is so much not herself that she can't get any perspective on her condition. She feels as though she is living in a black hole. Her expression doesn't change. She may whine in a colorless voice, or may be silent. She is most unpleasant to be with. Often it takes weeks before family, friends, even doctors recognize that she isn't acting in this seemingly self-centered, childish, inefficient, sulky way because she's a nasty person, but because she is very, very seriously depressed.

Postpartum depression can show up weeks and months after delivery. There is some evidence that links it to the onset of menstrual periods after giving birth—so don't assume that it's only a risk in the first few days and weeks after delivery.

Support sites include:

**Postpartum Support International:**
805-967-7636
*www.postpartum.net*
**Depression After Delivery:**
800-944-4773
*www.depressionafterdelivery.com*

Fortunately, severe postpartum depression is very uncommon, though not as rare as most people think. It always requires therapy, and often medication. The earlier it is recognized and diagnosed and dealt with, the easier it is to treat, and the faster the woman (and the baby) recovers.

## The Two Week Well-Baby Visit

The biggest worry for most parents is sitting in the waiting room with a zillion runny-nosed toddlers, all of whom are bug-eyed with desire to see, touch, and sneeze on the new baby. Call beforehand and ask if you could sit in the car until the exam room is ready. Or try to get the first appointment of the day. Some offices have sick and well waiting rooms, but don't take these titles too literally: most viruses are most contagious just *before* the child gets sick, or are passed on by children carrying the virus without ever becoming ill.

At this visit, every part of the baby gets looked at or listened to. Nearly all babies should have gained back the weight lost after birth. It isn't uncommon for something to have become noticeable since birth, such as a scalp swelling from the birth process (*cephalohematoma*) or hips that haven't quite rounded out their sockets or a heart murmur. Most often, these are findings that either go away by themselves or are easy to treat or just need watching.

Most pediatricians also discuss schedules, colic, safe sleeping positions, and maternal diet for nursing mothers. Many encourage starting an occasional bottle. Most pediatricians try to assess how the parents are doing emotionally, as well.

As for shots, some pediatricians start the series for hepatitis B immunization at this visit. (Others gave the shot at birth, or will start it at two months.)

## Looking Ahead

At the two-week birthday, most newborns will be settling into a schedule, and parents and pets will be adjusting to the idea that the baby is here to stay. Everyone will be exhausted except the baby.

The most foreign idea at this point is that the baby will change. You feel as if you're just getting the hang of this new escapade, feeding and diapering and surviving in general. Ah, but the baby WILL change. Look, you even have a landmark: the umbilical cord is separating. Look, there it goes, it's off.

A big event, a signature moment.

My advice is to keep that hard little nubbin someplace treasured, even though it looks like an old apple stem. I have always felt very bad that I didn't. Sara went to an enlightened Montessori preschool and at the age of three and a half, inquired, after a session on the facts of life: Where had we kept her umbilical cord? Not *if; where*. A bad moment in parenting. Learn from my error.

# Two Weeks to Two Months

## Cuddling and Colic

## To-Do List

- Continue to put Cherub to sleep on his back, day and night.
- Introduce a nursing baby to a Reminder Bottle, just in case.
- Make a two-month Well-Visit Appointment, and ask for the American Academy of Pediatric Guides to the immunizations that will be given.
- Read about the two month immunizations ahead of time.

## Portrait of the Age: It's My Party and I'll Cry if I Want To

Jason apparently likes these things: catching sight of his own hand, biffing himself in the nose, and lifting his head drunkenly from the prone position to stare at dust molecules.

While Jason entertains himself (is that what he's doing?), Sally and Joel have made progress also. They no longer feel quite so strongly that they are about to drop him, or break something. They have learned that when he has an erection he is about to pee in their, or his own, face. Burping him is a breeze.

Indeed, life is becoming normal. A new normal, yes, but normal nonetheless. Sally and Joel have even exchanged entire sentences in conversation. Sally has learned to sleep in two-hour stretches, and while she feels at times as if she were functioning under water, the sensation is not unpleasant. Thank God, however, for Joel's mother, who shows up every day at five-thirty with a casserole. Life is calm and manageable.

Up until four days ago.

Upon turning seventeen days old (who, Sally wonders, would ever have thought it to be a milestone date?), Jason entered a new phase. It starts twenty minutes after Joel walks through the door, which, Sally feels, is both flattering and tactless of Jason. Whatever it is, it isn't pretty. Shrill, inescapable, galvanizing, Jason's cries assault the house-hold.

"It must be six o'clock," the three loving adults say simultaneously.

Sally thinks she knows the reason Jason cries. It is that after their quiet day alone together, he is overstimulated. Too many people are holding him, and only she, Sally, knows how to do so correctly.

Joel worries that Jason is spoiled. Sally holds the baby too much. No wonder Jason cries the minute he's asked to spend a little time by himself, while he and Sally and his mother have five minutes together.

Jason's grandmother is irritated because she has been forbidden to brew chamomile tea for the baby's obvious colic. She is sure that Sally's diet is at least partly to blame. Here she's made all these marvelous dinner casseroles, and then Sally goes and has pizza for lunch. Yesterday, it was a salad with broccoli in it. No wonder the baby goes off like clockwork every evening.

Joel tries first, bouncing Jason on his shoulder and singing "Tiptoe Through the Tulips." That had worked the first night.

Then Sally takes over, offering the breast, a pacifier, her finger, and a bottle of warm sugar water. None of these has ever worked.

Then Grandma takes Jason, places him belly down on her capacious knee, and rubs his back while Sally turns on the vacuum cleaner. Unlike last night, this has no effect on Jason. Sally calls a friend. At the friend's suggestion, they decamp to the laundry room. Sure enough, held securely on the running clothes drier, Jason ceases his screams and doses off, vibrating peacefully.

The period from two weeks to two months almost always features this type of scenario. On the one hand, parents feel as if the routine of feeding, changing, bathing is getting to be second nature. The sensation of the early days, when every aspect of care was fraught with danger and difficulty, is past. It now becomes possible to think about some-thing else, talk on the phone, read a book while the baby sleeps. On the other hand, there is this new kind of crying, quite unlike the cry of the brand-new baby.

There are two aims for parents of crying babies. First, to be sure that the baby is healthy and thriving, and that therefore this crying, whether we choose to call it fussiness or colic, is normal for her. Second, to deal with the crying itself.

Most babies will have a two-week visit to the pediatrician. Typically, the baby is found to be thriving and perfectly healthy, and begins to scream her head off three days later. Is it fussy crying, or is it colic? Or is something really wrong?

## THE FIRST CRYING EPISODE

The first episode of really strong frantic cry-ing can make you panic. First, be sure the baby isn't hungry. Many nursing babies, and some on formula, have to "tank up" to get through a few hours of sustained sleep. They may eat nicely every two to three hours until about six P.M., then demand to eat every hour or so until eleven. Go ahead and feed, if that makes the baby happy.

If feeding, cuddling, and trying all the usual tricks don't work, I feel you should have some contact with the pediatrician to be sure that it IS normal crying. And you should feel free to do so for any episode thereafter that seems unusual. If the cry sounds different (very shrill) or the crying is accompanied by any other symptoms (see *Serious Illnesses* in the previous chapter), do call the pediatrician.

Before you call, take the baby's rectal temperature. Jot down anything unusual: runny nose, spitting or vomiting, abnormal stool. What is her color? Babies normally turn red or purple while crying, not pale, blue, or mottled.

If the baby is nursing, could there have been anything in the mother's diet to upset the baby—especially medications or caffeine?

Also, note the baby's activity before she started crying (and after, if she stops while you're waiting for the doctor to call back). Is she comfortable, and able to be briefly attentive? Or lethargic, or irritable? If there's any doubt about her basic health, she should be seen.

## SEPARATION ISSUES AND SETTING LIMITS: HANDLING THE BABY'S CRYING

This is the first occasion in which new parents are forced to regard their baby as a separate human being over whom they are, in some respects, powerless. (Though parents of sick or very premature babies learn this at the moment of birth, or earlier.)

It's a scary, humbling experience. You are so attached to the baby that when Cherub cries, it's an affront—it's impossible not to take it personally. How could such a cherished being produce such a wail? You want to tell Cherub firmly "No Crying!" and lo, have the crying just stop. And at the same time, you are overwhelmed with the need to make Cherub's world entirely perfect and comfortable, so that there will be no impulse to cry. That is, you want Cherub to assume full responsibility at the same time you want to assume that full responsibility yourself.

In the world of parents and children, Separation Issues and Setting Limits always, inevitably, clash. This is one of the most important challenges of parenthood. For babies in this age, what parents need to face up to is this: It is important to go to Cherub and pick him up, hold him, try different soothing ploys. It is NOT important, however, that he stop crying right away.

What you are doing by ministering to him is not stopping the crying, but inducting him into the society of human beings. You are telling him that his distress signal is meeting loving concern. And by not allowing yourself to become too distraught, you are letting him know that he is permitted to be the baby, and that you will stay separate and in control. You can't spoil a baby this age by holding and talking to him or her, but you can become a prey to guilt and anger if you feel that you have failed by not being able to soothe him.

### Normal Fussy Crying Versus Colic

Once you have been assured that the baby is healthy, how can you tell the difference between normal crying and colic? And is it important to differentiate the two?

I think it is. Dealing with a baby with colic is an all-encompassing exercise, and parents need the comfort of, at the minimum, a "diagnosis" even if the diagnosis doesn't mean much.

Many pediatricians define a colicky baby as one who cries inconsolably for an added-up total of more than three hours a day at least three times in a week (Wessel, et al., "Paroxysmal fussing in infancy," *Pediatrics* 1984:74:998). Yet study after study shows that many babies cry nearly that amount. At six weeks of age, the average baby cries two-and-three-quarter hours a day—if you keep a diary and add up all the crying spells. More than 35% of normal babies have at least one day in which they cry longer than three hours.

However, there really is a subclass of babies who cry much, much more than this. Some cry as much as six hours a day for weeks. Therefore, there is a difference between crying and colic—at least from parents' point of view.

### NORMAL FUSSY CRYING

Jason's crying is normal fussy crying. It occurs in the early evening and lasts about two to three hours. He pauses, unpredictably, after various soothing modalities, either because they work or because he has come to a good stopping point. Nobody knows.

His family's theories about the crying are all popular ones.

- **Someone is holding him too much**.

  No. Absolutely not. In fact, one of the few modalities found in research to really help normal crying is to hold the baby two to three hours more during the times in the day when he is NOT crying. (Unfortunately, this doesn't help true colic, as seen below; only normal crying.)

- **Something isn't agreeing with him**.

  Rarely, crying really can be due to an ingredient in the breast milk, or from some other irritant the baby ingests or inhales.

  What about the broccoli and the pizza that Sally ate?

  Theoretically, the broccoli (and cabbage, beans, etc.) shouldn't cause a problem to the baby. The person who actually eats these foods may become gassy because the complex starches in them break down in the intestine to form gas. However, neither the gas molecules nor the starch molecules cross over into the breast milk.

  Most nursing mothers worldwide live on a diet heavy in vegetables such as cabbage and beans and spices such as garlic. Their babies have no greater incidence of colic.

  Tobacco smoke contains nicotine, and babies get cramps from nicotine, which is, after all, a poison. Smoke in the environment is inhaled by babies, and high nicotine levels are found in their blood samples. If a nursing mother smokes, her baby gets nicotine both from her breast milk and from inhaling the smoke.

- **Somebody isn't handling him properly**.

  True, some babies can be over- or understimulated, but generally you can tell because they respond to a change in the way you handle them. Jason, like other babies with normal fussy crying, goes on crying no matter how he is handled.

Some studies have tried to pin the blame on maternal tension and stress, or on maternal personality. There is absolutely no evidence that this is the case.

- **The baby has acid reflux, also known as gastroesophageal reflux, or GER.**

    When acid refluxes up the esophagus, it burns and the baby cries. You might suspect this problem if your baby spits up and then cries, or if it looks as if he's regurgitating and then reswallowing, or if he arches his back with crying, or if he starts to feed eagerly and then, after a few minutes, starts to cry.

    This suspicion is worth checking out with your pediatrician. Treatments for this problem include thickening feedings with rice cereal, positioning the baby upright or on her right side after feedings, antacids, and medications that reduce acid secretions and/or help propel the stomach contents down the intestine rather than up the esophagus.

- **He's allergic to the milk.**

    Barely possible; more likely if there is a family history of allergies, asthma, eczema. It is true that a few babies seem to be sensitive to the cow's milk allergens, and sometimes to the soy allergens, that cross into the mother's milk. But any change in maternal diet should be undertaken only after medical advice.

Changing formulas is a frequent response to crying, but even if it works briefly it usually doesn't have a lasting effect—at least in terms of decreasing the crying. One study does indicate, however, that babies who cry a lot and then have their formulas changed tend to be regarded *for years* by their mothers as vulnerable and intrinsically fragile. This was true, though not so intense, even with mothers who were told explicitly that their babies were normal ("Perceptions of Vulnerability 3½ Years After Problems of Feeding and Crying Behavior in Early Infancy," Forsyth et al., *Pediatrics* 1991 88:4).

The one intervention scientifically shown to decrease normal fussy crying is to carry the baby for an extra three hours a day during his nonfussy periods. This is certainly worth a try. The study didn't examine whether skin contact made a difference, but you might try snuggling him between your breasts or to your chest with both of you topless.

Home remedies, like the ones Jason's family try, are useful, and sometimes they will soothe the baby. Who knows why? It is often helpful to try them in sequence:

- See if his diaper is dry.
- See if he wants his pacifier.
- See if he wants to nurse (or have a bottle).
- See if he wants to be held, rocked, walked, or swung in his swing.
- See if he wants soothing warmth on his belly: a receiving blanket that has been warmed in the microwave (carefully), a hot-water bottle, Grandma's knee.

- See if he wants a distracting soothing noise: the teddy bear with the tape of the intrauterine heartbeat, a vacuum cleaner, a White Noise machine, Mozart.
- See if he wants some kind of vibration: A rest on the clothes dryer, the vibrating attachment on the crib that simulates a car ride, or a ride in the real car.

All these are worth a try, in order of complexity.

Unless your pediatrician urges it, most medication won't help and may hurt. (Simethicone drops to break up gas bubbles seem to be very safe and may be worth a try.)

But what about colic? Real, true colic?

### COLIC

**Call your pediatrician if you have any doubts that your baby's crying is not due to colic.**

"You want to know why colic doesn't start until two weeks of age?" P. growls, glaring at the six-week-old bundle in her arms. "I'll tell you why. Darwin. The ones that started screaming at birth got put on the mountaintop. They didn't get to survive long enough to reproduce and pass on their colic genes. By two weeks you're bonded. You're stuck."

Now P. is a wonderful mother, devoted, sane, intelligent. She is also a pediatrician, and until the arrival of Bundle had given excellent serene advice to hundreds of parents of colicky babies.

Colic can do this to people.

Colic has been variously defined, but YOU KNOW when your baby has colic. Here is the usual description: a baby from two weeks to three months old who is thriving and otherwise healthy[1] who cries more than three hours a day as a regular thing. Colic peaks at about six weeks of age. The more than three hours of crying each day can all be at one excruciating time, usually the late afternoon or evening, or can be spread out to enliven the entire twenty-four hours.

How can a baby who is thriving and otherwise healthy cry so much? Trust me, she can. Why does she cry so much? I don't know. Nobody knows. There are five theories:

- The baby is allergic to something in the milk.
- The intestinal tract is immature so that the waves of peristalsis are erratic and painful.
- The central nervous system is immature, and the baby has a low threshold for any stimulation, external or internal.
- The mother (or caregiver) gets tense and passes the tension on to the baby.
- It's normal. Statistically, crying in babies follows a bell-shaped curve like other natural phenomena. Some babies hardly cry. (Forget you read this sentence.) At the other end of the bell-shaped curve are the babies with colic, who cry practically all the time.

---

1. If your baby has a heart murmur, or sweats while feeding, or has prolonged jaundice (more than two weeks), or frequently assumes an arching position, or has green or watery or painful stools, or has any other deviation from normal, I would insist on a full work-up, or a second opinion, before assuming that fierce incessant crying is due to colic.

The last explanation makes the most sense. Why would the central nervous system or the gastrointestinal tract be immature at two weeks but not at birth? Why do colicky babies suddenly "outgrow" their "allergy" consistently by three months? Why do very tense mothers have perfectly calm babies, and very calm mothers have diabolically crying babies?

If you have a baby with colic, you may be at the point where you couldn't care less about the cause. All you care about is fixing the colic.

Alas. You can't fix something that is a normal variant. All you can do is wait until the bell shaped curve changes, at about three months, and your baby starts crying less.

Now, many remedies have been tried for colic. Changes of maternal diet and formula. High-tech soothing techniques from White Sound machines to tapes of the intrauterine pulse to crib attachments that replicate the motion of a moving car. Infant water beds. Innumerable low-tech ploys: Carrying the baby on your arm belly down. Hot-water bottle on the belly. Placing the snugly wrapped baby on the clothes dryer, running the vacuum cleaner, putting the receiving blankets in the microwave, and swaddling the baby.

All these remedies tend to follow Gluck's Law. This is a law named after Dr. Louis Gluck, a well-known neonatologist. It has complicated origins, but in essence it says: *Everything works at first, for a little while.* Nothing works to stop the crying for very long except the baby's getting older. Many things will work briefly, once or twice. Your job is not to stop the crying. Your job is to be there for the baby while he cries, and to help yourself and your coparent maintain your sanity.

The suggestions for coping with colic, such as they are, address these two independent problems. They aren't great, but they're all we've got.

For helping the baby, see all the suggestions above. Medications for colic either work only temporarily (simethicone for breaking up gas bubbles) or are risky (sedation like chloral hydrate, which can cause a terrible pneumonia if the baby chokes on it; atropine-containing drops, which no longer are prescribed because of several cases of sudden death).

For helping the parents, the best remedy is to get away from the situation for a bit. Get help. Take a walk, a hot shower. Or get away without leaving physically: plug into your Walkman, turn on the *Eroica* symphony, meditate yourself into a transcendental state.

Pray.

Call everybody you can think of and describe the colic loudly, in detail. If you have run out of friends and relatives, start on 800 numbers.

Write down all the bad words you can think of and burn the paper. Write a reproachful letter to the baby, your spouse, the hospital, the stork, and burn that.

Don't even THINK of trying to keep a sense of humor.

## MILESTONES

From two weeks to two months, babies learn a great deal, though not nearly as much as their parents do.

### Vision

Babies can make out a face (live or drawn simply) at about eight to twelve inches away, and prefer to be looking at it eye-to-eye, not crooked or upside down. Until about four months of age, babies tend to look cross-eyed some of the time, and when they look at you may seem to be staring at something just off to the side: this is because they don't have central focus yet. However, it is not normal for one eye to be stuck looking in one direction. Taking flashbulb pictures will startle, but not injure, the baby. When you look at the flash pictures, the pupils should both be red (unless you have one of those newfangled cameras that eliminates the "red reflex"). If one is white, take the picture, and the baby, to the pediatrician to make sure the eye structures behind the pupil are normal.

### Hearing

Babies should startle to loud, unfamiliar, sudden noises. If they've been hearing the dog bark all through pregnancy, they won't startle now; but the noise of a lid falling or the rip of wax paper may evoke a look of terror and a full-blown startle response. If very carefully tested when awake and alert, a baby will slowly orient towards a high-pitched voice, turning toward the sound.

### Social

It's the smile, of course, that outshines all the rest. By two months, nearly all full-term babies smile. Some smile all the time; others bestow their favors rarely. Cooing ("talking") starts now.

### Motor

When placed on their tummies, two-month-old babies can lift their heads up and turn them from side to side. They move arms and legs at random but equally, showing no preference for right or left. When stood up, two-week-old babies will march right along in a reflexive manner; at two months, most will enjoy bearing their own weight.

A rare two-month-old will be able to roll from tummy to back, and some have been known to roll off changing tables and other surfaces.

## SLEEP

Most babies who weigh eight pounds can sleep four hours at a stretch. Many babies who weigh eleven pounds can sleep as long as eight or more hours. However, a very long or tall baby who weighs eleven pounds may need to eat more frequently in order to sustain the greater height, so don't count on it!

If it is important to you to have the baby sleep through the night in late infancy and toddlerhood and later, it's a good idea to help the baby learn at about two months to get used to falling asleep at night in the bassinet or crib, not in your arms. Of course, this means that the baby gets lots of holding

## BACK IS BEST

Nearly all babies should sleep only on their backs, day and night: the back position is highly associated with a reduction in the risk of SIDS, sudden infant death syndrome. (Rarely, a baby must be positioned on her tummy due to underlying medical conditions, but your pediatrician will guide you on this issue.) Parents often say that their baby sleeps so much more restfully and deeply on the tummy. Well, that may be exactly the reason that tummy sleeping is associated with SIDS! A little wakefulness may be an important safety feature.

When Cherub is awake, and you are right there watching to make sure he doesn't fall asleep, he can play on his tummy, as long as the umbilical cord is off and healed. When Cherub is able to roll from back to tummy spontaneously, without any assistance, you can leave the baby in that position. This skill requires neurological maturity and muscular strength that help protect against SIDS and usually isn't developed until after four months of age, the end of the highest risk period for SIDS. Always put Cherub to bed in the back position, however, and let the baby do the rolling.

and cuddling all during the day. Then you put him or her into the sleeping place awake. Most babies will take this equably, dozing off within a few minutes.

But a few higher-needs babies will get frantic, and they need to be held and cozened and allowed to relax and doze before being put to bed. I wouldn't let any baby this young "cry it out." You still have the next few months to encourage this habit of falling asleep in the crib, a few months for the baby to learn self-soothing devices. (After about six months of age, it becomes much more difficult because the baby will have grasped the concept of separation and absence, and won't want you to leave the room.)

## GROWTH

By the age of two weeks, most babies will have regained any weight lost after birth. From now on, until about six months of age, most will gain about an ounce a day. Some very long babies need to gain more and very petite ones less. Nursing babies often gain as much as formula-fed ones, but may only gain two-thirds as much.

## TEETH

Please see this section in the previous chapter, *Birth to Two Weeks.*

## FEEDING AND NUTRITION

*Schedule*

Most babies have evolved into a schedule by about two weeks of age. This doesn't mean they are sleeping through the night, but that the feedings are fairly regular. By the time the baby weighs eight pounds, nursing babies are usually spontaneously eating every three hours; formula-fed babies about every four hours. Most babies at around eleven pounds are able to sleep six or seven hours a night.

Once in a while, a baby needs to feed much more frequently.

- A small baby may still be catching up on growth. If the baby weighed less than six pounds at birth, she may need to eat every two to three hours because she needs to gain rapidly but her stomach is still small and her energy for sucking limited.
- A very tall or long nursing baby may need to eat frequently to stimulate the increased milk required to keep up with her growth.
- Occasionally, a baby gets into a pattern of nursing or bottle-feeding every two hours in the daytime, but sleeping six or seven hours at night. Most parents value that long stretch. It's a good trade-off.

### THE TOO-FREQUENT EATERS

But sometimes a baby who doesn't *need* to eat frequently does so anyhow. One such pattern shows up in an occasional baby who seems to get confused about his internal signals. Such a baby eats to the point of overflow, then within two hours cries again, and seems hungry. Unlike the truly hungry tall or long baby, this one is, not to put too fine a point on it, fat.

It's not subtle. You don't have to stare at him to try to decide whether he's maybe just nicely filled out. He has three chins and deep creases in the thighs and arms. Sometimes he is so chubby it's hard to find his penis. Strangers come up to pinch his thighs.

If you are really in doubt, weigh the baby on your doctor's baby scale; a gain of much more than an ounce a day supports this theory.

This kind of baby seems to interpret the intestinal feeling of "very full and digesting" as "uncomfortable and therefore hungry." He gets into a true vicious cycle, trying to cure the full feeling by nursing some more.

It is remarkable how grateful such a baby is when straightened out. The trick is to teach him what his feelings really mean. To do this, when he cries sooner than three hours after a feeding, give him a pacifier, a bottle of water with perhaps a pinch of sugar, a ride in the car—anything to get him to at least three hours, better yet three-and-a-half or four, after the beginning of the last feeding.

If this is truly what was going on with this baby, only two or three such feedings are enough for him to catch on. Wow, he seems to say. This is what it feels like to be hungry. And this is what it feels like to be full. Then the baby settles contentedly into a three- to four-hour schedule. Without his ever losing weight, he gradually slims down as he grows into his chubbiness.

## Giving a nursing baby a reminder bottle

Whether a nursing mother is returning to work soon or never, whether she is planning to nurse for three months or six months or a year or until the child weans herself, whether there is a father or grandmother or friend eager to offer a bottle or she has to invite one to do so, I feel it is important to introduce an occasional bottle now, given that nursing is well established and the baby is back up to birth weight.

At two weeks, such a baby is not likely to suffer "nipple confusion" and learn to prefer bottle to breast. Nor is the supply/demand rhythm of nursing going to deteriorate when a bottle is introduced. Engorgement should be a thing of the past. The daily routine is calm enough to withstand mother, father, and baby's learning another new skill: expressing breast milk, giving a bottle, and sucking from a rubber nipple.

If you wait until three weeks of age, however, many babies stubbornly refuse an offered bottle, whether it contains breast milk or formula. If the nursing mother suddenly can't nurse for whatever reason, such a determined baby can make life very difficult. Even if you are planning to take the baby to work with you, or need not return to work at all yet, you still could have an unexpected interruption of nursing: a trip, a crisis, a need for medication that crosses into the breast milk and is contraindicated for the baby.

So this is the best time to introduce an occasional bottle.

At this feeding, you don't need to give a full bottle of breast milk. The reason you're giving a bottle is to develop Cherub's ability to go from breast to bottle and back again. So an ounce or two would be fine, although you can offer up to one ounce of milk for every two pounds Cherub weighs—a rough estimate of a baby's stomach capacity.

If you give the bottle too frequently, she may start to prefer it over the breast; too infrequently, and she'll forget how to go back and forth. My advice is not more often than once a day; not less often than three times a week. If she starts showing a marked preference for the bottle, drop it for a couple of days and nurse like crazy; if for the breast, introduce the bottle at least once a day.

Once in a while a baby even at this age will refuse the bottle, especially if it contains all or mostly formula.

If this happens: first, get her used to the bottle containing just breast milk. Have her father or a friend or other relative who is patient, calm, and supportive give the bottle while the nursing mother leaves the house, if necessary. Make sure the baby is good and hungry. Turn on calm music.

Once she will take a bottle of breast milk daily, start adding, if you wish, a little more formula each time. Sneak up on her with it. When she will take the bottle you wish when you wish it, don't let her forget how. Keep giving a reminder bottle at least three times a week.

1. Support breast in one hand. With the other hand, gently work downward from above the breast.

2. Working all the way around the breasts increases milk flow through areolar ducts.

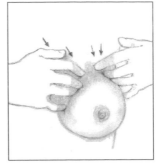

3. Gently stroke and draw your fingertips down toward the areola several times, avoiding any pressure on the breast.

EXPRESSING MILK BY HAND

4. Gently apply downward pressure on areas behind the areola, using the thumbs and fingers.

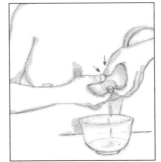

5. Squeeze thumb and forefingers together, pressing backward against the chest wall— back to the ribs. Don't pinch the nipple! Move all around the areola to empty all the areolar ducts.

*Weaning, totally or partly*

If it is necessary to wean the baby from the breast, it's important to do so gradually. Many mothers returning to work early can continue to nurse two or three times a day, before and after work, even when they don't express milk during the working day.

Weaning takes about two or three days for each feeding. Count how many feedings are going to be missed, multiply by three, and start the weaning process at least that many days before returning to work. For example, if you are working the three to eleven shift, and usually nurse twice in that period, start weaning the baby at least six days before you return to work.

Start with any one of the feedings. For three days, at that time of day, nurse the baby only until the breasts feel slightly decompressed and don't cause you discomfort, certainly not more than a couple of minutes. OR, if you find that impossible, just express that much manually or with a pump. If you empty too much, you will simply stimulate milk production. Apply cold packs afterwards.

Of course, you then finish with, or substitute, a bottle of stored breast milk and/or formula for that feeding.

### Wet nurses

Sometimes nursing mothers wish to nurse each other's babies, a grand tradition in royalty and novels. Those who wish to should be aware that viruses can be transmitted through the breast milk, notably AIDS, hepatitis B, and CMV. So can medications and pesticides. Ask your pediatrician, but I'd be cautious. No one REALLY knows someone else that well.

### Formula

As the baby grows, her stomach capacity enlarges. A good rule of thumb is to offer in each bottle about one ounce for every two pounds the baby weighs, up to a maximum of eight ounces.

Formula-fed babies may need extra water in very dry areas or during hot spells. Many babies will refuse plain water at first, and it is tempting to add a little sugar. I would try not to. If the baby is thirsty, she'll drink the water. Even babies this young can develop a hankering for sugar for the sake of sugar. An exception would be a little brown sugar in water for the baby with hard stools; it's not very sweet, and has a mild softening effect.

#### PREPARING FORMULA

Preparing formula is usually straightforward: just follow the directions on the can or box. However, there are a few important warnings to heed:

- Read the directions carefully. Should the formula be diluted? If so, how much water to how much formula? Make sure that everyone preparing the formula understands the directions.
- Make sure the water supply is pure: ideally use filtered or bottled water. Check well water for any impurities.
- Make sure that *everyone* preparing formula washes hands first—and washes again if the hands become contaminated stifling a sneeze or rubbing a nose or eye or going to the bathroom.
- If baby is on a specially prescribed powdered formula for a medical condition such as severe allergy, PKU, or galactosemia, be careful where you purchase it. Try to find a store that has in place safeguards against the black market in this commodity—thieves steal ordinary formula and repackage it as the more expensive brands, selling it back to the markets at much more than its market value. Talk to the store manager. The wrong formula can be very dangerous ("A New Formula for Fraud," by David Cho, *Washington Post*, August 4, 2001, p. A01).

### Juice

Don't offer juice unless your pediatrician recommends a bit of juice for the formula-fed baby with hard stools. Juices add salt and sugar unnecessarily, do not add nutritionally anything the baby isn't already getting, and may contribute to an "addiction" to sweets even at this age.

# EXPRESSING, STORING, AND REHEATING BREAST MILK

## EXPRESSING MILK

### Getting the breasts to cooperate

It's usually easiest to express milk after a feeding (because the baby has already stimulated the "let-down" reflex) early in the day (when you have the most milk). You may not get much at first, but if you express in a sterile fashion you can add all the little bits together to make a respectable amount: close to the four ounces needed for that first bottle feeding.

If you can't express after a feeding, stimulate the milk by gently massaging the breasts.

You may know some "triggers" that make the milk flow: music, sitting in your usual nursing chair, a tape of the baby crying hungrily. A bit of self-hypnosis sometimes works; visualize and "hear" the baby in your mind. Use some of your birthing class exercises to relax.

### Containers

You will need a receptacle to pump the milk into. If you are pumping by hand, you probably need one that has a wide opening. Pumps have their own receptacles. A good receptacle for hand pumping is a measuring cup, so you can see how much you get.

If you plan to feed the expressed milk right away, the receptacle need only be clean, washed with hot soapy water and dried. If you plan to store the milk, the receptacle must be sterile. Rinsing it out with boiling water and letting it air dry will suffice.

You will also need a sterile storage container if you plan to refrigerate or freeze the milk. Unfortunately for ecology, the only really efficient way to do this *in quantity* is to use the little plastic milk bags that come with the plastic nursers.

### Pumping by hand

Wash your nipples. Wash your hands for ten seconds (don't use a scrub brush) and dry them thoroughly on paper towels. To pump the breast, encircle the areola with your thumb and fingers and press back against the chest wall, right back to the ribs. Don't pinch the nipple! As you press and spurt milk, move your thumb and fingers around the areola so you empty all the little ducts. Experiment a bit; it will come. Expect the first few times to be a bit frustrating and/or hilarious.

Don't pump forever. Stop when you start getting driblets instead of jetting streams.

Carefully pour the milk from this sterile container into the sterile storage container. Label the storage container with the date and the amount you expressed.

You can add new amounts to the sterile storage container if you are sure not to touch the inside of that container. You can update the label showing amounts, but don't change the original date, as this indicates how long you can store the milk.

### Using a pump

Do not be intimidated by the weird appearance of pumps. When you have a little time, after baby and work and all are back on a routine, you can design a stylish and elegant pump, a true fashion accessory, and patent it and make a zillion dollars.

Pumps need to be able to be sterilized. They need to be comfortable, and efficient, because the longer you have to spend pumping, the more irritated the nipples become, in contact with a rigid surface. Especially if you are planning to pump on a daily basis, or at work, I urge that you invest in renting the very best electric pump possible. Newest ones can be plugged into the cigarette lighter of your car, so you can pump in privacy, listening to tapes or whatever. (Be sure you have parked in a private and secure area.)

### Manual pumps

These vary considerably in quality. Some are so poorly made they can actually hurt your breasts, and these are also usually very inexpensive ones. Don't be tempted. You definitely do NOT want one that works by suctioning milk back into the rubber bulb. You want one that works by a kind of piston action.

### Electric pumps

These also vary widely. Battery-operated pumps usually can't deliver enough moxie, and like a poorly designed manual pump may hurt more than they help.

## STORING MILK

Either feed the baby right away or put the milk in a cold place so bacteria won't grow.

If you will use it within the next twenty-four hours, store it in the refrigerator. If it will be longer than that before you use it, store it in the freezer.

If you store it in the freezer, label the baggie with the date and the amount of milk.

You can add more milk to a baggie. When you do, you can add the new amount to the label, but don't change the original date.

Many people store milk for up to two months in the freezer; I'd make it two weeks.

By the way, every freezer should have in it a Sentinel Ice Cube so that if there is a power outage you know about it. The Sentinel Ice Cube sits alone on the floor of the freezer. If it has melted, you know that the freezer suffered a hot spell, and that everything else has melted and refrozen itself again, and is potentially toxic.

## REHEATING MILK

### Thawing the frozen milk

You can insert the tightly sealed baggie in a cup of hot water and wait until it thaws and achieves body temperature.

You can also use the microwave. Don't put in the nipple. Beware hot spots.

If the fat separates and the milk looks flaky, keep swishing the bag around until it comes into suspension again.

### Water

Breast-fed babies get extra water built into the breast milk; it's their mothers who need to drink extra water to meet unusual needs from exercise, heat, or dryness. Formula-fed babies may need extra water when the environment is hot or dry. If you have any concern about lead in the water pipes, you are best off running the water for a few minutes before filling the bottle. If you are using well water, be sure it has been carefully tested for lead, fluoride, nitrates, and other minerals and contaminants. Bottled water is probably safe, but bear in mind it is not at this time regulated by any federal monitoring agency.

## BATHING

At some point during this period, most parents will feel impelled to bathe the baby. Not just a sponge bath, but a real one. The umbilical cord has fallen off, the circumcision has healed, they have had a good three hours' sleep, and the voice of conscience is upon them.

Before you have lined up the camcorder and a volunteer, the sink, basin, or special baby tub, and a hygienic armament of soaps, creams, towels, think again. Is the baby really dirty? No. What are you trying to wash off? The baby has not sweated, nor played in the mud. His skin is delicate and his soul doesn't like sudden changes and jarring activities.

## HONEY AND KARO WARNING: INFANT BOTULISM

Botulism is a disease caused by spores of a bacterium that lives in soil. Because these spores (seeds, kind of) are heat resistant, they may survive food processing and get into honey and corn syrup. In infants, but not in children or adults, these spores can sprout in the intestinal tract and make toxins. (Adults get botulism when they ingest the toxin itself, in foods that haven't been canned properly.)

So avoid honey and Karo syrup under a year of age!

It is a devastating disease in babies, appearing first as floppiness, weakness, and constipation. All the muscles can be involved, even the breathing muscles.

If you ever suspect your baby has the symptoms of infant botulism, get help immediately.

Even if he never had honey or Karo, there are other rare sources of the spores. Early treatment and intensive care can give an excellent prognosis.

---

Consider waiting a while. A nice introduction to warm water in the form of gentle cleansing of his face and diaper area might be a happier solution.

### Shampooing

The scalp does require shampoo. Babies have a brief surge of hormones that increase oil production on the scalp. Often this results in flaky scalp and a rash on the scalp, face, and shoulders called baby acne: see *Common Minor Symptoms, Illnesses, and Concerns* below. Preventing and treating baby acne means a daily or every-other-day shampoo.

## DIAPERING, GENITALS, PEE AND POOP, DRESSING

Circumcised baby boys may rub the end of the penis against the diaper and irritate the urinary opening (meatus). You can avoid this by putting a glop of Vaseline on either the penis or on the part of the diaper he's likely to rub.

The pooping habits of breast-fed babies change dramatically during this period. They outgrow the "gastro-colic reflex," or the automatic release of poop when the stomach is full. Instead, they have to learn to recognize that the poop is building up and needs to be pushed out. Breast milk stool is so liquid that it takes a lot of it to inform the baby that pushing is now required. So a breast-fed baby may only poop every few days, but then stand back! I knew one young lady whose schedule was every eleven days, like clockwork. Her parents planned ahead to devote the day to her needs.

Of course, this is normal only when the baby is growing and thriving beautifully. If a thin baby goes only once every day or two or three, it's because the baby isn't getting enough milk, not because the reflex has been outgrown.

Formula-fed babies need to be watched for both loose, green, watery stools (which can indicate formula intolerance) and hard stools, formed when the digested protein makes a thick, firm curd. If babies think a stool is too hard to push out fairly easily, they may hold it back. You'll see them grunting and straining as if they were trying to push, but all that energy is really being put into withholding the stool. Ask your pediatrician for help with this if you see it happening. Formula-fed stools should have the texture of toothpaste, thick sour cream, applesauce. If they are harder than very soft peanut butter, the baby's on the way to constipation. If they are of Tootsie Roll hardness, you definitely need help.

## ACTIVITIES AND EQUIPMENT

Even if you got your baby no toys or equipment at all, you would still be supplying the best stimulus, entertainment, and excitement of all: your own sweet self. So if money is tight, don't spend it now. Mostly, talk to and giggle at and flirt with the baby.

Mobiles are well received. Of course, the mobile should be hung according to what the baby can see of it, staring upward, not what you see of it looking in from the side. Black and white designs interest babies, and so do big (four- to six-inch diameter) simple faces, black on white and white on black.

As the baby becomes older, reaching and grabbing the mobile brings great joy. See the next chapter for suggestions.

Infant seats are wonderful for entertaining a baby by changing the scene. Beware the unanticipated lurch, however; I see a lot of babies, usually large active ones, who tip the seat off the kitchen counter or washing machine or whatever.

Automatic swings are fine if the baby isn't subjected to whiplash; the seat should support neck and head. A cradle swing is safest. In any kind of seat, the baby shouldn't coil up in the fetal position with chin on chest: this can obstruct the airway.

Cloth carriers should be constructed so that the baby isn't coiled up in this kind of carrier, either.

## SAFETY ISSUES

Please see the corresponding section of the previous chapter. If you haven't "baby proofed" as suggested there, I urge you to do so now. We're only adding on a few things to that basis.

*Medicine Chest:* Have some acetaminophen infant drops on hand for the two-month-old immunizations. Don't use them for fevers or colds! If the baby gets sick, you still need prompt medical attention. If the baby is on formula, you might also want to have on hand glycerine suppositories for babies, or the little bulbs that contain glycerine, to ease hard stools. (Breast-fed babies do not have hard stools.) At this age, don't use them without calling the pediatrician first.

Beware direct sunshine. Babies love and need fresh air, but the sun is stronger now than it used to be. It can reflect from sidewalk, beach, water.

## SUGGESTIONS FOR A BRAND-NEW VOCABULARY

- *Air from mouth:* burp.
- *Stomach contents come up out of mouth:* Vomit, throw up, spit up. (I'd avoid barf and puke as kind of vulgar and juvenile.)
- *Air from rectum:* Gas, or passing gas. (I'd avoid the word "fart." "Whiffle" and such nicknames can confuse old-fashioned pediatricians.)
- *Urinate:* Urinate or pee. (I'd avoid "piss," as some find it vulgar; terms such as "spray" may be confusing.)
- *Bowel movement:* Stool, bowel movement, poop, poo-poo, or ca-ca. (I'd avoid shit and crap, as being interpreted as vulgar. I particularly discourage the euphemism "Going to the bathroom" as it is inaccurate, confusing, and conjures up a truly mind-boggling mental image.)
- Even when the baby has diarrhea, and what comes out looks like water, pediatricians still call the stuff stool, poop, ca-ca. So if the pediatrician asks about the stool, say it is watery, smelly, loose, or whatever. Humor the guy. Or gal.
- Also, when the pediatrician asks how often a baby has a stool, the parental answer is often "Every time I change his diaper." But what the pediatrician wants to know is the frequency of the stool—as in "every three or four hours" or "gee, nine times between six o'clock and midnight."

If you do need to be outside for more than five minutes, apply hypoallergenic sunscreen to Cherub's exposed body parts. Use a hat with a brim and neck and ear flaps. If Cherub starts looking even a little pink, leave the sun: it takes twelve hours for a burn to develop fully.

## Health and Illness

*Part II of this book discusses symptoms and illnesses of babies and children older than four to six months of age. Newborns up to two or three months old are different: more fragile, prone to different conditions, and still adjusting to life outside the womb. They are still, medically speaking, newborns. Symptoms and illnesses for this age group, two weeks to two months, are discussed at length in the* Health and Illness *section of the previous chapter, "Birth to Two Weeks." The following section merely updates that one by adding certain conditions. If your baby has any worrisome condition not covered here, or if you do not understand or are uncomfortable with the discussion in this chapter, get in touch with your pediatrician at once.*

## EXPECTATIONS

Under two months of age, babies are still considered newborns. They are medically fragile in a number of ways (though also surprisingly tough and resilient in others).

However, there are a few items that change at two weeks of age.

In terms of medical significance, the two most important differences between a baby this age and a much younger newborn are:

- Jaundice beginning now is not normal newborn jaundice. It may turn out to be due to an insignificant problem, but it always needs prompt evaluation.

- It is not unusual for a heart murmur to be heard for the first time in a baby this age. This doesn't mean that the pediatrician "missed it" earlier. The heart and circulation mature greatly during the early weeks, and the changing pressures in the heart and large vessels can make it possible to hear a murmur—the noise of turbulence in blood flow—that wasn't there before. Most often, such a murmur reflects a fairly innocent condition—one that may even resolve itself over time.

For most babies, though, these won't be an issue. Nearly all babies, however, will manifest one or more of the following minor problems.

## COMMON MINOR SYMPTOMS, ILLNESSES, AND CONCERNS

### Skin

Of all the categories of minor problems in very new babies, skin tops the list.

#### BABY ACNE

Baby acne is a red bumpy rash on the scalp, face, and shoulders. Sometimes it is flamboyant and covers the whole baby. The baby usually is perfectly happy with no signs of illness. If there is any sign of illness, including crankiness or a temperature 100° or over, call your pediatrician.

Baby acne is caused by hormones generated by babies of both sexes in the first weeks of life; the hormones make the oil glands overreactive. Baby acne is harmless and of cosmetic significance only. It eventually disappears by three or four months of age.

Treatment is aimed at getting rid of the oils, particularly on the scalp.

- Often, the only treatment needed is to shampoo the scalp once a day with baby shampoo (not baby bath or bar soap).
- If this doesn't work, try applying the shampoo, leaving it on for three or four minutes, then rinsing.
- And if this doesn't work, check with your pediatrician about the following remedy, which is safe (if you use the right cream) and usually effective. This involves applying a cortisone cream to the scalp, leaving it on for ten or fifteen minutes, and then shampooing it off. The cortisone cream

must be a mild and poorly absorbed one; it's best for it to be prescribed. It should be thinly but evenly applied. Make evenly spaced dots of cream over the scalp, then blend together. Don't rub it in.

### DIAPER RASH

Most pediatricians call a rash a diaper rash if it appears in the diaper area but not elsewhere. This doesn't mean that the rash is due to diapers, or to diapering technique. I guess we ought to say "genital and buttock rash" but I've never heard anybody do so.

Most babies will have some diaper rash in this period, and most cases will be benign. However, if a baby this age has diaper rash along with diarrhea, the rash won't get better until the diarrhea is cured, and a call to the pediatrician is indicated—and often a visit.

There are three common kinds of diaper rash:

1. *Irritant diaper rash:* The baby with this rash isn't happy about it. This is a red rash that usually spares the creases where the hips are. It comes from moisture, irritants in the stool, or from substances in the diapers (perfumes in disposables, soaps in cloth diapers). Prevention and treatment are the same. Cleanse very gently, using Vaseline on a soft tissue, not baby wipe or washcloth or water. Leave a layer of it on after you finish cleaning. Keep the baby as dry as possible. Don't use plastic or rubber pants; use the "breathing" cloth overgarments instead.

2. *Yeast rash:* The rash looks terrible, but the baby couldn't care less. This is a big splotch of a rash with a border of red raised dots; the rash advances as the dots merge together. It does attack the creases. It is caused by a yeast called candida or monilia. Yeast is in the environment, and loves babies' bottoms and their mouths (see *Thrush,* below). Prevention includes keeping the baby dry and washing hands before as well as after diaper changing. Treatment is a medicated cream containing an antifungal agent. These are both prescription and over-the-counter. The over-the-counter cream contains clotrimazole; a brand name is Lotrimin.

3. *Seborrhea:* This is the bottom's version of baby acne. It is also present in the creases, and usually doesn't bother the baby. Treating it like the irritant rash, above, usually works. Sometimes this needs a cortisone cream, but this should be on doctor's advice only. Applying a cortisone cream to this area, which is then covered by a diaper that seals it in, can make the rash much worse and can even be dangerous because the baby absorbs the cortisone.

### BIRTHMARKS

Many birthmarks aren't present at birth but appear at this age.

*Brown spots:* These are usually small, flat smooth frecklelike spots. They aren't freckles, but, usually, moles (also called nevi). Nearly always they are innocent and permanent. If they are large, or there are more than

## STRAWBERRY MARKS THAT SHOULD BE SEEN BY A PEDIATRIC DERMATOLOGIST

- Located on the eyelids or just below the eye;
- Located on the tip of the nose;
- Any large one on the face (rarely these become aggressive and damage underlying structures);
- Triples or more in size over a few weeks;
- Likely to bleed: on the lips, diaper area (genitals or anus), hands or feet.

two of them, your pediatrician will want to take a look.

*Strawberry marks:* These are raised red bumps that look like ripe, succulent strawberries. They start out as flat reddish areas surrounded by a white patch; they then grow to fill out the borders of the original white patch. Sometimes a baby has more than one of them. Usually they are on parts of the body that are either easily concealed or "noncosmetic," like the belly or scalp. Even though they can become very large, they do all go away by themselves in time.

If a baby has many of them, or they are disfiguring or impair a function (for instance, block a baby's vision), there are many techniques for dealing with them, and new ones all the time, including safe laser treatment. However, most of the time all that is indicated is to watch them grow, then watch them disappear. Most of the ordinary ones go away by the time the baby is about two.

### *Eyes*

#### TEARY EYES; GOOPY EYES

You may see tears welling up or rolling down the cheek, or find goo matting the baby's eyes. Check to see that the baby is otherwise well, without redness of the white of the eyes, or swollen lids, or signs of pain on looking at a bright light, or fever. Then call your pediatrician. A happy baby with an eye that that looks perfectly normal except for the tears usually just has a temporarily blocked tear duct, a usually quite harmless condition.

Parents sometimes think that the tearing eye is normal, and that the other one doesn't have tears yet, but this is not the case. Tears are formed by birth, and are being constantly manufactured by the tear gland under the upper eyelid. They float over the eye and then drain into a tiny tube that carries them into the inside of the nose, where they are absorbed by the lining mucous membrane. Sometimes there is a clog in this system.

The pediatrician will probably want to check to be sure this is all that is going on. If the eye is red, or the baby is squinting at bright lights, or seems to be in pain, the baby should be seen very promptly. These can be a sign of rare problems like glaucoma (increased pressure in the eye) or a serious infection.

Once it is clear that a blocked tear duct is the problem, the treatment is to massage the duct so as to force tears past the obstruction. Some pediatricians recommend a few drops of very clean breast milk into the eye; many prefer prescription antibiotic eyedrops. Massage is done with the short-nailed, clean, pinky finger of the adult.

### FOCUSING

Babies don't focus straight on until about six months of age. That is why the baby seems to be gazing so hard at somebody invisible standing just behind your shoulder: she is really focusing on you.

Wandering eyes still occur at this age. They are only likely to be a problem if one eye is always fixed in the same direction.

### Nose

The baby who is otherwise healthy but has a noisy nose probably has schnurgles, as described in the Newborn section. Remember, with schnurgles you don't see anything coming out of the nose; you just hear the noise. If the baby has a runny nose, that's more likely a cold, and at this age the pediatrician should be called, though a visit may not be necessary.

### Mouth

Thrush is by far the most common mouth concern. This is a yeast infection, causing a cottage-cheesey white paste on the tongue, insides of the gums, and cheeks. Thrush is not usually dangerous unless it becomes so extensive that it impairs eating; or unless it signifies, rarely, an underlying problem with the immune system (this might be considered if the baby were otherwise sick, not growing well, or had thrush exceedingly resistant to treatment). But it can make sucking uncomfortable, and a nursing baby can pass it on to the mother's nipples, which is not at all pleasant.

A call to the pediatrician will probably call forth a medication containing nystatin. It is important to know that just dropping the medicine in the mouth with the dropper won't get rid of the thrush. The medicine needs to be rubbed all over the inside of the mouth; put the medicine on a gauze pad and rub that around. If the thrush is on the tongue, putting the medicine on a sterilized pacifier will help do the trick.

Do not put anything back in the bottle of medicine that has touched the baby's mouth; it can contaminate the medicine and produce a stubborn case of thrush. It is a good idea for nursing mothers to put a little on nipples after feeding, also.

The medicine is not absorbed by the baby, and therefore is not dangerous; but it can irritate the lining of the stomach.

To keep thrush from coming back, wash hands well, frequently cleanse all rubber nipples, and avoid refeeding with a partly taken bottle.

## Chest

The breasts of babies of either sex can still be enlarged from the mother's hormones. And that little bump at the end of the breastbone (or sternum) called the xyphoid process is still prominent.

## Abdomen

The belly button may still be an outie. There has to be an opening between the paired, vertical abdominal muscles for the umbilical cord to exit, after all. No wonder it takes a while to close, and in the meantime the area where the cord was protrudes. As the muscles get stronger, most outies become innies.

When the opening is fairly large, a part of the abdominal contents can protrude. This is called an umbilical hernia, though it is not truly a hernia, since the opening is a normal, not abnormal, one. Preemies, African-American babies, and constipated babies often have these as a normal body accessory. As they get older, they and their friends like to diddle them, and that's fine. Nearly all of these disappear in the first two to four years of life. Bandages, coins, and other home remedies to constrict it can be too binding and can obscure infection, not a good idea.

Rarely, if the baby isn't doing well otherwise, the hernia may signify an underlying problem, such as thyroid insufficiency.

## Legs and feet

It is usual for the lower legs to be bowed. They still remember their intrauterine position. The feet are often curved, as well. If the curve is prominent, the pediatrician will want to see if the baby requires massage of the foot or perhaps shoes to wear to straighten them out. Since some babies with very curved feet may have a hip not firmly in the socket, the pediatrician will check that carefully also.

Rarely a parent rather than a pediatrician will discover that a baby's hip is out of the socket. If one leg looks shorter than the other, or the buttock creases don't line up, or the hip goes "clunk" when you change the diaper, let your pediatrician know (not in the middle of the night, though).

The feet of breech babies still tend to wander up towards their ears. It is hilarious to watch such a baby confused about which is his hand and which is his foot.

## SERIOUS OR POTENTIALLY SERIOUS SYMPTOMS, ILLNESSES, AND CONCERNS

### Pyloric stenosis: recurrent vomiting with force

You've probably heard about this one. The pyloric muscle is the gatekeeper muscle that allows milk (and food) to pass from the stomach to the small intestine. If it gets tight (stenotic), then every time it tries to pump food along it pumps it right back up out of the stomach instead. And the baby vomits with amazing force. As the muscle gets tighter and tighter, the vomiting becomes more and more frequent.

The typical time for a baby to start showing the signs of pyloric stenosis is about two weeks of age. Don't pay any attention to

A baby who vomits yellow, green, or brown material may have an obstruction lower down in the intestine. This can be caused by a twisting (called a *volvulus*) and is a more urgent situation than pyloric stenosis. Call if your baby vomits "colors" more than twice, or if it's only once but the baby seems unwell or in pain.

people who tell you that it only occurs in first-born males; that's not true.

What is true is that the diagnosis can be made very easily, by an experienced old pediatrician feeling the belly, or by ultrasound, or by the baby swallowing a little dye under X ray, or any or all of the three.

The treatment is surgical, but remarkably minor as surgery on babies goes. The surgeon makes an incision in the muscle in one direction, then sews it up in the other. (Imagine a horizontal smile sewed up to make a vertical smile.) If the diagnosis is made while the baby is still thriving and well hydrated, usually all goes so well that the baby is in and out of the hospital and eating again in short order, and is left with a practically invisible scar. Usually no blood transfusion is needed, either.

## RARE AGE-RELATED PROBLEMS

*Crib death (also known as sudden infant death syndrome, or SIDS)*
It is tragic. It is also rare: 2 to 8 per 1000 live births, depending on sex, race, season, geography, and other factors. Nearly every causal factor has been examined, from imbalances of trace metals to occult infections, from reactions to the pertussis vaccine to maternal smoking, from immature brainstem development to being overdressed. The latest statistical study at this time has to do with sleep position. We don't know a whole lot more than we did when we started.

Our best advice concerning SIDS is this:

Three factors seem to correlate strongly. The first is maternal smoking both during pregnancy and after birth. So don't smoke. The second is nursing babies. So nurse if possible. The third is sleep position. Full-term babies (without underlying medical conditions that make it inadvisable) should always sleep on their backs, both at night and for naps. Side sleeping is not as safe; tummy sleeping is the most dangerous.

Bed-sharing with adults is cozy but comes with its own dangers. Adults can "overlie" a small baby in their sleep; babies can become wedged between bed and headboard or wall; get tangled in bed clothes or pillows. Sleeping on a sofa with an adult is particularly hazardous, statistically.

Careful and thorough studies have exonerated the pertussis vaccine. It may even help to protect against crib death. Your child should have the DPT shot unless your pediatrician vetoes it for specific medical reasons.

### Near-SIDS (or apparent life-threatening event)

Sometimes a baby will have a spell where she doesn't breathe and turns blue or pale or mottled; she may need to have stimulation or CPR to start up again. Such a baby may be having apneic spells, in which the normal irregular newborn breathing results in a pause long enough to deprive her of oxygen. This may signal some immaturity of the breathing centers of the brain, or it may be an indication of some other cause of apnea or not breathing. One of the most common causes is the comparatively innocent "reflux" in which a baby regurgitates milk from the stomach, and for poorly understood reasons this causes her to hold her breath too long. This can be diagnosed and treated. **A baby who suffers such a spell should be fully evaluated, usually with a brief hospitalization.**

Some parents want a monitor for their baby because of a family history of SIDS, or because of generalized worry. Many pediatricians will go ahead and prescribe a monitor just to ease parental worry; who knows if it really does? Such monitoring has not decreased the incidence of SIDS, and parents can be driven to distraction by oversensitive or malfunctioning monitors.

### INJURIES

#### Shaken baby syndrome

Sometimes a baby this age cries so much there is an overwhelming temptation to shake some sense into him. This can be tragic.

A baby's blood vessels are very fragile, especially those surrounding the brain. A baby who has been shaken can die of a hemorrhage in the brain, or suffer anywhere from mild to severe brain damage. A shaken baby may look pale and mottled, be lethargic or irritable, have a seizure, or go into a coma. The diagnosis is made by examining the retinas of the eyes, performing a lumbar puncture, and by various imaging procedures such as an MRI or CT scan.

Shaking in anger, shaking in play, bouncing, tossing—none of these for babies this age.

### Burns

Since babies get their first baths around this age, make sure your water heater is turned down to 120° and that you test the water with your wrist.

Since babies get bottles at this age, with either formula or breast milk, beware the microwave. Hot spots can severely burn the baby's mouth.

## Window of Opportunity

Learning about trust and reciprocity are the two main tasks of the very young baby. Fortunately, they are almost impossible to get wrong. The main thing to bear in mind is the more attention the better; you can't spoil a baby this age by too much holding, conversation, singing.

Practically speaking, this is the age to make sure a nursing baby can go from breast to bottle and back to breast again with relative ease.

Any baby who experiences such a "near-SIDS" or "ALTE" (apparent life-threatening event) should be hospitalized without delay. She should be monitored while studies determine why such an event occurred, what if anything to do about it, and whether or not she should be monitored electronically at home.

From the medical point of view, the main easy-to-prevent problem is constipation in the formula-fed baby. A hard stool that is uncomfortable to pass may inspire the baby to try to hold back on pooping; this results in an even larger, harder, less comfortable stool. While this vicious cycle usually starts later, when foods are introduced, it can begin this early. Ask your pediatrician if you think this problem is brewing.

## What If?

### STARTING DAYCARE

Many mothers return to outside work between six and eight weeks after delivery. This is often very painful, sometimes unexpectedly so. Hormones are still asserting themselves. Fatigue is normal. And the bonding of parent and baby is reaching a crescendo.

"Will the baby remember who I am?" is the question most asked. The answer is an unqualified yes. Babies bond, too, and can pick their mothers and fathers out of a crowd by voice, fragrance, sight, and touch.

If you don't believe it, watch how a young baby responds to the father who's been away all day.

The best daycare for a baby this age is as close to one-on-one as possible. If this is not possible, the next best is to share an adult with one, two, or at the very most three other babies in the same age range. This is in part to protect young babies from the infections that are a normal part of larger daycares, and in part to make sure they get the cuddling and prompt response to their needs that they require.

If this is the arrangement, the adult should be charged only with baby care, not with other housekeeping chores. While babies this age don't climb on the coffee table or tease the dog, they still require high standards of cleanliness, patient and loving attention, and lots of cuddling and smiles.

In any daycare situation, the provider, besides having personal qualities of warmth and patience, should understand the importance of hygiene. This means that diaper care should be relegated to one area, as far as possible from bottle preparation, and that careful handwashing gets done after each diaper change, before feedings, and before handling the baby.

## EXTENDED SEPARATION FROM THE PARENTS

### Physical separation

There are two kinds of physical separation that can occur during this period: a parent must be away from the baby (and usually the other parent) because of illness, outside demands (such as military obligation), or marital strife; or the baby must be hospitalized or rehospitalized.

The first situation is highly complicated by family dynamics, and is too complex to discuss here. However, I think that there is one universal aspect that any parents in such a situation need to consider. It is very difficult for a parent left with the baby to become attached and bonded to the baby while at the same time mourning the absence of the other parent. It is as if nature allows only one of those emotions at a time.[2]

Because of this enormous demand on emotions, the parent left with the baby may unconsciously make one of three choices: bond to the baby and not mourn the absence of the other parent; mourn, but not bond to the baby; or opt out of both, throwing emotional energies into other endeavors. Of course, the choice of response impinges on the parent who must be, or chooses to be, absent. Anger, deprivation, rejection, worry all increase.

Obviously, one solution is to try to avoid such a separation whenever possible. When such a separation is inevitable, both parents need to be aware of the burden not only on the "away" parent, but also the dilemma for the one who stays with the baby, and how this absence may affect future marital and family dynamics. It is my conviction that any family in such a situation at this time needs counseling.

When a baby must be hospitalized or rehospitalized, especially if parents have been told that the baby is seriously ill, or even that the baby may die, they are put into the same bind: they are asked to bond and to mourn at the same time. Most neonatologists and newborn ICU staffs are very aware of this enormous emotional burden, and strive to help parents through this difficult time. Experience has shown that parents can help themselves by trying to:

- Participate as much as possible in the baby's care. If parents can visit frequently, the highly skilled NICU nurses often can involve them in even the sophisticated aspects of care.
- Become as informed as possible about every aspect of the baby's condition and care, and involve themselves in decision making as much as they can. Having a sense of being in control over what happens to the baby is a crucial factor in feeling like the baby's parent.
- Communicate about major decisions, and about the baby's prognosis, with one physician, rather than going from one to another. Ideally, the chosen physician should be one who emphasizes positive

---

2. Marshall H. Klaus, M.D., and Avroy A. Fanaroff, M.B., *Care of the High Risk Neonate*. W. B. Saunders, 1991.

aspects about the baby, who explains clearly what is going on, and who welcomes parents' participation. Parents who have an excellent relationship with the baby's nurses often can find out from them which physician or physicians might be the best choice. Since physicians rotate, it is particularly important to anticipate such schedule changes and to request that the physician to whom they are bonded "sign them over" to a hand-picked physician on the next rotation.

- All these steps are greatly facilitated if parents can establish excellent relationships with the neonatal ICU nurses caring for the baby. Even highly stressed parents may be able to recognize that the nurses, too, constantly carry the same emotional burden of trying to bond and mourn simultaneously. Nurses can model such strength, but sometimes even the most professional staff members burn out. Recognition from parents helps greatly, and helps parents and nurses to bond. If a parent can find a moment, a bit of energy, to write a note of thanks to a staff member, or a note of commendation sent to the hospital administrator, senior physician, or nursing supervisor, this is much appreciated.

*Emotional separation: postpartum depression (PPD)*
Even when no physical separation occurs, rarely a mother may be the victim of severe depression and become emotionally separated—not just from her infant, but from her family, the world, her past, herself. For-

tunately, this is a rare condition. But the faster it is recognized and treated, the better for everybody.

Postpartum depression is quite different from the mild "baby blues" of feeling a bit self-pitying and teary after the baby's birth. PPD may declare itself only gradually over these weeks. A mother who can't stand the baby's crying, and doesn't enjoy holding the baby or like the way the baby looks or smells or sounds, probably has a strong element of postpartum depression. If she can't find the energy to care for herself or the baby, she surely does. And if she has impulses to hurt herself or the baby, either by neglect or by violence, she needs immediate professional help.

Often both family members and physicians, even obstetricians and pediatricians, miss this diagnosis until it practically shouts its name, or until a tragedy or near-tragedy occurs. If any friends or family members suspect that a mother has postpartum depression, they must obtain help, even at the risk of offending and alienating the mother and her mate and relatives. An urgent but confidential call to the obstetrician or pediatrician is one option.

## TRAVELING WITH THE BABY
The main concerns of traveling with a baby this young are feeding considerations and exposure to infectious illnesses. The latter consideration is most difficult to deal with, and if an extended trip, especially air travel and especially during the winter months, can be avoided or postponed, I would urge you to do so. If the trip is to an isolated or

medically unsophisticated area, I especially urge you to reconsider.

## Feeding

### NURSING

It's far safer to nurse a baby while traveling than to try to provide sterile formula in bottles. Nursing mothers should realize how easy it is to become dehydrated while traveling. Air conditioning works by pulling moisture out of the air, and thus out of the bodies of travelers. No matter how you are traveling, bring your own water supply along. Avoid diuretics such as excess sugar, caffeine, chocolate, and, of course, alcohol. Be prepared to nurse the baby frequently, because the dryness makes the baby more in need of frequent fluids as well.

### FORMULA

Thanks to a very generous formula company, Chuck and I took a very young Sara on a tropical vacation with a suitcase full of ready-to-feed four-ounce bottles of formula. That was back in the days before we (the medical profession, not just Chuck and I) realized it wasn't ethical to accept such gifts. It was great not to have to worry about providing sterile formula. Unfortunately, we hadn't considered the water problem: she was thirsty, not just hungry for formula. Not realizing this, we kept giving her more formula. No wonder she was rather cranky much of the time, and no wonder she gained an enormous amount of weight in a short period. We were lucky she didn't get into real trouble from lack of extra, nonformula water.

My advice to parents of formula-fed traveling babies is to find a reliable and sterile water supply, one that is generally available wherever they travel, and to use powdered formula. The water source I suggest is any brand-name water purchased in factory-sealed bottles. Boiled tap water is a next-best choice, but tap water can be contaminated with lead and other minerals.

Start well ahead of time giving the baby powdered formula mixed with distilled water, so that she is used to it.

Remember that once the baby has put her mouth to the bottle and sucked the milk, the milk is contaminated with her mouth germs. If she doesn't take the whole bottle, it must be discarded, not refed.

Make up the bottles yourself, rather than asking a babysitter to do so, so that you can be sure the proportions are correct.

## Infectious illness

Besides avoiding intestinal upsets by attending to sterile feedings, parents of traveling babies need to protect the baby from air-borne and hand-borne illnesses. During this period, the baby outgrows maternal immunity and starts to produce his or her own immunity. But the baby is still very prone to colds and other viral illnesses, such as RSV bronchiolitis and rotavirus diarrhea (see Glossary).

- Keep the baby in a front-carrier, and don't let anyone, no matter how doting and curious, handle the baby.
- Avoid secondhand smoke, which irritates the airway and mucous membranes and prepares the way for infection.

- Keep a container of premoistened wipes handy, not for the baby's bottom but for your hands. Wash your hands with them each time before touching the baby or touching your own face, especially eyes and nose. You will constantly be touching contaminated surfaces. If you catch a virus, you are likely to pass it to the baby.

## ADDING TO YOUR FAMILY

Having two babies in rapid succession has advantages and disadvantages for all participants.

- Cherub will be around a year of age when the new baby is born. That means he will not as yet have established territorial rights (which occurs around the middle of the second year). Outright jealousy may not be very apparent. The older child often ignores the baby for weeks, and then may be quite affectionate. Moreover, a child that age isn't very ambivalent yet about wanting to grow up and remain a baby simultaneously. He's still pretty much *just* a baby. So the sight of a new baby in diapers, nursing or with a bottle, isn't likely to cause him deep personal doubts and longings.

  However, when the older child is about a year-and-a-half old, the new baby will just be starting to become mobile. For both children, these are ages that require intense adult attention. If there is one harried adult caring for both, especially if that adult is charged with housework, errands, or other responsibilities, life is likely to be strained. The older child will be working out the period of *rapprochement*, which is something like adolescence. She wants to be independent and infantile simultaneously and intensely. The older one is also going through a very intense language-learning stage. If most of the conversation he hears consists of orders, demands, and reprimands, his own language development will be restricted. Toddlers need lots and lots of the kind of adult talk that fosters speech. See the chapter *"Eighteen Months to Two Years."*

  All these develomental challenges are difficult enough, even when there is not a crawling, inquisitive infant to compete with. At the very least, the older child is going to be no reliable help whatsoever in supervising or entertaining the younger one. And when the older child is going through the possessive and negative aspects of the second year, the younger one will be at the age of imitation and aggression.

  The older child cannot be expected to share, either possessions or attention. It may be difficult for the younger one to garner the adult attention that is required for consistent limits and for verbal skills to develop optimally.

- For the mother two pregnancies in close succession can be exhausting. A nursing mother needs to pay special attention to vitamins and calcium intake.

- For the caretaking adult, this age difference can be a wonderful exhilarating challenge, or an unbelievable strain; this depends a lot on temperament, both the baby's and the caretaker's. Nobody in this position should count on doing more than baby care, right from the birth of the second child. If a crying infant demands attention and energy now, imagine how you'd cope when you also had a one-year-old climbing up the stairs, eating the dog food, taking off her own diaper, and regarding the new baby as open game for attack and exploration.

## DIVORCE AND CUSTODY

From the pediatric point of view, many of the issues surrounding divorce in such a brand-new family are those of adjustment to the new role of parents. These are discussed both in the previous chapter, and in this section above, under *Extended Separation from the Parents*. The process of bonding to a new baby may exacerbate all kinds of issues in parents. As discussed above, it is nearly impossible to both bond and mourn at the same time. A parent who has a mourning burden from the past, from childhood scars, may find that the bonding process painfully invokes that burden. Moreover, each parent may feel somewhat rejected by the other's preoccupation with the baby.

All I can say is, if a couple seemed to be happily suited before the arrival of the baby, and then everything comes apart after the birth, seek the help of a therapist before that of an attorney.

If divorce is inevitable, strongly consider mediation rather than litigation. A baby this age needs parents who can work together for the baby's best interests, and a situation in which custody is not seen as an issue for bargaining for material assets or as a means of marital revenge.

## SURGERY AND HOSPITALIZATION

These are dealt with in the section *Extended Separation from the Parents,* above.

## MOVING

My, some people do get stuck making several mind-boggling life changes all at once. The parents I see who have to move just after the arrival of the baby usually also have five or six other things going on simultaneously: career change, elderly grandparent coming to stay, final exams, whatever. And in the midst of all this, they usually decide that this is the right time to get a pet—often a large, untrained, energetic puppy.

The only advice I can possibly give is to simplify everything as much as you can and bear in mind at all times that your main priorities are: The baby and your relationship to each other, and that sometimes the order of these priorities is reversed. Get help from friends and family. Call in all your chips.

Once you've moved, do two things right away: Find the best pediatrician on your health plan and make an office visit, even if there isn't any special reason to do so. Get established in the practice. Make sure you understand your health insurance.

Then turn your new water heater down to

120°. Find out the fluoride content of your new water supply. Don't give your baby extra fluoride in drops if she is drinking fluoridated tap water in her formula.

## The Two-Month-Old Well-Baby Visit

The two-month well visit is a fun visit. Most babies are smiling and cooing by now. They will alert to a sound, but not actually turn towards it, and a loud noise makes them startle. They can follow an interesting object, such as a grinning pediatrician, but often not past the midline. Placed on her tummy, a two-month-old can lift her head and shoulders. Once in a while, a two-month-old can even roll from tummy to back. She'll be moving both arms about equally, and both legs will have the same strength. She may even bear her own weight for a moment when held standing on a surface.

Most of the time the examination holds no surprises. Most babies are gaining an average of an ounce a day now, though a few gain more and some nursing babies may be gaining only two-thirds of an ounce a day.

The rare surprises on the examination are most often the discovery of a heart murmur or a previously unsuspected dislocation of a hip. It is helpful to know ahead of time that a murmur discovered at this age is frequently insignificant, and that a hip dislocation discovered at this age is usually treatable without any surgery.

Two months is the age by which babies have outgrown much of maternal immunity, and the age at which their own immune systems start to function, so it's time for the first immunizations. (Though some babies will have had a shot for hepatitis B at birth.) At this time, all the vaccines given at this visit are inactivated, killed vaccines: they cannot cause the disease for which they are being given. They all come in the form of injections; however, some may be able to be combined together into one syringe.

The DPT immunization is given to prevent diphtheria, pertussis (whooping cough), and tetanus. It is an updated version of this vaccine and has dramatically fewer and milder side effects than the older version. The inactivated polio vaccine prevents this paralyzing, sometimes fatal viral disease. The HIB vaccine is given to prevent infections with the bacterium *Hemophilus influenzae,* which in prevaccine days used to be the biggest cause of serious meningitis in young children, as well as the main cause of the life-threatening disease called epiglottitis. The hepatitis B vaccine prevents this virus, which can cause incurable liver disease. Finally, the vaccine called Prevnar is given to prevent serious infections with pneumococcus bacteria, such as meningitis, sepsis (infection of the blood), and pneumonia. It also has a less significant preventive effect against middle ear infections caused by pneumococcus.

Some of these vaccines (like the DPT) may be able to be combined into one "shot," but it is likely that babies will continue to get at least three or four pokes at this visit. Fortunately babies (unlike kindergartners) can't yet count. To my observations, they used to cry just as lustily and long from two pokes as they do now from four.

A baby may have fever as high as 102° and sore muscles after the immunizations. If he is uncomfortable and feverish, acetaminophen fever drops will make him much happier.

Many parents dread Immunization Day. Part of this dread is not wanting to see the baby in pain. Part of it is not wanting the baby to feel betrayed, as you stand by allowing some hard-hearted person to poke a needle into him. Many parents ask to leave the room during the shots. Part of it is worry about the immunizations themselves, especially the DPT.

For the first dreads, reorienting one's attitude is helpful. Try balancing the pain of the needle against the pain of suffering, say, tetanus. As for leaving the room, I am convinced the baby feels more betrayal at being left. She looks to her parent for comfort and for reassurance that this assault has her parent's okay and therefore can't be as frightening as it seems. On this note, a confident and reassuring voice is probably more helpful than one that is too pitying and worried.

As to the latter dread, worry about the immunizations, it is exceedingly rare for a baby to have a severe reaction to any immunization. In fact, it is doubtful if any serious, permanent damage from killed vaccine immunizations ever occurs. Some have been reported connected in time with immunizations, but this doesn't mean the immunization caused the problem, even though such isolated tragedies have been assumed to be immunization-caused. This is most unfortunate, because the diseases prevented by immunization are themselves frequently fatal or severely handicapping.

If a baby does become ill after these immunizations, the same guidelines for calling apply as would if she had not had an immunization. See *Medical Concerns* in the chapter "*Birth to Two Weeks.*"

## Looking Ahead

Now it really becomes fun. Over the next two months, most cases of colic resolve, nearly all babies sleep all the way through the night and are on a regular feeding schedule. They chortle and smile and gaze with love. Get ready for a good time.

{ F O U R }

# Two Months to Four Months

## The Meaning of Drool

## To-Do List

- Continue to put Cherub down to sleep on her back, not her tummy or side, and make sure this happens at daycare, too!
- Make an appointment for the Four-Month Visit.
- If you are breastfeeding, and are not likely to start your baby on solid foods by four months of age, ask your pediatrician about supplemental iron and vitamin D drops at the Four-Month visit.
- Never put a baby this age on anything she could roll off of: you could be surprised!
- This is a big habit-forming age. Make sure a breastfed baby gets an occasional reminder bottle with breast milk in it; help Cherub learn to fall asleep in the crib rather than in your arms; and start rubbing the toothless gums daily, so that Cherub becomes accustomed to you fooling around in there.

- Try to help Cherub separate the act of nursing or bottle giving from the act of falling asleep. Getting Cherub accustomed to going into the crib (or bassinet) or family bed awake, and learning to fall asleep there, will help assure a good night's sleep later on.

## Portrait of the Age: Me and Not-Me

Melissa has been perfecting the push-up for weeks now. Prone in her portacrib, she pushes up all the way until she balances, wobbling, on her hands, a goofy, slightly crazed expression on her round face.

Her mother, Janice, keeps expecting the game to turn into peekaboo, and for Melissa to seek Janice out each time she pops up. But no. When she does spot her mother, she's thrilled; she bursts into a gorgeous grin just as she wavers and collapses again. There's no sign at all that she expects to see, or looks for, or even remembers from last time, her mother's smiling face.

Janice respects such dogged persistence on her three-month-old daughter's part. But she is aware of a certain impatience, a wish that Melissa would, developmentally speaking, get on with it. Not that she's complaining. It's wonderful to have the schedule regular, to have the baby sleep at least eight hours a night, the colic over.

And she has reached one milestone: Melissa has learned to stare. She loves to look at objects in the near to middle distance—faces, toys, key chains, TV, whatever. She'll watch them as they move with an intensity that seems beyond her control. But can you put that in the baby book? "Today Melissa stared at my sunglasses." Oh, yes, and she's learning to kick. In fact, lying on her back, she performs a kind of bump and grind that is quite vigorous.

But there hasn't been anything *really* new for—oh, weeks now, it seems. Janice feels underchallenged. Melissa doesn't eat solid foods. She can't crawl or pull to stand or even roll over yet. She's still making the same cooing noises she uttered at three weeks, though they've recently been interspersed with an occasional shriek, Bronx cheer, or fake cough.

When, Janice wonders, is It all going to begin? She isn't quite sure what she means by It, but she knows that It will require more parental input than she's called upon for right now.

Suddenly she catches a gleam, not from Melissa's eye, but from her chin. Melissa is drooling! Janice sweeps the baby up and runs her finger along the ridged pink gums. Nothing there, but still, Janice is thrilled: Melissa is teething! Quick, the baby book!

She's not, actually.

But the first drool is still a landmark, and I believe it ought to be right up there with first teeth and first words and first steps. There is a purpose to drool. Drool isn't just overflow because the baby doesn't swallow efficiently: the fountain of saliva that appears at between two and four months of age wasn't there before. And it appears at this age consistently, no matter when the baby's first tooth shows up.

In fact, drooling shows up just about the time that a baby learns to reach out and grab something and put it in her mouth. Obviously, then (to my mind), nature has timed drooling to coat that object. Saliva contains all kinds of immune factors to counteract dirt, plus a good deal of slime that smoothes the rough edges. No doubt, drool is a legacy from Neanderthal times, when a baby's teething ring was a grundgy mastodon metatarsal, and the baby who drooled more survived to pass on her superior drool genes to her progeny. Drool is a sign that the baby is about to take charge of part of her world.

For the first time, Melissa can rely on herself, rather than on her parents, to change the world. She gets a whole new universe every time she does a push-up.

Even more startling, Melissa is discovering that where her body ends, something else begins. After staring at her hand for days and weeks, she first started biffing at things, then grabbing them and gumming them. She seems utterly engrossed, as if trying to figure out why, when she gums her hand, she feels it getting a slobbery massage, but when she gums her teether, she doesn't.

As with Janice, this stage of development often creeps up on parents who are looking for something a little more specific. It can seem as if the landmarks of this age—the push-ups, grabbing an object—are mere preparation for what lies ahead. As if the baby herself hasn't really changed, but just acquired a few new skills.

But to the baby herself, the discovery is fraught with meaning. It is, in fact, the beginning of the discovery of a self, which starts with the discovery of a boundary between self and other

That boundary can be a scary place or it can be trustworthy and loving. It's the task of this age to determine which is the case. The baby thus needs loving, consistent, handling. She also needs the kind of stimulation and response to her overtures that fit her particular temperament and style.

This exploration of the boundary between Self and Other is a game, a dialogue; the beginning of real play. It sets the tone for learning in general: learning about other people, about the real world, and about what kind of a person the baby is in the eyes of others. People of all ages seem to know this instinctively. I remember watching Big Sister Krista soothe Baby Paige. Slowly, gently, Krista would stroke one

(freshly washed) finger down Paige's forehead and nose, again and again, hypnotically. Soon Paige's eyelids would begin to close, and Krista would stroke those, too.

You have to learn Self and Other before you can learn the difference between Familiar and Strange, before you can understand that something familiar can disappear and reappear, before you can understand what a possession is, and how to share it.

Playing the game, exploring the boundary between Self and Other is the most important work of this age.

Some babies are ringleaders. They'll do anything to engage you in play, catching your eye, chortling, performing bumps and grinds shamelessly, making raspberries and biffing you in the face, and pulling your nose. Some are coy flirts, inviting you to make the first move and rewarding you with gleeful chirrups and funny faces. Some are scholars, letting you know only by a sudden silence and stillness and stare that they are ready to venture into the realm of very gentle play.

Babies very much appreciate parents who don't have their heart set on a style that the baby doesn't express. They also reward a parent who knows when enough is enough stimulation. A ringleader may suddenly collapse at the height of the game. A coy flirt may turn away just when you think you've pleased her the most. A scholar may retreat into himself, or howl, before you think the game has even begun.

## SEPARATION ISSUES: THE BONDED BABY AND DAYCARE

It takes a heart of stone not to become attached to a baby this age. Working parents both suffer and benefit from this fact.

Who wouldn't want those special smiles and belly laughs all for oneself? This is the reward for the past few weeks, and it feels brand-new. For one thing, most people are getting close to a good night's sleep. The colicky crying has ceased, and the schedule is fairly regular. And then just when it all starts to be fun, when the baby starts selecting you out for his smiles and belly laughs, when he's the most adorable being in the world—you have to leave him with someone else!

On the other hand, the daycare person is equally vulnerable to the beaming enticements of the baby. This may cause pain, seeing the Other Mother engaged in deep flirtation with the baby, but when viewed in another light, it is most reassuring.

As Burton L. White suggests, it is almost as if Nature makes babies irresistible at this age so that adults around them will be fully bonded, their hearts in thrall, by the time Cherub develops a will of his own—and ambulation (the ability to get from one place to another) to go with it—in a mere three or four months. It's like the honeymoon that lays a strong foundation for the inevitable disagreements and tribulations ahead.

Babies this age do seem to need not so much one, sole, special person only but for the special people in their lives to be physically nurturing, promptly attending to cries of distress.

When you see a baby this age adored by an adoring mother, father, a grandparent or two, and a daycare nanny, you see a circle of love. This is not a deprived baby.

## LIMITS

What possible limits could there be for a baby this age?

Such a young baby can't be "spoiled." He doesn't yet act by intent much of the time, so he can't disobey. Even when he does act by intent, he hasn't many choices. Shall I push up one more time or roll over on my back so that I'm stuck like a turtle staring at the ceiling? Shall I hit daddy in the eyes or the nose? Nor is there much he can do that is really forbidden.

The only limit to be set, then, is on the adults, and it is a very mild, good-humored, fun limit. That is: be alert to a change that occurs at the end of this period. There will come a time when the best response to the baby's cries may not be to automatically pick him up and cuddle him close.

In the first weeks of this period, that's a vital and appropriate response. The baby learns that every part of his or her existence is adored: appearance, vocalizations, odor; pee, poop, spit up. The baby learns the first and most important steps in self-esteem and in trust: if I cry, the universe responds.

Around three-and-a-half months, the baby makes an enormous intellectual leap. She discovers that not only does the crying bring an adult response, but that she can choose to cry. It is not an involuntary utterance all the time anymore. When a baby cries voluntarily, the cry sounds different, and the baby watches you differently. And she wants something a bit different from you: she wants to be stimulated and entertained. She has started to use her intelligence!

Picking up and cuddling the baby will indeed cure the boredom. Anybody will tell you it's a great way to be stimulated and entertained. And this may satisfy both baby and adult. But this response makes little demand on the baby's new skills.

Of course, the three- to four-month-old baby who cries from boredom does need attention, and promptly, to show the baby that this need is respected. But when you get to her, you let her know that you understand what she needs: something new to look at and slobber on, a new noise to listen to or create, a flirtatious interlude. Not merely a cuddle and kisses. Those are wonderful, but they don't let the baby discover why she cried.

It's a good idea to make sure the daycare person, even as she is bonding with the baby, takes pleasure in entertaining him as well as in cuddling him. A baby who is used to being held and carried *all the time* by the caretaker may start to cry when the parent returns. Just so, a baby may rebel when handed from a constantly cuddling parent to the other parent. This can be falsely interpreted: "She doesn't like her daddy," or "She loves her daycare mom more than me!" or "She has stranger-anxiety already!"

Highly unlikely. What is probably happening is that she is firmly establishing the idea of what is normal, and has been persuaded that normalcy consists of being held and carried all the time.

# Day to Day

## MILESTONES

### Vision

By four months, babies can see and follow an object in an 180° arc. They may still look very cross-eyed on occasion, but one eye should not be stuck in one position, and most of the time the eyes should focus.

### Hearing and vocalizing

All babies should alert to a novel sound, and by four months many will turn in its direction, especially to a high-pitched adult voice. Babies like to squeal, laugh, scream, make raspberries, and perform a social cough that sounds like the chairman of the board: "May I have your attention, PLEASE."

### Social

Distinguishing Self from Other is the very basis of all social interactions, and thus is the primary task of this age group. Individual styles don't really matter. Some babies regard anything that is novel, strange, and new as a real threat: they are said to have developed Stranger Anxiety. "How *dare* you exist!" they seem to be thinking. One false move, and a howl ensues. Other babies, especially those whose world is filled with the unpredictable (siblings and pets, for instance), take things more in stride. Some will even welcome newness with glad cries and joyful lurches, bouncing and kicking and lifting their arms.

Both extremes, and all the gradations in between, are normal. However, if by Four

Months of age a baby genuinely doesn't seem to care about the discovery of Familiar versus Strange, the pediatrician needs to investigate. A baby who consistently does not respond to the closeness of her loving adults with pleasure, and to the close approach of strangers with some kind of awareness of the novelty (from a moment's focused assessment, to excitement, to an offended howl) may just be at one far end of the spectrum. But it's important to be sure that this aspect of her development—the awareness of self, and the development of bonds to the rest of the world—is monitored, just like other aspects of her growth.

*Motor*

The most noticeable development during this period is the ability to choose an item, grab it, and put it into the mouth. The grabbing should be done equally well with right and left hand. Protect your earlobes if you have pierced ears.

Many babies hate being on their tummies, because they can't get a wide view of the world in that position. It's a good idea to have the baby do these prone floor exercises every day to encourage upper-body strength. Many will need inspiration and encouragement: an adult face to see every time they push up.

By four months, babies will learn to roll from tummy to back. Now, this doesn't have to be exactly by the four-month birthday, but close. (I remember giving Sara a little nudge with my toe on December 19.) Usually, the baby who doesn't roll from tummy to back isn't abnormal in any way, but insists on being placed on her back in the first place

and thus has no incentive to roll. However, a few energetic babies will learn to roll not just from tummy to back, but from back to front. When they can roll both ways, beware: a rolling baby can move with startling swiftness and get to, or off, or under, places you might not think of.

When put in a sitting position, Four Months can't hold the pose but can keep the head steady. Many like to be put in a standing position, and are proud of it. Sometimes merely being allowed to stand up will cheer a crying baby for a bit. Some babies won't be bearing weight quite yet.

SLEEP

Hang on to your hat. By three months many babies, and by four months most babies, should be sleeping uninterruptedly through the night.

Well, that's not quite accurate. Nobody really sleeps through the night. Everybody has a sleep cycle that features normal, brief, arousals to the waking state. Very young babies need these arousals so they can eat; their stomachs are small, they need frequent feedings. But older babies don't need these frequent feedings, and if they know how to get back to sleep after a brief normal arousal, they'll do so. They won't cry and wake their parents.

Does it matter to you whether your young baby "sleeps through the night" now? Perhaps not. It's a joy, if you aren't exhausted, to get up in the night and cuddle and feed the baby. Perhaps you have yet another reason for wanting to continue night feedings: if you are nursing, and if you are relying on the hormonal changes of nursing for your one and only effort at birth control, you had better keep on nursing at night. A large study showed that it wasn't frequent daytime nursing but night nursing that postponed ovulation in nursing mothers. (They had a hard time controlling whether or not the mere act of getting up at night to nurse itself deterred conception—fatigue and absence and all that—but they were pretty sure it was the hormones.)

However, if you look down the road of life and see yourself exhausted by an older infant or toddler who demands feedings every three or four hours through the night; if you are not eager to join the Family Bed movement; if you are a single parent, or a two-career couple, or are planning a second pregnancy very soon (please read the section *What If?* first), or have thin walls and edgy neighbors, this is your chance to insure a good night's sleep later on by your actions now.

First, try to schedule the daytime feedings to every three or four hours. Snacking and topping off can lead to a baby so orally engrossed that the need to suckle overrides the ability to go back to sleep after a brief arousal period.

**Second, put the baby in the crib awake and let the baby learn how to fall asleep without adult help.**

Third, make sure the baby gets loads of cuddling and attention during the daytime, not at bedtime or the middle of the night.

If you do all these, the baby won't wake you up in the middle of the night, and you won't be faced with the problem of whether or not to let him or her "cry it out." The reason that this is such a difficult and painful question is that it is the wrong question. The right question is, How can I help the baby learn to go back to sleep on his or her own after a normal brief arousal at night?

## GROWTH

Most babies continue gaining about an ounce a day through this period of time. Some babies gain more—a lot more. Really fat babies have a hard time rolling over and pushing up from prone. If a plump baby has a family history of obesity, if there's a tendency to stop crying with food, or if the baby is really eating a huge amount (nursing more often than every three hours in the day and five at night, or taking more than thirty-two ounces of formula in twenty-four hours), the pediatrician ought to be consulted specifically about this issue.

Thin babies are also a concern. If the baby comes from a sylphlike family, if her weight gain maintains the same growth curve without crossing lines, and if her physical and developmental exams are normal, there probably is little reason for concern; however, your pediatrician may wish to do basic lab tests, especially for occult urine infection, to be sure.

## TEETH

Most babies have no visible teeth yet. Please see the section on *Teeth* in the chapter "*Birth to Two Weeks.*" If your baby has already sprouted a tooth, see the same section in the chapter "*Six Months to Nine Months.*"

## NUTRITION AND FEEDING

For guidelines on eating fish, see page 7.

### Nursing

Nursing babies may act as if they are trying to wean from the breast at this age. This is almost never what is really going on.

They pull away from the breast, eagerly casting their eyes about, and have trouble getting started on again. But this isn't breast refusal. They are merely discovering an amazing thing: the breast is Other. Up until now, they had assumed that the breast and they were one and the same. In starting and stopping, they are experimenting with and exploring this discovery. Look, I can stop! Look, I can start!

This isn't a case of the baby being distracted by the outside world, so it doesn't usually help to go into a dark quiet distractionless room. (However, it's worth a try.) What the baby does need is understanding of what in the world the baby is up to, and patience while she works it through.

### Bottles and formula

On the average, a baby drinks at each feeding one ounce for every two pounds she weighs. A twelve-pound baby will take five six-ounce bottles a day, on the average. A sixteen-pound baby will take four eight-ounce bottles. It's unusual for a baby to need more than thirty-two ounces of formula a day. Make sure the formula is high-iron, even if the baby may seem to have trouble digesting it. If this is really a problem, talk with the pediatrician rather than just changing to a low-iron formula. Both intellectual and physical growth are highly iron-dependent.

### Foods and juices

Off and on, there seems to be a competition to see who can start a baby first on a) foods and b) juices. I cannot figure out why this is, except that I can understand why parents (like Janice) get impatient at this time for some developmental breakthrough.

Up until about four months of age, many babies actively reject a food-laden spoon, thrusting it out of the mouth with the tongue. Good for them. Breast milk and formula provide all the baby needs.

(Don't forget to give the occasional reminder bottle if you're nursing. Babies this age are involved in sticking to the familiar, and reacting with suspicion to the new. If you don't give a bottle for a couple of weeks and then have to do so, you may meet with incredible resistance.)

So wait on the foods. Foods at this young age:

- Take up the room that milk ought to occupy. Gram for gram, ounce for ounce, milk (formula or breast) is more nutritious than food.
- Impair the absorption of iron from the intestine into the blood, making iron deficiency more likely.
- Perhaps set the stage for food allergy later on, as the baby's intestine is more porous, allowing larger, allergy-sensitizing molecules into the bloodstream.

Juices, on the other hand, are generally well received by young babies. No wonder:

## THE TROUBLE WITH JUICES

- They don't add any nutrition not already in human milk or formula.
- If the baby is chubby, they add more calories he doesn't need.
- If the baby is thin, they suppress the appetite for milk and replace it in the diet.
  (Why, you may ask, don't they *suppress* the appetite of the chubby baby and *add calories* for the thin one? For the same reason that chocolate has the parallel effect in adults. There is no justice.)
- Sugar increases thirst, so the baby becomes more thirsty and drinks more juice, which makes him more thirsty, which brings on more juice. Very nice for the juice companies.
- Juice is often used as a pacifier or as entertainment, giving babies the firm conviction that sweet oral activities are a solution for anxiety and boredom.

babies are programmed to like sugar and salt. Both sugar and salt are present in all juices. Just because they are natural sugar and natural salt doesn't make them any better for babies.

### BATHING AND SHAMPOOING

Now that the baby is sturdier, the umbilical cord safely off, and the circumcision healed, you might want to try a real tub bath. This is also a good time because the baby can tell you what gives pleasure, not just what gives discomfort. However, babies this age usually don't get really dirty (except the diaper area) unless they are confirmed spitter-uppers. So don't feel obliged to do any but sponge bathing if you don't want to.

Bathing is not mysterious. You will find your own tricks and holds easily. The main concerns are safety, warmth, gentle cleansing and thorough rinsing, and pleasure.

Of course, a baby must never ever be left alone in a tub or basin, even if there is only an inch of water. That is the very moment she will roll over and inhale the water and panic. Equally, be sure electric appliances are a good distance from the water. When you're bathing a small baby and the phone rings or you're distracted, sometimes you grab something you don't mean to. Don't put more than a couple of inches of water into the basin. Finally, you have already, I hope, turned the water heater temperature down to 120°. Be sure to test the water in the tub with the inside of your wrist. It should feel just warm, not hot, not cool. Keep the room warm, and have soft towels on hand. Not just one for drying, but one for wrapping after the bath, and a spare for emergency wrapping if you have to interrupt the bath and start again.

As for supplies, I advise using a small,

towel-lined basin. A towel-padded roasting pan is just fine. Wet babies are unbelievably slippery, and you want a very controlled bathing environment. As for supplies, I suggest using a mild, white, scentless moisturizing bar rather than liquid cleanser (even labeled for babies) or a bar of pure soap. You want to keep the bath water as soap-free as possible, for rinsing. Put a little soap on the wet washcloth or your hand, and rinse with clear water. For the scalp, the name-brand tearless baby shampoos are designed to help cut down on cradle cap (seborrheic dermatitis). I strongly suggest no powder. Petroleum jelly or a brand-name diaper cream is useful to apply after the bath, giving the baby's bottom a Teflon-like coating, which makes it easier to clean.

Count on it, the baby at some point will either pee or poop while in the bath. Urine is sterile, so just regard it as a bit of extra water. Poop means you have to start all over again, so you'll need that extra towel and safe place for the baby to wait while you clean her bath basin.

## DIAPERING AND DRESSING

Diapering should be second nature by now, and I have nothing to say about it.

As for dressing, though, the primary concern is safety. There is a strong impulse I see to adorn babies of this age with jewelry, headbands, and fancy footwear. I'd be very concerned about necklaces, even religious medallions. A baby who can roll can catch the necklace chain on something and stran-

gle. This becomes more a danger as the baby starts to sit, roll around, and pull to stand, but are you going to remember to remove it then? Better not to start now. At the very least, keep the chain securely under the clothing, and watch for rashes due to metal contact.

Piercing the baby's ears? Usually safe, especially if you ask your pediatrician to do it or for a referral. If it's a strong cultural tradition, well, you probably are going to go ahead anyway. But a few cautions. Make sure the needle is sterile. Never trade, or allow the older child to trade, pierced earrings with someone else: there is always the chance that a drop of blood clings to them, and can transmit disease such as AIDS. Look at the earlobes frequently for signs of infection or allergy, and get the earring out right away if you suspect either; otherwise the lobe will swell and getting the ring out will be a nightmare.

If it's not a cultural ritual to pierce the ears, I strongly suggest not doing it. Given the nature of teenagers, your child is likely to object, in about thirteen years, to your having allowed such a "mutilation." Moreover, if children decide at eleven or thirteen or whatever that they love pierced ears, you can't use the piercing as a reward or bribe if you've already had it done in babyhood.

Headbands certainly are a fashion statement, but I can't help worrying that they will slip down to encircle the baby's neck and act like a necklace as a potential strangling hazard.

Fancy footwear includes polyester socks which for some reason seem to rub on toes and produce reddened areas of cellulitis. It also includes real shoes, which are very cute but which are not necessary or desirable for "forming the foot" unless your baby has had special shoes actually prescribed by your pediatrician or the orthopedic surgeon. I'd spend the money on something else, but it's your choice. The baby, over the next few months, will simply regard them as something else to chew on.

## Activities

Babies this age are eminently portable. They also tolerate moderate temperature changes fairly well. But there are two important cautions:

- Midday sunshine should simply be avoided. Reflected sunshine and sunshine on cloudy days are particular dangers.
- If you do spend more than five minutes outside, apply hypoallergenic sunscreen to Cherub's exposed body parts. Use a hat with a brim and neck and ear flaps. If Cherub starts looking even a little pink, leave the sun: it takes twelve hours for a burn to develop fully.
- Babies are still susceptible to wind and cold. Most temperature regulation occurs through the scalp, as the head is still the biggest part of the baby's body. A warm hat is a must. If the air in the house is cool, both a hat on the baby and a humidifier in the room will help to keep him safe and comfortable.

Shaking, bouncing, and tossing can damage the blood vessels surrounding the brain and spinal cord. It's too early for joggers, bicycle seats, or the game Flying Baby, in which you hold the baby either horizontally or vertically and toss her. Not a good idea. Even if the baby loves it. Jiggling seems to soothe many babies, but the arc of the jiggle should be such that the motion is smooth and gentle. When in doubt, don't.

This is an age that likes to watch changing light patterns through windows or tree branches—or on TV. If the latter, the baby should be at least four feet from the set. This is about the only age a baby really interested in TV doesn't need to have the watching censored. (Or so we think.)

## Toys

A baby who hasn't yet learned to grab and hold still loves toys to look at and bat.

A mirror is fun. When Janice wasn't right there to give Melissa a smile every time her push-ups allowed her to surface over the top of the portacrib, she placed an unbreakable mirror close to the head of the crib. Melissa seemed to enjoy this very much, holding her push-up at full extension to watch her own reflection.

Toys to kick can be entertaining. Taping a squeaky toy to the foot of the crib and putting the baby within kicking distance is likely to give the baby a great deal of joy and exercise and the parent a nice little vignette to video. Tying (very securely) a bell to the baby's bootie will entertain the baby and eventually drive you bonkers.

At around three months many babies will start to put things in their mouths. Smooth, sturdy objects, like plastic key rings, are graciously accepted. Parental noses and chins are particularly fine toys. Beware the developing ability to grab earrings, especially if yours are the pierced variety.

## SAFETY ISSUES

The next few months will whiz by faster than you can believe, and you will discover several phenomena: that babies climb before they walk, for instance; and that the baby's hand-to-mouth coordination is far superior to an adult's grab-it-out-of-the-baby's-mouth coordination.

When you baby-proof the house and develop baby-safe habits and skills now, you are not merely cutting down on the chances of tragic accidents. You are also setting the basis for effective and happier discipline later on. Setting limits is a lot easier when there aren't that many limits that have to be set.

## Health and Illness

*Part II of this book discusses symptoms, illnesses, and problems with body parts and bodily functions of babies about three to four months of age and older. Since babies this young are still more fragile than older babies, there are special concerns that are discussed in this chapter. With these exceptions, however, Part II can be consulted for babies this age. If you have any questions not answered to your satisfaction in these sections, of course, consult your own pediatrician.*

## EXPECTATIONS

Even though babies are better able to deal with illness at this age, the ideal would be to avoid sickness. The baby's immune system is still immature, and often doesn't make useful, long-lasting immunity to infectious diseases—so there's no real point in having any. Besides, a baby this age is still fairly fragile, easily dehydrated or exhausted.

However, many babies will have a first illness now, either a cold, with or without an ear infection, or a bout of intestinal upset. Fortunately, babies this age can withstand minor illnesses pretty well. A temperature of 101° or 102° in a comfortable, cheerful infant beyond the newborn period doesn't automatically mean laboratory tests and likely hospitalization. Any baby this age with a temperature over 100.4° does need pediatric attention, though (see below).

To help the baby avoid unnecessary illnesses, some of the old newborn guidelines still are useful.

- Everyone should wash hands before handling the infant and after diaper changes.
- Avoid exposure to sick people and to crowds.
- As far as is possible and tactful, limit exposure to toddlers and preschoolers. They are at the age when their "job" is to catch and develop immunity to infectious diseases. Most of these illnesses are most contagious just before the child

## MEDICINE CHEST

- Acetaminophen drops (ask your pediatrician for dosage at each well-visit)
- Electrolyte solution such as Pedialyte
- Electronic rectal thermometer

## BABY-PROOFING

- Set water heater at 120°.
- Toss out all pacifiers with a base that is smaller than Cherub's mouth. Check that the nipple is securely attached to the base—tug on it.
- If Cherub's in an infant seat, make sure he's strapped in, and don't carry the seat by its handles: his weight can make it tip and even spill him out of it. Never use an infant seat as a car seat.
- Change diapers and dress Cherub either on the floor or always belted in on the changing table.
- Never leave Cherub alone on a surface it's possible to roll off of, not even the middle of the bed. Not even for "just a minute."
- When Cherub is in his car seat or infant seat, never put it on a counter or appliance; it's too easy for it to tip over.
- Cherub must always be in the car seat for plane or vehicle travel, never on a lap.
- Cherub's car seat must be installed correctly, facing backwards, ideally in the backseat. If there is a front or side air bag for that seat, it must be disabled.
- Make sure there are no loose objects in the car (including pets) that could become airborne in an accident and hit Cherub.
- Think ahead: take a peek at the baby-proofing suggestions in future chapters to get a head start—particularly if you have more than one child, guns in the house, pets, or a water source such as a pool, pond, or fountain.

shows symptoms and remain contagious after he is well.

This is an especially good idea during the winter months. If there is a community outbreak of chicken pox, measles, or bronchioli-tis (see below, and Glossary), it's a good idea to be even more careful. These diseases can all be transmitted through droplets in the air, rather than just on the hands.

A baby this age isn't quite as fragile and mysterious as a newborn. The heart and lungs have fully adjusted to the jolt of being born and having to breathe and deliver blood to working lungs. Jaundice should have disappeared long ago, because the intestine has stopped reabsorbing bilirubin. Most crucial, the immune system is now active: the baby doesn't depend anymore on the maternal defenses obtained through the placenta.

So in medical terms a baby this age has made enormous strides.

But perhaps the most important difference between a newborn and a young infant isn't physical and biochemical maturity. It's that now the baby can smile!

When she is sick, the disappearance of that smile is a worrying sign. Parents now know their baby's vocabulary of cries: a shriek that isn't hunger, annoyance, or boredom, but rather that sounds shrill and frightened, gets their attention. A baby who stops playing the Baby Game of interaction and cooing and wriggling, parents know instinctively, is a baby with a medical problem that needs attention.

## Common Minor Symptoms, Illnesses, and Concerns

### Colds

Even in the best-regulated families, babies will catch colds. A baby this age with a cold can be perfectly cheerful, as mucus mixes with drool for her to paint her face with; or miserable, because she hates the feeling of not being able to breathe through her nose.

A true cold is characterized only partly by the symptoms present (runny or stuffy nose,

extra sneezing, an occasional cough) and greatly by symptoms that are absent.

Signs that the baby could have something worse than a cold, and needs prompt pediatric attention, are:

- Poor color. He should be pink, not pale, bluish, or mottled.
- Trouble breathing that originates in the chest. His nose can be a real mess, but when he breathes you shouldn't see the muscles in his neck, between his ribs, or in his belly working to force air in and out. His nostrils shouldn't flare.
- Fever over 100.4°. (If this is the only thing out of the ordinary, a phone call in the night and a visit the next day may suffice, especially if the baby is over three months old.)
- Signs of pain. Fits of crankiness are to be expected, but a real pain cry, especially if the baby is restless and unable to sleep lying down, suggest an ear infection or other complication.
- Persistent cough. The cough of a cold is sporadic, and doesn't cause the baby to pause in what he's doing. It doesn't interrupt sleep.
- Eye(s) that are red, oozing, or have red swollen lids.
- Refusal to eat, sleep, or smile.

Even if your baby just has ordinary cold symptoms, you'll probably want to notify your pediatrician if it's the baby's first illness, or if your doctor has asked you to do so.

Good home remedies for a simple cold are:

- Elevating the head of the mattress to make breathing easier. A few books under it will do the trick.
- Unless you live in a very damp climate, try a cool-mist humidifier. A baby breathing through her mouth dries out the mucous membranes quickly, and moisture is needed to keep them healthy. Don't let the room get uncomfortably cold, or so damp that you can practically hear mold growing along the carpet.
- The aspirating bulb you took home from the hospital can help to get rid of excess nasal mucus. Before you go to suction, you can try instilling in each nostril a couple of drops of Saline Nasal Solution, a buffered salt water preparation available over-the-counter in pharmacies. Then count to three, then suction (by squeezing the aspirator with your thumb on the round base before you enter the nose and then releasing it for suction once you're inside).
- Only if OK'd by your doctor, acetaminophen can make the baby more comfortable. But you don't want to mask real pain or the onset of a new fever.
- Extra nursing, if you're nursing, and extra water by bottle if you're not.
- Over-the-counter oral decongestants and antihistamines aren't a good idea in babies this age most of the time. The former have never been shown to "work." The latter dry the secretions, when nature wants them to flow. Both ingredients can cause behavioral side effects, which can confuse you as to how bad the child is really feeling.

- For a stuffy nose, rather than a runny nose, some pediatricians recommend medicated nose drops such as Neosynephrine. These can help symptoms, but also can be addicting, as the membranes of the nose swell up even more after the drops wear off. If you do use them, most pediatricians suggest that you put one or two drops in one nostril only. Do so when needed, before, say, a feeding or sleep. Alternate nostrils. And don't use longer than three days.
- Cuddling and distraction. This is no time to vigorously encourage the baby to rely on his own resources.

Try not to catch the cold yourself.

Colds are passed on primarily through hand contact, not through particles floating through the air. You can avoid catching your baby's cold by washing your hands before you touch your own nose or eyes. But if you think this is easy, put a little colored chalk on your fingers and then look at your face in the mirror half an hour later.

*Intestinal Upsets*

Many babies will have an episode of diarrhea or vomiting during this time. Make sure that if the baby is breast-fed, he isn't getting "secondhand" caffeine or medications. If the baby is in only mild distress (see below), the problem is likely to be transient. If you do suspect caffeine, medications, or other maternal indiscretion, it's a good idea to pump and discard breast milk and give stored expressed milk (or formula) for a feeding or two.

If the baby is bottle-fed, check that whoever is mixing up the formula is diluting it properly. If you suspect that the water added is too little or too much or contaminated, this can be very serious; call the pediatrician immediately.

### SIMPLE DIARRHEA

Simple diarrhea in young babies shows up as loose or watery stools, without blood, in a baby who has no fever, vomiting, respiratory symptoms, or rash, and who is in a cheerful state of mind with good appetite. If all these conditions are fulfilled, most pediatricians recommend:

- Nursing babies continue to nurse, as often as they wish.
- Bottle-fed babies be given one or two feedings of an electrolyte solution such as Pedialyte or Ricelyte, and then return to formula. Some physicians recommend a formula such as Prosobee (see Chapter one) that contains the simple sugar glucose rather than milk sugar (lactose) or table sugar (sucrose).

For any baby with diarrhea, it is most important to be sure the baby stays well hydrated. The eyes ought to be bright and shiny, the mouth flowingly wet, the baby cheerful. If in doubt, stick a clean finger in the baby's mouth; it ought to encounter actual wetness. If the roof of the mouth is sticky or tacky, the baby is dehydrated and you need to get medical attention right away.

Counting wet diapers can be deceptive, as the stool can be as watery as urine. If you happen to catch the baby in the act of peeing, and the urine is as clear as water and there's a good amount, that is truly reassuring.

### VOMITING

Most babies this age will spit up, some fairly frequently. True vomiting usually means simply that she ate too much and then burped or jiggled it all up again.

If the vomiting looks like an isolated episode in an otherwise cheerful, thriving, vigorous baby who has no fever or other symptoms, it is safe to try the following:

- Let her stomach rest for one hour after the vomiting.
- Then give her small, frequent amounts of a liquid containing sugar. If she is nursing, nurse her one or two minutes a side every ten minutes. If she is bottle-fed, give her a measured tablespoon of an electrolyte fluid (Pedialyte or Ricelyte) every ten minutes.
- Do this for an hour or so. If she hasn't vomited again, gradually increase the amounts over the next two hours.

Some vomiting is not simple, but complex (see below).

### Poop problems

Infrequent or too-firm stools most often are innocent in cause, but can become a major management problem if ignored. Rarely, they can indicate a serious underlying cause.

### INFREQUENT STOOLS

Infrequent stools in a breast-fed baby this age are usually normal. The baby has outgrown the reflex that triggered a poop every time his stomach got full. (We all have to outgrow it sometime.) Now the baby only poops when the rectum becomes distended enough to inform his brain that he's got to push the poop out. In a breast-fed baby with very soft stools, this may take two or three days.

### TOO-FIRM STOOLS

Breast-fed babies have only soft stools, never hard ones (unless they're taking rice cereal or more formula than breast milk).

Formula-fed babies sometimes have hard stools, with a consistency that varies from thick peanut butter to Tootsie Rolls and rabbit turds to downright rocks.

Constipation often becomes a vicious cycle if not treated. When a baby this age has a stool that is uncomfortable, painful, or even merely difficult to pass, he tends to not want to pass any more stools. You'll see him grunting and turning red and making all kinds of fuss, but it's all in the cause of holding back the stool, not pushing it out. And the more stool he withholds, the harder it gets, so the more he withholds.

So address the issue promptly and with dispatch.

- If your pediatrician says it's okay, relieve the hard stool by using either half of a pediatric glycerin suppository, or the little bulbs containing glycerin, called Babylax.
- Again with your pediatrician's consent,

add a tablespoon of adult prune juice to the baby's bottle once or twice a day, and give an extra bottle of water between formula feedings.

- If this is a persistent or recurrent problem, ask your doctor about changing formulas.
- When the baby is trying to poop, sit him up and bring his knees up to his chest rather than letting him do the job lying on his back. Remember what it's like to try to go on a bedpan?

## SERIOUS OR POTENTIALLY SERIOUS SYMPTOMS, ILLNESSES, AND CONCERNS

### Fever

Fever is a normal response to infection and inflammation; it is a symptom, not a disease. In the newborn period, it is a very powerful symptom. Very young babies can have a serious infection with only a slight elevation of temperature to show for it. Moreover, other behavioral signs of serious illness can be very subtle. The younger the baby, the truer this is. However, as babies mature, the signals gradually become as important as, or even more important than, the degree of fever.

In a baby from two months to four months, a rectal temperature of 101° or greater should prompt a visit that day or a phone call to the pediatrician even if there are no other symptoms. In a baby who is three months or younger, a temperature of 100.4° deserves a visit. And a baby this young with no or even a slight fever who acts sick needs prompt attention.

The exception is the fever that babies sometimes have after their immunization. This can be as high as 102°, but if it is higher, a phone call is in order. And if the baby is highly irritable or lethargic, or showing other signs of illness, don't assume that this fever is simply a normal "shot" reaction. Now and then a baby who seems well at the office visit is really coming down with an illness, and the fever is not a shot reaction but an infection. Fever from an immunization shouldn't last longer than forty-eight hours.

*Spells*

### JERKING OR TWITCHING SPELLS

All babies this age jerk when they are startled by a noise or sudden movement, especially a sudden falling backwards from sitting. But if a baby does a "startle" type movement repeatedly, without provocation, there's a possibility that this is a seizure.

Rarely, a baby in the first year of life will have a kind of seizure called infantile spasms. These are spells in which the baby's head drops, and her arms and legs jerk repeatedly, more than once. They typically occur as a baby is waking up, just starting to fall asleep, or as she is eating.

Since babies do so many weird things that are actually normal, such behaviors may turn out to be perfectly innocent. However, it's important to make sure. If these truly are infantile spasms, Cherub needs *urgent* diagnosis and treatment.

If your pediatrician so much as suspects a chance that the movement is a seizure, the next step is an EEG (see Glossary), a recording of "brain waves."

### NOT-BREATHING SPELLS

Every now and then a baby is brought in because her parents discovered her not breathing. This is every parent's nightmare, of course. However, most of the time, when the baby gets to medical care she is fine: pink, vocal, outraged at all the fuss. There are several possibilities when this happens.

- **Misinterpretation:** Since babies normally breathe irregularly, she may have been in the long "sighing phase" between breaths. Your pediatrician will go over the incident with you very carefully, and examine Cherub very carefully. If there is any doubt, tests are in order.

- **Reflux Behavior:** Many babies with gastroesophageal reflux (see Glossary) will pause in their breathing.

  Your pediatrician may entertain this diagnosis, especially if your baby has other signs, such as arching her back, crying, and spitting up or regurgitating frequently.

- **True ALTE:** An ALTE is an apparent life-threatening event; that is, a near case of sudden infant death syndrome.

  If your pediatrician feels this might be a possibility, it's a good idea to admit the baby to the hospital for tests to determine possible underlying causes, for a discussion of monitoring the baby's breathing, and for a parental course in CPR.

### Breathing problems and persistent coughs

When we think of more serious respiratory problems in babies, the mind tends to jump to pneumonia. While pneumonia can certainly occur in babies this age, the most frequent and sometimes epidemic respiratory disease is called bronchiolitis. (It differs from bronchitis because it is the tiny tubes of the airways, the bronchioles, rather than the larger ones, that are infected.)

Bronchiolitis is caused by a virus called respiratory syncytial virus, or RSV. A baby with RSV makes an effort to breathe, especially to push the air out. Her chest, neck, and stomach muscles may suck in and out as she breathes, and you can often hear a high-pitched noise, a wheeze, as she exhales.

A baby who wheezes must be seen, either urgently or for a same-day office visit depending on how hard she is working to breathe. Until she is seen, humidified air and being held upright may give some relief. A baby this young with bronchiolitis may need hospital care.

Persistent coughing or violent coughing, even without trouble between the spells, can also mean trouble in such a young baby. The most worrisome illness causing this symptom is whooping cough, or pertussis. The two-month immunization for the disease gives strong, but not 100 percent, protection. Any baby with this symptom also needs a trip to the physician.

### Complex diarrhea

There are two major concerns for a baby with anything more than simple diarrhea: Is the underlying cause serious? Is the baby becoming dehydrated?

A baby who has blood in the stool, fever, or vomiting has complex diarrhea that may have a serious underlying cause. These symptoms need prompt attention.

A baby with very frequent, watery stools can easily become dehydrated; she needs medical supervision and close watching. The most frequent cause in babies this age is a virus called rotavirus, which often appears in the winter, in daycare centers. (See Part II, under *Common Illnesses,* for more details.)

Finally, a baby whose stools are always loose black or very dark or foul-smelling, needs pediatric assessment, even if the baby is otherwise doing well.

And a baby who has simple diarrhea, but who also has respiratory symptoms and is much more cranky than usual, also needs a visit; middle ear infections often give this set of symptoms.

### Persistent or projectile vomiting

Isolated episodes of vomiting in a baby this age aren't unusual. Most often, the baby simply has a very full stomach. The sphincter muscle that holds food in the stomach isn't well developed yet, so the milk refluxes back up. This can appear as spitting or as occasional vomiting.

But sometimes vomiting heralds a serious problem: an overwhelming infection, an obstruction in the intestine, or a disorder of metabolism.

## VOMITING THAT NEEDS AN URGENT VISIT

- Vomitus that is yellow-green, brown, or that contains blood.
- Cherub seems to be in severe pain.
- Cherub becomes pale, very quiet, or limp between vomiting spells.
- If Cherub is behaving very unlike herself, with either irritability or lethargy (in either case, not smiling or interested in attempts to entertain her).

*Complicated Constipation: Hard or Infrequent Stools not Caused by Diet*

The time to worry about infrequent stools, even though they are soft, is:

- If the nursing baby isn't gaining weight well. Some "happy-to-starve" babies will have a small, loose, green stool every three or four days. They will urinate frequently, however. But they are still starving. If this goes on for long, such babies really look thin. Their arms and thighs are sticklike, their eyes look huge, their ribs stick out, their bellies aren't rounded but flat. Often their skin is loose. If you aren't sure about how plump your baby looks, either visit the pediatrician or weigh the baby on a reliable baby scale. A baby this age usually should be gaining one ounce a day.
- If the baby is not thriving or developing well, infrequent stools may rarely mean that she has either a thyroid problem or an intestinal disorder of motility called Hirschsprung's disease. In any case, if the constipation isn't an isolated, short-

term problem, medical advice is in order.

- If the baby is floppy, sleepy, and can't suck well, there is a chance she has a disorder called infant botulism. The chance is greater if she's been fed honey or Karo syrup, which once in a blue moon contain spores of botulism. Such a baby needs medical help urgently, as the breathing muscles can be affected.

## INJURIES

### Rolling off surfaces

Nathanson's First Law states that the first time a baby rolls over, it is likely to be off of something. I would suggest not leaving a baby this age unguarded even for an instant on any elevated surface. Nor would I leave a baby in an infant seat on any such surface: they can rock themselves right over. Especially big babies.

### Shaking

Tossing, shaking, or even vigorously jiggling a baby this age can cause brain damage and even death, because the muscles of the neck are weak and the blood vessels surrounding the brain are fragile.

### Smothering

Even the strongest baby should not be left on a soft surface like a water bed or pillow. And beware any playpen with mesh sides; they have been known to collapse on a baby and smother.

## Window of Opportunity

### TRUST

This is the age in which babies start to establish their ideas of the Normal. They're finding the answers to the questions: Is this a responsive and welcoming universe? Am I entirely wonderful and worthy? Is life predictable—and therefore do I have a degree of control over events? Am I a participant or a passive bystander?

They develop trust and the foundation of self-esteem most readily when they are in close contact with people who think they are absolutely wonderful, delicious, and perfect. And the contact needs to be consistent.

This doesn't mean that every day has to be exactly the same as every other, or that a baby's environment has to be exceedingly consistent. On the contrary. A baby in such a consistently unchanging world may have such a constricted notion of what is normal that change can induce panic.

"Normal" to a baby means primarily the people and intimate items he depends on for care and entertainment and stimulation. He does need loving, consistent adults; and a few consistent loved items, so that adventures into the world have a solid core of familiarity. With these on hand, he could travel round the world and still learn about Familiar and Strange in a healthy way.

For a baby to feel that normal means participating and trying to control the world, playing the Baby Game is key. The Baby Game is any game of reciprocity. His or her coos and smiles and gurgles call forth your responses, back and forth, back and forth. Parent and baby imitate each other's funny faces.

### SLEEP

Towards the end of this period, babies start to become quite firm in their ideas of what is normal and what isn't. A particular area to be aware of is what the baby considers to be a normal going-to-sleep routine.

At this age, you can rock the baby, or nurse the baby, or give a bottle until the baby is sound asleep, then put him in the crib (on his *back*, remember!), and count on a good night's sleep for all.

This will change.

Later this year and on into toddlerhood and beyond, babies become more like adults in their sleep cycles. That is, they awaken briefly several times a night as they come into the light stage of sleep. If the older baby or child is used to falling asleep being rocked, held, nursed, or with a bottle, he'll need to repeat that ritual every single time he wakes up in the night.

By six or seven months, she'll have turned this ritual into a habit. She'll be more aware of separation—she won't let you leave her. And by nine months, often earlier, she'll be able to pull up to a stand and shake the crib rail at you and scream. Several times a night.

So now, at close to four months, try to help the baby learn to fall asleep in the crib rather than in your arms or while nursing or taking a bottle. This doesn't mean plopping the baby into the crib to cry himself to sleep, but getting him relaxed and drowsy, putting him gently in the crib, and, if necessary, patting him and singing until he falls asleep. Gradually put him in more and more awake. It's usually easiest if the mother is not the one putting the baby in the crib.

Along the same lines, don't let the nursing baby forget how to take an occasional bottle. If she gets it down pat as normal now, you aren't likely to have a problem later. If you have a real refuser, though, don't despair; by five months, many babies can learn to take significant amounts from a cup.

Finally, many babies prefer to lie on their backs and watch the world. Make sure such a one gets time on the tummy as well, to strengthen arms and neck. Try giving her a reward for her push-ups, as Janice does with Melissa.

## What If?

### STARTING DAYCARE
This is the most frequent time for mothers to return to work. For many, leaving the baby is tougher than the birth itself. It feels as if you've just started reaping the rewards: no more colic, cooing and chortling, sleeping through the night—the baby is a delight. And now you have to leave!

The main thing that helps is having a daycare person you trust and like and with whom you can communicate. One of the nicest gestures I have encountered in a daycare nanny was her practice of writing out a report each and every day, for even the youngest babies. "Today I went for a walk in my carriage, took two six-ounce bottles of breast milk, and had a very nice nap. I giggled three times."

One thought that may help is that this is a very bondable age, in which adults find it easy to adore the baby. While it may be painful to see the baby and daycare nanny having a wonderful time, it is just what you would wish. Also, babies this age are just establishing their idea of what is normal. It is a nice thing for a baby to understand that other adults besides mother and father adore them. What a lovely normalcy.

Both for adoration and for physical hygiene, the ideal arrangement for such a young baby is one-to-one with a loving adult. Next best is a group of not more than four babies the same age. If the other babies have older siblings, it's best if these siblings are older than about three; toddlers are meant to catch many viruses and bring them home to share with their baby brothers and sisters, who then take them to daycare with them. Viruses are most contagious just before the child shows symptoms of illness.

The home should be smoke-free. If there are pets, they should be wormed regularly.

One trusted mature dog may be unlikely to nip and bite a baby, but two or more dogs may find their wolf-pack instincts aroused by a baby and actually attack. Even one dog can have an ungovernable impulse. Cats, when teased, can claw and bite.

No matter how bereft the mother feels, and how engrossed in the baby when she is home, it's best for the baby to spend time regularly with each parent, one-on-one. That means alone with the father, unsupervised by the mother. Some studies show that early infant-father bonding with girls makes the father-daughter tie stronger all the way through adolescence.

And, ahem, it's a REALLY good idea for parents to spend time alone together without the baby every now and then, certainly once a week. Just to remember what they liked about each other before the baby was born.

## EXTENDED SEPARATION FROM PARENTS

Separating from a parent for a prolonged period at this age can be done without upsetting the baby too much, but it tends to be very hard on the parent, and may temporarily distort the baby's attachment to that parent. From the baby's point of view, she needs to have continuing loving care from a person she knows well. This relationship should be well established before the parent takes off.

Of course, the baby will attach herself and give her best smiles and rare belly laugh to the person taking closest care of her, so it's crucial that that person think she's the most wonderful baby in the world.

If separations can be avoided, they should be; but if a parent MUST leave it is crucial that guilt not be added to the pain of separation. A sense of doom and gloom is certainly inappropriate.

The returning parent will need to spend some time "courting" the baby again. This may take several days. At first, such "courting" should be in the presence of the stay-at-home caretaker, but after the first couple of days, baby and newly returned parent ought to have time alone together without "gatekeeping" (monitoring and supervision) by anybody.

## TRAVELING WITH THE BABY

This is a fine age for travel. Babies this age love watching new sights, including busy travelers. They don't suffer from restrictions on their new skills; and they don't scream in frustration very much.

Try to take the following items along with you:

- A pad on which to place the baby for diaper changing.
- Wipes not just for her bottom but for her and your hands. Use them not just after diapering, but after you've gotten settled in your plane or train seat, before you touch the baby. Most illnesses are picked up from touching other people and then your, or your baby's, face. You can also use the wipes to clean baby toys should they fall, because the older baby in this range is going to put them in her mouth.
- Bottled water to drink. Travel is dehydra-

ting, and you'll need water for yourself if you're nursing and for both of you if you're bottle-feeding. Don't trust airplane water from a tap, even a "potable" one.

- A small first-aid kit with a rectal thermometer, Vaseline, acetaminophen drops, and a bulb suction and salt water nose drops.

Try to avoid crowds (including planes) during times when "everybody's sick." This is generally the winter months.

## ADDING TO YOUR FAMILY

Parents have all kinds of reasons for reembarking so soon. Biological clocks; wanting to "get it over with all at once" and have the pregnancies and diapers and so on fitted into a block of time; wanting to have the children be close in age and thus close as friends; being lulled into the idea that recent pregnancy and nursing are a form of "natural" birth control.

A baby conceived now will be born when your present baby is a very young toddler. The older baby will need an enormous amount of adult attention, in these forms:

- Protection from danger. A one-year-old is not so much fearless as unacquainted with the physics of life. Heights, falling objects, heat and cold, sharpness—all these are mysteries he'll still be exploring and figuring out. All day. Every day.
- Language stimulation. A one-year-old has very good receptive language; by fifteen months a baby is often described as "understanding everything." If what she hears most often is "NO!" and "Mommy has to take care of Baby first," and "Daddy is too tired to play," she may get discouraged about language. If adults are too fatigued to engage her in nonverbal, and then gradually verbal, play, she may not have much incentive to learn and use words.
- Separation practice. A one-year-old is still clingingly attached to his beloved adults but makes brave forays into independence. These are braver and happier when he has a loving, consistent, in-control adult base to return to. Parents of a brand-new baby are apt to be tired, a bit irritable, and not as patient as they usually are. The skills learned in separation are necessary as a basis for learning about limits, a relationship that is often overlooked. But in order to understand that a limit on behavior is not a judgment on one's own worth as a person, a toddler has to feel good about himself as a separate person.
- Ego-enhancement. A one-year-old has a great need to think of herself as the center of her world, the most wonderful person in it. (She needs to feel this way in order to be able to confront limits on her behavior without thinking of them as diminishments of her self-hood.) She won't be sophisticated enough to be *jealous* of the new baby; she won't make any reasoned connection between the baby's arrival and diminution of her status as a Wonder of the World, but she will sense any such loss of worship.

- Modeling behavior. A one-year-old is an imitator par excellence. He'll "fold" laundry and "brush" his teeth. He'll "drive" the car. He'll replicate faithfully nursing his dolly, changing Dolly's diaper, kissing Dolly, being gentle. He'll replicate just as faithfully yelling "NO" at Dolly, slapping Dolly's hand, stamping his foot and making a big noise. As he learns to talk, he may announce that he's "fed up" and "sick and tired." And worse.

Of course, a One with a brand-new sibling can have all these characteristics turned to his advantage. All it takes is forethought, luck, and, I fear, resources—including, usually, financial ones. Such a One merely needs:

- A consistently available beloved adult who is not primarily caring for (and exhausted by) the new baby. This adult should think that One is the most marvelous person in the world, but should also be willing and able to set limits and model and encourage language skills.
- An environment for play that is pretty much unrestricted and safe.
- Limits that are few, reasonable, and well enforced (without slapping, spanking, or demeaning language, but consistent and with follow-through).

Attention to these needs can bring about something very close to the idyllic state parents often envision, of two very close siblings. Such parents often have more psychological "distance" from their children,

thinking of them as "the kids." This is subtly different from parents who don't have a second child, or whose next child is born after the first one is three or older. They tend to think of each child as more of a singular, unique person. I am not saying that one psychological approach is better than the other; just that they are different.

## DIVORCE, SEPARATION, AND CUSTODY ISSUES

Custody decisions for a baby this age are very difficult. It's important that both parents stay involved with the baby, but an infant this age has a short memory. Since her primary developmental task is to bond to a very small number of special adults who think she is the most wonderful person in the world, to establish a base of the Normal and Familiar, parents should try to find a solution in which someone remains a constant presence, warm and loving, in the baby's life. This may be a grandparent or a caretaker, or it could be either parent if the other is willing to make frequent visits to the baby, rather than removing him, for a stretch of time.

A baby who is ping-ponged back and forth between parents is at a developmental disadvantage. She stays with one just long enough to integrate the ping universe, and then is shuttled to pong, where the same process must take place. She can't hold the thought of ping during pong, or pong during ping. Her emotional task of establishing an idea of what is Normal Life, loving and serene, is disrupted.

## Surgery and/or Hospitalization

When a young baby needs hospitalization, it is either to rule out a serious illness, such as overwhelming infection, or it is to treat such an illness or a condition, such as dehydration or respiratory disease. Surgical hospitalizations in this age group are rare, but when they occur are often for intestinal obstructions, and even more rarely, for kidney or for neurological surgery. Very rarely, other problems arise. No matter what the reason, parents in love with their new babies are going to find hospitalization and the inevitable procedures terribly difficult.

Most hospital admissions for babies this age are not planned but are performed on an emergency basis. With this in mind, here is my advice:

- Whenever possible, admit the baby to a children's hospital, or at least to a children's ward. If surgery must be done, ask for referrals to pediatric surgeons and anesthesiologists.
- If your child is admitted to a teaching hospital, many procedures are likely to be performed by doctors in training, medical students, and nurses. Be assured that it is these people, not the senior and attending doctors, who are best at these procedures because they do them all the time. Don't insist that the Chief of Whatever start the IV; she or he probably hasn't done one in years or decades.
- Be as active as you can in understanding what is going on with the baby and why

procedures are necessary, but try not to get in the way, either. Many necessary procedures involve restraining the baby and inflicting pain or discomfort, though usually this is brief. Most of these procedures are done without local or general anesthesia: local anesthesia involves a needle itself, and carries its own risks; general anesthesia is indeed risky and reserved for major procedures.

- Unfortunately, many of these procedures must be done in a hurry, either because the baby requires them rapidly or because the staff is rushed. You can help the baby most by concentrating on explanations so that you don't have to ask for too many repeats; by staying calm during the procedure, even though several unsuccessful attempts may be necessary; and by doing what you can to soothe the baby afterwards.
- If your baby is admitted to rule out infection, the baby will probably need blood drawn, a catheterized urine specimen, and a spinal tap, and then an IV started for antibiotics. He or she may need a chest X ray or other procedures as well. After all this, it may seem enraging to discover that the underlying cause of the fever and illness was just a virus. It may seem as though all this was "unnecessary." It wasn't. When you are ruling out serious infection, naturally most babies will turn out not to have it. But such care is lifesaving for the ones who do. Nobody has a crystal ball.
- Your best chance for a hospitalization that leaves you feeling good about your-

self and the experience afterwards lies in having a positive relationship with the nurses, medical students, and house staff. Bear with them if they seem tired or brusque; offer compliments when you can, and make sure to let a superior know when one of them performs especially well or kindly. Gifts of food, such as cookies or brownies, especially on the night shift, are much appreciated. Often the staff are of a different racial, ethnic, or language background from your own, or of an unexpected gender or sexual preference. This does not make them any less competent, by any means.

- After the ordeal is over, bear in mind that you may be inclined to regard your totally recovered baby as specially vulnerable—not just for a little while, but forever. Be alert to this tendency. If you suspect it in yourself, seek help from your pediatrician, and perhaps from a counselor.

### MOVING

My advice here is exactly that of this section in the previous chapter, with one addition.

A baby this age is learning to roll over and grab things. Moving day is the day she will learn to roll off the bed you settled her into the middle of, just for a moment while you direct the movers. It is the day he will learn to grab the baby powder can and put it in his mouth and squeeze. Beware moving day accidents; get a trusted babysitter.

Two things to do as soon as you catch your breath: turn down the water heater to 120° to prevent accidental burns, and find out the fluoride content of the tap water if the baby is using tap water–mixed formula.

## The Four-Month-Old Well-Baby Visit

Babies this age usually prefer to be examined on a parent's lap or on a parent's shoulder. It is always cute to see how strongly bonded a parent and baby are at this age. Often when I ask to see the baby's "other ear" the parent turns so that *his* other ear, not the baby's, is the one presented. Even so, try to keep in mind that it is the parent's job to restrain the baby during delicate parts of the exam—like removing ear wax when necessary. For the safety of the baby and for the safety of the pediatrician (corneal abrasions from tiny fingernails, etc.) concentrate on holding the baby still rather than on watching the baby's expression or the procedure itself. Also, try not to give the baby a full nursing or bottle before the examination. The milk may be confused with, or mask, thrush; and the tongue depressor may produce spectacular results.

Much of the focus of the four-month-old examination is on the new integration of reflexes and skills. Babies this age should be focusing smoothly; any suspicion of a lazy eye or of abnormalities in the eye structure needs further evaluation. They should be able to orient towards a sound, though perhaps not look directly at its source yet. Arms and legs should move symmetrically, and most babies now can bear their weight on their legs when you stand them up.

Bellies during infancy and toddlerhood are always examined carefully for enlarged organs or other masses, and hips continue to be checked for dislocation, as sometimes this occurs as a developmental or acquired rather than as a congenital problem.

Any deviation from the growth curves of last time is noted.

Finally, there are the immunizations. At this writing, these are the same as at two months of age. See Part III for a discussion of immunizations in general.

## Looking Ahead

Over the next two months, babies will increasingly call for attention rather than cry merely from distress and need. Everything goes in the mouth. Babies will learn to get from one place to another by squirming or rolling or inching, seemingly for the sole reason of putting whatever they can reach into their mouths.

But the main delight of four to six months lies in the new fun of the Baby Game. If you think it's a pleasure now, just wait!

# Four Months to Six Months

## Reach Out (Way Out) and Touch (and Fondle and Mouth) Someone

## To-Do List

- Reward babbling with your attention and conversation.
- Baby-proof the house.
- Insist all caregivers take a CPR course.
- Introduce a cup containing breast milk or formula (not just juice or water).
- Discontinue the Pacifier.
- Make sure Cherub goes into crib or family bed awake, and falls asleep without needing to nurse or take bottle or pacifier.
- Continue to put Cherub to sleep on his or her back. When the baby learns to roll from back to tummy, without any assistance, Cherub can stay in that position.

## Portrait of the Age: A Mouth, Surrounded by Baby, Assisted by Hands

Zeke's gaze is a joyful searchlight, pinning adults to the spot with the delight he radiates. His grin is overwhelming; his approval without reservation. He reaches for everything. Every item is fascinating: a nose, a key chain, a phone cord, his own foot. It must be mouthed, examined, mouthed again, and squealed about.

His parents, Hope and Marcos, his grandma, and Carmella, his daycare nanny, are all having a wonderful time.

Until Zeke cries.

It's not the shrill, painful, colicky cry of weeks ago. When Zeke cries now, though, it somehow seems even more affecting. He is such a delightfully happy baby, with a grin and a coo for everybody. When he cries, you feel an urge to fix it quick.

And everybody has his or her own theory about how to do so.

His mother, Hope, who misses Zeke sorely during her work days, and feels guilty that he must miss her even more, picks him up and whuzzles his bellybutton. She cuddles him close and carries him for rides around the house.

Carmella, having been forbidden to give him apple juice, offers a toy and a teething biscuit, and then goes about her business, chatting loudly to Zeke and "rounding" on him every so often.

Marcos resists the impulse to pick up his son. He restrains himself and ruffles the baby's hair instead, making faces and showing him items from his pockets.

Grandma, of sterner stuff, responds with conversation from afar: "Who's that little rascal calling me? What could he want?" Indeed, Zeke will stop crying momentarily and look in the direction of her voice; then take up the wail exactly where he left off.

She wants to get Zeke a walker, now that he's nearly five months old. He holds his head steady, likes to kick, and seems ready for adventure. Grandma is eager for Zeke to develop his leg muscles. But Hope and Marcos are adamant: their pediatrician has told them that walkers are dangerous, and that the American Academy of Pediatrics has even issued a statement warning against their use. "A walker is an excuse for parents to ignore the baby," Hope says with a sniff.

So while they're all delighted with Zeke, the adults are less than thrilled with each other. Hope worries that Zeke is being deprived; Marcos that he's being babied; Grandma that he's being overprotected; and Carmella that he's being spoiled.

At around four months, babies arrive at a fabulous discovery. They can act on the world and make things happen. They can reach and grab and get that necklace into their mouths. They can roll over and change the way the whole universe looks. They can cry not just because they are in the throes of pain, but because they would like a bit of attention, please. Right now, please. I said NOW, please. And fairly often somebody comes.

This kind of crying, however, can set loving adults at odds.

You have to sympathize with babies this age. They are filled with eagerness for experience. Enchanted, for a bit, with a novel item to examine, chew, lick, sniff, they eventually exhaust all its possibilities.

There they are, being very polite really, entertaining themselves with all kinds of poor substitutes for human contact—their sock, a rattle, plastic keys, measuring spoons, their own feet. And when that gets a bit tiresome, what other way do they have to relieve boredom? They can't get anyplace. They can squirm a bit and roll in one direction, but that doesn't get them very far. They can't say, "Excuse me, Dad, could you bring your nose over here?"

Crying is a magnificent way of manipulating their world. It almost always works: somebody comes. Unfortunately, that somebody may misinterpret the cry, and think it is a distress cry. If this happens often enough, and the response is to pick up the infant and cuddle her like a younger baby,

Four Months may be persuaded that this is really what she had in mind in the first place. After all, such a response relieves her boredom. There's lots of tactile, olfactory, gustatory, visual, auditory, and all-round sensory stimulation in a cuddle.

Once persuaded, infants are rather hard to dissuade.

There are two solutions. One is to come to Four Months every time she cries, bearing new and interesting items for her to shake, chew, examine, and discard. Another is to get a safe walker. The old, unanchored walkers were dangerous. There used to be about 29,000 walker accidents a year bad enough for a baby to be seen in an emergency room. In an untethered walker a baby can hurtle downstairs, get to the cup of hot coffee, reach the tail of the dog asleep on the sofa. One Four Month I know attained the coffee table, where he devoured most of an entire tuberous begonia. It was nontoxic, but had been the cherished project of his gardener father. If the walker is poorly designed and the baby is strong and determined, he can tip it over or even launch himself out like a projectile and land on his head. He can pinch his fingers, if the walker has springs within his reach.

So if you do get a walker, get the kind that is anchored onto a giant saucer so that Cherub isn't let loose to roam the house and get into trouble. It will make Cherub happy, and give her some exercise, which will make naps and bedtime easier.

The only downside is that the walker may be so enchanting it's tempting to leave her to her own devices much of the time. But Cherub needs your attention and your conversation. This is when language learning starts to take off: that's why she's making all those weird vocalizations. So talk to her. Here is the red rattle. It makes a rattling sound. Hear it rattle? That is the red rattle, rattling.

## Separation Issues: The First Inkling of Separation and Stranger Anxiety

The first great accomplishment of babies this age is to discover that they not only have a separate self, but that this self can act on the world. To discover this, there needs to be a bit of delay and inconsistency when the baby asks for a response. Every slightly different reaction to his cry teaches him a little bit more about his universe. If he were to get picked up and cuddled as soon as he cried, he really wouldn't learn much.

Delay and variation in response also teach him that he can rely on himself to satisfy or at least to temporize his needs. And he gets the glimmer of a perception that while he is incredibly important and powerful, his beloved adults may have other things to attend to before they respond to him. This is a valuable lesson for him to learn before he takes off into the creeping, crawling, exploring, demanding stages coming up.

The second great accomplishment of Four Months is the development of an idea of what is Normal and what is Strange. Sometime along here, most babies develop some degree of what is unfortunately called Stranger Anxiety. The first sign of this phenomenon is The Stare. When the baby's attention is caught by a new person, she responds with a gimlet gaze. I have seen a baby stare for nearly sixty seconds without blinking. If the adult makes the error of staring back, or, worse yet, of approaching the baby, the stare turns into a frown and then often a frantic howl. Some babies, especially those whose days are spent with only a few very consistent adults, have an intense reaction. Others, especially those surrounded by people and noise, stare but briefly and warm up without elaborate "foreplay." But in every baby there is a sign, albeit subtle, that the baby is integrating new information.

An adult confronted by The Stare does best to ignore it—not easy, believe me. Look anywhere but directly at the baby. Many babies seem reassured if you conduct a quiet conversation with her trusted adults. When the baby finally has stared her full, she will break the gaze (sometimes with a little sigh) and turn to something else to play with. Her computer has logged you in. Now you may approach, humbly, eyes cast sideways, placating her with a toy. After a few minutes you may be able to make brief eye contact and even touch her. It will be easier to pick her up if you do not take her from the arms of one of her beloved adults, but from a neutral turf such as the floor.

Fortunately, most babies this age still don't usually have separation anxiety, perhaps because they don't yet have the concept of object permanence. Parents, however, do. Bonding is so intense, I believe, in large part because babies inspire little ambivalence at this age. They don't ask for much except attention and entertainment, which they reward with dazzling smiles and laughs. They don't get into much trouble. They are interested in anything and will smile at anyone.

While it may be very difficult for an attached parent to leave the baby with someone else (and the baby's normal "stranger anxiety" may reinforce this difficulty), it will be easier now than in the next three months, when separation anxiety appears. It may be helpful for parents to look upon the introduction of new people as the enlarging of the baby's social vocabulary. It's like learning new words, or trying new foods.

For a discussion of bedtime separation anxiety, see *Sleep*, below.

## Limits: Fitting Adult Responses to the Baby's Needs

Beginning at about three months, babies start to cry from boredom and/or for attention. By four and five months, it becomes increasingly important that such bids are responded to appropriately.

For the baby, it's empowering for him to know that responses are appropriate to his needs. A hunger cry brings food, a distress cry brings a cuddle, a bored cry brings a new activity and interaction. Tolerable delay between the cry and the response teach him a little about his own resources for self-soothing and independence.

For the loving adults, this age provides a

good introduction to limit-setting later on. It's a natural transition from the baby as bundle to be cuddled and adored to the baby as adventurer, to be anticipated, pursued, and outwitted.

Zeke is lucky that he has more than one devoted adult, and that their opinions about limits differ so widely. If he had only one of them, he would learn less about his own separate self.

If he had only Hope, and she *always* picked him up and held him close, he would learn to be satisfied with nothing less, and would be proceeding down the path to being, indeed, "spoiled."

If he only had Marcos, and his father *never* picked him up, Zeke wouldn't learn the spectrum of joys, from self-initiated activity to being cuddled.

If there was only Grandma and she *always* forced him to wait for gratification, he wouldn't grow up spoiled, but his capacity for joy in intimacy might be diminished.

And if there were only Carmella to comfort him, and her idea of comfort *always* included an oral treat, he might grow to depend on his mouth being occupied, with something to drink, eat, chew, or smoke, in times of boredom or anxiety.

The adults are lucky to have each other. As they see how Zeke responds to each, they can adapt and become more flexible themselves, incorporating each other's styles and enriching their own.

# Day to Day

## MILESTONES

### Vision

By six months, babies should be focusing directly on what they're looking at. No longer do they seem to be staring at something over your shoulder. When you flash a light at a six month old, the reflected glint should be in the same place in each eye.

At the same age, babies can follow an object moving all the way around them, from right to left or left to right. Once the object disappears off to the side, however, they won't yet turn in expectation of its reappearing at the other side.

If one eye seems to turn in or turn out, this needs to be evaluated by a pediatric ophthalmologist. (An ophthalmologist is a medical doctor, unlike an optometrist. This is important because a baby this young needs the whole eye examined; it's not enough to determine nearsightedness or farsightedness.)

Once again, a pupil that shows up as white in a flash photo, or a pupil that doesn't match the other one in size or shape or color, needs prompt investigation. So does the baby whose head is consistently tilted to one side. This may indicate a wandering eye (strabismus).

### Hearing

During this period, babies learn to alert to where a sound is coming from and turn in its direction. Since the most frequent word a baby hears is his own name, it may seem as if that's what he's responding to. Babies this age babble a lot, but you're not likely to hear much in the way of consonants yet. They belly laugh, blow raspberries, make grunts, bark, cough oddly, and often screech.

If you suspect your baby isn't hearing well—he doesn't babble much, or doesn't turn to the sound of your voice—insist that he have a hearing evaluation. Babies who have a hearing loss detected and treated before the age of six months are likely to have far superior language skills (and success at school) than those diagnosed later.

Casual hearing tests, such as seeing if a baby jumps or startles to noise, are completely useless. A baby with no hearing at all can sense vibration and react.

### Social

A four-month-old baby is adept at playing the Baby Game, responding to overtures with a smile or sound or motion and then waiting attentively for the other person to do something else.

She probably won't start playing peekaboo until her six-month birthday. That's OK, you can use the time to practice.

### Motor

Most Four-Months-Olds can bat at and sometimes grab objects and bring them to their mouths. They use both hands pretty equally, and kick both legs with equal vigor. By Six Months, most can also transfer an object from one hand to the other.

Most babies like to bear weight when stood up on a firm surface. If your baby doesn't do this by six months of age, bring it up at the six-month Well Visit.

Sitting with support starts to develop over this period. By six months, Cherub ought to be able to sit in a high chair, and many babies will be able to sit by themselves on a surface and play with a toy with their hands.

Rolling in either direction is not a very stable milestone. If your baby is doing everything else, likes to bear weight on her legs, holds her head steady when pulled to sit, and tries to squirm on her tummy to reach a toy, she is most likely doing fine. If in doubt, ask your pediatrician for an assessment.

### Growth

Most babies during these two months continue to gain about an ounce a day, and follow their early growth curves.

Fat babies (those whose weight for length is at or above the 90th percentile—see *Chubby Or Not, Here We Come!* in Part III) with double chins and deeply creased thighs and fat bellies may start to slim down as they become more interested in activities other than eating. It's important to give them the opportunity to do so. Fat babies on formula may have a harder time slimming down than breastfed fat babies, as the composition of the fat is different.

Rarely is there an underlying medical problem. It is more likely that Cherub is just a very "good eater" and never says no to an extra few ounces of milk or juice. At this age, it's easy to help a very oral baby both slim down and learn healthy eating habits.

It's a good time to start many of the following good habits:

- When she cries, don't assume it's hunger unless it's mealtime.
- Keep feedings on a regular schedule. Don't be lured into "snack-nursing" when Cherub is bored, or into taking a bottle of formula (Juice I won't even mention! Out of the question!) along on errands to stick in her mouth when she cries.
- When you feed her, respect her pace and stop feeding when she slows down in her eagerness. Wait patiently to offer her more, to make sure she really wants it.
- Make sure she's getting exercise—down on the floor on her tummy, doing her push-ups; bouncing in her bouncer; playing in her saucer-anchored walker.
- Don't let a feeding become part of the going-to-sleep ritual. If she's still nursing or taking a bottle during the night, ask your pediatrician if her size is such that you can help her stop that habit.

Thin babies, too, may need medical evaluation. Most of the time, a slender Four-to Six-Month-Old is healthy but more interested in occupations other than eating. But sometimes such a baby has a hidden illness, or is "happy to starve," or is in the care of someone who has misguided ideas about fatness and thinness.

## Teeth
Most babies won't show a tooth until about six months at the earliest. (All that drool is still protective; see the previous chapter on *The Fountain of Drool*.)

Whether or not you see or feel a tooth, still keep massaging those little gums twice a day to keep him and you in the habit. If your baby already is getting a visible tooth, see the *Teeth* section in the next chapter: *"Six to Nine Months."*

See the chapter *"Birth to Two Weeks"* for a discussion of fluoride supplements.

This is the best age to help break the bedtime-bottle or nursing habit, not to mention the eating-at-night habit. At this age most babies have few if any teeth, and it won't hurt for them to have a mouthful of sugar (milk sugar) to fall asleep with. However, in a few months you really won't want to think of those pearly whites coated with decay-inciting sugar for hours and hours. And by that time, the baby may regard the feeding as a necessary part of the prenap, prebed routine. A routine that is spoiled, somehow, by any attempt to clean the teeth afterwards.

## Sleep
Most babies this age sleep about fifteen out of twenty-four hours, about nine at night and six in the daytime, divided into two naps. Babies still should be put to sleep on their backs. *If* they can roll onto their tummies unassisted, they may stay there.

This is the age beyond which it will be very hard for babies to learn to fall asleep in the crib. And if they don't learn to do so, they won't be able to put themselves back to sleep when they wake up in the night.

All babies go through sleep cycles in which they awaken several times a night. And nearly all babies, when they thus awaken, will demand the same means of falling back to sleep that enabled them to fall asleep in the first place. If bedtime means rocking, nursing, bottle, or three choruses of "A frog, he would a-wooing go," then that's what it will take at two A.M. and three-thirty A.M. and five A.M.

This doesn't mean you can't hold and rock the baby and sing to him. But most of the time try to put the baby in the crib awake, when he's just gotten relaxed but his eyes are still open. At this age, he's likely to roll around a little, coo and talk, put hands, feet, and toys in his mouth, and doze off on his own. If he's a little fretful, singing and patting him usually help him quiet down again.

If you get beyond this age and the baby hasn't learned to put himself to sleep a large part of the time, you will have passed the easiest time in which he can do so. Between six and nine months, most babies develop separation anxiety; they cry when left by themselves. Fortunately, habit is even more powerful than separation anxiety for most babies. The baby who has become accustomed to putting himself to sleep usually comes to enjoy it. He may howl when his mother puts him in the crib in order for her to take a shower, but not when he knows it's time to go to sleep.

Moreover, between six and nine months, a baby learns how to pull to a stand. When he doesn't like the idea of being left in the crib, he can thus shake the crib rail at you, hold his arms up yearningly, and fall heavily backwards, hitting his head on the crib side. He can also drop his toys off the edge, and scream imploringly as they disappear. He may be able to scream until he throws up.

Trying to teach him to fall asleep on his own at that age requires a heart of stone or a state of desperation that can only be acquired after at least two weeks of constant parental wakefulness.

Besides teaching him to fall asleep on his own after he's been relaxed with a bedtime ritual, it's a good idea to teach him that he doesn't need one specific adult to perform that ritual. Alternate who calms him before bed. If you are a single parent, invite a trusted friend or hire a babysitter. This is a good idea even if the baby spends your working day with someone else, and you cherish that bedtime closeness. It doesn't need to be often, but a few experiences will pay off later when you leave him with Grandma.

### Naps

Most babies this age take two naps a day, of two to three hours each.

Just as a baby can be persuaded to sleep through the night, so can he be persuaded to nap at pretty much the same time each day. The key is to keep his schedule as regular as possible in other regards, and to condition him to the napping environment. You don't usually have to do the elaborate night routine and "lullaby" for naps. Just having the day punctuated by meals, bath,

Honey should not be given to children under a year of age, due to the risk of the disease infant botulism. For fish-consumption guidelines, see page 7.

outings, and play in the same order at the same time is enough; fit the naps in at the same time each day also. A rocking and cuddling just before the nap is a wonderful cozy idea; just make sure the baby doesn't really fall asleep during it.

So I'd try not to get anybody into this habit.

## FEEDING AND NUTRITION

### Milk: Breast milk, formula, and others

Both breastfed and formula-fed babies may be ready to experiment with foods now, but most of their nutrition should still come from human milk or formula. If you have a strong family history of allergies, your pediatrician may urge you to keep the baby solely on breast milk or high-tech formula until the age of six months. No baby should start cow's (or goat's) milk out of the carton until about a year of age. It can produce iron deficiency, which can cause impaired development. The balance of calcium and phosphorus, and the range of fats, is not appropriate for infants.

Start to introduce milk (expressed breast milk or formula)—not juice—in a cup at about five months. I suggest a plain cup

without the "learner" spout; many babies find that spout confusing, and don't know whether to sip or suck.

### Solid Foods

By solid foods, I mean Baby Foods: smooth purees of vegetables, fruits, cereals, and meats. Some books refer to these as "beikost." It's best to use pure foods, without additives (except for iron supplementation) rather than mixtures until you're sure how Cherub is going to react.

Since most of the nutrition for a baby this age comes from milk, regard this as an "introduction" to solid foods.

Babies likely to take well to solid foods can hold their heads up. They often look very interested indeed when watching other people eat, and may open their mouths in mimicry. When you offer a spoonful of food, they accept it without thrusting it back at you with their tongues.

Start a new food every three or four days to see whether Cherub tolerates it. Signs that Cherub has a problem with the specific food just introduced include facial or diaper rash, actively refusing the food when offered, vomiting, watery or painful stools. It is OK (though disconcerting) for the food just given to appear in the stool apparently unchanged, as long as the baby shows no dis-

This is a good time to internalize the Golden Rule of the Good Eater, as formulated by Pediatric Nutrition Guru Ellyn Satter: Adults decide **What** choices of food are offered, and **Where** and **When** they can be eaten. Babies and children decide **How Much** of each food they want.

*Nathanson's First Amendment*: If a baby seems to be forming finicky tastes (taking only yellow veggies, for instance), simply stop offering the addictive food as a choice.

*Nathanson's Second Amendment*: It is easier to avoid introducing seductive foods (such as fruit juices or Cheerios to munch on, catsup on anything) than it is to try to limit or remove them later.

tress about it. Some of it probably did get absorbed.

Guidelines for which foods and when are based as much on commonsense and tradition as they are on scientific findings.

## BATHING

Many babies this age adore the bath. For the few who are frightened, it often helps to take the baby into the bathtub with you, so that bath and cuddling are combined. As babies become older and fascinated with bubbles, it's tempting to make every bath a bubble bath. If you do, bathe the genitals with clear water before drying the baby to avoid irritation of the glans of the penis and of the vaginal area. Powders are truly a hazard at this age: it's so easy for the baby to grab the container, put it into her mouth, and squeeze.

## DIAPERING AND DRESSING

The only news on this front at this age is the universal tendency of babies to roll and grab

while you're trying to get them dressed and diapered. Either perform these acts on the floor or belt the baby onto the table. Never turn your back for an instant. Keep baby powder out of reach, as well as anything else the baby could grab and eat or choke on.

Please see this section in the previous chapter for a brief discussion of jewelry, headbands, and footwear. No shoes are the best shoes for babies whose feet don't need prescription footwear.

## ACTIVITIES

I tend to think of babies this age as a Mouth eager to explore. The other body parts of the baby—hands, feet, eyes—are just accomplices.

Hands are for reaching, grabbing, and putting things in the mouth. Any toy small enough to handle but oral-safe (not smaller than two-and-a-half inches in any diameter, no sharp edges or detachable parts, and nontoxic) is good for at least a few minutes' exploration. Toys that float away when you

# BABY FOODS SHOULD BE SAFE AND HEALTHFUL

- Avoid sticky, stringy, chunky, crunchy foods; avoid anything that comes in pieces that can be choked on. We have seen some pretty severe choking accidents on crunchy teething biscuits.
- Don't add, or buy foods with added salt, sugar, fats, or artificial ingredients.
- Avoid foods naturally high in salt, sugar, and fats: ham, pork, apple juice, egg yolk.
- A few varieties of home-prepared vegetables can be dangerous. They can grow in areas where the soil is rich in nitrates and these can cause anemia in babies. It's best not to home-prepare carrots, beets, turnips, and collard greens.
- To set a good example for the future, offer a diet *rich* in green and yellow vegetables, fruits and cereals, *low* in sugar, refined starches, fats, and red meat.
- If Cherub is not eating iron-rich foods (such as enriched rice cereal) and is not taking iron-rich formulas, ask your pediatrician about supplemental iron drops.
- Make sure you're still giving Cherub the same amount of milk you have been. Don't decrease it. You are using foods right now just as a supplement to fill him up as he grows and to introduce Cherub to new flavors.

  At this age, most of the baby's nutrition still comes from breast milk or formula. A breastfed baby still should be getting about five good nursings a day; a formula-fed one, about 24 ounces depending on size. (A good rule of thumb is one ounce for every two pounds he weighs, at each of five feedings a day. The limit to formula would be 32 ounces a day; going over that means both a very stretched stomach and a very stretched household budget.)

  Some babies like to take solids before nursing or taking the bottle; some like it after; some in the middle. And some may vary from meal to meal and from day to day. Go with the flow, is what I say.

- Give the milk first and then the food, staying alert to how much Cherub really wants.

  Don't try to finish the jar, or aim for a certain quantity, or "top him off" so he'll sleep longer. Even a four month old can take control of much of the process—just wait until he spits it back and your world-view turns spinach-green. Whoever is feeding the baby can hold the spoon in such a way that he needs to turn or lean towards it; can time the approach of the spoon tactfully; can exchange comments with coos throughout the process. Respect personal tastes, e.g., don't force the mashed lima beans, and stop when the baby turns away.

- Start with rice cereal, then move on to a green vegetable, then yellow veggies, then fruit.

    Rice cereal is bland. You can mix it with a little breast milk or formula, so that it tastes familiar. It's a good introduction. It is also fortified with iron, so that breastfed babies who eat it do not need iron drops.

    The other foods are introduced in order of sweetness, from least sweet to most sweet. This is based on the theory that once a baby knows that there are sweet potatoes and peaches out there, he'll turn up his nose at green peas.

    Green vegetables are indeed a big adventure in flavor and texture. You may need to offer the same green vegetable at every meal for three or four days before Cherub consents to take more than a taste.

    Baby food meats are quite a challenge in terms of taste and texture. Since Cherub is already getting sufficient protein and iron, there's no rush about introducing them.

- Give Cherub three or four days to show there's no reaction to a food, and then add a new one from a different food group than the first.

    Reactions to food certainly can occur, but they aren't necessarily allergic: there may be more fiber, or sugar, or fat than the baby is able to digest. Very acid foods can give an acid stool and acid drool, which give rise to diaper and face rashes. But a spirit of fun, tempered by caution, is a healthier attitude than one of apprehension.

- **Expect mess.**

    Glopping food on the fingers from dish to mouth may start as early as five or six months of age. The more participation, the better. The more control the baby has the happier he'll be, and the more he'll learn.

grab for them, such as cradle gyms suspended by flexible cords, aren't very good play items.

Eyes are for catching sight of something mouthable. Mobiles aren't very good; you can't grab them. Mirrors are fun if they are close enough to kiss and lick and drool on the reflected baby.

Feet are good for putting in the mouth.

Legs are good for kicking something that makes a noise, because you can do that while you are also putting something in your mouth.

Ears are good for listening while you are putting something in your mouth.

## TOYS AND EQUIPMENT

Babies this young don't need any real help developing the skills they are destined to learn. That is, you don't really need special toys and equipment if you have a loving, play-oriented caretaker and ordinary objects to explore, such as measuring spoons, mirrors, nonglass baby-safe coffee tables, adult spectacles, hats, and noses.

After all, a Four Month is most happy finding, staring at, mouthing, and going on to the next thing. Mirrors, things that squeak or move interestingly when shaken or grabbed (like an adult nose) are excellent toys. Change in type of motion is fun, but gives most pleasure when the baby is active in producing the motion. That is, a **safe** "jumper" or a walker will be more fun than sitting in a swing or being pushed in a carriage.

Babies are still prone to damage from whiplash-type injuries and shaking. I know it is tempting to use a wheeled baby jogger, or to toss the baby for a giggling reward, but the concerns may outweigh the benefits. Don't toss. Try to find a jogger with the baby sitting backwards to avoid whiplash, and so that she can see you rather than just the unpeopled passing scene. Make sure that you aren't rattling the wheeled vehicle over bumpy terrain, and that she isn't a direct hit for cars' exhaust.

## SAFETY ISSUES

If you haven't baby-proofed your home and stocked your medicine chest as described in the previous chapters, please do so now. We're just adding on to those basics.

This is the moment to think ahead to the next few months. Normal babies show an enormous range of motor development from now to a year. Your baby may be a placid soul who would prefer to sit and figure out small objects, or she may be someone who walks independently at eight months of age. Or he may be one that is vertically inspired, like a certain nine-month-old I know who could engineer a climb from pillow to chair to table to counter to cupboard. It may be hard to believe that the easily corralled Cherub blowing bubbles is going to make such leaps in such a short time. Believe it.

Take the time now to make your home baby-safe for the mouthing, mobile, climbing, exploring baby. Not only may you prevent a tragedy, but you will make the art of discipline much more sophisticated, interesting, and effective. Setting a few consistent limits is much less taxing and more educational than "no" repeated ad infinitum.

*Actions to take*

- Take a CPR course, and make sure the other parent and all caretakers do too. Concentrate especially on how, and when, to use the "baby Heimlich" maneuver to help a baby choking on a small object. If your caretaker speaks a foreign language, make sure she takes such a course in her native tongue, and knows enough English to dial 911 and get help.
- Check the batteries of your smoke alarms.

- Always use a car seat. Make sure it is installed properly, and that the shoulder straps are always used correctly. And keep the car seat in the middle of the backseat, facing backwards, as long as the baby fits that way; it's much safer in every way.
- Move all breakable, sharp, edible, or chokable items off any surface lower than four feet, and any surface reachable by climbing. Pay special attention to the bathroom, and form the habit now of closing the door after use and of not leaving electric appliances near sources of water. And watch out for the curling iron, tempting and easy to grab: make a special hard-to-reach place for it.
- Beware all sources of water: pools, Jacuzzis, fountains, toilets, tubs, buckets. Babies can lean over the side, get pulled over by the disproportionate weight of their head, and drown in just a few inches of water.
- Affix safety latches to cabinets, and put safety covers on electrical outlets. Equally, if not more important, try to make plugged-in electrical cords inaccessible. Severe mouth burns from biting a cord are more common than electrocution from electrical outlets.
- Check the lead exposure in your household: chipping paint or plaster in a house built before 1960, lead from exhaust emissions in dirt, lead in water pipes. If you're suspicious, discuss with your child's pediatrician.
- Move under-the-sink chemicals in bathroom and kitchen to a higher, safer location. Get rid of any you can, especially pesticides. In most cases, the pests are safer than the chemicals. Also, get rid of or hide well the liquid plumbing aids that open clogged drains.
- Place by each phone the number of your pediatrician, Poison Control, and a reliable neighbor.
- Don't turn your back for an instant if the baby is on a surface other than the floor. He or she is sure to fall.
- Don't put chains, strings, or necklaces around the baby's neck; it's too easy for her to get caught and strangle. Beware playpens with netting, which can catch the button on a shirt and "hang up" the baby; and beware any curtain draws or other cordlike objects near the crib.
- Never leave even for a second when the baby is in the bath.
- Baby swimming classes are NOT advised. The American Academy of Pediatrics urges no immersion swimming lessons until about three years of age. Reasons:
  1. Some babies immersed in water will have a reflex that closes the upper airway and chokes them.
  2. Some babies when underwater involuntarily and automatically drink huge quantities, diluting the blood and throwing off the chemistry. This can make the baby very sick, even with seizures and coma.
  3. Babies who inhale water are prone to ear infections and pneumonia. Their anatomy and immune systems are immature. If they are in a class with other diaper-clad babies, it is a rare,

huge, very well maintained pool that can control all those germs.

4. Swimming lessons don't drown-proof a child. A baby who can swim will still panic if he falls in the water unexpectedly. Indeed, such lessons may endanger a child who has come to love the water and may have no inhibitions about heading towards an unguarded pool.

# Health and Illness

*Part II of this book covers illnesses in babies this age and older. It covers how to assess symptoms, such as diarrhea or a cough; how to tell whether a child is sick enough to require urgent or prompt attention; and how to analyze what is wrong with a specific body part, such as eyes or legs. The younger the baby, the more careful we should be in getting help promptly for any condition that could be serious. I have tried to be very conservative in this area, but if you have any doubts or questions consult your own pediatrician promptly.*

EXPECTATIONS

Medically speaking, the four- to six-month-old is like the two- to four-month-old in many ways. His immune system is immature still, and he derives no benefit from contracting or being exposed to infectious diseases. However, minor illnesses like colds and gastrointestinal upsets are a bit less threatening, simply because the baby is stronger and bigger.

But there is one big difference between this age and the last. Now the baby is reaching, grabbing, and mouthing objects. How likely are such objects to pass on diseases? Should you boil or scrub or eliminate anything that has fallen on the floor? Do you do more damage to the soul by being fastidious than you do to the body by being casual, or vice versa?

Both viruses (like rotavirus) and bacteria (like strep) can be passed on via surfaces of toys and furniture. If the baby touches a contaminated object and then puts her now contaminated hand (or her foot or the object itself) in her mouth, she may get an intestinal illness; if she touches her eyes or her nose, a respiratory one.

So what can you do?

Forget the idea of keeping the baby from touching her eyes, nose, mouth. And you can't annoy the baby by constantly washing her extremities. As I see it, the most helpful ways of keeping the baby healthy are:

**Boost the immune system** as much as possible. Breastfeeding, keeping up on immunizations, providing sufficient rest, exercise, and pleasure (I am one who believes, to an extent, that taking pleasure is an important natural function) are ways to do this. Keeping the baby's mucous membranes sufficiently moist is another; consider a cool-mist humidifier.

Keep the environment supportive of the immune system. No, we don't catch colds by going out in the wind and rain without protective clothing. However, such stimulus does tend to generate increased mucous production and bodily discomfort, and these may decrease the ability of the body to fight off a cold.

**Avoid obvious sources of contamination.** Don't let your baby mouth or handle toys in the pediatrician's waiting room. Keep her on your lap, and discourage the attentions of other children. A good way to do this is to talk to the *baby*, loudly: "No, no, Miranda, you can't play with the nice little boy; the doctor says you are VERY contagious until next Tuesday." Make a point of holding Miranda up to your shoulder, facing away from the curious youngster. Smile apologetically at Little Boy's parent. "We've all had it at our house. We call it the curse. I'd hate to pass it around the whole community." Then sneeze (discreetly into a hankie), or at least clear your throat huskily.

Try very, very hard to find a daycare situation in which there are no more than three other babies. Babies this young need not share toys. Even if they do, or if they are always touching each other, they are much less likely to be in a contagious phase of an illness than are toddlers and preschoolers.

If the baby's toy does fall in an area with a high potential for infection, wash it. If the baby does touch something with a high potential for infection, wash the hand or foot. This is especially true if there is a seasonal epidemic going around. Use a diaper wipe if you need to. Since the baby can't get around on her own yet, this shouldn't happen very often, and nobody should accuse you of being fanatical or obsessive-compulsive. You really don't want your baby to catch rotavirus diarrhea or RSV-caused bronchiolitis at this age.

**Wash your hands** after you touch something highly suspect (including diaper changes) and before you touch the baby. Out in public, keep those premoistened wipes handy. Wash your hands always before you touch your own eyes or nose; most respiratory viruses are transmitted in this fashion. If you think you never touch your eyes or nose without knowing it, put blue chalk on your fingers and look in a mirror half an hour later. For the same reason, try to wash your hands after touching your eyes or nose and before touching the baby.

Finally, however, some exposure is inevitable because the respiratory viruses can be transferred through the air. Avoiding crowds and gatherings of young children, especially during times of widespread illness, is a perfectly valid way of protecting a baby this age.

I don't think these measures are overprotective. A baby this age doesn't receive any benefit from being sick. She doesn't make very good immunity yet, and she's at risk for more severe side effects than is an older child. Moreover, while babies this age may show enjoyment when they're exposed to crowds, sick adults, birthday parties, and young children, they equally love plastic keys, their own feet, a shiny mirror, and their own family.

## Common Minor Symptoms, Illnesses, and Concerns

### Colds

Babies this age usually can be made comfortable and stay remarkably cheerful during uncomplicated colds; they regard the nose run as an extension of drool, perhaps. See the advice in the previous chapter for assessing and comforting the symptoms of a cold.

### Intestinal upsets

Uncomplicated vomiting and diarrhea can be watched and treated by the same methods as for younger babies. See this section of the previous chapter.

### Simple constipation

A baby taking formula or solid foods can easily become constipated, and this can become a frustrating vicious cycle if not interrupted promptly. See the section in the previous chapter on avoiding this very common problem.

## Serious or Potentially Serious Symptoms, Illnesses, and Concerns

### Fever

Fever is a normal response to infection and inflammation; it is a symptom, not a disease. In the newborn period, it is a very powerful symptom. Very young babies can have a serious infection with only a slight elevation of temperature to show for it. Moreover, other behavioral signs of serious illness can be very subtle. The younger the baby, the truer this is. However, as babies mature, the signals gradually become as important as, or even more important than, the degree of fever. In babies this young, it isn't a good idea to give acetaminophen for a fever unless you know what's causing the fever; you don't want to mask an important symptom.

In a baby from four to six months, a rectal temperature of 101° or greater should prompt a same-day contact with the pediatrician. If the baby is acting sick (see *Acting Sick, Smelling Sick, Looking Sick* in Part II of this book) or if the fever is as high as 104°, obtain urgent attention.

Most fevers are due to viruses, but occasionally there is a surprise, such as a urinary tract infection. Teething, I am sorry to say, has never been shown to cause fever. This legend started because babies this age and through age two are always teething, so that there is usually a tooth coming in whenever a baby has a fever. Why is it that nobody ever suggests that fever causes teeth?

### Breathing problems and persistent coughs

Assessment and action for these problems is the same for this age as for younger babies. See this section in the previous chapter.

### Complex or prolonged diarrhea

Once again, concerns for a baby this age are the same as for a younger baby. Please see this section in the previous chapter. Also, see *Intussusception*, below.

*Persistent or projectile vomiting*
The concerns about vomiting are, with the exception of the condition called intussusception, the same now as they were in the previous age range.

*Intussusception: An intestinal obstruction with various symptoms*
Intussusception (IN-tuh-suh-SEP-shun) is a fairly rare problem, occurring in only about four babies out of one thousand, and it peaks between the ages of three- and thirty-six months.

The problem is that many babies have symptoms that suggest this obstruction, even though few actually do have it; and in every single one of them, the diagnosis needs to be considered and set at rest. This means that your pediatrician may mention this diagnosis as a possibility if your baby shows any one of several symptoms.

In this condition, the small intestine "telescopes" on itself like a wrinkled shirt sleeve. This causes the circulation in the bowel to be cut off, as by a tourniquet. Left untreated, the bowel will "strangle."

A baby with this disorder often, but not always, screams every time the bowel kinks and squeezes. She often, but not always, vomits. She often, but not always, has bloody stools that look like current jelly. She sometimes has no vomiting, no bloody stool, and no sign of abdominal pain, but recurrently goes limp and pale and very lethargic, then wakes up again and seems "almost OK."

When the diagnosis is made early in the course, the intestine can often be unkinked with a special enema under X-ray guidance. Later on, surgery is sometimes necessary. If the condition goes untreated, the baby can go into shock, with a very high fever and lowering of blood pressure.

## MINOR INJURIES
These are now related to the increase in mobility of the adventurous young baby.

*Rolling off surfaces*
Fortunately, when they do fall, babies this age usually roll off low surfaces such as couches and beds, and it is unusual for the baby to be severely injured. After the screaming (both parental and infant) stops, look at the baby for obvious bruises: undress her completely. Make sure she is using both arms equally; sometimes a baby will crack a collarbone, arm, or wrist in a fall. Then, even if everything seems OK, call the pediatrician anyway, partly so that your careful and honest reporting can be noted in the chart if bruises develop and somebody accuses you of child abuse. And partly because you may notice something in the next day or so, and forget to relate it to the fall.

*Falls while being carried*
I put these under Minor Injuries because I've never seen a major problem ensue, thank heavens. But there's always a first time. Here are common scenarios:

- Adult on the stairs, gets distracted, no handrail;

- Too-young or too-impulsive child carrying the baby;
- The infant seat has a structural problem with the handle, or the parent is distracted changing its position—usually while changing hands, especially if there's a blanket covering the seating area.

### Burns

The most common kind of burn in babies this age is probably sunburn. It's so lovely to have a portable baby that it's easy to forget the sun is always there, whether there are clouds or shade or not. Reflected sun and filtered sun still burn. Since a burn takes six to twelve hours to fully develop, be alert to the slightest pinkness and get the baby inside (not just into the shade). Any burn with blisters needs to be seen and dressed by the pediatrician. Don't break the blisters! And with sunburn goes overheating. A baby who is hot and flushed can be cooled by just changing environments, giving extra fluid, and perhaps sponging extremities with tepid water. A baby who looks sick and has poor color may have heat stroke (see *Fever,* Part II, Chapter 12).

### Overheating

With sunburn goes overheating. Or the baby is bundled up for a cold day, and then put into a heated car, or in a window with sun pouring through.

A baby who is hot and flushed can be cooled by just changing environments, giving extra fluid, and perhaps sponging extremities with tepid water. A baby who looks sick and has poor color may have heat stroke: see *Fever,* Part II.

## MAJOR INJURIES

### Shaken baby syndrome

Shaking, tossing, or vigorously jiggling a baby this age can damage the blood vessels in the brain, causing seizures, impairment of vision, hearing, intelligence, or even death. It is so much fun to toss babies, and so tempting to shake some sense into them when they cry, that this terrible possibility must be foremost in the mind of all caretakers. Make sure yours is vitally aware of this. A baby with the syndrome may show no external signs of trauma, but may act sick, lethargic, and irritable.

### Choking

A baby this age doesn't have a pincer grasp with which to pick up coins, batteries, peanuts, and other objects so likely to cause airway obstruction. But she can pick up a two-inch block, or a Ping-Pong ball, or a marshmallow, or a doll with a head that comes off in her mouth. Or, given a container of small candies to shake, the lid can come off while it's in her mouth.

## Window of Opportunity

### Bedtime

See the section on *Sleep* in *Day to Day*, in this chapter, to ensure a good night's sleep for the next year or two.

### Starting to drink from the cup

Everything else goes in the mouth, after all, and the baby is less interested in sucking and more interested in other oral adventures. At this point, she can't hold the cup herself, but she can get the hang of sipping. I'd avoid those spouted tops; I am sure many babies are confused by them. A small Tupperware cup or even an unbreakable shot glass with a little breast milk or formula works well. Don't put just juice or water in the cup.

### Giving up the pacifier

Sucking needs are so overshadowed by exploring needs, this is a good time to "lose" the pacifier. Most babies won't miss it—or will fret only briefly. But if you don't do this now, "binkie" will turn into a beloved security object.

### Talking to the baby

If anybody is shy or self-conscious about talking to a baby frivolously, this is a good time to overcome it. Babies are interested at this age in anything you say. And you don't have to censor your speech much; you can talk politics, analyze your friends, propose unlikely scientific theories. You can get used to naming genitals out loud, if you have an innate shyness about doing so. This is good preparation for imparting sex education later on.

Just talk. He loves the sound of your voice.

Right now, a baby will turn towards any new voice, no matter what that voice is saying. But she is probably learning a bit about inflection and the kind of music her own first language makes.

## What If?

### STARTING DAYCARE

It can be terribly hard to leave this enchanting baby to go back to work. It can be even harder to return from work and find the baby having a wonderful time with the caregiver. There is one overwhelming consolation, however: from the baby's point of view, making the change now may be easier than it will be later on.

Most babies don't develop separation anxiety for a couple more months, at which age their attachment to their parent, usually the mother, will be more like Superglue than Velcro. Now, however, after a short period of making friends, they'll "go to anybody." Moreover, their schedules are generally regular, they delight in simple entertainments, and they aren't yet mobile on their own.

These characteristics not only help the baby transfer to the nanny, but help the nanny to bond with the baby. This may be of some importance, especially if the child's personality is a very active, determined one. It is easier to cope with a rambunctious nine-month-old or fifteen-month-old if you have already fallen in love with him at this endearing stage.

It remains important to take enormous pains with the type of child care. A nanny who cares for children of varying ages may seem perfect now: toddlers often find this age baby irresistible, and the baby chortles in delight at their attentions. However, in just a couple months, the baby will be mobile and will start to test limits, and the patience of both Nanny and toddlers will wear thin.

Moreover, those toddlers are engaged in important immunological work. Their job is to catch a variety of infectious diseases, so that they will develop a base of immunity. They are most contagious before they develop symptoms. And the baby will catch these illnesses, guaranteed.

It does the baby no good to do so. He can not usually make permanent immunity to them, so his suffering is for naught; and he learns less well when sick. If he is prone to ear infections, the repeated viral infections are likely to develop into otitis. During the episodes, his hearing is likely to be diminished. This would be too bad, as receptive language appears to take off around this age.

Also, a baby this age may be amused by other children, but doesn't require them for full development. Loving adults and an explorable environment are his needs.

Finally, a nanny busy with toddlers may not have the time, energy, or incentive to be sensitive to the rather subtle needs of a baby this age. She might well answer every "attention" cry with a bottle of juice, or by picking the baby up and carrying him around the house. This can set the stage for spoiling problems later on. Or she may not have time to really play with him.

Of course, one-to-one daycare is impossible for most parents. But I strongly feel it is worth trying to find a situation in which the other daycare children are either agemates of the baby or several years older. And one in which the total number of children is not more than four, not only from the point of view of attention but from that of infectious disease spread.

## SEPARATION FROM PARENTS

While babies haven't developed separation anxiety yet, they do appear to have some sense of continuity and routine, and they certainly now recognize the specialness of their parents.

Separation at this age is much easier than it will be in a few months, and may even be easier than it was a few months ago, as the baby can do more to entertain herself and isn't so dependent on being closely held and cuddled. A separation of a week or so may be easier on the baby than on the parent. However, it is *crucial* that the adult the baby is left with has already strongly bonded with the baby beforehand. Even when all this is said, don't expect instant recognition and "forgiveness" when you return. It may take a day or so before the old ease is reestablished.

## TRAVELING WITH THE BABY

Expeditions at this age are even easier than earlier, as the baby is such a crowd-pleaser. Again, the only caveats have to do with infectious illnesses (carry wipes for your hands, the baby's hands, and anything the baby handles; don't let small children come up and sneeze on him) and secondhand smoke. Avoid it. If anyone rudely persists, say greenly that you are very sorry but the smell makes you throw up. They'll stop.

See this section of the previous chapter for further suggestions.

## ADDING TO YOUR FAMILY

Parents of a baby this age, especially of an easy-tempered, easily scheduled baby, may be sorely tempted. The days of crying or colic are receding in memory. It is very difficult to envision this delightful baby as anything other than acquiescent and undemanding.

If you conceive now or start adoption proceedings, by the time your new baby arrives, this one will be climbing and running, testing limits vigorously, interested in hitting, biting, pinching, and pulling hair. He'll require sustained adult patience to set limits appropriately and adult attention to develop language skills. Parents taken up with the demands of a newborn tend to be sleep-deprived, and thus slightly impatient and preoccupied.

Moreover, when the new baby is starting to test limits at about nine months of age, the older one will be at the climax of the separation work that characterizes the nearly two-year-old. Parents can feel quite desperate when they are torn between such strong emotional needs of their children.

Of course, all this is exacerbated when there are other demands on the parents, such as an ill parent, two careers in the family, or financial problems.

A special health concern applies if the older baby will be in daycare during the pregnancy and while the new baby is under six months of age. The older one may bring home viruses (especially CMV, cytomegalovirus—see Glossary) that can affect the fetus, should the mother become infected. After the baby is born, the toddler in daycare will want to sort of engulf the new one, kissing and mouthing and hugging, and passing on, thereby, any and all viruses she's acquired that day.

This is made all the more difficult because most of the time toddlers pass on such viruses before or without getting sick themselves; you have no warning that it's happening.

## DIVORCE AND CUSTODY ISSUES

When parents of babies this age tell me they are contemplating divorce, I always wonder: Is Cherub's age-related behavior a factor? Is one parent so entranced with the baby that there's no room (or incentive, or time, or energy) left for adult conjugal affection? At Four to Six months, Cherub is designed to bond very intensely with a special loving adult. Unfortunately, this often goes along with Cherub's giving every appearance of rejecting other adults, including—no, especially!—the other parent.

When you are ignored by your adult partner and rejected by your beautiful baby, it's easy to feel as if this whole relationship is a big mistake. Talk about unrequited love and lack of appreciation!

If you are in this kind of predicament, it's crucial for both parents to behave like adults: that is, to think about the situation rationally, get a little perspective, and work on transforming this asymmetric bonding. To do so, it's important for *both* parents not to take Cherub's behavior personally. Cherub is merely engaging in an important, normal, temporary, universal developmental learning phase.

The "cure" is for the Rejected Parent to be left alone, one-on-one with Cherub on a regular basis. This will work best if you don't hand Cherub from the one parent to the other, but perform the transfer with Cherub in a neutral territory, such as a blanket on the floor, where the Rejected parent can make overtures using a wheeled toy and making The Car Noise.

Divorce when a baby is this age usually involves a combination of the above issues, plus those discussed in *Starting Daycare* and *Extended Separation from the Parents.* On top of which:

- It may be very tempting to turn to the baby for closeness and intimacy during the bitterness that accompanies many divorces. In particular, as a single parent you may want to keep the baby in bed with you at night. This is very understandable, but can set the stage for problems later on. Not just sleep problems but problems with blurred personal boundaries. A baby this age is discovering the difference between Me and Not-Me, and if his closely attached parent is confused, he will be also.
- Conversely, a parent exhausted and embittered by a divorce may be tempted to take out these feelings on the baby, especially if the baby is of the same sex as, and looks like, the divorced spouse. When a baby this age cries for attention, such a cry may seem like an outrageous demand. It may be overwhelmingly tempting to shake or otherwise physically punish the baby. Being aware that

this can happen before it actually does can help a parent get preventive help in the form of counseling and alternative baby care. A "hot line" to a trusted friend, relative, or child-abuse group can be a lifesaver.

- Custody issues may be very difficult to deal with, as a very bonded and angry parent may not be able to tolerate the idea of the other parent having access to the baby. It is important to try to get a grip on the fact that for the baby, such continued contact is almost always in his or her best interest (unless the absent parent is psychotic or abusive).

## SURGERY AND/OR HOSPITALIZATION

The advice here is exactly the same as the advice in this section in the previous chapter.

## MOVING

Again, the advice here is exactly the same as that in the previous two chapters, but with an even greater caution about safety. Accidents happen on moving days. Have the baby absolutely elsewhere, with a trusted child care person.

# The Six-Month-Old Well-Baby Visit

My favorite part of the visit is when I go to listen to the baby's chest while he's on his parent's lap. The baby almost always leans down with an air of noblesse oblige, opens wide, and mouths my stethoscope-holding hand like a kitten. I always wash especially thoroughly before (and after) this exam.

The usual focus of this exam is on neurological and emotional development, the abdomen, the eyes and hearing, and the hips and feet. Babies should move arms and legs symmetrically, without a preference for right or left. They should have enough fine motor control to reach, transfer, and examine an object, and put it in their mouths. Most will bear weight on their legs if you stand them up. The abdomen shouldn't have any enlargements or unusual masses, and no hard stools. Babies this age should focus on an object steadily, and should turn to a voice. Hip sockets should be in place so that when the baby is lying on his back and you press the knees out in the "frog" position, they can both touch the surface he's lying on even though he doesn't like it and will yell for a moment. Feet shouldn't curve in very much, though the bone of the lower leg may curve in quite a lot (see Tibial Torsion in the Glossary).

At this writing, the baby will still get the DPT, polio, H. flu and meningitis, and sometimes the hepatitis B vaccines at this visit.

## Looking Ahead

Get ready, get set, go. Your baby is about to be mobile, adventurous, and unstoppable. You are about to encounter the real meaning of the concept of *limits*. But you won't really notice, because the Six- to Nine-month-old is such a personable, delightful creature. You also will discover the meaning of separation anxiety, and how it feels to carry around an eighteen-pound weight attached to your leg while you try to go to the bathroom.

{ S I X }

# Six Months to Nine Months

## The Drama of Disappearance

## To-Do List

- Introduce the joy of books (chewing, tearing, staring).
- Introduce a cup, finger foods, spoon.
- Don't be seduced into giving beverage juices or "handout" snacks.
- Baby-proof and take a CPR course.
- Read ahead about possible scary but innocent behaviors.
- Start to Set Limits: Become (temporarily) an Authoritarian Parent.

## Portrait of the Age: Discovering a Separate Identity

Katie doesn't just sit. She invests the verb "sit" with a new weight. Solidly grounded on the floor, she rakes toward her one object after another. She examines each one closely as if searching for clues. Then, suddenly, she loses interest ("That's not it," she seems to say to herself) and reaches for another toy. As she does, the first one drops. Katie watches her hand as she lets go. Then on to the toy in the other hand.

She likes to tour the house in the arms of her father, George, looking for objects to examine. George loves to hear her responsive babbles as he describes these objects to her. He would never have dreamed there were so many things you could say about a jar of pencils or a coffeepot. He cherishes these tranquil moments with Katie.

*That was at six months.*

Katie at eight months is a different story. Even when she sits in her high chair, now, she needs George. These days, she watches the objects as they fall, peering over the tray with rapt attention. Indeed, the primary appeal of all her expensive toys seems to be their ability to fall when dropped. She empties her high-chair tray and then yells for George to pick everything up for her to drop again.

And no more walker. Here comes Katie,

performing the military crawl on her belly. Sometimes she "inchworms" about. She's much harder to keep track of. Earlier this morning George panicked before realizing she had disappeared under the bed.

The things she finds while crawling fill George with terror. Over the last two days he has extracted from her hand a penny, a transistor battery, a walnut shell, two roly-poly bugs, and what appears to be a lump of dog food, which has totally mystified George because they have never owned a dog.

Limits? Katie doesn't know the meaning of the word. Shouting "No!" evokes from Katie a delighted grin, the same grin with which she rewards a retrieved clothespin or a Cheerios. George spends a lot of time picking Katie up and putting her down again in a safer spot. She reminds him of a wind-up toy: set her down and watch her go.

George is having a great time but he's feeling his age. His back hurts from lifting Katie, who is no lightweight; his knees creak from bending to pick up her dropped toys; and her fascination with ever smaller objects—can babies see molecules?—has forced him to think seriously about bifocals.

But he feels very special towards Katie, as do her mother, grandfather, and nanny. This specialness is reinforced by Katie's reaction to unfamiliar people.

When someone else approaches Katie, she first stares at them, and then her face crumples into a worried little frown. If the newcomer makes the error of staring back, the frown turns into tears, and in a moment Katie is clutching at George, burrowing into his shoulder and weeping. Suddenly brave, she turns, takes a fresh look at the other person, and howls anew.

George was flabbergasted the first time it happened. Now he anticipates it, and tends to call out to approaching adults (such as visiting aunts, the rabbi, friendly people in the checkout line), "Don't make eye contact!" Not a few of these adults think George is pretty strange.

Another thing about Katie: she insists on keeping Liz, George, Grandpa, or nanny in view at all times. Or at almost all times. Her special rule seems to be: I can leave them, *but they must not leave me.* A common adult distress call at Katie's house is, "I have to go to the bathroom. Can you take Katie?" The alternatives are to place Katie in a crib or playpen and listen to her frantic howls, or to take her along into the bathroom, an exciting and dangerous enterprise. George wonders whether Katie is paranoid or insecure, and what he and Liz have done to produce this distress.

He confides his fears and observations to Katie's mother, Liz. She agrees that life with Katie has changed, but is more romantic about it than George. "All of a sudden, she has awoken. She's a real person. It's like Sleeping Beauty."

## Separation Issues: The Velcro Baby

Separation anxiety coincides with the baby's new ability to get around on his own. It's as if the baby's pleasure in doing so boomerangs. The power and control a baby feels in being able to leave is outraged when the baby's beloved adult tries to pull the same stunt. An adult who leaves violates the baby's sense of autonomy. Put this into the modern context of travel, daycare, single parenting, and the dissolution of extended families and community bonds, and you have a volatile mix.

All of this is complicated and made even more fascinating by the fact that individual babies and families differ so markedly in temperament and style when it comes to these issues.

Katie's style is about mid-range. A few babies—Tony is one—pass through this period with total equanimity, greeting the wide world and its inhabitants with broad smiles, and allowing their parents to leave them with little or no fuss. And a few babies are as sensitive as Yolanda, who not only howls upon the approach of a newcomer or the temporary absence of her mother and father, but trembles and turns pale.

Parents always worry. Tony's are concerned that he hasn't really bonded with them; that everyone is equally special to Tony. Any observer, though, can see that Tony saves his most special smiles and chortles for his parents, that he is quieter and less effervescent with others. He sucks his thumb a lot when his mother leaves the room, and looks about for her. And while he consents to being held by someone else, he continually glances to his father as if for reassurance.

But Tony's parents still worry that starting him in daycare early has warped his capacity for intimacy. They make increasing

efforts to spend time with Tony, talking to him and showing him things.

Yolanda is way at the other end of the spectrum. Her parents are both embarrassed and genuinely upset. They have tried everything. The pediatrician cracks open the exam room door and sidles in with averted eyes, making reassuring coos. Visiting grandparents remove scary eyeglasses and grope their way towards Yolanda, dangling gift offerings in the hope of distracting her from the disturbing fact of their existence. Her mother and father play endless games of peekaboo and hide-and-seek, trying to get her used to brief separations.

They worry that they have warped her by holding her too much in early infancy and by not introducing her to people outside the family when she was younger. All their friends have babies who go to daycare; perhaps, they worry, they have overprotected Yolanda.

In my experience, *all* parents worry about first babies this age. I'd go ahead and give in to the tendency, but realize that you would worry no matter how your baby behaved in the face of strangers and separation. This bit of perspective will come in very handy over the next few decades.

Most of the time, learning to cope with strangers and separation simply comes with time and repeated experience.

It often seems to help babies in such straits to design a sort of tutorial in handling strangers and separation with repeated, calm exposure to other people.

Very shy babies like Yolanda, who may indeed have become "overattached" to their nuclear families, may benefit by introducing a gentle, healthy, social toddler or preschooler as a friend. The friend can visit. Yolanda can visit the friend at his house. The friend can come along on scary occasions, like pediatric visits. Often, a baby like Yolanda, after watching and smiling at her "friend," will become more adventuresome and social herself.

I know I've said that such social contact isn't necessary, and that babies can catch viruses even from healthy-appearing young children, but such a trade-off in this case is worthwhile.

Who knows why this works?

Perhaps some babies like Yolanda have inherited shy temperaments from their parents. Perhaps such shy parents give off subtle anxiety signals around strangers. Perhaps the toddler-friend acts like an animate security object, much the way a volatile thoroughbred horse is soothed by having a calmer animal, like a goat, as a stablemate.

For whatever the reason, I have seen again and again that such shy babies aren't shy around small children; in fact, they are fascinated and unthreatened. And when the preschooler fearlessly approaches adults, the baby seems to regard *them* as less threatening, as if the preschooler has decontaminated the adults.

Whatever the baby's temperament, the job of this age seems to be to integrate each new thing from a solid basis of familiarity. Thus, the needs of this age are twofold: that solid basis of familiar, regular life, with a predictable schedule and a few loving, well-known caretakers; and opportunities to explore and integrate new people and things.

Adults can help by being sensitive to the etiquette of stranger anxiety. A key sign that the baby is engaged in integrating a new person or thing is The Stare. It is terribly hard not to return The Stare. But if you do, the baby, utterly undone by the eye contact, will react. His stare will turn into a frown, and then tears.

The proper etiquette is to sit quietly, gazing elsewhere, until the baby stops staring and looks someplace else. Babies often signal, with a sigh or a blink before they turn away, that they have completed some kind of mental process. Then you can approach gently, still not making eye contact, and offer her a toy or object to look at. Once she consents to examine your offering, and then looks up at you, you may return her gaze.

On the other hand, changes in what a baby considers a Normal Human Being can upset him greatly. Glasses, beards, hats can all reduce a young infant to horrified wails. The solution: discard the hat and glasses when you are first being introduced, and then put them on again while the baby watches. Beards and bushy eyebrows just need a lot of working through.

Practice separations are also helpful. I have never been much impressed by the use of games of peekaboo and hide-and-seek to teach babies to cope with separation. What I do think helps is setting aside times for practice separations, where the baby is brought to the home of a (healthy nonsmoking) friend, relative, or neighbor, and allowed to get over any stranger anxiety while the parent stays with him. Then, with a brief, cheerful good-bye, the parent leaves for a brief period. When this is done regularly, and especially if more than one friend or neighbor volunteers, the baby gains skills in ability to separate.

For separation at bedtime, see *Sleep*, below.

## Limits: Setting the Stage for Saying "No"

Setting a limit on a closely bonded baby's activities can seem like a violation of the bond itself. It sometimes happens that parents respond either by not setting limits at all, or by deciding, often unconsciously, that when the baby violates a limit she is rejecting the parent. To a parent with such a conviction, the very act of setting a limit means that the bond has to be dissolved. Then limit-setting becomes punishment, with slapping or spanking, not discipline. Setting limits effectively for such a young baby may thus seem easy, but really implies a big step in maturing parenting skill.

A baby needs to get clear on one basic fact at this age. That is: there are certain behaviors that are not permitted. This needs to be so definite and clear that it becomes part of Cherub's world view—very similar to the force of gravity. So you are not in the business of teaching specific rules, but rather the concept that Cherub is *not* the person in charge here. She is not the one who makes decisions about what is permitted to happen. Parents (or appointed grown-ups) do that.

Within the array of permitted behaviors, Cherub gets to choose which ones to pursue.

The keys of setting limits at this age, then, are:

- **Help her to establish an ability to tolerate frustration.** When the baby becomes upset because she can't reach, fix, do, or manipulate something to her satisfaction, give her a chance to solve the problem herself. You can substitute your reassuring, guilt-free voice for your immediate presence.
- **Don't go overboard in soothing her expressions of frustration, but don't punish her for them either.** Think of your response as a teaching device about life. Try not to take her frustration personally, even though she glares at you as if the immobility of the footstool was your own nasty fault.
- **Get rid of obvious causes of recurrent frustration.** This means removing temptation. Baby-proof, baby-proof, so you don't have to be constantly clearing a path, rescuing either the baby or the Waterford crystal. Some "friends" may tell you that this is spoiling the baby, that "he's got to learn to live in a civilized way" right from the beginning, but they are setting you up for disaster.
- **Use distraction and substitution as much as you can *before* he gets into trouble or becomes frustrated.** There are two reasons for this. First, a baby's approach to the world and to learning is really taking off now, and the more accessible and interesting that world is,

## SAYING "NO!"

When she really has gotten into something absolutely forbidden, don't give a warning. Say just the one word: "NO." Be firm, but try not to shout. At the same time, immediately, move her. Or move the object. Do not give an explanation or go into detail. Just one word. Babies love to hear adults talk to them, and if you go on and on about what she did or why you said No, she'll go and do it again just to hear you talk.

the better he'll do intellectually. Second, if you have to keep setting firmer limits, moving him constantly, pleading with him, slapping or spanking, you will become annoyed and angry. The baby will sense this and will become less secure in his sense of being the Wonderful Center of the Universe. This may preoccupy him so he can't learn as well as he is designed to. It will also make it much harder for him to learn to separate, because physical punishment is too scary a reminder of the emotional separation he already feels.

• **Try saying "EH."** Well, yes, this is how you train a dog. As the baby approaches something forbidden, just say, "eh," in a calm but slightly warning voice. He's likely to stop and look at you for a signal, and you respond with a sympathetic but warning glance. A baby this age is often sensitive enough to detour off to something else.

One final word about limits. All babies this age explore their genitals—i.e., masturbate. This is normal and expected and should not be reprimanded. Parents might try gentle distraction and substitution when they have to change the diaper. By the way, I have *never* seen a baby boy actually pull them off, but I've seen plenty of worried parents. Try to relax about it. Another piece of advice: don't be too upset if the baby seems interested in his own poop. Don't overreact. Just wash his hands well. Don't give the impression that the poop is disgusting or that you are scared or repulsed. The baby may think that it is the Baby you are reacting to.

## Day to Day

### MILESTONES
Babies this age can see well, but still do not have 20/20 vision. A baby ought to be able to follow a small object the size of a pencil eraser. Once seen, such an object is grasped and then eaten. Or inhaled. Beware.

### Hearing and babbling
Normal hearing allows a baby this age to increasingly associate words with people and

things. By nine months, babies respond to their own names with recognition; they don't just turn in the direction of the sound. In the office, I see this when the baby sits on the floor, occupied with toys and drool, while the parents and I talk quietly. When her name is mentioned, she looks up inquiringly.

By nine months, a baby's babbling includes consonants as well as vowels. If it doesn't, she needs to be checked for subtle fluid in the ears, and have a hearing check. A very mild hearing loss (as happens with middle ear fluid) keeps babies from hearing consonants, and thus from reproducing them.

When talking with a baby this age, try to guess what she might want to say, then say it for her, and then respond. Here is a typical conversation George has with Katie. "Katie finds the ball. The big blue ball. Katie makes the ball roll. Roll, ball, roll! Katie wants Daddy to get the ball out from under the table. Here goes Daddy! Here, ball, come here ball, ouch."

### Motor skills

Babies learn to grab objects first by raking them up and holding them with thumb opposed to all fingers. Over this period, they perfect this grasp. Most babies learn to really "pinch" a tiny object between nine months and a year. Babies have trouble letting go of an object at this age. If you want to keep your baby from grabbing his poop or your nose or earrings, just give him two toys, one in each hand; he will stare at them, stymied, unable to let go to grab the forbidden object. Babies do not show a definite, persistent preference for using one hand over the other at this age; if your baby does so, bring it to the attention of the pediatrician.

By about seven months, most babies can sit solidly and use their hands to hold objects rather than to prop themselves up. They can also roll from back to tummy (as well as tummy to back) during this period, and can get places by rolling to them. By seven months, nearly all babies will bear their weight on their legs when you stand them up. By nine months, most can pull up to a standing position. Once the baby learns this trick, he'll pull up on anything or anyone, including the dog.

Babies in this age range have a great variety of ways to get from one place to another. Many, of course, crawl. Some crawlers go by the book, on hands and knees. Some prefer the Suspension Bridge, on hands and feet; others are like Katie, the Inchworm. A few perform the Crab, sideways, or scoot on their backs. Given that your pediatrician has checked out the baby and pronounced everything in good order, all these styles are perfectly normal. They don't have any predictive value about age of walking or intelligence.

A large minority of normal babies with fine intelligence and athletic abilities do not go through a crawling stage. They roll, or squirm, or just sit around until they get up and walk at between nine and sixteen months of age. If the baby is doing everything else just fine, and has had regular pediatric checkups, can sit unassisted while holding a toy, bears weight on his legs when you stand him up, and is getting plenty of attention and incentive to explore, don't worry about the absence of crawling. Despite any rumors you have heard, this does not mean he will have trouble reading later on, nor that you need to teach him to crawl so as to avoid academic problems later.

## SLEEP

Most babies this age sleep about fourteen hours out of twenty-four, about ten at night and four in the daytime, divided into two naps. And with all the new things going on, they and their caretakers need every minute of it.

If you have decided on the Family Bed, there are many spokespeople to help you along. If you value your own private bed, you may be having some dilemmas.

### Getting through the night

Separation is harder at night. Babies at this age, and in the next few months, who have not yet learned to fall asleep in the crib will find it much harder to learn to do so ·now. And they will make it harder on their parents as well. A baby who cannot fall asleep on his own is very likely to awaken the parents every time *he* awakens in the middle of the night. And starting at about seven months, some of these babies will not just holler; they'll stand up and shake the crib rail and holler.

Desperate parents often ask what to do when the baby awakens them in the middle of the night. This is not the right question. No matter what you do in the middle of the night, you won't solve the problem.

The problem is getting the baby to learn to fall asleep on his own in the first place, without being cuddled, and without bottle or pacifier.

And that is not going to go smoothly until the baby can separate calmly in the daytime. It is neither logical nor kind to ask him to do so, suddenly, at bedtime, when it is dark, parents are tired and slightly impatient, and he's had close attention from adults all day along.

Such a baby needs practice in separating. Part of this practice is not holding him or picking him up a lot in the daytime. Other physical caresses can be substituted for cuddling. He can learn a little delayed gratification by not having his attentional cries answered so promptly he hardly gets a chance to yell. He can learn that every whim doesn't get gratified. He can start getting more and more pleasure from his explorations of the world, rather than just from his interactions with adults. Another part of this practice is visiting other homes and being cared for by other adults on brief but frequent occasions.

When putting the baby to bed in her crib, some parents prefer to stay in the room; others go in every few minutes with words of reassurance, lengthening the time between "visits" each night. In my experience, neither of these approaches work. It just makes Cherub angrier and angrier. I suggest, instead, the "Ferber" method (Richard Ferber, M.D., is the author of *Solve Your Child's Sleep Problems*). That means three nights of letting Cherub discover that you are not going to come back into the room.

This is often called "Making the baby cry himself to sleep," which sounds cruel. I don't agree. Many babies this age have learned the skill of falling asleep on their own long ago: it is an age-appropriate skill. You are not abandoning Cherub, nor causing physical pain. Cherub is not terrified; Cherub is furious. Listen to him. Is that the cry of terror—the cry he gave when that big dog came up to him, or when you accidentally doused him with cold water in the shower? I think not.

Think of it this way. Cherub needs parents whose relationship is happy and fulfilling, and who are full of energy and good will. At the minimum, these achievements demand a good night's sleep. It is in Cherub's self-interest to allow that to happen.

To help set the stage, and give you a clearer conscience, however:

- **Start to have a ritual at bedtime,** with gentle play, a cuddle or song, and then into the crib without a pacifier or bottle, but with cuddly toys, a special blanket.

- **Communicate to the baby by tone of voice and body language that the day is over** and that sleep is desirable, not a punishment or deprivation. Also, that whatever you are going to do now on your own or as a couple is infinitely boring, and the baby won't be missing anything.

- **Don't allow her to fall asleep other than in her crib.** A baby even this young can soon become entranced by evenings in the living room or family room. If she doesn't know what she is missing, she is

less likely to get out of her crib at eighteen months searching for nightlife.

- **Consider having her start sleeping now in a sleeping bag.** There are many advantages. She won't get cold kicking off the covers. She won't get her pajamas twisted around uncomfortably, and a wet diaper is less likely to be cold and nasty and awaken her. And if she starts sleeping in one now, when she is a year-and-a-half she won't be able to get out of her crib in it! (Starting a sleeping bag at a year-and-a-half won't work; she'll rebel.)

- **If she does wake you at night, it's important to go in right away, as soon as she starts to cry.** If she is in real distress, sick or uncomfortable or frightened, she must be taken care of, of course. If she is crying for attention, and stops at the sight of you, whining or beaming and holding up her arms, try to be as boring as possible. Your voice is slow and not inflected, your eyes are heavy lidded. You move slowly, and pause before you move or speak. You are underwater. Think of yourself as being contagious with sleep. Hear in your mind the voice of Fred Rogers, but even slower. Pick her up if you must, but then put her down very slowly and pat her hypnotically, then leave the room.

- **Try helping her use a security object.** (See *Window of Opportunity,* below.)

### Night terrors

A few babies may have their first Night Terror now. This is a sudden, incomplete awakening from a very deep stage of sleep. The baby screams and thrashes and can't be consoled. These are different from bad dreams, in which a baby reaches out for comfort. Please read about *Night Terrors* in Part II of this book. At this age, I'd have the baby checked the next day for medical causes of partial awakening—a sudden pain, perhaps from ear infections or constipation, that may have awakened her.

*Naps*

It's a good idea to try to keep the sleep and nap schedule regular now. A baby who skips an afternoon nap and goes to bed early may well wake up too early the next morning, take the morning nap too early, and get meals and naps out of kilter from then on. If such a pattern is developing, anchor it by sticking to the meal schedule, and by waking the baby at his regular time even though he may want to sleep later, and not letting the afternoon nap last later than its old pattern—usually not later than two-thirty or three P.M. After a day or so, you should be back in the old groove.

## GROWTH

Growth slows down a bit now. Most babies will gain about half an ounce, rather than a whole ounce, a day.

Also, babies often start to show their genetic growth inheritance more now. It is especially common for babies who started out life very tall, but whose parents are average or short, to start slowing down. I have often seen this come as quite a relief to parents who may have wondered where those tall genes came from.

Chubby babies generally start slimming down a bit too. A chubby baby who is still gaining an ounce or more a day is probably eating from reasons other than hunger. Snacks and juice are a big reason, as it can be tempting to use these to comfort, distract, and amuse. Restricted physical activity can be another reason. As soon as the baby can creep, squirm, crawl, or roll from one place to another, she ought to be free to do so for large parts of the day.

If the baby is very chubby, and/or if there is a family history of obesity, I'd ask the pediatrician, and I would not be content with "it's just a phase; he'll slim down." See the essay *Growing in All Directions* in Part III.

Thin babies may be following a family trend, but if a baby is getting proportionately thinner at each well visit, the pediatrician will probably wish to do some investigation, both an analysis of food intake and basic lab tests.

## TEETH

Most babies get their first teeth about now (though some perfectly normal babies won't get any until a year of age, or rarely even older). These are usually the two central bottom teeth. They often look very raggedy at the edge and are set like a V instead of in a straight line, and that's usually just fine.

Now you have two kinds of teeth to deal with: visible and invisible, unerupted teeth

> Honey should not be given to children under a year of age, due to the risk of the rare disease infant botulism. For fish-consumption guidelines please see 7.

forming in the jaw. There are three keys to protect both kinds of teeth:

- **Make sure the baby is getting the right amount of fluoride.** If a nursing baby who has been receiving fluoride drops goes to formula, the need for fluoride may or may not change. If your community's water supply contains fluoride, your baby may get enough from taking formulas prepared with that water. Soy formula concentrate has fluoride built in. Babies on soy should not receive extra fluoride—they should not drink fluoridated water, nor should the formula be prepared with fluoridated water, nor should they take fluoride drops. Make sure that the baby isn't getting too much fluoride from the toothpaste you brush with. Use no toothpaste or a very small bit of fluoridated toothpaste or nonfluoridated toothpaste.
- **Start cleaning the teeth two or three times a day.** It's easiest and most fun to do this with the baby lying on two adult laps. Use a soft-tufted toothbrush, or the over-the-finger version. Be sure to get all surfaces of the teeth, not just the fronts. (It's like painting a chair; you keep finding surfaces.)

- **Respect the importance of the baby teeth,** even though they are going to fall out, eventually. They not only pave the way for the permanent teeth. If cavities become rampant in the baby teeth, infection can spread and damage the permanent ones. Fluoride and tooth brushing are only part of the tooth-watch. It's also important to limit the amount of time baby teeth spend coated with sugar-containing foods and liquids: see my diatribe against juice as a beverage, below and elsewhere.

For the issue of fever with teething, see *Fever,* under *Health and Illness,* below.

## FEEDING AND NUTRITION

### Milk

Most babies need about twenty-four ounces of breast milk or formula (iron fortified!) daily. Whichever they're drinking, put some of it into a cup. Don't let the baby get the idea that only juice or water comes from a cup, and that milk only comes from breast or bottle.

Learning to drink from a cup is a fun, messy skill; the baby won't usually be able to hold the cup until ten to fifteen months of

I hope you have already internalized the Golden Rule of the Good Eater, as formulated by Pediatric Nutrition Guru Ellyn Satter, and its amendments: Adults decide **What** choices of food are offered, and **Where** and **When** they can be eaten. Babies and children decide **How Much** of each food they want.

*Nathanson's First Amendment:* If a baby seems to be forming finicky tastes (taking only yellow veggies, for instance), simply stop offering the Addictive Food as a choice.

*Nathanson's Second Amendment:* It is easier to avoid introducing seductive foods (such as fruit juices, baggies of Cheerios to munch on, catsup on anything) than it is to try to limit or remove them later.

age, but if you hold it at least you're giving him the incentive to learn.

Some pediatricians suggest starting whole milk at six months of age in formula-feeding babies. It is certainly less expensive than formula. However, the Academy of Pediatrics recommends strongly that babies stay on breast milk or iron-fortified formula until their first birthday, for several reasons:

- There's no iron in milk, and the milk itself prevents other dietary iron from being well-absorbed. Iron deficiency can be present long before anemia is reflected in a blood test, and can cause behavioral changes and impaired development.
- The fat and protein compositions in cow's milk are farther from those in breast milk than those of formulas. Whether this is a subtle, important, or irrelevant difference for growth and development has not yet been well elucidated.
- A baby's intestine is still immature and lets larger protein molecules enter the

bloodstream than it will later on. Some evidence suggests that the earlier the exposure to cow's milk, the greater the chance the baby will develop an allergy to cow's-milk protein.

### Solid Foods

Think of this period as the age in which a baby discovers that she can tell when she is hungry, what she feels like eating, and when she is full. It's a time when she can come to tastes and textures with a comparatively open and innocent attitude. She has not yet been corrupted by sugar, salt, and fat. She hasn't yet been confused by "grazing" on crackers or juice.

On the other hand, she probably already has a basic orientation to the diet prevalent in your particular home. If she is nursing, she gets secondhand tastes of whatever her mother has been eating. Nursing or not, she smells food cooking or (if you don't cook, just heat up or whatever) being served. She has a rudimentary idea of the approach to eating at your home. Is it social, with people gathering at meals? Rather haphazard? Fun? Tense?

This might be a good time to consider in a relaxed way whether or not you want to make changes in your family's eating patterns before the baby becomes habituated to aspects you yourself don't like. If you have a family history of obesity or eating disorders, this is a good time to make a special appointment, maybe without the baby along, to discuss this aspect with your pediatrician or with someone, a nutritionist or nutritional therapist, the pediatrician recommends.

It's also a good time to confront your ideas about Mess. Feeding at this age has to be messy. A baby has to explore food's texture, smell, taste. She has to learn to get the right amount on her fingers. Eventually, to get the spoon in the mouth. She'll bang it on her tray and fling food with it at first.

She has to do it because she has to learn that she "owns" her hunger and the ability to satisfy it. Keeping this in mind may help you to bear with the mess. Put down newspapers on the floor. Rather than bibs, have her wear only a diaper, and take her in the shower afterwards.

Let her know you think eating is fun, and she's doing a delightful job at it. Don't coax her, urge her, or even praise her too much. Let her own it!

As to the nutritional aspects:

- Three meals a day of up to four ounces each. You can use prepared baby food or puree or mince cooked food you make yourself. Don't add sugar or salt to anything, but spices are fine if the baby likes them. (Of course, not bottled condiments like catsup and Worcestershire sauce and mustard. These are all pretty strong stuff and contain lots and lots of sugar and/or salt.)

- The texture ought to be smooth until you notice, at about eight months, that the baby is making chewing motions. Then you can introduce soft lumps for texture. But never anything chewy, crunchy, stringy, sticky; nothing in size/shape that she could choke on, like a slice of hot dog or a grape.

- The nutritional requirements of babies this age are, as far as we know now, slightly different from children who are older than age two. In most respects, you don't have to worry; if your child is eating the food groups described below, and taking breast milk or formula as outlined above, he is getting all those minerals and trace elements and so on. The main different feature about babies this age is that they are allowed much more fat in the diet than children over two, older children, and adults. A baby this age can have from 35 percent to 50 percent of his calories as fat. Much of this fat comes from the milk, so you don't have to worry about purposely adding fat to the foods.

- As to variety, think of four major food groups, and try to serve a representative from each every day. Cereals, bananas, and yellow veggies are often constipating, so you don't want more than one of each every day. Make up the rest of the "solids" intake with other foods.

- **Cereals:** Rice, oats, barley, wheat, white potatoes, tapioca, grits. Some pediatricians discourage wheat and corn products until nine months to a year: check with yours.
- **Fruits:** Applesauce, pears, peaches, bananas, apricots, plums, prunes.
- **Green veggies:** Peas, beans, broccoli, kale (steamed and pureed). Spinach belongs with the yellow veggies.
- **Yellow veggies:** Carrots, yellow squash, sweet potatoes. The yellow veggies and spinach are wonderful foods, but if a baby takes more than one helping a day the carotene in them may turn her skin an orange color (*carotenemia*; see Glossary). This isn't harmful, but can be an indication that she is preferring the yellow group to a degree that she's skipping other foods.

Meats and poultry are fine in moderation, but don't need to be one of the major food groups. The baby gets all the protein and B vitamins he needs in his milk, and doesn't have to have any food from this group. If you do serve this group, be sure each portion is boneless, fatless, unsalted, and minced very finely. If you wish, serve a portion each day, but don't feel you have to do so from a nutritional standpoint. Letting the baby subsist as a lactovegetarian, or using animal-based foods only as "condiments," is fine too.

Children should not eat raw seafood or shellfish at all, as they are particularly susceptible to possible food poisoning. Other fish consumption guidelines may be found on page 7.

At this point in our knowledge about food allergies, we are advising parents to avoid highly allergenic foods such as peanuts and nuts until the child is three years old. This may be entirely wrong, but this is where we are now.

Egg yolks are fun, but not very nutritious. Unfortunately the iron in them isn't very accessible to babies, and they are rich in cholesterol. On the other hand, you don't really have to worry about restricting cholesterol until the baby is two. One hard-cooked egg yolk a day would be fine, included as part of the Meat food group. Egg white contains the protein albumin, and some babies may develop egg allergy if you start this before nine months to a year.

Table foods from family meals are fine, again if they are pureed or minced, and haven't been sugared or salted. But avoid highly allergenic foods, even if you can get them into a safe-to-eat form. These include chocolate, peanut butter, and nuts. (They have lots of salt and sugar in them, too.)

*Snacks and "handouts" or "grazing"*

A planned mid-morning and/or mid-afternoon snack seems to fit some babies' personalities. They are light eaters at meals, and need a "pick-up" a couple of hours later. The ideal snack at this age is a portion of whatever part of their meal they didn't eat— but not as a punishment. It all depends on the attitude of the parent. If that sounds too unappealing, a protein-and-fruit snack is another good choice. In any event, serve the snack as a mini-meal, with baby in the high chair with adult company.

But I'd be sure the baby really wants and needs the snack, and that without the snack, she is whiny, tired, not herself.

"Handouts" or "grazing" is quite a different thing. Giving a baby crackers or dry cereal or juice as a constant accompaniment to other activities has nothing to recommend it. It cheapens the sensations of hunger and satisfaction, gets the baby used to constant oral occupation, and doesn't let her be in charge of her own needs. It encourages a baby to go "uh-uh-uh" in a demanding way when she sees the magic container or even hears the magic words "juice" or "cracker." This may be cute now, though I doubt it; but it will turn into whining in a few short months, and that isn't cute at all. Grazing is terrible for children's baby teeth. It makes crumbs. It is too easy a substitute for really paying attention to the baby's activities, needs, whims, and desires.

## BATHING

Many babies this age enjoy their baths, but some become phobic. Perhaps it's that so many changes are occurring during this stage, and the change from dry to wet is the last straw. Perhaps the bathtub is slippery, an upsetting factor for someone just learning to roll around or crawl or pull to stand. Perhaps the water going down the drain gives a sinking, scary feeling to someone preoccupied with the drama of disappearance. Whatever the reason, I'd honor that phobia, and try sponge baths, or baths on your lap, or even showers, for a while.

## DIAPERING AND DRESSING

The first trick for these enterprises, especially diapering, is to occupy the baby's hands before you start. At this age, babies have trouble voluntarily letting go of an object, so if you put a small nonchokable toy in each hand, the baby won't be able to grab handfuls of fascinating poop. (Of course, if there's no poop, let him or her explore genitals freely.) The second trick is to diaper and dress the baby on the floor. I have never known a baby to fall off the floor.

The Early Education Specialist, Burton L. White, has observed many babies in the act of being diapered, and thus in the act of escaping from being diapered. So he knows what works for the Diaper Escaper. Here are his suggestions, as given in his fine book: *Raising a Happy, Unspoiled Child* (Fireside, 1995).

- Before you begin, establish a "baby-game" atmosphere with the baby, talking and smiling. Give her a toy to play with, too. At the first sign she's trying to escape, discontinue the talking, smiling, and eye contact. Don't act angry or disappointed; simply remove all attention, all at once, on her signal that she's about to fling herself over and crawl away.
- If she stops rebelling, resume your social behavior. If she continues, simply get down to business, hold her down with your leg if need be, and get the job done. Don't talk—don't even say "No." And after you have finished, keep your attention removed for a minute or so.

If you do this consistently, there's a good chance the escaping will taper off. Good luck.

## Shoes

If the environment is safe for bare feet, that's the best state for the baby. If shoes are necessary for warmth or protection, the least expensive shoe that fits and is flexible is the ideal shoe (unless your doctor has prescribed special shoes for a special problem). You don't need expensive shoes to "form the foot" or "help the baby learn to walk."

Shoes for decoration just need to be vetted for safety. Babies this age love to eat their shoes, and you don't want any little parts coming off and being inhaled.

## ACTIVITIES

Once a baby becomes mobile, whether she is rolling, squirming, creeping, crawling, or cruising, there is an overwhelming temptation for many parents to enroll her in a gym class or an infant stimulation class or *something*. This temptation becomes stronger with each passing month.

The truth is, babies will learn the important skills of this age such as getting around, handling items, and exploring, without any special classes or equipment. Moreover, while they may become excited by the presence of other babies, they have no developmental need to be around other very young infants.

So such classes are fun for adults, and the babies like to see each other, but there are many down sides.

- It is impossible for parents, especially of first children, not to compare their child's abilities with those of others. At this age range, the normal span of motor competence is huge. Some babies crawl at five months, some walk at eight. Many crawl at eight months and walk at a year, or thirteen months. Some never crawl but sit around happily until they get up and cruise. All these patterns are not only normal but don't have a thing to do with intelligence or later athletic skill.
- Such classes may give parents the feeling that their child wouldn't develop normally if she weren't actively taught by experts with special equipment.
- Babies this age mouth everything, including everything that all the other babies have mouthed. This is a great way to pass on cold viruses that cause respiratory and intestinal diseases—especially in the winter, when viruses like rotavirus and RSV (see Glossary) are rife.

Swim classes have all these problems, and more. Once again, I hate to be a spoilsport, but I wouldn't do it. The reasons are the same as those mentioned in the previous chapter:

- Sudden immersion of the face in cold water can make a baby stop breathing, and make the upper airway spasm shut.
- A baby who does like to "swim" underwater can swallow a dangerous amount, automatically and invisibly. This can cause a serious, dangerous dilution of the

blood, resulting even in convulsions, coma, and death.

- Water up the nose can give rise to ear infections and pneumonia.
- Contaminated water (from the other diaper-clad babies in class) can cause diarrhea.
- It doesn't make the baby "water safe," just "water happy." This may make him more inclined to seek out fun bodies of water, but if he falls in, all the classes in the world won't keep him from panicking.

## TOYS AND EQUIPMENT

Though there is a myriad of toys for babies this age, babies can be perfectly happy with objects adults don't regard as toys but babies do. These include doors to bang, floors to inspect, drawers to open; pans, pots, and things to drop in or out of them such as old-fashioned, nonclamp clothespins; rolled-up socks; laundry baskets; spoons; and plastic dishes.

*Recommended toys:*

When you do buy toys, bear in mind what babies this age really like: anything that can help them work out the mystery of disappearance and reappearance. These include balls of any size to roll back and forth (and someone to participate in this), sturdy books with pages to turn, toys that change as you move them, things to hide behind or under, things to push about from room to room, and soap bubbles that pop.

A climbing toy made of smooth material that is low and sturdy and can be surrounded by cushions is a good toy, and may help you to keep the baby from climbing on the furniture. Some babies want to climb as early as eight months of age. If you have stairs, try putting a protective gate at the level of the third stair, and provide nonslip pads for these first three steps so the baby can practice going up and down and up and down and up and down and up and . . .

Some babies towards the end of this phase like to have an object with wheels on it to push around the house as they practice walking. (The wheeled stool in the doctor's office is just right, alas.) Such a "walker" should have a low center of gravity (so it doesn't tip over when he yanks himself up on it) and no finger-pinching coils or springs. If it stops suddenly and the baby keeps on going, you don't want a jarring bump of the mouth, either, so the height and the softness of the surface should be considered.

*Not recommended:*

Since babies this age are just getting a grip on the real world and cause and effect, fantasy toys don't give much joy. These include, alas, stuffed animals. Little hairs wind up in the mouth, nose, eyes. They get very dirty and dusty. The stuffed animal's primary play value is in being dropped, or as a security object.

If you don't have a playpen, I wouldn't get one unless you are desperate for a brief, occasional restraining device for the baby, or for a location for the VCR, et cetera, to keep it out of the baby's clutches.

## SAFETY ISSUES

If you haven't already baby-proofed your home and stocked your medicine chest as suggested in the earlier chapters, please do so now. Here are some important things to remember when baby-proofing:

- **Babies can pick up smaller and smaller items.** They can choke on foreign objects and swallow spilled pills, eat the dog food (or worse), become fascinated with electricity outlets and cords. Be vigilant, and get child-proof outlet protectors.
- **Some babies can pull to a stand before nine months of age.** So beware the toilet, buckets of water, low fountains. A baby can stand up, lean over, and have the weight of her head pull her down into the water. Babies can drown in only two inches of water.
- **Some babies start to cruise around the furniture before nine months of age.** Put rubber corners over the sharp ones on the coffee table so when she falls and bumps her head it won't bleed.
- **Babies climb before they walk.** This is the time to anticipate the attractions of climbing from the coffee table to the sofa to the bookshelf, or from chair to table to counter to top of refrigerator. Any furniture rearranging that cuts down on the more tempting dangers is warranted.

If your baby is already climbing or walking, see baby-proofing in the next chapter, *"Nine to Twelve Months."*

## Health and Illness

*Part II of this book covers illnesses in babies this age and older: how to assess symptoms, how to tell whether a child is sick enough to require urgent attention, and what kind of home treatment may be appropriate. It also examines problems by body part. The Glossary is an alphabetical listing of medical terms with pronunciations and informal definitions.*

### EXPECTATIONS

It is a rare baby that gets through this period without at least one cold or episode of diarrhea, and many have their first ear infection at this age. Even babies who aren't in daycare, even babies with no older siblings, can get sick at this age, simply because they are beginning to explore on their own. Viruses on surfaces can stay infectious for hours, and the baby who touches the virus and then touches her own eyes or nose can inoculate herself and get sick.

If the baby does get sick, she will probably handle it better than she would have earlier. Many of the same infant illnesses of concern earlier are still a worry: bronchiolitis and diarrhea due to rotavirus are two of them (see previous chapters and Glossary). But at this age the baby is a bit less likely to get into serious trouble, as she's less likely to become exhausted or dehydrated.

More important, a baby this age has established consistent norms of healthy behavior. You can tell when she's cranky, or eating less than usual, or doesn't have her usual energy.

## COMMON MINOR SYMPTOMS, ILLNESSES, AND CONCERNS

### Ear infections (otitis media)

Many babies will have their first ear infection now. Fortunately, statistics suggest that if a baby's first infection occurs after six months of age, it's less likely to herald severe recurrent infections than if it occurs earlier.

Babies with an ear infection are likely to have had a cold first, then awaken in pain at night. The baby may grab or rub or poke her ear, or bat her head, or pull the hair on that side. However, ear infections can present without anything more than behavior changes in sleeping, eating, or mood. Have a high level of suspicion.

Once an ear infection has been treated, it's important to have a recheck to make sure fluid isn't lingering behind the eardrum. The subtle and temporary diminished hearing that occurs when there is an ear infection or fluid starts becoming more important now because receptive language learning begins at this age. See the essay in Part III on *Trouble in the Middle Ear.*

### Croup

Croup is a respiratory illness with a cough that sounds like its name: "Crooooop! Crooop!" The upper airway contracts, and the baby makes a rasp as she inhales. The raspiness is called stridor. It is different from a wheeze, which is a spasm of the lower airway and causes problems exhaling.

Typically, a baby with or without a previous cold awakens in the night. She is frightened, and frightening, struggling to inhale. When she coughs, she sounds like a seal barking or a goose honking.

It's important to make sure that this truly is croup, not something else such as a foreign body obstructing the airway or one of the rare diseases called epiglottitis and bacterial tracheitis. In ordinary croup, the child is vigorous, pink, and upset. She doesn't resist being held—doesn't fight to assume one position of breathing comfort. Nor does she thrash about frantically.

She'll calm down markedly if you give her soothing steam or cool moist air to breathe. You could take her outside into the cool night air (dressed warmly). Or you could take her into a bathroom made steamy by the hot shower. (Remember steam rises; don't sit on the floor or the toilet with her. Stand and hold her.) If you can't do either, you can carefully hold her over a steamy kettle of water, but this is potentially a very dangerous enterprise.

You might want to read the discussion of *Croup* in the *Common Illnesses* section of Part II under *Trouble Breathing.* So that when your baby awakens you at three A.M., frightened and noisily crouping and rasping, you will not be completely thrown for a loop.

### Roseola

Roseola is a virus (Human Herpes Virus [HHV] 6) illness in which the baby gets a very high fever without any other signs of illness—well, maybe a slightly runny nose. The temperature may go as high as **105°**. Typically, the baby remains pretty cheerful.

After three or four days, the fever simply disappears. Often the baby breaks out into a rash after the fever disappears. This is usually a pink flat or bumpy rash on the trunk and face, and it lasts for about twenty-four hours. During the time the rash is there, or for twenty-four hours after the fever disappears even if the rash doesn't appear, most babies are incredibly grumpy.

**The first main feature of roseola is that it is a diagnosis of exclusion. That means that a baby with a high fever, but without other symptoms such as runny nose or diarrhea, still needs to be seen by the pediatrician. What has to be excluded? A serious infection in the blood, urine, or spinal fluid.**

The second main feature is that 10 to 15% of babies with roseola will have a fever convulsion. This is a very scary (though innocent) event. I recommend you read about it in *Frightening Behaviors*, in Part II.

### Common Virus Diseases

There are a number of very common but unpublicized viral diseases of young children, and young babies can catch all of them. See *Illnesses, Both Common and Uncommon* in Part II of this book. In particular, babies are still prone to catching rotavirus diarrhea and RSV bronchiolitis. These are also discussed in the "*Two Months to Four Months*" chapter and again in the Glossary. These are likely to be somewhat less serious in a baby over six months of age.

### Viral diseases with familiar names that are becoming, or are about to become, rarer and rarer

These are discussed in Part II, Chapter 16, under the heading *Uncommon Illnesses with Familiar Names* and include measles, mumps, chicken pox, and German or three-day measles (rubella). All of these diseases are preventable by immunization.

## Common Scary but Usually Innocent Problems

Ah ha! A brand-new category!

It is kind of babies, I think, to wait until this age to indulge in the scarier medical performances. Most of the time, they wait until even later, twelve to eighteen months of age, before manifesting any of these alarming but essentially benign conditions. But now and then a baby this young will show up with one of the following, presumably in order to keep his or her parents fully alert and paying attention at every moment. You can be one up on the baby by reading about these now. Even if your own child never manifests any, perhaps you can be a hero when some other parent calls you in a state of shock and horror.

The three entities in this category at this age are fever convulsions, breath-holding spells, and fused vagina or labial adhesions.

### Fever or febrile convulsions

Fever convulsions are frightening but harmless seizures that occur at the onset of fever in about five children out of a hundred. They do not occur much before six months or much after six years of age. The average age is eighteen to twenty-two months. If anyone else in the family had fever convulsions, the chance of your baby having one is a little higher.

They don't mean a child has epilepsy. They do not cause and are not caused by brain damage. Their primary effect appears to be to scare parents to death.

They usually occur when the fever first comes on, and especially if the temperature goes up rapidly. Often it is the first sign that the child has an infection, so that controlling a fever once the child is sick has little or no role in preventing fever convulsions. The baby's arms and legs jerk, and he is unconscious, holds his breath, and turns blue. The seizures last only a few minutes, not longer than fifteen at the very longest, but it seems like an eternity. See *Frightening Behaviors* in Part II.

### Breath-holding spells

Breath-holding spells can occur any time from six months to six years, but most happen between twelve and eighteen months of age. About 5% of all children will have one. In a breath-holding spell, a minor event (like a scolding or a fall) triggers an involuntary spell of not breathing. The child may cry out or not, may turn blue or pale, and may even have a brief seizure. They are almost always completely innocent. See *Frightening Behaviors* in Part II.

### Labial adhesions or fused vagina

The inner lips of the vagina are very sticky in babyhood. If they get irritated for any rea-son, they tend to stick together and fuse. This can be a very alarming sight: it looks as if her vagina has disappeared! Overnight!

Don't panic. Even if you were to do nothing, the vagina would open up eventually. However, a pediatric visit is in order, partly to determine what irritation caused the labia to fuse. If the baby has any signs of a urinary tract infection, or a history of one, or seems very uncomfortable, the pediatrician will want to treat the adhesions. This means either separating the labia gently in the office or applying hormone cream at home.

## SERIOUS OR POTENTIALLY SERIOUS ILLNESSES AND SYMPTOMS

### Fever

Fever all by itself is still of concern in babies this young. Most pediatricians want to see a baby this age with a rectal temperature over 103° or if a lower temperature is present for more than three days, even if there are no other symptoms whatsoever and the baby is happy, eating well, and sleeping normally.

Most such unworrying fevers are due to viruses, but occasionally there is a surprise, such as a urinary tract infection. Teething, as I have said earlier, has never been shown to cause fever. This legend started because babies this age and through age two are always teething, so that there is usually a tooth coming in whenever a baby has a fever.

A fever in a baby this young that goes as high as 104° requires a prompt, even middle-of-the-night, phone call and a same-day visit, even if there is no other symptom and the baby is happy. If there is any other symptom, the baby certainly needs a very prompt, if not urgent, visit.

### Intussusception: An intestinal obstruction

This is a rare problem. However, it is hard to diagnose at times because many other illnesses have the same symptoms. It's a good idea to have a nodding familiarity with the condition, as your pediatrician may at some point mention it. See the complete discussion in the previous chapter.

### Urinary tract infections

Bladder and kidney infections can appear at any age. They are less common in boys than in girls, after the newborn period, because boys' anatomy protects them: the plumbing and the stool are widely separated. Girls and women, though, are very prone to UTIs.

In my experience little girls are more likely to show up with them starting at this age. This is because it isn't that easy anymore to clean the baby when you change her; she's fighting you every step of the way, and wants to handle everything in that area, and finger-paint too, if you get my drift. So it isn't unheard of for germs from the rectal area to enter the urethra, and so infect at least the bladder.[1]

Often a child with a UTI will have fever and show signs of abdominal pain and/or vomiting. But sometimes she will merely have fever, or be cranky without fever, or not thrive. This is why urinalyses and cultures are performed so frequently on babies with nonspecific symptoms like these.

### Undescended testicles

A testicle or testicles that haven't descended fully into the scrotum yet are not at all likely to do so in the future, and many pediatricians and pediatric urologists feel that surgery can and should be performed between six months and a year of age, both because of psychological factors and to maximize the future health, fertility, and freedom from malignancy of the affected testicle. If the surgery is done this young, it is particularly important that the urologist be very skilled in handling delicate baby tissue (ideally, a pediatric urologist) and that the anesthesiologist be one who works primarily with children or ideally is a pediatric anesthesiologist.

### Iron deficiency

Iron deficiency is potentially serious not just because it can cause anemia but because it can damage development and impair the potential intelligence of the child, and can diminish physical growth as well. A child can be iron deficient without being anemic (see Glossary), though a test for anemia usually will detect iron deficiency as well. It is also possible for a child with an inherited

---

1. It is rare for boys of this age and older to get urinary tract infections because Nature has made the stool relatively inaccessible to the urethral opening, which is pointing in the opposite direction. Moreover, most little boys this age prefer grabbing their penises to finger painting with their stool. Urinary infections are rare in uncircumcised baby boys, and are ten times more rare in circumcised ones.

variant of the usual hemoglobin (such as *thalassemia* or sickle trait) to be anemic without being iron deficient. Most pediatricians check for this condition at the nine-month visit by dietary history and sometimes by blood test.

This concern is one of the main reasons for continuing a baby on iron-fortified formula rather than switching to cow's milk, until a year of age.

### Lead poisoning
A baby who is crawling and mouthing can be counted on to find and ingest exactly what he shouldn't. Chipping paint and plaster can be dangerous if any layer in it might contain lead. So even if your house is newly painted, if any underlying coat of paint could contain lead you have a potential problem. Lead poisoning can occur even at low blood lead levels, affecting intelligence and development, and these effects are only partially reversible with treatment. Discuss testing, prevention, and treatment with your pediatrician.

## MINOR INJURIES
Most minor and major injuries for this age reflect the increased abilities of babies to get around, reach, and grab.

### Burns
Spilled hot food, splashes of hot bath water, and touching something hot (often a curling iron) are unfortunately fairly common at this age. Prevention, of course, is best. For Treatment, see the *First Aid* chapter in Part II.

### Rolling off surfaces
Falling off a slightly raised surface usually is an innocent accident in a baby this age. A fall from four feet or higher, though, always requires a prompt pediatric visit, even if the baby seems fine—statistically, such falls are more associated with occult fractures. Of course, call the pediatrician, but do not have a frenzy of guilt. Or have one, but don't let it haunt you. These things happen. See *Bonks on the Head*, in Part II.

## MAJOR INJURIES

### Choking
Choking is the worst. Avoidance is paramount, and, as noted above, taking a CPR class is vital. Make sure anyone who cares for the baby regularly takes such a class. Review the steps in your mind frequently, and practice occasionally on a doll.

### Water immersion
Babies this age are driven to seek out receptacles of water, from pools to toilets to buckets. When they lean over to splash or admire their reflections or whatever, the weight of their heads, which are the heaviest parts of their bodies, pulls them in. A baby this age can drown in two inches of water.

Prevention and an updated CPR course are your major weapons here. Not just *your* prevention and CPR, but your caretaker's as well.

# Window of Opportunity: New Importance of Toys and Challenges

### GEARING UP FOR LEARNING

In the first few months, babies learn about their relationship with their own bodies, with their beloved adults, and with a few objects. By the time they reach six months, they are aware, on a basic level, that:

- They are separate people from their beloved adults.
- They can control their environment. If they bat at something, it moves; grab it, it comes to them. If they cry, an adult materializes.
- There is a Familiar world and a Strange world, and time and experience and play can turn a Strange into a Familiar.

With these concepts under their belts, the six- to nine-month-old is ready to take on the universe. The more he can enjoy separateness and cope with its anxieties, the more he can control his environment. The more she turns strange into familiar, the more eager she will be to learn and try new things from now on.

Parents who want to gear up for the amazing expansion in development that takes place from now through age two need a very special caretaker for the baby. And that's true whether the caretaker is mother, father, grandparent, other relative or friend, or someone they choose and employ. Here's what to keep in mind:

- The person who's with the baby most of the time has to think the baby is wonderful and be "in love" with the baby. The baby needs to have that security as she ventures into exploration and mastery.
- That person needs to talk to the baby a lot.
- That person needs to *design* an environment rich and safe enough for the baby to explore without either boredom or real danger. She needs to act as a resource for the baby, showing and explaining how parts of the world fit together and work.
- That person needs to be willing and able to set limits appropriately and to help the baby learn a bit about delayed gratification and other people's needs.

### COMFORT (OR SECURITY OR TRANSITIONAL) OBJECTS

How old was Linus when he became attached to his thumb and blanket? Probably between six and nine months of age. A soft, strokable something that can be used instead of a person is a very handy comforter for somebody just discovering separateness and strangeness. It's a way station between having to have someone else comfort you and being able to comfort yourself.

When a baby has lots of love and attention and adventure, a security object is a healthy, helpful item, allowing routine separations such as for nap or sleep to be friendlier, and to foster a little bit of independence and maturity. Many studies have shown that such security objects can endure all the way through childhood and even adolescence, without any sign that their owners are emotionally insecure or deprived or otherwise warped.

When a child undergoes a great deal of unusual stress, such a security item can be of immense importance: hospitalization, parental discord, illness—these all can be survived a little bit better with the security object.

If parents feel embarrassed or self-critical or guilty about their child's security object, it may help to remember that such an object is to the child "a piece of Mommy" or "a piece of Daddy," and if the attachment to Mommy or Daddy weren't pretty terrific, the object wouldn't have much use.

To help a baby having trouble with separation attach to a security object, try using the same blanket or whatever to stroke and soothe him when you're cuddling, feeding, nursing. Then leave it in the crib with him. Just a couple of warnings:

- Safety is paramount. Nothing in the crib should be big or stringy enough to smother or strangle. Or fragile enough to pull off a piece off, which could then be inhaled or swallowed. No fur.
- Don't let it get too grungy, because part of its "magic" may come to be its particular, poignant, and all too pungent smell.
- Try to coax the baby to like an item that can be cut into pieces so that there isn't catastrophe if one piece gets lost.

- Try to guide the choice. One little girl in the practice became very attached to a pair of her mother's silk underpants. Until they became so worn their identity was concealed, we were all faintly embarrassed by the waiting-room scene.

## What If?

### STARTING DAYCARE

If a baby needs to start daycare for the first time during this period, it may be better to do so early on, since stranger and separation anxiety peak at about eight months.

The best situation, the ideal, is still a one-to-one relationship. Next best is a small group, a total of four or fewer, in which the other children are either age-matched infants or older than three, so that the caregiver isn't totally involved in setting limits and monitoring toddlers. As always, the caretaking situation must be safe. Get personal references. Do not tolerate secondhand smoke. Make sure the caretaker is up-to-date on CPR. Hygiene and handwashing are crucial.

- Pick the caretaker carefully: see *Gearing Up for Learning* in *Window of Opportunity*, above.
- A baby this age needs to be introduced slowly to a new caretaker. See the section on *Separation*, below, and the proper etiquette of introduction.
- Make sure the caretaker agrees with you on how to set limits in an age-appropriate way. See *Limits*, above.

- From now on, it is more and more likely that your ideas will be at odds with the daycare person's ideas. Open communication is vital.
- Babies thrive best in very stable situations where the schedule is regular and life is predictable.
- The environment needs to be truly safe, not only to keep the baby from being injured, but to avoid too many restrictions on exploration. (This may mean finding a caretaker who has neither a house full of precious knick-knacks nor a pet. Even the kindest dogs and cats can lash out impulsively at a tail-pulling infant.)
- The caretaker should understand that children under the age of two-and-a-half have no social skills and need constant supervision; she shouldn't expect toddlers and older infants to "work things out on their own."
- She should avoid the snack trap. A planned snack, nutritious and scheduled, is fine; the indiscriminate offering of juice, crackers, and dried cereal leads to dentist bills, weight problems, and ingrained snacking habits. See Feeding, above.
- If she transports the children, every single one needs a car restraint that they can't get out of.

## Extended Separation from Parents

This is a very hard time for babies to be separated from their best-loved adults. Like Katie, babies this age seem to have a rule: It's OK if I leave them, say, to crawl into the bathroom, but it's not OK if they leave me. If at all possible, I would try to defer a lengthy separation until this intense phase resolves a bit over the next few months.

If the separation is necessary, the next best move is to make sure that the baby is well established with a loving, patient adult who thinks the baby is absolutely wonderful, and who is not conflicted about setting limits in an age-appropriate way. Even then, parents need to expect that the baby will be very angry with them when they return. Some babies will have a tantrum and act out anger, but it is more common for a baby to turn away, reject, and ignore the returning parent. This can be most upsetting, but on the other hand it is a compliment and a statement that the bond is the strong one it should be. Patience and wooing and time will win the baby back, but it may take more than a week.

## Traveling with the Baby

This is likely to be easier now than it will be in a few months, when the baby insists on being walked, walked, walked with you holding his hands up and down aisles, pausing only to eat old chewing-gum wrappers.

A variety of novel toys that do things when manipulated, indestructible books with pages to turn, and soap bubbles are good traveling items.

## ADDING TO YOUR FAMILY

If you have a new baby in nine months, your present one will be eighteen to twenty-one months when the next is born; and twenty-seven to thirty months old when the new one starts to explore and test limits.

An eighteen-month-old is very parent-intensive. He is going through another peak of separation anxiety, which may make the mother's second delivery tough on everybody.

Eighteen Months is working on very important social and intellectual skills. These skills should be well developed *by age two*, so this is an important six months for him.

- He needs to learn how to get an adult's attention in an appropriate way, to ask an adult for help without whining or screaming, and to express annoyance, anger, and affection in ways that others can accept and respond to.
- He is developing language at a great rate, and the language he uses will reflect exactly what he is used to hearing. A fatigued, impatient, harassed parent may pass on highly undesirable and embarrassing language.
- He has a basic idea of compassion that needs to be actively encouraged. He doesn't yet really understand that another person has a separate and different perception of reality. He may comfort a toddler who is crying, but then turn around and pinch another friend fiercely and not understand why that friend becomes upset.

- Before he can learn to share, he has to figure out the concept of Possession, and will hold on to objects for dear life while consolidating the idea of Mine. He is not easy to negotiate with at this time.

Not only does Eighteen Months need a great deal of patient adult attention to acquire such skills, but he has little experience to tell him what is dangerous and why.

With a newborn sibling, then, the toddler may have a hard time getting the intensive parenting he needs over the next nine months. He may learn to scream and whine and have temper tantrums, and his newly developing language may reflect parental inflections of annoyance, and be heavy on such phrases as "Stop that this minute or I'm going to smack you one!" Or worse.

By the time this older child is Two, the difficulties with the new sibling may have made these undesirable traits engrained. Moreover, the older child may not have had the guidance necessary to play on his own and with other toddlers, so he'll have little to distract him from the new baby. And he may not feel very good about himself. Unfortunately, just at this time, and just when he ought to be emerging as a pleasant, sociable person, good company to be with, the younger sibling becomes exploratory and mobile. Since the older child is still highly possessive (see the "*Two Years to Three Years*" chapter), forages into his toys and possessions are likely to wind up in violence and tears.

Parents with excellent health, inner resources, and outside help can certainly work through this period and give both children the parent-intensive care they need. But single parents, parents who will need full-time daycare for both children, or who cannot afford part-time daycare for rest and recreation, may be very sorry they spaced the children so closely.

## DIVORCE, SEPARATION, AND CUSTODY

While children are never ultimately responsible for their parents' divorce, sometimes developmental aspects may play a role in initiating the issues that force separation. The newly developed separation anxiety shown by a baby this age may be the last straw in making the less-attached parent feel rejected, both by the baby and by the more-attached parent. This feeling of rejection, of course, can become a self-fulfilling prophecy. I say this in the hope that if a parent recognizes this phenomenon, communication and marital counseling may be given a try.

If separation and divorce are a reality, parents can help a baby most by recognizing that such a dramatic and sad separation by the two adults can make it more difficult to deal with the baby's separation anxiety and testing of limits. Certainly, being conscious of this can assist both parents in obtaining help with these issues.

Custody for babies of this age is tricky because attachment to parents is strong, separation is so threatening, and the baby has no way of understanding the arrangement. Today is all that exists for a young baby; the idea that tomorrow, or next week, will bring

a new home and the other parent is utterly incomprehensible. It is much better for the baby to stay in one place, with the primary caretaking parent, and have visits from the other parent (with the first parent out of the house).

## SURGERY AND/OR HOSPITALIZATION

The happy flip side of separation anxiety, and of being anchored in the Here and Now, is that babies this age can undergo procedures and hospitalization with a degree of equanimity if their beloved adult is constantly with them, if they have familiar loved objects to hold on to, and if their need for pain medication is respected. If elective surgery needs to be done, this is a good time to do it if these conditions can be met.

## MOVING

Similarly, babies this age are not likely to be unduly upset by moving if their attached adults are consistently with them, and if their beloved objects are close at hand. The main concern is safety. From now on, if you move, on the day that you move, have the baby absolutely elsewhere. In the hustle of the move, babies undergo rapid developmental acceleration. They learn to dive into the toilet, or out the second floor's newly unscreened window, or to close themselves into the refrigerator. At the very least, they learn to utterly disappear and throw everyone into a furor looking for them.

Make sure, in the new home, that you

turn the water heater down to 120° and check the fluoride content of the tap water.

## The Nine-Month-Old Well-Baby Visit

This is a difficult visit, as the baby has reached the age of stranger anxiety and won't let herself be touched (or even approached sometimes) by the pediatrician, who has developed all kinds of weird tricks to win the baby over. No matter how wonderful both baby and pediatrician are, brace yourself: there will be screams and tears. Moreover, the baby will want to crawl all over, picking up and mouthing previous patients' Cheerios, squeaky toys, and abandoned articles of clothing. Bring her own mouthing toys along, and try to keep her on your lap with a bunch of new (they can be cheap and tearable) picture books.

The focus of this visit is on developmental milestones, vision and eyes, hidden ear infections, the abdominal exam for enlarged organs or masses, and the genital exam, checking for labial adhesions in girls and enlarged scrotum from fluid or hernias in boys. Hips are checked carefully, as they are at every well-child visit, for dislocation.

At this time, it is routine for most pediatricians to perform a blood test for iron deficiency and/or anemia. Many do a test for lead exposure as well. And there certainly may be a new vaccine or two recommended at this age by the time you read this.

## Looking Ahead

The next three months are truly remarkable. Babies flourish in three directions.

- Locomotion. They crawl, or cruise along the furniture, or walk (often pushing a wheeled item along) to get themselves from one place to another. But whatever their style of getting around, they all climb. Stairs. Cushions on the floor. From coffee table to couch. From chair to counter to refrigerator top.
- Language. No, they don't talk until about eighteen months, usually, but over the next three months, you'll have clues that your baby understands a pretty good proportion of what you say.
- Limits. A biggie. For the first time, your baby will actually *challenge* a limit rather than acquiescing to it or objecting to it. There's a difference. You'll see.

# Nine Months to One Year

"Yours is the Earth and everything that's in it"
Rudyard Kipling: *If*

## To-Do List

- Last chance to give up pacifier with relative ease.
- Start phasing out bottles and putting both breast milk and formula in cup.
- Encourage finger-feeding and trying out the spoon.
- Avoid between-meal juice, "handout" snacking, and addicting foods.
- Consider a baby-harness or leash: this is the age for Cherub to get used to it.

## Portrait of the Age: Language, Locomotion, and Limits

Matty has discovered the fireplace.

And the ants marching in through the crack in the hearth.

Nancy has come upon Matty *flagrante delicto* four times in the last twenty minutes. The nine-month-old lies on her belly, gazing fixedly at the ants. She squashes them with one finger deliberately, almost elegantly. She gazes at the finger solemnly, then licks. Her expression is thoughtful, almost glazed. Her mother's is not.

Nancy can't spray the ants. Poison Control has already informed her (four times now—she keeps hoping to reach a different expert) that the pesticides are far more harmful than the insects. She can't switch rooms either, because her word processor is in this room and, while watching Matty, she's working on some data. She can't use the playpen. It was given up eons ago, and now houses folded laundry. Besides, Matty's great joy these days is to stick closely to her beloved adults, peeling herself off to make forays into the world and returning again.

"Matty is squashing the ants," Nancy says the first time. Matty watches her attentively as if taking in every word. "The ants are looking for water to drink. But Matty is squashing them instead. The ants need to drink their water. Mommy will help find Matty something else to do."

Over the last few weeks, the adults around Matty have started talking in this fashion, almost automatically. Matty doesn't exactly understand the words, but they seem to flow over her and some tend to stick. Matty now will turn to her mother at the word "mama" and to her father for "dada." Without being aware of it, her mother puts into practice all the suggestions for encouraging language skills:

- She *recognizes* Matty's interest, and *names* it for her.
- She regards Matty's gaze, action, or babbling as Matty's part of a "dialogue," as eloquent as speech.
- She uses full, short sentences.
- She avoids confusing pronouns such as "you" and "I."
- She uses simple, or slightly advanced, language.
- She uses the same nouns repeatedly.
- She keeps the interchange brief, acknowledging Matty's still short attention span.

Matty consents to be distracted with a shape toy. She stuffs a square block into the round hole. Nancy smiles and hands Matty the round block. Matty bangs the toy, chortling, a few times, and finally spits on it. Then she returns to the hearth.

The second time, Nancy lures Matty away with an activity center. Matty switches the on and off toggle thirty or so times, but soon becomes bored. "Matty can spin the ball, too. Mommy spins. Now Matty spins the ball." This lasts just until Nancy settles herself at the word processor. Matty is instantly back at the fireplace.

In desperation, Nancy brings out the *pièce de resistance:* an old kettle containing half a dozen rolled-up socks and eight nonclamp clothespins straddling the rim. Crowing, Matty empties the kettle, and then replaces item after item, examining each one carefully and mouthing it.

The fourth time, just as Matty gets to the hearth, she turns back to look at her mother. There is a new expression on Matty's face. It contains challenge, interest, and a spark of mischief. Before Matty can even raise the death-dealing fingers, Nancy is on top of her.

"No ants!" her mother yells. She dashes to Matty, picks her up firmly, and strides out of the room with her. "Yuck! Yucky! No ants!" She draws breath. "Matty must not squash the ants. Ants are nasty! Ants are not food to eat! Poor ants!" She makes a horrible face. "Yuck!"

Matty absorbs this maternal display with a charming smile, her pearly new teeth smeared with black dots.

This drama contains many crucial older-infant ingredients. Matty is *acting intentionally,* with an idea of cause and effect. She displays *physical skills* in getting to the hearth, *fine motor skills* in squashing the ants, *oral curiosity* in tasting them. All estimable traits. On a conceptual level, her fascination with the ants springs from her recent appreciation of appearance and disappearance. Look, the ants get squashed—and new ones appear! The concept of the *permanence of an object*, an ant or a mother, its disappearance and reappearance or replacement, is a source of constant experimentation.

The whole episode has taken only a few minutes, and neither Matty nor her mother has marked the event as momentous, but it is. Matty tested her mother intentionally, and a limit was set. Matty seemed pleased rather than frightened or offended. It is likely that Matty now will use the ants as a way of testing her parents, to see if they always will enforce this new rule about eating insects. It is important, for the whole family, that they do so.

By the way Nancy handled this challenge, she showed Matty that she:

- Values her curiosity, skills, and impulses, and shares her interests. She allows Matty to roam the house, and provides her with small, safe objects to examine and experiment with.
- Treats her body lovingly and with respect. No slapped hand or spanked bottom.
- Is not afraid of the emotional distance, i.e., separation, implied in setting a limit, and thus neither backs off nor overreacts to Matty's testing her resolution.

However, when it came to responding to the eating of the ants, Nancy has sent Matty some mixed signals. She allowed Matty to get away with it at first. When Matty challenged her, Nancy responded with exciting, novel attention: she gave Matty her eye contact, her voice, and her touch. She talked a lot about those ants. Wow.

Want to bet Matty goes right back to them, and this time challenges Nancy right away!

Over these three months, Nancy is going to need to hone new skills in setting limits. Her goal is to preserve those important insights above, but also to teach Matty another important one:

There are behaviors that are forbidden, or "NO," and there is no "gain"—no pleasure, no excitement, no satisfaction—in challenging a "NO."

It will take Nancy several heart-stopping experiences to discover the key to success. That is: if you talk too much about an activity to Matty, Matty will engage in that activity more and more and more, relishing your attention.

So if you want her to *stop* an activity that is dangerous, destructive, or aggressive, you should use only one lone, lorn, word: "NO!" At the same moment you utter it, you move Matty or the object. And then you turn away from her for half a minute or so, pointedly removing your attention from her. It was Disraeli's secret in setting limits on foreign and domestic troublemakers, and it works equally well when setting limits on babies this age: *Never apologize, never explain.*

## Separation Issues: Wait Until Dark

Everybody in the Franklin household is experiencing separation anxiety—except the cat, who has taken to disappearing.

Matty can't be transferred from one adult

to another without holding up her arms and weeping piteously. She won't let her mother, or her father, or her caretaker, out of her sight—except when she leaves them under her own steam. When she does so, after a few minutes she discovers that she is "abandoned" and shrieks with dismay.

When Nancy leaves Matty to go to work, she feels an awful pang, as if part of her were being ripped away. She has strong moments of regret that she went back to work so early and "missed" Matty's babyhood. She rails that now she is likely to "miss" Matty's first word, first step, first everything. She finds herself interpreting Matty's tears as reproach. She has even found herself weeping, filled with guilt, and with anger at Larry for having "allowed" her to return to work, at her parents, her colleagues, her clients, Mother Nature, and the feminist movement.

At other times, she is delighted to see Matty exploring on her own, bravely terrorizing the cat and eating ants, making ties to Larry and to her caretaker. She is a bit grateful for having an adult life not entirely defined by her role as mother, because she's getting an inkling that Matty won't always be defining herself solely as her parents' baby.

Larry spends a lot of time engaged in an internal dialogue with himself. Partly, he is contending with his own feelings about Matty's new mode of both clinging and testing limits. It makes him feel, he admits silently, rather godlike. He enjoys the exploring more than the clinging, and finds himself impatient with his daughter's tears—has caught himself wanting to tell her not to act like a baby. What has surprised him, however, is the change that has come upon his marriage. He can't help but realize he and Nancy are jealous of each other for Matty's attention, and jealous of the caretaker too. Worse, he is jealous of Matty for Nancy's attention.

Beyond anything, both Larry and Nancy feel pressed. Both feel that all their time and energy should be divided between career and Matty. The only time they talk with each other is to issue directives, plot to get out of social and family functions, snap at each other about the way each is responding to Matty's behavior, and hunt for the cat.

## Bedtime and sleeping through the night

If you think separation is a challenge in the daytime, just wait until dark. Most babies this age spend an average of ten hours in nighttime sleep, with two additional daytime naps of about one to two hours each.

Often, their parents aren't nearly so well rested.

Refusing to go to sleep and waking up at night are very frequent problems at this age. And this behavior often leads to a vicious cycle: the less sleep the parents get, the less they are able to cope with the baby in a patient, consistent, loving way. This makes the baby upset. The more anxious the baby gets, the less she's willing to separate at bedtime.

If she can't put herself to sleep at bedtime, she won't be able to do it in the middle of the night when she comes to the light parts of her sleep cycle and awakens briefly. So she'll shriek and stand up shaking the crib rail, until an exhausted parent appears to put her back to sleep.

Here's what it takes to have a nine-month-old fall asleep on her own and rely on her own devices to put herself back to sleep when she wakes, as she normally will, during the night. She needs:

- Enough one-on-one time during the daytime with each parent.
- Some experience relying on her own devices. If she cries from frustration, a minor scare, such as going "down boom" on her bottom unexpectedly, or boredom, she needs sympathy and attention—but not RIGHT away, and not overly intense.
- Some experience in separating from her primary caretaker and spending time with other warm, responsible, comfortable adults, with her caretaker out of sight and out of hearing.
- A good, friendly relationship with her crib. Sometimes it helps if her beloved adult spends some time with the baby in the crib with the rail down, playing quietly.
- To get in the habit of falling asleep in the crib, without bottle or pacifier, it's often easiest for the parent who is not the primary caretaker to do the bedtime ritual, because separation is easier. This may take time, as a baby this age may cling to the one parent and reject the other, which is all the more reason to cultivate that relationship.

- To know that her parents are comfortable with the separation implied in setting limits on her behavior. If they aren't panicked by this most threatening kind of separation, she won't be scared either.
- A bedtime ritual, one that is the same each night, of maybe four to seven steps, that melds into a sleep mode. The parent can emulate Mr. Rogers on a particularly tranquil day.
- For some Nine Months, a transitional object. (See *Window of Opportunity* in the previous chapter.)

## Limits: The Real Challenge Begins

Once you say the word limits, you automatically evoke two opposing hazards of parenthood. You worry about spoiling. And you worry about breaking his spirit.

If no limits are ever set, and the baby isn't forced to choose an acceptable over an unacceptable behavior, he'll start to become spoiled. On the other hand, if the baby isn't allowed to make most of his choices of behavior on his own, he won't get the satisfying feeling of autonomy—of setting a goal and carrying it out. That is what is meant by "breaking his spirit."[1]

---

1. The worry about breaking a child's spirit becomes more intense as the child enters into social activities later. This worry tends to focus on whether or not to limit physical aggression and to encourage courtesy. Many parents worry about raising a "wimp." This may make parents unwilling to set any limits. In this situation, the baby can't become comfortable with the idea of emotional separation from parents. He'll be preoccupied with it, and will still be dealing with this basic issue when he should be learning other things. He may even turn out to be a wimp after all, because the idea of challenging limits is so fraught with anxiety. Or he may simply seem spoiled and obnoxious, which is not the same thing as having an unbroken spirit.

It's not so complicated. Think of it this way. This is the age when a baby discovers that limits exist, and develops, at a primitive level, the skill of keeping some behaviors within limits. Babies make this discovery and develop this skill just the way they do everything else: by constant practice.

- **Most activities should be encouraged or at least allowed**. Babies thrive on exploration, infant adventure, and play. Be sure you are updating your baby-proofing, as described in the *Day to Day* section under *Safety Issues* (see below). Don't worry: no matter how baby-proof you think your home is, your baby will find at least several utterly outrageous or dangerous activities for you to set limits on. I promise.
- **Limit the limits.** If everything that is faintly annoying or disruptive is limited, the baby will have a hard time learning just what a limit is. Parents and caregivers won't be able to teach her effectively: they won't be able to catch her just as she starts each and every one of a hundred "limited" actions, move her or the object, and follow through. They'll be inclined to be exhausted, impatient, constantly alert for accidents, inconsistent, and not much fun. Babies don't learn limits in this environment; they just learn that life isn't very enjoyable.
- **Avoid confrontations as much as possible.** Undesirable behavior should be prevented by distraction and substitution as much as possible, ideally before the

baby actually takes up the forbidden activity. Head her off at the pass. Nancy might have tried her seductive distractions before Matty actually got to the ants.

When you do set a limit:

- **Don't issue a warning before you act on a true limit.** A baby this age doesn't understand warnings. If she is asked not to do something, and then she does it, and THEN a parent takes action, she is confused. She'll need to perform that action again and again and again, trying to figure out what in the world her parent has in mind.
- **When you set a true limit, say one word only: "No." Move her or move the object of her attentions at the same time.**
- **Follow through.** No matter how many times the baby heads toward the forbidden behavior, move her or the object or both, without a warning. Use your firm voice.

# Day to Day

## MILESTONES

### *Vision*

A baby this age doesn't have true 20/20 vision yet, but you'd never know it. She can't tell you that the street sign is still a blur, or that she can't see individual leaves on the tree. But she certainly can spot tiny objects such as lint, and follow the flight of a bird or a helicopter or a hot-air balloon.

*Hearing and babbling*
The best sign of good hearing is the amount and style of babbling. At six months, most of what you hear will be vowels; by nine months, there will be consonants as well. Many babies specialize in b or d syllables. If your baby appears to say Dada but not Mama, try not to take it very personally (whichever you are). I suspect that wise mothers heard what their babies were babbling, and then informed the fathers that Dada was the babies' term for the paternal figure. Not until about a year of age do most babies assign Dada to father and Mama to mother. Some babies will say several words besides Mama and Dada.

## SOCIAL AND COGNITIVE SKILLS

- She should be very interested in gaining and sustaining adult attention. She should like Peek-a-Boo, and imitate waving Bye-Bye or blowing a kiss.
- When you point at something, she should usually turn to look at what you are pointing at.
- When she isn't sure if what's going on around her is OK, she should look to you for reassurance.
- By One Year, she should point and grunt (or screech) to what she wants, or take your hand and drag you there.

Cherub should be getting the idea of What Is Expected. You'll find ways to notice

this: she may sit up and hold her arms out as you approach, or start getting excited when she sees her bottle.

*Motor skills*
The range of normal motor skills is very great during this period. How early a baby walks has little to do with future athletic ability and nothing to do with intelligence, morality, or parental excellence. True, some normal healthy babies will walk as early as eight months; but others don't until nearly sixteen months. By kindergarten, an observer won't be able to tell which was which. And, as pointed out in the previous chapter, some normal healthy babies do not go through a true crawling stage, but get about with alternative styles until they learn to "cruise" around the furniture. Most babies are pulling to a stand by at least a year of age, often much sooner.

Most babies learn to climb before they can walk, at whatever age they do walk. This can come as a major surprise. Once a baby can pull to stand, the lure of verticality becomes a major preoccupation. Babies under a year get up onto couches, low tables, anything they can pull up on and boost themselves onto. They are fearless about heights, and do not understand that when you come to the end of a surface you will fall off.

*Hands*
Some babies have a well-developed pincer grasp by nine months; others, not until a year of age. Babies should still be able to use both hands just about equally. Though they

may show a preference for one hand over another reaching for a toy or feeding themselves, if that hand is otherwise occupied the less preferred hand should seem just as effective.

Most pediatricians want to check the development of a baby:

- If the baby loses an ability—not a behavior, but an ability—she used to have. An ability is a neurological skill such as cruising around the furniture, using a pincer grasp, using both hands equally well, transferring from one hand to the other. A behavior is a willed act, such as blowing a kiss. No longer performing a behavior is normal, as other behaviors become more intriguing.
- If the baby seems not to use one arm or hand or leg as well as the other.
- If by nine months the baby isn't able to grab toys with a grasp in which the thumb opposes the fingers, and by a year isn't able to pick up a small item, such as a Cheerios or a potato bug, between thumb and forefinger.
- If the baby doesn't babble in an age-appropriate way. Most babies this age make a lot of sounds and noises. They babble when alone in their cribs, and they babble, flirt, and squeal as a reciprocal game if you talk to them.
- If the baby's muscle tone seems floppy. By nine months, babies should bear weight if you stand them up. When on their bellies, they should be able to push way up and roll over in both directions. They should be able to get from lying down to sitting up and sit, holding a toy, without assistance. They should be trying to pull up to a standing position.
- If the baby doesn't seem social, interested in gaining and sustaining adult attention. A baby who seems to lack separation anxiety entirely but is social and affectionate in every other way is probably fine, and of a very easygoing temperament; but the baby who couldn't care less whether a loving adult is there with him needs to be checked.
- If the baby doesn't seem able to anticipate events to a degree. For instance, at ten months, she should find a toy that she sees you hide under a cloth.
- If you have any concerns about vision or hearing. Babies whose eyes look crossed, who squint and tear in sunlight, or who don't seem to follow a visually exciting but silent object as you move it from far right to far left (or vice versa) may have vision problems. A baby with recurrent ear infections needs special attention about hearing; so does a baby who doesn't turn to the sound of her name or who doesn't take pleasure in babbling and making noises.

## SLEEP

Most babies this age sleep about fourteen hours out of twenty-four, about ten at night and four divided into two daytime naps. Sleep problems are usually related to separation anxiety; see *Separation Issues*, below, in this chapter.

> Honey and Karo Syrup should not be given to children under a year of age, due to the risk of the rare disease infant botulism. Fish consumption guidelines may be found on 7.

## GROWTH

The exceedingly rapid growth rate of the first six months has slowed to about half what it was. Most babies will gain about a pound a month, half an ounce weight gain a day.

Many babies whose initial height curve indicated they would be much bigger than their parents slow down now, manifesting their genetic destiny. This may be reassuring to parents who have envisioned standing on tiptoe to pat their five-year-old on the head.

A baby whose weight-for-height curve changes dramatically at this time may need special attention. See the essay *Growing in All Directions* in Part III.

Harvey, for instance, used to weigh just the average amount for a baby his age: his weight for his height (or length) was at the fiftieth percentile. Now it has jumped to the ninety-fifth percentile.

This is partly because Harvey has learned to go to the refrigerator and whimper piteously for juice and snacks. It is partly because he is so very active and persistent in his explorations that his parents often count on food as a way to distract him and gain a moment's respite.

Unfortunately, the habits and weight gain started now can become self-reinforcing, and Harvey is in some danger of obesity. He needs adult help in changing his lifestyle. Harvey needs to learn the difference between the feelings of, and solutions to, hunger versus boredom versus anxiety.

With Melody, it's the opposite. Her weight for height used to be at an acceptable level at the twenty-fifth percentile. But now she's gaining less and less each month, and her weight is below the fifth percentile for her height. Since she seems perfectly healthy, and since she looks so much like her aunt, a ringer for Audrey Hepburn, her parents aren't really worried. Nevertheless, they sensibly allow Melody's pediatrician to perform a complete physical exam and basic blood and urine tests to make sure that no lurking medical problem is causing the deviation.

## TEETH

Teeth sometimes bewilder parents of babies this age. Some babies have six teeth, others none. If the baby is growing and developing normally, each pattern is fine. Some babies cut their two upper lateral incisors first, and almost always temporarily acquire the affectionate nickname "Fang."

Since most babies have at least two visible teeth by now, this is an important time to set habits for good care. Turn to the previous chapter for specific suggestions.

I hope you have already internalized the Golden Rule of the Good Eater, as formulated by Pediatric Nutrition Guru Ellyn Satter, and its amendments: Adults decide **What** choices of food are offered, and **Where** and **When** they can be eaten. Babies and children decide **How Much** of each food they want.

*Nathanson's First Amendment:* If a baby seems to be forming finicky tastes (taking only yellow veggies, for instance), simply stop offering the Addictive Food as a choice.

*Nathanson's Second Amendment:* It is easier to avoid introducing seductive foods (such as fruit juices, baggies of Cheerios to munch on, catsup on anything) than it is to try to limit or remove them later.

*Nathanson's Third Amendment:* By fifteen months, Cherub not only decides **How Much**, but should be feeding herself entirely. So start her early, at Nine Months, learning how.

One new factor for teeth is giving up the bottle and pacifier. Most babies this age don't need to suck a great deal. They want to mouth everything and chew, though, and if given a bottle or pacifier they will give every sign of "needing" it.

Sucking on a pacifier most of the time may seem like a harmless entertainment, but it can become a habit that is increasingly hard to give up; it can also interfere with talking, since it's difficult to engage in both enterprises simultaneously. Bottles are a bigger problem, since they often contain juice or formula, which coats the teeth for long periods, producing "nursing bottle cavities," which means rampant tooth decay. Another good reason to introduce, with jolly optimism, formula in the cup. (As for juice, see the section on *Juice* in *Feeding and Nutrition*, below.)

## FEEDING AND NUTRITION

Make sure you're offering breast milk or formula in a cup, and give Cherub a spoon, which at first will function more as a catapult.

Babies this age often prefer grown-up foods to baby foods in jars. You can feed your baby dishes that you are eating as long as the following conditions apply:

- Cherub has eaten all the ingredients before and tolerates them.
- The food is free of highly allergic ingredients such as peanuts, nuts, and chocolate. (Now, I have to say, this statement is based on current guidelines. There is a possibility that in fact we should be offering these foods early in life. See the essay on Allergy in Section III!)
- You have mashed, ground up, steamed, or otherwise modified the texture so that

it is easy to "gum". Never offer chewy, pellet-like, stringy, or otherwise "chokable" foods. No foods shaped like a circle (slices of hot dog) or a sphere (grape, frozen pea).

- Don't use juice for thirst or recreation. The juice beverages (apple, pear, cranberry, grape) have barely any nutrition that doesn't have to be added at the factory. They are very sweet; some babies start to prefer juices to water or formula. Babies may cry for juice, and drag you to the refrigerator and point and grunt for it. Not wholesome behavior, no matter how "wholesome" the juice is advertised as being.

- If you offer a snack, do it because you are absolutely convinced your baby needs one, rather than from culturally-imposed routine. Most babies don't need snacks.

- There is not one baby in the entire universe who needs "handouts": those little baggies with crackers, raisins, whatever that inhabit the diaper bag and get hauled out whenever there's a lull in the action. A baby is not a parking meter, who needs frequent "tokens" administered so that an alarm doesn't go off.

Feeding should be messy, nutrition should be easy, and both should be fun. There are only a few guidelines for mealtimes at this age. Cherub only needs about 750 to 900 calories a day at this time, and half those calories come from breast milk or formula. So expect Cherub not to eat a whole lot. Intake is supposed to be pretty erratic from meal to meal and day to day. Keep in mind:

- Salt, sugar, and fat in excess are not only unhealthy, they are downright addicting. That means chips and fries, catsup and mayonnaise, sodas. Once you start them they are hard to stop. Don't add salt, sugar, or fat unless you have to, to make the food edible.

- When you buy processed foods, check the labels. You *do* want foods, such as cereals, fortified with iron; you *don't* want foods containing artificial coloring or sweetening. Preservatives are necessary for food safety, but try to limit how much you give your baby. Canned and frozen foods don't require preservatives.

### Food dangers

- Make sure that you scrub *and* peel all fruits and vegetables. Any food can be contaminated with pesticides or with food-poisoning bacteria, such as E. coli. Never give anybody, especially a baby, unpasteurized juice.

- Each meal ideally should represent at least two food groups. You can keep track by evaluating the color scheme of each meal. If each food is a different color (a white cereal, green or yellow vegetable, yellow egg yolk, brown meat, red or blue or white or yellow or purple fruit) you are probably doing fine.

- Keep portions small, about the size of Cherub's fist. Offer one serving each of three different foods, and see what he does with them. Weight gain starts to slow down now, and you may see his appetite decrease rather than increase. Be alert to signs that he's had enough.

## More subtle aspects

1. **Too much of a good thing**

    *Food "addictions":* If you serve more than three helpings from one food group each day, you and your baby may be getting into a nutritional rut. Babies often prefer the starchy cereal foods over everything else, or become addicted to one particular food.

    Too many helpings of yellow veggies will turn the skin orange (carotenemia), not a medical problem in itself, but a potential sign that the baby isn't getting enough variety. (Rarely, carotenemia can be a sign of other problems; see carotenemia and jaundice in the Glossary, and the discussion of *Skin* in Part II.)

    *Fat:* Babies this age need up to forty percent of their calories as fat. (Life isn't fair. Fatty foods, like youth, are wasted on the young.) Only after age two, babies need some active, planned, fat restriction in their diets. At this age, the restriction is just passive: don't go out of your way to give fats. Something that is larded or dripping grease or is just and simply fat, like the border of a steak or the skin of the roast goose, is not good baby food. The breast milk and formula give him a good amount of fat, and he'll pick up the rest he needs from a normal, "no fat added," diet.

2. **Too many constipating foods**

    One of Nathanson's laws is: a foods isn't popular unless it's constipating. In older children, this law covers pizza and chocolate. In our present age group, the deliciously constipating foods are rice cereal, bananas, and cheese. At the first sign of a stool firmer than Play-Doh, back off these foods and bring out the prune juice. (A tablespoon in each bottle of formula, or mixed with peaches or other fruit for the breastfed baby.)

## BATHING

**NEVER EVER** leave a baby alone in the bath even for the few seconds it takes to get a towel or answer the phone. A baby can drown in only two inches of water, because she can panic and inhale if she puts her face in suddenly and by accident. Make sure you have no electrical appliances she can pull into the bathtub. To avoid hot-water burns, test first, of course, but also keep your hot-water heater turned down to 120°.

Baby girls, like older girls, should avoid as much as possible sitting in soapy water—bubble bath, shampoo water, lotion soaps dumped in the bathwater. These can all produce vaginal irritation. The best "baby soap" is a moisturizing white unscented bar soap. Use a little on a washcloth, and rinse with clear water. Baths are fun by themselves, with constant adult company and water to pour and splash, they don't need bubbles.

Powders are dangerous at this age. A baby can grab, squeeze, and inhale in the blink of an adult eye.

Bath phobias are not uncommon, and can be handled by avoiding the bath itself but encouraging gentle water play with a basin and interesting toys to splash. Gradually, a baby is likely to consent to sit in a shallow container partly filled with water, then on a parent's lap in the shallowly filled bathtub.

Some babies like to shower with a parent. This is fine.

## DIAPERING, DRESSING, AND CLOTHING

See the section in the previous chapter for help with the Diaper Escaper. As Cherub grows older, try to resist the temptation to ask Cherub for permission to undergo a diaper change. That is, don't say "Want to get a clean diaper on?" or "Time for a diaper change, OK?" You are not really offering a choice, and it's not a good idea to get into the habit of phrasing orders as choices.

The main question about clothing is shoes. Unless a baby has feet that need special shoes (say the sole of the foot curves in so that the baby's feet look like the parentheses marks surrounding this phrase), no special shoes are needed. High-priced, leather baby-jogging shoes and professionally fitted bronzable baby shoes are both very nice indulgences, but they do not form the baby's feet or make her walk straight or do anything but make everybody feel elegant but poorer. Find an inexpensive shoe that is flexible, that fits at the heel and doesn't cramp the toe. Don't bond emotionally with this pair of shoes; they'll be outgrown in a whiz. Let the baby go barefoot whenever it's safe and warm enough.

In these ozone-depleted days, clothing provides not merely decoration and warmth, but sun protection. Even one blistering sunburn in childhood increases the chance of skin cancer. Read the label! **Some sun lotions do NOT protect from the most harmful (UVA) rays as well as the tanning (UVB) rays. These are worse than no sun lotion at all, because you lose the early warning of sunburn.** Hats with brims, long sleeves, and pants are important.

Moreover, sunglasses with UV protection are an excellent idea. Because babies are so busy with their hands and with locomotion, they have ways of entertaining themselves other than sitting around removing their spectacles. Give it a try. Sun damage to the eyes is also accumulated over a lifetime, and is not nearly so obvious as skin damage, though just as, if not more, worrisome. **Sunglasses without UV protection are WORSE than no sunglasses at all.** The dark lenses allow the pupil to dilate, so more unprotected rays can enter and damage the eye.

## ACTIVITIES, TOYS, AND EQUIPMENT

The joys of this age include getting from one place to another, learning about small objects, enacting the drama of appearance and reappearance, and exploring the concept of Self and Other.

Successful toys are enablers of these activities.

- Wheeled toys a baby can push as he "walks" are real winners. The wheeled stool in the pediatric office is a perfect example. The child's own stroller is another.

- Babies love toys that involve appearance and disappearance, or repetitive change. Toys that babies love to experiment with include beach balls to push to and fro, switches to turn on and off, handles to push up and down, and hinged items. Hinged items include doors to bang: cabinets, cupboards, room doors. They also include books, which are appreciated for this feature and for being mouthed before their literary qualities are discovered. A toy telephone with a rotary dial is another delightful toy.

- Related to this fascination with the permanence of objects are toys that can be emptied and filled. These include water toys, and, with reservations, sand toys.[2]

- Small, smooth safe objects to examine, drop, and mouth are endlessly fascinating. These need not be real toys, though big plastic teething keys are fine. Measuring spoons, large spools empty of thread, clean smooth clothespins, rolled up socks, anything in a handbag, prescription expensive glasses, salt shakers—these are all toys, some more appropriate than others. Mouthing and investigating these help a baby understand the difference between her body and a possession.

This is a difficult concept and isn't well established until after the third year of life.

Toys to avoid include:

- **Dangerous toys:** small toys that can be choked on, sharp toys, toys that come apart into smaller or sharper pieces; toys with strings to choke on; poorly balanced, wheeled push toys; and BALLOONS. Balloons, when popped, can injure eyes with the explosion, and when choked on do NOT respond to the Heimlich maneuver.

- **Toys that promise inappropriate teaching:** flashcards and counting toys, for instance. Reading and counting by printed numbers are symbolic acts. To engage in a symbolic act, one must first understand the object the symbol stands for! If you don't mind your child eating the flashcards or ripping them up and dumping them in the bath, these are fine; the child is using them as toys in an age-appropriate fashion, and discovering the properties of cardboard and plastic.

- **Expensive toys with limited appeal:** expensive dolls, stuffed animals, puppets, building blocks, wind-up toys, music boxes, and many "activity centers."

2. Sand can get into eyes, noses, mouths, vaginas, rectums, and under intact foreskins, and cause pain. Sand can be contaminated with animal feces and even asbestos. Sand at the beach is worth it, usually; sand in a sandbox should be asbestos-free and covered at night. Container toys like Matty's kettle are very successful, as are nesting and stacking toys. Shape toys are frustrating for adults, because children this age don't match the shapes but play with them and force them into the "wrong" openings.

## Safety Issues

If you haven't already baby-proofed your home and stocked your medicine chest as suggested in the earlier chapters, please do so now. These are suggestions to update, not to start afresh.

- **Take a CPR course again. The guidelines for treating a choking baby change when the baby turns a year old!**
- Consider a Harness or Leash for the future.

  If you foresee that you will be doing errands and shopping with Cherub and once he becomes a toddler and tires of his stroller or backpack, I suggest that you consider a baby harness or leash.

  Such a device is much more safe and reliable than trying to grab his hand, or chasing after him, or relying on friendly strangers to head him off at the pass. It avoids the occasional embarrassment, such as my little friend who recognized "her" potty in the display window of a store. And a leash helps to decrease oppositional behavior, because Cherub won't be able to keep testing you by running off.

  If you anticipate the need for a leash or harness, now is the time to introduce it, so that by the time he's Eighteen Months to Two, Cherub thinks it is normal and friendly.
- Make sure the baby is properly restrained and can't get out of the car seat. Until Cherub is 20 pounds *and* one year of age, he should face backward in the backseat.

- Beware choking. The worst: peanuts, popcorn, anything round or spherical, like a slice of hot dog or a grape.
- Beware ingestions. Particularly: chemicals, medications (ask visitors to put handbags on top of the refrigerator; they often have medicines with non-child-proof tops), poisonous plants, alcohol, and drugs. Also, beware the small batteries used for toys. Especially be careful of sugar-coated iron pills or any medication containing iron.
- Beware guns, even this young. If your child can manipulate a light switch, he can accidentally pull a trigger.
- Beware burns. Especially watch out for curling irons, clothes irons, motorcycle exhausts, and hot fluids.
- Beware water immersion. A baby this young can inhale water and panic and stop breathing, even in shallow water such as a fountain, a barrel or bucket of water, a wading pool.
- Every child this age encounters bumps, bruises, cuts, abrasions. Read *First Aid* or *Body Parts, Bodily Functions, and What Ails Them* in Part II. Tell yourself you won't be surprised.

## Medicine Chest

To your already stocked chest, add:

- "Kling" bandages so you can wrap a sterile gauze pad onto a cut. (Don't use adhesive strips; children this age tend to take them off and mouth and choke on them.)
- Acetaminophen elixir.

- Acetaminophen and ibuprofen. Make sure you check dosages with your pediatrician! And remember that the Infant Drops are much stronger than the Children's Liquid. If you give a teaspoon of the Drops, instead of a teaspoon of the Liquid, you may give your baby a truly dangerous overdose. BEWARE!
- Gatorade or other **nondiet, sugar- and salt-containing** soft drink for the vomiting baby who may refuse the infant rehydrating solutions.
- A sterile stool or urine container in case you have to bring in a sample.
- A "barf bucket" for peace of mind when your child is vomiting and you have to drive.
- Tweezers, to get out splinters and foreign bodies.
- A flashlight and extra batteries. To find her under the bed, to look up his nose for the pencil eraser, to check for pinworms in the middle of the night.

## Health and Illness

*Part II of this book covers illnesses in babies this age and older: how to assess symptoms, how to tell whether a child is sick enough to require urgent attention, and what kind of home treatment may be appropriate. It also examines problems by body part. The Glossary is an alphabetical listing of medical terms with pronunciations and informal definitions.*

## EXPECTATIONS

Most babies will have something during these months: a cold or two with a runny nose and mild cough, an ear infection, some constipation or loose stools, eczema, and/or diaper rash. Bruises, bumps on the head, and mildly smashed fingers from falling, climbing, and swinging all those hinged doors are pretty much inevitable.

However, recurrent illnesses at this age should be considered a real problem. There are several reasons for this:

- The immune system doesn't fully mature until about age two. Many parents have heard that a child who gets sick a lot as a baby at least will have built up a lot of resistance by school age, and will be tougher. Recent studies have shown that this isn't the case. There isn't any advantage to having diseases at this age.
- Because the immune system is immature, illnesses may be more severe now than they would be when the baby is older.
- Babies this age can't tell you where it hurts, and it may be hard to judge the intensity of illness. The pediatrician may be required to perform laboratory tests to decide just how sinister an illness is. These may be painful and expensive.
- Serious illnesses requiring hospitalization may be very rough at this time of intense feelings about separation on the part of both babies and parents.
- Babies this age have such basic skills and concepts that even mild illnesses may

interfere with learning. A slightly feverish and irritable two-year-old can still engage in fantasy play, being read to, or (alas) watching TV. A nine-month-old in the same boat can only be cuddled.

- Receptive language learning is a key skill for this age. Many viral illnesses in this age group turn into middle ear infections, which can temporarily diminish hearing and delay language and speech development. If these are frequent, permanent learning and behavioral problems may result.

Babies in group care have been shown to acquire, on average, many more infectious diseases than do babies who are at home with no or only one sibling. If the group care is large, more than four children, the number of illnesses is even higher.

The onset of such recurrent illnesses may come as a shock to parents whose babies have been healthy up until now. They haven't changed caretakers; their lives are otherwise the same.

Why do illnesses start now?

First, babies this age actively pursue, and actively attract, toddlers and preschoolers. Bodily fluids are freely exchanged as they rub and reach and mouth toys and each other.

Toddlers and preschoolers are going through age-appropriate illnesses: their immune systems are mature, and they are supposed to experience multiple viral illnesses (at much smaller risk for complications) in order to build up a repertoire of immunity. Most annoying of all, these ill-nesses are most contagious just before the child shows signs of illness. The apparently healthy two-year-old may be much more likely to pass on a disease than the very apparently sick one.

Second, babies this age handle everything. Then they rub their eyes or noses (transferring to themselves respiratory viruses) and put objects in their mouths (transferring to themselves intestinal viruses).

What can parents do about child care to counteract these problems?

- If it is at all affordable, try for as close as possible to one-adult-to-one-baby care. Most learning at this age is dependent on the close attention of a loving adult, not on contact with other children. A wonderful exception is much older children, those seven or older, who are often enthusiastic and protective teachers, friends, and guardians of babies this age.
- If this is impossible, try to find a daycare that limits the number of very young children to four, even though there may be several older children.
- Sometimes parents find it possible to form a select daycare group of four or five age-matched infants. Families may chip in to hire a daycare person, or parents may work out a schedule so that a different family takes the group each day. A good resource is parents who attended the same childbirth classes.

In any event, the caretaker should be aware of which illnesses—the majority, in fact—can be contained by careful hand-washing: and should perform such careful hygiene. See the chapter on *Viruses and Bacteria* in Part III.

## COMMON MINOR SYMPTOMS, ILLNESSES, AND CONCERNS

### Colds and Ear Infections

The main problem with the upper respiratory viruses is that they often turn into middle ear infections. It is likely, though not conclusively proven, that the temporary and mild hearing decrease that occurs with recurrent ear infections can impair cognitive, language, and emotional development in young children. There have been more than sixty studies trying to determine how much to worry about this. While there are conflicting results, it is certainly clear that recurrent ear infections are never beneficial and potentially very worrisome. See *Ears* in Part II in *Illness and Injury* and the essay on the same topic in Part III.

### Conjunctivitis (pink eye)

A baby this age with tearing or goopy eyes is more likely to have an infection than just a blocked tear duct. This is especially likely if the whites of the eyes are pink or red. Alas, a baby with conjunctivitis has a pretty good chance of having a concomitant ear infection, and should be checked for this (even if the pediatrician has consented to prescribe eye drops over the phone) if she is cranky, not sleeping well, or digging at her ears.

If only one eye is tearing, there is some chance that he has scratched the cornea or has a foreign body in the eye, particularly if the baby squints at a bright light, seems to be in pain with the eye, or won't open it. This scenario always requires a pediatric examination.

If you or your pediatrician have any suspicion that the baby could have herpes in the eyes, the baby should be seen very promptly by an ophthalmologist. This could happen if the baby has oral herpes (see *Common Illnesses with (More or Less) Unfamiliar Names*, Part II) or if someone with obvious oral herpes has been taking care of him, or if the "pink eye" symptoms are particularly dramatic and painful.

### Croup

A cough that sounds like a goose honking or a seal barking; a frightened child who goes "croooop! crooop!" when she inhales; and all of this in the middle of the night. Croup is a respiratory illness usually caused by a virus. It is discussed fully in the *Common Illnesses* section of Part II under *Trouble Breathing*.

### Common viral diseases with unfamiliar names

There are several very common but unpublicized viral diseases of young children, and babies this age can catch all of them. They are discussed in Part II. They include those with spots: fifth disease and hand-foot-and-mouth-disease; those with sore mouths and throats: herpangina, oral herpes or gingivostomatitis, and infectous mononucleosis (the "kissing disease" of adolescents); and the specific high-fever-then-rash disease roseola (the most familiar one).

*Viral diseases with familiar names that are becoming, or are about to become, rarer and rarer*

Even though you may suspect your child has chicken pox (Varicella), measles, mumps, or rubella (German or three-day measles) this is unlikely because these diseases have been pretty well controlled by immunizations. However, outbreaks in areas where immunization hasn't taken hold, for one reason or another, are not altogether rare. See Part II for descriptions of these illnesses.

## COMMON SCARY BUT USUALLY INNOCENT PROBLEMS

Fortunately, babies haven't added much to their repertoire in this category between the last chapter and this one.

*Fever convulsions*

These very frightening but innocent seizures can start as early as six months of age, or as late as five or six years, but are most common between twelve and eighteen months. They were discussed in the *Six Months to Nine Months* chapter, and you'll find another discussion in Chapter 12, Part II in *Seizures, Convulsions, Fits, and Breath-Holding Spells.*

*Breath-holding spells*

Lots of young children, when furious, seem to take forever to get the first cry out, and turn purple and glare at you. But a true breath-holding spell is different. These start out with anger and frustration, fear, or a minor injury. The toddler lets out a cry, then either turns blue or pale, goes limp, and may have a convulsion. The age range for these is exactly the same as for fever convulsions, and these too are discussed fully in the *"Six Months to Nine Months"* chapter and again in the above-mentioned chapter of Part II.

*Labial adhesions or fused vagina*

The apparent "disappearance" of the vagina due to this condition can be startling but is not serious. See the description in the previous chapter.

## SERIOUS OR POTENTIALLY SERIOUS SYMPTOMS OR ILLNESSES

*Fever*

Fever all by itself is still of concern in babies this young. Most pediatricians want to see a baby this age with a rectal temperature over 103° or if a lower fever is present for more than three days, even if there are no other symptoms whatsoever and the baby is happy, eating well, and sleeping normally.

Most such unworrying fevers are due to viruses, but occasionally there is a surprise, such as a urinary tract infection. Teething, I am sorry to say, has never been shown to cause fever.

A fever in a baby this young that goes as high as 104° requires a prompt, even middle-of-the-night, phone call and a same-day visit, even if there is no other symptom and the baby is happy. If there is any other symptom, the baby certainly needs a very prompt, if not urgent, visit.

## SERIOUS INFECTIONS

### Sepsis and meningitis

The two most common kinds of serious infections in this age group are *sepsis*, which is an overwhelming bacterial infection in the blood, and *meningitis*, an infection (usually bacterial) in the spinal fluid.

(The term *spinal meningitis* is really redundant; there isn't any other kind of meningitis. The really important thing to know about meningitis is whether it is caused by a virus, in which case it is usually but not always fairly innocent, or by bacteria, in which case it is always potentially very serious.)

A baby this age who has a temperature of 104° or over has a somewhat higher chance of having such an infection than one with a lower temperature, but most fevers this high are from more innocent infections. Nevertheless, such a high fever demands a phone call and either an immediate or same-day visit.

More crucial is the presence of other symptoms like vomiting or trouble breathing, and abnormal behavior: not able to smile or play, limp and lethargic, or highly irritable. If there is a rash, don't assume that the illness is just a virus. See *Fever* and *Not Acting Right* in Part II.

### Cellulitis

Cellulitis is another serious infection. This is an infection of the deep tissues of the skin, often a cheek or an eye. Usually the child acts sick, has a fever, and there is a swelling that is red or reddish blue.

### Urinary tract infections

These occur most frequently in little girls. But either sex may have a UTI. See the discussion in the previous chapter.

### Undescended testicles

A testicle or testicles that haven't descended fully into the scrotum yet are not at all likely to do so in the future, and many pediatricians and pediatric urologists feel that surgery can and should be performed between six months and a year of age.

### Iron deficiency

Iron deficiency is potentially serious not just because it can cause anemia but because it can damage development and impair the potential intelligence of the child, and can diminish physical growth as well. A child can be iron deficient without being anemic, though a test for anemia usually will detect iron deficiency. It is also possible for a child with an inherited variant of the usual hemoglobin (such as thalassemia or sickle trait) to be anemic without being iron deficient. Most pediatricians check for this condition at the nine-month visit by dietary history and by blood test.

This concern is one of the main reasons for continuing a baby on iron-fortified formula rather than switching to cow's milk, until a year of age.

If your child, or you, takes iron supplements, do remember that iron taken in overdose is poisonous. Ironically, children's iron preparations taste bad enough that they are rarely ingested in quantity. It is the adult sugar-coated iron tablets that babies get into. Beware!

*Lead poisoning*

A baby this age is just tall enough to reach window sills, and has excellent fine motor skills for picking off paint chips and eating them. Since layers of old lead-containing paint can hide in these chips, a young toddler is a candidate for this way of getting lead poisoning.

Lead poisoning can be present without any symptoms at all, and it takes only a low amount of lead in the body to effect development and intelligence, so the best way to diagnose lead poisoning is with a screening test on all toddlers. To test for lead, it's necessary to take blood from a vein in the arm. In many pediatric practices, this is a routine test performed at least once in early childhood. In other practices, however, very few children turn out to have excess lead levels in their blood. These practices may test only those children who have a particular risk factor for lead poisoning.

Here are some risk factors:

- Spending time (home, daycare, relatives) in a house built before 1960 that has peeling or chipping paint, or in which construction is going on or planned.
- Having a crib that might have been painted, or repainted, with paint containing lead.
- Having a sibling or playmate diagnosed with lead poisoning.
- Being around an adult whose job or hobby involves exposure to lead, such as colored pool table chalk, arts and crafts with lead paint, working at a battery plant.
- Living near an active lead smelter, battery recycling plant, or other industry likely to release lead into the environment.
- Well water that might be contaminated with lead.

Symptoms of lead poisoning, which appear after damage has already been started, include constipation, abdominal pain, headaches, anemia, slowed growth, convulsions without fever, and mental retardation. Of course, all these symptoms are much more likely to arise from other causes, unless the baby has been exposed to lead.

*Gluten-Sensitive Enteropathy (GSE)*

This is a condition, also called sprue or celiac disease, in which a child can't digest an ingredient in wheat, rye, and barley (and maybe oats). People who never eat anything with gluten in it can have GSE and never know it.

Most children don't start on these grains until after a year of age, but some do. And if they have this inherited condition, they can have symptoms ranging from very mild to severe. If a child is very gluten-sensitive and is exposed to these grains, she will have abdominal pain and foul, pale stools. She'll have a nasty temper, won't grow well, and her muscles will be weak and underdeveloped. She may be anemic. She might have a smooth tongue and sores in her mouth.

The possibility of having GSE is highest in those whose ancestors originated in Western Ireland: one in three hundred people from that region have sprue. It's least if you hail from Sweden (one in five thousand). Everybody else is in between.

The diagnosis can be suspected if you take the child off all gluten-containing foods and she does better. Blood tests may increase suspicion. But the only way to make a definite diagnosis is by having a biopsy of the small intestine. It is worth doing if GSE is suspected, because the condition doesn't go away, and because you may be able to warn relatives and siblings if you know that one person in the family has the condition.

## MINOR INJURIES

Babies are just starting to hurt themselves on a regular basis at this age, so this is a good time to gird one's loins and decide in what spirit you are going to take the inevitable bumps and bruises. Would you treat a baby girl differently from a baby boy? Why? How did your parents treat your childhood escapades, and what do you think about that in retrospect? What parenting approaches have you seen and admired, or seen and been repulsed by?

This is the age at which babies are learning to come to an adult as a resource. If the injury is obviously a minor one, letting the baby decide whether she needs help is certainly appropriate.

Moreover, babies this age are learning which behaviors are considered "cute" and attention-getting. Withholding sympathy is hard-hearted and mean, but parents might wish to consider how much and what kind of fuss to make over a bonk on the head or a fall from the couch to a carpeted floor. "I told you you'd fall!" is a conceptually useless statement to the child, no matter how much better it makes a parent feel.

Preventing major injuries is ideal, but minor contretemps are inevitable. In fact, it's good to plan ahead, assuming that they will occur.

### Head bonks

Most babies will hit their heads at least half a dozen times during this period. Most such Bonks will be innocent. Your major challenge will be putting something cold on the area in an effort to prevent swelling. The scalp is endowed with lots of blood vessels, and the bruise can look like a goose egg. If there is a cut, therefore, expect lots of bleeding.

The best first aid is a cold pack—the moldable kind you buy at the athletic store and keep in the refrigerator. If you're really stuck, make your own with frozen veggies in a plastic baggie. Never put ice, or the plastic bag, directly on the skin: put a thin cloth over it.

Never put an ice pack or your vegetable baggie on body parts that can get frostbite. Use a cold cloth, instead. These body parts are:

Fingers and Toes
Penises and Noses,
And ears,
My dears.

When do you need to have Cherub examined for a bonk on the head? (See Part II.)

**Not much worry:** If Cherub had a fall, but was never unconscious, looks fine, and is acting perfectly normally, she is probably OK. Make sure you check her all over, not just her little bonked head! It is important to watch her for at least two hours, to make sure nothing is developing.

She should be seen by the pediatrician if:

• There is a bruise on her scalp (where the hair grows) that you can see or feel.
• Bruises on the forehead are usually not significant, but if it is a big, "boggy" one, it could indicate an underlying skull fracture, and must be checked out.
• She fell from a height of three feet or more, or fell with unusual force or onto a particularly hard surface, such as concrete, linoleum, or wood.
• You didn't see what happened, just heard a thump and there's Cherub, wailing, at the bottom of the stairs, for instance.
• There is any fluid from the ears, or blood or persistent fluid (not from crying) from her nose.
• There is vomiting after the injury.

In any of these situations, your pediatrician may order a skull X ray. A skull X ray showing a fracture is merely a sign that there could be bleeding into the brain. The presence of a fracture in a baby this young is an indication for obtaining a CT scan to see if there is any bleeding or risk of brain injury.

Don't bother checking pupils *if the baby is otherwise fine.* Dilated pupils are common after a shock such as a head bonk, and in an alert, yelling baby who calms down and acts fine, are not a sign of trouble. Unequal pupils after a head injury are a fairly late sign of bleeding into the brain, and you would have plenty of reason to worry, and to get help, before they occur.

**Severe worry:**

*Signs that there could be bleeding inside the skull, or a brain injury, require urgent help: call 911 or go right to the emergency room if:*

• The child is unconscious, even for a few seconds;
• The child acts irritable or lethargic or behaves in any other worrisome way, holding her head in pain and moaning for instance;
• She has a seizure; or
• She is not acting quite right and her pupils are of obviously different size.

*Bonks that bleed*
With head bonks that bleed there is always much more blood than you would imagine, so much that you might be terrified. Almost always, it looks like much more than it really is. Try to keep this in mind when that inevitable crack on the head happens.

**What to do:**

*First assess the likelihood of serious injury— see above. If there are any signs you need immediate help, set that in action.*

If there are not, then address the bleeding.

- Most often the baby will be fully conscious and screaming, without any trouble breathing or impaired consciousness at all. In this case, wipe the area off with the cleanest cloth available (usually your own shirt) and see where the bleeding is coming from. When you find the cut, apply firm pressure for five minutes. Don't peek under the cloth. Use whatever soothing noises you can to calm and distract the baby. Pretend that blood is your favorite bodily fluid, and admire the baby's. Make your tone of voice like that of Big Bird when he sees a particularly gorgeous flower.
- As you do this, try to assess the rest of the child's body. Arms and legs working OK? Any teeth knocked out of place?
- After five minutes, the bleeding will have stopped and you can apply your cold pack off and on for another five minutes.
- In a baby this young, any bleeding head injury deserves at least a phone call to the pediatrician, whether or not the cut needs to be stitched. So call and get instructions.

Check that Cherub is up-to-date on DTP vaccines, since diphtheria and tetanus germs live in dirt.

## MAJOR INJURIES

*These should be avoided at all cost. Nevertheless, this is a good time to be sure all the child's caretakers are up-to-date on CPR training and speak enough English to summon help effectively.*

### Choking

This is perhaps the most serious and dreaded accident. Prevention is crucial. All caretakers should take an up-to-date CPR course, and rehearse its lessons mentally every so often.

### Poisoning

Most babies this age can't unscrew any container tops of medications, much less the child-proof variety. Their danger comes rather from plants, household chemicals, and tablets or capsules or open bottles lying around loose. The local Poison Control phone number should be posted by each phone.

### Water immersion

The danger in this age is not limited to pools and hot tubs, though these are a major problem. Pools with soft pool covers are particularly treacherous, as the baby can become trapped invisible beneath the cover. Baths, toilet bowls, buckets of water, and fountains are all irresistible and potentially lethal.

### Burns

Hot liquids, stoves, and fireplaces are what parents usually, and rightfully, worry about. Consider also: curling irons, frayed electric cords, exhaust pipes.

Syrup of ipecac, used to produce vomiting in poisoning incidents and overdoses, can be dangerous. If you intend to use it, you need specific instructions to do so. Do not rely on a product label for instructions!

### Animals

The German shepherd who loved the sedentary baby may become quite annoyed at the tail-grabbing nine-month-old. Even cats can inflict major damage.

### Falls

Serious falls are quite rare in this age group, as babies don't usually climb higher than a couch. Any fall from a height of more than four feet should result in a medical assessment promptly, even if the baby seems fine. Beware windows above the first floor.

### Car accidents

These are a special problem now for several reasons. First, babies sometimes can't fit easily into an outgrown car seat that faces backwards. If that's the case, invest in a larger one! Facing forward during a car accident can cause serious damage to the spinal cord at this age, as the neck isn't yet stable. Second, babies are more social and active, and it is tempting to put the car seat in the front passenger seat rather than in the safer backseat. Third, some babies become very bored and distracting to the driver. Fortunately, hardly any baby this age can undo the clasp and liberate herself. Babies must have shoulder straps, not lap belts. Accidents with lap belts can damage internal organs and even the spinal column. If your car has an air bag, its deployment could smother a baby. Consult your car manual above having an air bag deactivated. Drive carefully and soberly.

If Cherub hates her car seat and screams when she's put into it, try this. Bring the car seat inside the house, put it on the floor, and let Cherub explore it. Then put her in it, and give her a food treat. Once Cherub seems to be attached to her car seat indoors, put it in the car again while Cherub watches you, then immediately put her into it and give her a food treat. Pediatric medicine and veterinary medicine have a lot in common, don't they?

### Older children

Older siblings, especially those in the two-to-four range, can become extremely jealous and annoyed by mobile, curious, attention-getting babies this age. Even young children can inflict serious injuries in a jealous or frustrated rage.

## Window of Opportunity

### GEARING UP FOR LEARNING: A NEW ROLE FOR ADULTS

In many respects, this is a very important three months. For babies, it is the beginning of independence, language, and the ability to use adults as a trusted resource. For parents and caretakers, it is the beginning of a new role, in which they not only nurture, but also teach, impose limits, and tolerate separation and independence.

To summarize, parents and caretakers can assist their babies best by:

- Frequently engaging the baby in a dialogue of play and language.
- Setting appropriate limits without using punitive restrictions or physical punishment.
- Rewarding curiosity and exploration with attention and interaction, not just praise. Busy adults may be tempted to give attention mainly to problem behavior, and the intelligent baby will respond by identifying problem behavior and producing more of it.
- Thinking of masturbation (self-stimulation; "playing with oneself") as a normal pleasure to which the baby is entitled, and as a first step in protecting the baby against molestation and sexual abuse later on. Babies need to learn that they are allowed to touch themselves so that they feel "in charge" later, when told other people aren't allowed to do so.

Specific skills that are easier to learn now than later are:

- Self-feeding, with a spoon introduced towards a year of age, to use like a shovel.
- Substituting cup for bottle increasingly.
- Reinforcing sleeping in the crib without

waking parents. See the section in this chapter on *Separation Issues.*
- Relinquishing the pacifier. This is the last easy time to accomplish this. After a year of age, such a habit is overdetermined and very hard to give up. Excessive pacifier use can affect teeth and jaw development and even interfere with learning to talk—it's hard to babble with your Binkie in your mouth all the time.

To help a baby discontinue the pacifier, it helps to make the pacifier less attractive. Give it a little help "wearing out." That is, take a clean darning needle and poke one hole in the pacifier every night or so. The pacifier gradually becomes less appealing as its surface becomes rough and the voluptuously air-filled nipple collapses. Bye-bye Binkie.

## LEARNING TO LOVE BOOKS

Books are not valued primarily as literature at this age. They are hinged toys, and they also can be eaten. The seventeenth-century essayist Francis Bacon must have known some nine-month-olds when he wrote, "Some books are to be tasted, others to be swallowed, and some few to be chewed and digested." ("Of Studies," 1624.)

Towards the end of this period, however, language begins to take hold. The words on the page won't mean a thing, but the concept that the picture reflects an entity gradually will evolve.

If cuddling up and reading to your baby becomes a habit established now, both adults and baby will not be tempted away from it as the baby becomes more active and negativistic later in the second year of life.

### Drinking from a Cup: Giving up Bottle

Drinking from a cup is exciting now, and this is a good age for many babies to wean from bottle to cup. Be sure you're putting milk in the cup, not just water or juice. But you know how I feel about juice.

## What If?

### Starting Daycare

Because of the sometimes exhausting activity and strong attachment of babies this age, many parents suddenly yearn with new force for a few free hours a day. Others may have planned from the beginning to return to work at this time, since the baby—now mobile and adventurous—seems much less a baby now.

If you do start daycare now, you will be grateful to yourself if you do so intelligently. The baby needs to feel very attached to the primary caregiver if she is going to be able to feel comfortable separating from that person—and that includes accepting the setting of limits.

This is also an extremely important time in the child's learning life, when her relationship with her caregiver helps determine many crucial traits.

- The care arrangement should be as close to one-adult-one-baby as possible. The adult need not be highly educated. Ideally, the adult ought to speak grammatically. This is not so vital, however, as that she or he should retain a bit of childlike delight in simple play and "baby chatting," and actually take pleasure with the baby in simple games like switch-the-lightswitch and drop-the-socks. Frequent, pleasant interactions are crucial. Babies who do well have been shown to participate in nearly twenty such interactions with a loving adult every hour.
- Spanking, slapping, or inappropriately advanced strategies such as "time out" must be absolutely out of the question. However, the caretaker must be able to intervene appropriately when the baby tests a limit.
- The caregiver should share the parental attitude towards feeding and snacks, scheduling, and naps.
- The caregiver should understand the importance both of necessary hygiene and necessary mess. It's crucial to have the diapering area and the food preparation area separate, and to wash hands thoroughly and often.

*Safety Concerns at Daycare*
- If there are older children, there need to be special rules and arrangements for keeping little toys, such as Barbie slippers, out of the reach of babies and toddlers.
- Every caretaker must pass a yearly CPR course, and be able to summon help in English, unless the local medical setup is bilingual.

- The ratio of children to caretakers should not be greater than 4:1, unless the "extra" children are much older and able to take care of themselves.
- Secondhand smoke and aggressive pets should disqualify a caretaker.
- Pools and Jacuzzis should have better security than Fort Knox.

## EXTENDED SEPARATION FROM THE PARENTS

By this, I mean leaving the baby with someone with whom she has not formed a deeply loving and trusting attachment (for example, not with the other parent).

To do so is not a good idea at this time, unless it's absolutely necessary. Babies are just forming their attitudes towards disappearance-reappearance, separation and loss. It is within this context that effective, kind limit-setting can take place.

If parents must leave the baby with another primary caretaker, if at all possible the caretaker should come to the baby's home, and spend at least a day or two interacting with the baby and participating in the schedule, especially the bedtime ritual. It's helpful if the baby already has established a transitional object attachment both to the crib and to a toy, blanket, or piece of parental clothing.

If this simply cannot be done, parents must ensure that their substitute caregiver is patient, kindly, and hasn't a mean bone in her body, because she—or he—is likely to have a rough time of it. If the baby does NOT show displeasure and anxiety loudly, the caregiver should be alert to signs of infant depression: not eating, not playing, withdrawing. These are urgent signs that every effort should be made to get the parents back as soon as possible.

When the parents do return, they must be prepared for either anger or, what feels worse, rejection; the baby may turn away, appear not to recognize them. It may take days or weeks to win the baby over, and even then the process of making brief separations easily may have to be learned all over again.

Given all this, parents should not feel unduly guilty unless this was a purely-for-pleasure indulgence. A second honeymoon to save a very troubled marriage, a necessary though not urgent hospitalization, a business trip in which a career hangs in the balance, don't fall into the category of indulgence.

The vital aspect is not to take the baby's point of view lightly. This is a crucial time, and the preparations made, the care with which the substitute parent is chosen, and the time spent afterwards in reestablishing trust and ties may ultimately enrich the entire family.

## TRAVELING WITH THE BABY

The advantages of traveling with a baby this age is that she is easy to please, interested in everything as long as she doesn't have to be worried about being separated from her beloved adult. She is more likely to make friends and flirt than to scream, unless she is tired or sick or hungry or uncomfortable.

The disadvantages are that she will want to

be "walked" everywhere, with a bent-over parent grasping her upstretched hands, and will investigate everything at your knee level and below. She will find things to scale that were never meant for climbing. Since everything in the world is brand-new and exciting, your oohs and aahs over adult-oriented spectacles will leave her unexcited; her need for a cracker supersedes your interest in the cathedral.

I would discourage trying to sedate a baby this age for travel. The usual sedatives often backfire. They make the baby just sedated enough to lose her primitive abilities to adapt to new situations, and she is more likely to be outraged and irritated by strange events and people. Some babies have this happen to such an extent that they become paradoxically agitated and "hyper." It's a long way from New York to Paris. It's even longer if you have to walk the whole way, up and down the aisles of the 747.

## ADDING TO YOUR FAMILY

Oh, brave new world, that has such people in it!

Here is one way to think about such an enterprise: how would it feel to have a demanding newborn right now, right this minute? Would you be able to spend the time you do now dialoguing, distracting, substituting, gently roughhousing, allowing mess, and enjoying this wonderful nine-month-old? Would you have any time at all to nurture your marriage, sustain a career, brush your teeth?

If the answer is no, then think again about conceiving another child. Because when that second baby is a newborn, your present child will be infinitely more demanding than she is now. She will be eighteen months old or so, and going through the demanding clinging-and-then-rebelling, possessive, negative phase called Rapprochement. She will be needing enormous amounts of attention to foster independence, to prevent "spoiling," to encourage crucial language skills, and to keep her from endangering her, you, and the newly crawling and investigating second child's life and limb. She will have an occasional temper tantrum, and if her life is too frustrating and the adults around her are too fatigued, short-tempered, or expect too much from her, these will increase.

Of course, anything is possible! If your biological clock is ticking rapidly, if there is an unplanned Event, or if you are convinced from personal experience that closely spaced siblings are better friends, you may want to undertake this adventure.

However, it is best to understand ahead of time what you're getting into, and plan for it. Two children with this spacing need twice as much one-to-one adult input, not half as much as one child of either age alone. Not only will the older one be incapable of "helping," he or she will devise ingenious means of diverting a great deal of attention away from the new baby. It is a little like having three, not two children: the older child, the younger child, and the "evil invisible child" that is the relationship between the two siblings.

"Thank God they both take naps," Georgia says, hair, eyes, and hands all flying about distractedly. "I close myself in the bedroom and just whack the bed with a tennis racquet. It keeps me from murdering them. Or their father."

How to reconcile spacing dilemmas?

The first step is to recognize they exist, before you contemplate, or conceive, child number two. If you fantasize a wonderful dream world in which the two-year-old happily entertains the one-year-old, you are in for a terrible shock.

If the older child is to attain optimum language and social development, he or she will need:

- At least 15 minutes alone with either (preferably, with each) parent every day, without anything interfering: errands, phone call, sibling, coparent. Bedtime rituals count.
- An enormous amount of language stimulation, in which brief conversations about pleasant activities occur ten to twenty times an hour. Saying "No!" doesn't count as a brief conversation.
- Patient help with daily activities, such as feeding, dressing, bathing, as he or she learns to perform these tasks independently.
- Close adult guidance as he or she enters the social world of other two-year-olds, a world of territoriality, aggression, and just the beginnings of interactive play. Sharing does not occur until age three-and-a-half to four!
- Understanding that since his or her language skills are still rudimentary, the child can't talk about negative feelings

to discharge them, nor very effectively use symbols (such as flogging a doll instead of a sibling). Instead, she will act them out.

As Burton L. White says, "Though parents can do much to alleviate the problem of closely spaced siblings, there is simply no way of making the situation as easy as dealing with a first child only or with widely spaced children. Both parents must understand and accept that fact." *The First Three Years* (Prentice Hall Press, 1990), p. 248.

## DIVORCE, SEPARATION, AND CUSTODY ISSUES

I'm not a marriage counselor, but I have seen lots and lots of families.

Very often, marital discord at this age springs in large part from the "triangle" produced by a very strongly attached baby. One or the other parent is so in love with the infant that the other feels rejected and jealous, and miserable and guilty for feeling this way.

Such an overwhelming attachment may spring from the parent's own undernurtured infancy or childhood. Bitter feelings may resurface at this time, and may seem to be comforted by this new strong attachment to the infant.

When this is the case, the very attached parent may actually resent participation by the other parent, which is perceived as diluting these extremely strong feelings. Usually, discussion is of limited value here, and what is needed is therapy. Some therapists feel they can help the situation in only one or two sessions.

If things do progress to divorce, the custody issues can be brutal, with the overwhelmingly attached parent demanding total rights, and the rejected one demanding overnight visitations, usually to an abode, and often with a lover with or without other children, that the baby has no ties to.

If strenuous marital therapy cannot turn the marriage around, then new rules come into play. From the point of view of the baby, it is vital that they be strictly enforced.

1. No intramarital violence, word or deed, with baby as witness.
2. No abrupt change in primary caretaker. If daycare must be changed or instituted for the first time, follow the guidelines above.
3. Visits from the departed parent should take place on the baby's turf, at least initially. Brief visits, without an interruption of schedule, may gradually be instituted to the noncustodial parent's new abode. However, it is too much to expect a baby this age to swiftly become accustomed to spending the night or a weekend. If the new abode includes toddlers, all parents must bear in mind the caveats above on closely spaced siblings.
4. Grandparents still have rights, unless they have forfeited them by choice or behavior, and the baby has rights to them.
5. Questions about feeding, clothing, visitations, social contacts, and so on may

certainly be brought up with the pediatrician. However, the smart pediatrician will not allow himself or herself to be quoted, out of context and angrily, by one parent to the other. And the smart parent will avoid doing so, as this will result in an angry phone call from the offended parent back to the pediatrician, who will be justifiably annoyed.

## SURGERY AND/OR HOSPITALIZATION

Many surgical procedures, some minor and some major, are technically best done during this age range. These may include hernia repair, bringing down an undescended testicle, placing "tubes in the ears" for chronic otitis media, procedures for a cleft palate, and some kinds of corrective heart surgery.

In many respects, there are some psychological advantages to having such procedures done at this age. The baby is not old enough to anticipate the occasion, is still orally focused enough that such distractions as sweets and sucking can soothe, and is heavily dependent on the presence of a parent, rather than on a familiar environment, for security.

Of course, these can only work in a baby's favor if the parent can be with the baby until anesthesia takes over and after it begins to lighten, and if the parent can maintain as calm and supportive a demeanor as possible.

Bring your own sleeping bag and pillow, besides the usual toothbrush and book, to the hospital if you are staying overnight. Bring more than one pillow so that you can prop the baby and your arms when you hold her.

Soap bubbles are a fine hospital distraction at this age. Noisy, large toys are not. Books are not likely to be appreciated. Balloons are, or should be, forbidden.

For urgent hospitalizations, the most important task is to insist on finding one medical professional who can interpret, harass, coax, and otherwise deal with all the others. This is probably your own pediatrician. It isn't necessary or even desirable, frequently, that your own doctor run the show, but he or she should fulfill the other roles. Make sure that this is clear when you choose a pediatrician.

## MOVING

As with all big changes in this period, babies will take their cue from parents, and parental calm and pleasure over a move will smooth the way for the baby. A few precautions will help:

- Keep the baby's schedule as regular as possible.
- Any introduction of even a casual, brief babysitting arrangement should be prepared for ahead of time.
- Have the baby absolutely elsewhere during the move itself. Statistics for accidents during a move are astonishing and scary.
- Keep one room at least of the new home baby-proofed from the beginning.
- Don't change the furniture in the baby's room at the same time you move.
- After you move in, turn down the hot water thermostat to 120°, and check the fluoride content of the local tap water.

## The One-Year-Old Well-Child Visit

I love this visit. The baby is usually happy to flirt, and most of the exam can be performed with the baby in the parent's lap. The baby is interested in, but no longer tries to eat the stethoscope and otoscope.

As with the nine-month-old visit, it's best to bring along the baby's own toys, so he doesn't mouth and handle everything in the waiting room. Snacks should be crumbless and spillproof unless you want a bad reputation with the office staff, and should be used only if you know there will be a wait. Giving a baby a snack before the throat exam invites gagging to the point of vomiting. Moreover, the pediatrician will have to decide whether that stuff in the throat is thrush, pus, or merely cottage cheese.

Most babies will receive a test for exposure to tuberculosis at this visit, as a needle injection under the skin (PPD). Remember that this is a test for exposure only; if the test is positive, it does not mean that the baby is sick, but that further testing is needed.

Immunization schedules tend to change as vaccines are developed, improved, or combined with other vaccines. Some possibilities at the One-Year Visit: the MMR, the live vaccine for measles, mumps, and rubella; the chicken pox vaccine; the vaccine for strep pneumococcus—a cause of bacterial meningitis; a booster of the hepatitis B vaccine.

## Looking Ahead

These coming six months set the tone for the whole second year. This is the year when you guide your baby towards being a delightful person to be with by the age of Two. The delightful two-year-old, as opposed to the "Terrible" two-year-old, is someone who is fun to be with. He listens and talks, he can express negative emotions without (most of the time) having a tantrum or physically attacking, he is attached but able to separate. He feeds himself, sleeps the night (mostly) without waking you up, and is starting to play with other children, though he may be a long way from wanting to use the potty and even a longer way from being able to truly share.

Already you have a One who knows that she is a beloved and wonderful person, that other people have some rights that occasionally take precedence over hers, and that physical and emotional separation and limits exist and are not scary. One knows that the world is a fascinating and trustworthy place to explore, and that people can be counted on to treat One with respect and consistency. Building on this fine foundation makes the journey from One to Two an exciting, engrossing, demanding, and enlightening adventure.

# EIGHT

# One Year to Eighteen Months

## Where There's a Will, There's a Wail

## To-Do List

- Engage One in lots of chit-chat.
- Have daily picture book sessions.
- Expect a slow-down in appetite.
- Encourage complete self-feeding, cup, spoon, and fork.
- Be Firm about Limits.
- Change the batteries in the smoke detectors (each year on Cherub's birthday).
- If you plan on moving Cherub from a crib to a bed before Age Two, see the next chapter—consider erecting a barrier to her bedroom door *now*.
- If you intend to use a harness or leash later in the year, introduce it now, while One is still in a stroller.
- At one year of age *and* 20 pounds, Cherub needs a new car seat.

## Not-To-Do List

- Stay away from juices, "handout" snacks, ritual snacking, and catsup.
- Don't get into a TV habit at home; rule out TV at daycare; vow NOT to put a set (or a computer) in Cherub's bedroom.
- Don't move from crib to bed yet.

## Portrait of the Age: Twelve to Fifteen Months: Revving Up

Kyle loves to almost-give things.

Arms out, feet wide apart, he toddles across the room, spies his beanbag frog, and tries to stop. He overshoots, turns, squats to pick it up, and heads towards his mother like a wind-up toy. "Thank you," says Emily, reaching. Just as her fingers close, he grabs it back with an enchanting smile, goes to his book corner, and returns with an ancient

*Time* magazine. "Thank you," says Emily, as Kyle snatches it away.

On his way back to the book corner, Kyle spies a crayon on the floor. As he goes to pick it up, he lets go of the magazine and it falls on top of the crayon. Kyle steps onto the magazine. He bends and yanks at it. It won't budge. He pokes at the lump the crayon is making under the magazine he is standing on, and tugs the magazine again. He frowns in frustration. Again and again he tries, his brow ever more furrowed. Finally, he stands up and points at it and grunts, staring at Emily with a command and a request written on his face.

Emily, trying not to laugh out loud, goes to help him. "Look, Ky, you can't lift it up because you're standing on it."

As Kyle turns to figure out what she is saying, his feet slide on the slippery cover and he sits down, hard. Success! In doing so, he has displaced the magazine from the crayon. Pleased, he grins at Emily's applause and picks up the crayon. On the way to give it to his mother, however, he is detoured by a tempting chair seat and pauses to rub the crayon on it. You couldn't call it scribbling, exactly.

"Chairs are not for crayoning on," Emily tells him. "Here is paper for Kyle to crayon on."

Kyle isn't interested. He has paused by the book cabinet, where, dropping the crayon, he opens and shuts the door a few times, staring at the hinge as if it contained the secret of the universe. Seeing him engrossed, Emily dashes to the laundry room before he can notice and object to her absence.

She returns with a basket of clean clothes. Kyle sighs with satisfaction. His favorite thing. He finds a pair of his father's shorts and places them over his head, inhaling noisily. "What a nice hat! Kyle made himself such a nice, fragrant hat!" Kyle peeks out from under the hat and sniffs with greater vigor. "I wish I had my camera. What a great ad for fabric softener!" Kyle giggles at her, making flirtatious faces.

Casting off his unconventional headgear, Kyle heads to the laundry basket, grabs its rim, and tries to pull it over. "The laundry basket isn't for tipping over!" Emily says, detaching Kyle and righting the basket. She has used this sentence structure so many thousands of times that it has carried over into her adult life. "Order pads are not for doodling," she tells her subordinates. "Newspapers are not for putting coffee mugs on," she informs her husband.

Kyle stares at Emily, and returns to the basket. Again, she detaches him and rights it. And again. And again.

Finally, **"The laundry basket is not for pulling over!"** Emily seizes Kyle and deposits him in the playpen, which is standing in the corner. "Please wait in the playpen until Mommy finishes with the laundry."

Outrage! Kyle shakes the net siding as he builds up for the Yell. As it bursts forth, he stops shaking the siding and concentrates on his howl. He pauses a fraction of a second to assess his mother's reaction. Clearly finding it displeasing, he inhales again, goes to shake the siding, loses his balance, and sits down hard. In this position he yells, eyes scrunched shut, for approximately thirty seconds, at which time he opens them and looks around the playpen. Still yelling, he finds his wire-and-wood-ball construction toy, and settles down to squish and stretch it.

Emily works furiously, folding laundry. Her inner timer tells her that she has thirty seconds, now twenty, now ten, until Kyle comes to the end of his attention span.

## Portrait of the Age: Fifteen to Eighteen Months: Under Way

Kyle loves his beach ball.

He loves his collection of bald tennis balls, rubber balls, sponge balls, He loves to roll them, drop them, stare at them, caress them, and just carry them around. In fact, he appears to love everything that is round and moves spinningly: bicycle wheels, toy-car wheels, rolling crayons. He watches and watches and watches them spin. Sometimes he makes his car noise; sometimes he just gazes as if hypnotized.

If Kyle were not so winningly social, Emily would worry that he is autistic. She thinks that under electron microscopy, the Y chromosome probably can be viewed rolling a ball. Ted, on the other hand, predicts a career in pro sports or race-car driving.

Today, Kyle has decided to carry his beach ball with him to the store.

"Kyle won't be able to hold Kyle's big ball in the car seat, or in the shopping cart," Emily explains. "Let's take Kyle's race car instead."

She holds out the shiny green Corvette to her son. Kyle merely grabs the ball more tightly and sticks out his chin. Emily, in a hurry, pries at the ball with one hand as she swoops the car around enticingly with the other, "Vroom, vroom! See that car go!"

Kyle sits down on the floor, now gripping his ball with white-knuckled fingers. He begins to roar, but without tears, glaring at her.

Emily glances at her watch. "OK, OK, take a ball, but take a smaller one." Cunningly, she fetches the Nerf ball. "Here we go."

The roar gets louder.

"All right, then, here!" Despite visions of toppled grocery displays, she gets the six-inch red rubber ball, Kyle's usual favorite.

No go.

"Honestly!" At the end of her rope, she reaches for the beach ball and grabs it from his hands. Kyle topples backwards. There is an ominous second of silence and then fury. Screams and kicks. Emily simply stares. Why, she realizes, Kyle is having a tantrum! He's had outbursts before, but this is a true-blue tantrum (blue, indeed; purple, more like it. Will he burst?).

She knows the rule. Never let the child succeed in getting what was forbidden by throwing a tantrum. But time is passing rapidly, and what's the big deal about a beach ball after all? Maybe he can contrive to hold on to the thing.

"Kyle, here!" She offers him the giant ball. Kyle stares at it, stares at her. They both hold their breath.

Then Kyle strikes out, pushing the ball furiously away with both hands and feet, so vigorously he bounces, then picks up his interrupted shriek exactly where he'd left off.

Why? Why refuse an offered gift, again and again and again? Why reject the very object he had been fighting to hold on to?

## POSSESSIONS AND SHARING

Connected to the idea of becoming an individual is the idea of possession. The two are inseparable: every individual human being is defined by his possessions, whether they are material or in the realm of spirit.

When One-Year-Old Kyle "almost gives" something, it isn't because he has an impulse to share. On the contrary. When he "gives" an adult an object, he *knows* that the adult will give it right back—it's a game you can count on. Like peekaboo. Kyle is proving to himself over and over that he can make a possession disappear, sort of, for a second, and reappear. When it comes back to him, it hasn't been changed by its brief disappearance.

When Fifteen-Month-Old Kyle holds tight to his beach ball, and rejects all other balls, he is demonstrating his progress in understanding the meaning of a possession. It is not any one particular ball, but the fact that he is holding on to it, that defines him as a person who possesses—and, therefore, as an individual. When his mother grabs it and then tries to make it all right by giving it back, she is (from Kyle's point of view) devastatingly missing the point. She is challenging his status as individual, as much by giving the ball back to him as by taking the ball away. It would be different if *he* grabbed it back.

This exasperating phase is the very first step in learning to share. To share, a child has to be aware of himself as a person who possesses something and who therefore can give it away. So naturally the first step is to work out the idea that "I am a person who possesses something." Just working this out takes up a large portion of this whole year. Here are the puzzles Kyle must solve before he can share:

- The concept that words can "stand in" for objects or people, so that you don't always have to keep hold of the thing or person physically.
- The realization that someone else has a point of view different from one's own, and therefore a sense of fairness.
- A sense of future time, so that giving up your beloved beach ball doesn't seem like abandoning it.

That Kyle will acquire all these steps, in rudimentary form, by age three or four seems to Emily unlikely. Or at least amazing.

I guess it is amazing.

And sometimes things go awry. Bobbi, at a year, had twin brothers age three and a new baby sister. Her brothers wouldn't let her hold on to anything, and her parents were too busy with the baby to intervene.

On her third birthday, her friends gave her presents. Bobbi unwrapped the first one, her eyes huge. "It's from your friend Andy," her mother said. Bobbi's expression changed to horror, anger, grief. "Not Andy's! Mine! Mine! Mine!" Tears streaming, she hugged the gift to her chest, staring at the stack of remaining presents with a look composed half of anticipation and half of dread. As if, her mother says, she were looking at a fortress she was going to have to defend.

Bobbi is still working out the idea: What is a possession? She is miles from the normal three-year-old concept: A possession can be given up and gotten back. She has a lot of catching up to do before she'll be in any condition to share.

## Oppositional Behavior—Part of Growing Up

During the year from One to Two, a child carves out his personal niche of identity in the world. Much of the time it seems to parents as if Cherub is doing so by opposing himself—no, no, let's say by *sculpting* himself, against anything he runs up against. Against other people, the external world, immutable physical laws, and even against his own passions.

Another person's plans must be rejected simply because they are another person's plans. Stairs and tables are to climb, animals to attack, boxes to get into, objects to throw or kick, simply because they are there. Immutable physical laws, such as gravity, or the concept that you have to let go of one object to grab another, must be defied.

Getting from One to Two is thus a crucial journey. Over this year, a child discovers that some self-defining activities are rewarded, some condoned, and some forbidden. Yes, forbidden. Really and truly NO—along the lines of, No you can't drink a glass of milk lying flat on your back.

Getting from One to Two is a crucial journey for parents, too. Most parents discover that some of their discipline strategies work, some have no effect, and some backfire.

For instance, it's natural for parents to feel that if they say "No," and explain their reasoning, Cherub ought to stop. Not so. Even their firmest, strictest, interdictions—against biting or running into the street or pulling the cat's tail—are defied again and again and again. Even when they shout! Even when they *spank*!

Parents don't realize that successfully setting a limit for Mid-One is a skill with a technique, and that it can be learned. This is too bad, because if setting limits doesn't "take" by age Two, life with Cherub can become a lot less fun.

So be sure to look carefully at the section on *Limits*, below.

## Temperament and Individuality

Normal one-year-olds differ from each other mostly by temperament. They can't plan ahead, so they can't yet really choose how to behave. They are more or less at the mercy of their feelings.

Two-year-olds have incorporated temperament into individuality, into a personal style of relating to other children, adults, and the challenges of the world.

Given the same environment, an easy temperament makes the road from One to Two smoother; an intense temperament gives a more challenging ride. Willy-nilly, by Two a child has developed attitudes about and ways of dealing with self and the world that can make Two either delightful or terrible.

Over this coming year, a child discovers:

- Whether his autonomy—his need to set a goal and achieve it—is respected by the adults closest to him. Can he get help when he needs it—or is he constantly interrupted and interfered with, or ignored? Is he able to explore and investigate freely—or is his day filled with negative commands or with restrictions or with physical punishment?
- How to get adult attention. Does charm do it or a simple request? Does she have to misbehave to get it? Or compete to such a degree that this activity takes up all her energy and focus? Or is the adult so closely bonded that Cherub is never free of adult attention and is lost without it?
- What language is for. Is it for conversing in a dialogue? Or is it used by adults only to give commands and orders, to complain and whine? Or is it shot through with anxiety about performing?
- Whether she is allowed to own her own emotions. Do her tears upset her parents too much—or do they get no sympathy at all? If she has a tantrum do her adults try to coax or threaten or argue

or placate her out of it? Or is it treated calmly, as though she can learn to prevent and stop tantrums on her own? Are her parents able to allow her her own emotional struggles, or do they seem to appropriate them?

Every child, no matter what temperament, what lifestyle, what culture, what family, what methods of parenting, develops his or her own unique answers to all these challenges this year.

The role of parents throughout this year is to help the child maintain a balance between her bonds to her parents and her mastery of the world. This means that parents need to be sure themselves of the boundaries between toddler and parent, and can allow the normal frustrations and negativism and possessiveness and tantrums of this period without taking them personally, as an attack on the parent's own worth and autonomy.

There aren't any simple recipes for accomplishing this.

Every child has a unique temperament. Many are easygoing, not easily frustrated, fairly easily pleased. They make transitions, from one caregiver to another, from one activity to another, pretty calmly. They venture eagerly into the world to explore and investigate. They succumb to tantrums when severely frustrated, not at every little disappointment or difficulty. A somewhat smaller group has a bumpier time of it. These Ones are easily frustrated, and have more tantrums. They are more cautious about exploring the world, and they have a harder time with transitions. And a very small group, mostly boys, take everything hard. When presented with something new, they cling and cry. They are easily frustrated by the world—the laws of physics seem to be their personal enemies. They have many, many tantrums.

Fostering independence and autonomy in an easygoing relaxed child is quite a different challenge from doing so with a high-strung, intense one. The main guide for parents is to bear in mind the goal and not take temperament and normal developmental challenges personally.

Not that a one-year-old starts off with **only** a temperament.

Kyle, just turned one, has already learned a whole range of things about himself, his world, and his family. The first and most crucial is how lovable he is. His appearance, odor, taste, feel, and bodily products are all delightful to his parents, especially to his mother. His accomplishments, particularly walking, are fabulous achievements.

The world, as he discovers anew each day, is a fascinating place. He can make things happen, and happen again, and again: it is a predictable place. When he almost-gives something away, he can retrieve it or it is given back. It's not gone forever. When something vanishes, like a crayon under the *Time* magazine, it will stay there, retaining its identity, until Kyle can uncover it. He can have a goal, an aim—pulling over the laundry basket, say—and work to achieve it. If he can't get it the first time, maybe he will

on the sixth, or the thirtieth. It is dawning on him that cause and effect are related, and, even better, are related through his own efforts.

In his explorations of the world, Kyle is pretty much in charge of his own activities. Nobody forces him to sit down and look at a book when he is enthralled by a truck with wheels. Instead, his father will come and help him run the truck on the floor and make truck noises and talk about what trucks do.

In fact, Kyle is discovering, speech is pleasurable. Most of the time it consists of praise, assistance, instruction, and narration of activities, with Kyle as its focus. He has all kinds of incentives to understand what people are saying.

When he does point and grunt, a grown-up will come and help. This is a new step in a true dialogue, with Kyle acting out questions and comments and the adult responding with actions and words.

Adults are clearly of an advanced species. Not only are they capable of amazing technical feats, like folding laundry, but they are able to determine what Kyle is permitted to do and to stop him, when necessary, by changing his location or the object of his interest.

Even though Kyle is clearly the center of their universe, his mother and father have lives of their own. Sometimes Kyle has to wait. Sometimes he has to stop an engrossing activity. Sometimes other adults, his daycare Nanny and his grandfather, take the place of his parents.

Fortunately, Kyle's universe is governed

by adults who are reassuringly in control. He has not been spanked, hit, or yelled at in a scary uncontrolled manner. Of course sometimes adults talk loudly and get an unpleasant annoyed tone in their voices now and then, and that's interesting also, and just scary enough to push beyond and see what will happen next.

When Kyle does push them, adults may remove their glorious attention from him in response. Not being the center of their universe for a period of time is desolating. It is the nadir of existence. Exile to the playpen or the crib is about as bad as it gets.

Emily and Ted have learned a few things as well.

First, it is much more fun and less wearing to remove temptation from Kyle than to remove Kyle from temptation. The Kyle-safe rooms of the house have been stripped of plants, fragile objects, and remote-control devices. This leaves plenty of off-limits items for Kyle to test them with: electric cords and sockets, the fireplace, the stairs, the dials of the TV. They have solved the stair problem by putting a gate three steps up so that he can practice going up and down to his heart's content.

When Kyle does venture into the realm of the truly forbidden, poking at the sockets, twisting the dials, they say "NO," and move him. Some days they may have to do so fifteen or twenty times, or even double that. This can drive someone crazy, especially if that someone is trying to accomplish his or her own agenda: fold laundry, make a few calls, write checks, deal with a report. On

occasion, only their powerful vision of themselves as wonderful parents keeps them from completely losing it.

In this, they are actually having a very easy time of it. Kyle is an easygoing person, open to distraction, seduced by novelty, responsive to applause and expressions of displeasure. Moreover, his easygoing temperament is matched by that of his parents. They are not disappointed that he isn't more aggressive, intense. He is just right for them. They are also blessedly free from external stresses: marital conflict, money worries, poor health, fatigue.

They and Kyle both have a firm foundation to take on some of the more demanding aspects of this six-month period. Kyle has a very nice balance among the three enterprises that engross him:

- Exploring the world of objects.
- Practicing his motor skills.
- Developing his relationship with his parents and caregiver, especially with his mother.

If one aspect overwhelmed the others, Kyle and his parents would be in a for a tougher time than necessary, because at some point during this period each of these three enterprises goes through a transformation:

- Instead of just manipulating objects in the physical world, he will wish to possess them.
- Not content merely to practice his motor

skills, he will increasingly use them to accomplish goals that he sets for himself.

- His passion to achieve this autonomy will put him into conflict with his parents and caregiver. He will particularly push away the adult closest to him because that is where he feels the strongest pull toward dependence. Every such pull demands an equal and opposite push away by the child.

## Separation Issues Twelve to Eighteen Months: Push-Me-Pull-You

Ted and Emily have started to notice another odd thing about Kyle during this six months. Sometimes he seems to cling to them like a barnacle; then, without much in the way of transition, he's pushing them away. Sometimes he can't seem to make up his mind which to do, and dissolves into frustrated howls.

"He needs two contradictory responses from us at exactly the same moment, and he wants them both with exactly the same incredible intensity," Ted says. "I remember how that is. I felt just the same way as a teenager. I wanted to conquer the world and be a safe little kid, both at the same time."

The problem isn't so much sympathizing with the toddler as figuring out how to help him. The whole tone of separateness, and of relationships with adults, is established during the period from about fifteen months to about two years of age.

During this time, caretakers are charged with several tasks:

- Allowing the child to work through his establishment of personal identity without either overindulging him or "breaking his spirit."
- Encouraging the development of language, so that the child can make use of symbols (words). Otherwise, he's pretty much restricted to body language, such as kicking, grabbing, and tantruming.
- Keeping a balance in the child's life between his intense relationship with the adults closest to him, his exploration and mastery of the world, and his development of motor skills.

It isn't the job of the adult to solve the child's ambivalence; only time and experience can do that. It's the adult's job to maintain balance in the child's life between the baby's intense emotional bonds to the parents and his need to move out into the world.

This seems to be hardest when the child is so racked by tantrums that the caregiver or parent feels attacked and harassed. It is also difficult with the occasional very tranquil, early-talking baby. This child may seem so much a companion that the relationship is exceedingly rewarding to the adult, who must consciously encourage the baby into independence and exploration. Any problems parents may have had in their own lives with separation and loss may resurface now, and not necessarily as conscious thoughts. Sometimes babies this age act out with devastating candor exactly what the adult may be feeling about his or her own place or history in the world. Extreme problems in marriage and parenting that arise during this time may be partly a response to this "echoing" of unresolved loss. Parents who are aware of this in themselves, and who seek help, lay a foundation for their own growth and for their children's sanity.

## NIGHTIME SEPARATION

Sometimes separation problems are vividly enacted all day long. Sometimes they only appear at bedtime, when the end of everybody's rope has been reached. Both parents and toddlers may feel that bedtime is, indeed, The Twilight Zone.

Some families choose, or collapse into, the Family Bed during this period.

For most, however, sleeping through the night, away from the baby, is a high priority. Parents need intimate time alone together, and not just for sexual activity. They need to debrief, give each other support, recall that they are adults. They need to use long words and complicated grammar and pronouns instead of proper names. They may even need to say a few words they don't want to hear repeated, loudly, by their child in public.

And everybody, including the baby, needs unbroken rest with which to tackle the next day.

Moreover, sleeping *en famille* may erode the balance between bonding and independence many parents and babies find so precarious right now. This is particularly likely in families with only one or two children.

Troubled bedtimes can be tackled by realizing the heavy issues being dealt with by everybody, and focusing on the underlying reasons for all this distress.

- **Daytime separation issues:** If the parents are having trouble keeping a balance between the baby's close attachment to them and his need for independence, bedtime and night will exacerbate the issue. Is the bonding too tight? Is the baby spending time on her own, exploring and mastering her world? Is she able to come to the adult or call him as a resource, rather than spending most of her time clinging to him and demanding reassurance and attention? Is she happily developing new motor skills, climbing and starting to run? Or is she in the care of someone who consistently understimulates and underresponds, so that she is desperate for parental closeness at night?
- **Daytime stresses:** Events that make separation more difficult include anything that threatens balance between independence and bonding. Illness; physical separation through travel; divorce; moving; new daycare, or a change (such as of caretaker) in the old; a new sibling; a slightly older sibling who resents the toddler; poorly supervised socialization between the baby and other children—these are some common stress points.

- **Self-comforting:** If the child can't fall asleep in the crib in the first place without bottle or pacifier, nothing will help middle-of-the-night awakenings!

- **Bedtime slowdown:** Babies love to rough-house at this age, and often a parent who hasn't seen the baby all day finds this a delightful activity. But it does churn things up right before bed. A slowdown period culminating in a session so subdued that Mr. Rogers looks hyperactive by comparison is helpful.

- **Bedtime rituals:** Furthering the transition to separation and sleep, these can now start to consist of three or four steps, which are always the same and which occur always in the same order. If the last step is a song, it helps for it always to be the same song, even if one parent is tone deaf and the other opera material. "Morningtown Ride" or another slow, repetitive dirge with a less than one octave range is recommended.

- **Putting to bed:** The parent who is less intensely bonded is the ideal putter to bed. The other parent may take part in the bedtime slowdown and the early steps of the ritual.

- **Middle-of-the-night waking:** When/if the child calls to the parents in the night, that same less-intense parent should be the one who goes in to the child, and who says good-night in a firm but boring voice. He should emanate boredom and lassitude, shuffling in, eyes half-lidded, big yawn, scratching. Everything slowed way down. Big pause before registering what the child is asking for. Barely rec-ognizing that someone is upset. Why on earth should someone be upset? He can give a pat or ruffle the hair, but in a distracted, sleep-walking fashion. He can wander in and out like this every five, ten, or fifteen minutes until the child "catches" his somnolence.

- **Bedtime is cribtime:** Make sure the child does not taste the delights of joining the parents beyond his bedtime and falling asleep not in the crib. Such memories of paradise take a very long time to vanish, and add fifteen or twenty decibels to the child's objections.

- **Sleeping bags:** If the baby is still sleeping in the sleeping bag (as suggested in the last chapters) don't change now. Within the next six months he'll be able to climb out of his crib, given the opportunity. Lower the crib mattress. A crib extender, purchased or crafted, is a good idea.

  Also don't talk about the concept "climbing out of the crib" where the child can overhear you. Certainly don't tell him or her not to do so! Maybe it won't occur to him or her. Don't let anybody else talk about it or ask about it either. If you have to do so, spell out the words.

# LIMITS: The Temporarily Authoritarian Parent

Included below: • What is an Ideal Discipline, and why • Aggression • Tantrums •

Whining • Trying to run the family by rejecting one parent and preferring the other • Thinking about your strategies • Counter-opinions • The role of temperament.

Every child in the first six months of this year has to test limits. The challenge of parenting a child of this age is to define clearly what is to be forbidden, and then to find a discipline that works. By "working," I don't just mean making a forbidden behavior stop. I mean something larger.

Every time you set a limit at this age, you are trying to teach something more important and far-reaching than that it is "wrong" to bite, or to pull the cat's tail. What you are trying to convince Cherub is: I am the Boss of You. When I say not to do something, That Thing Will Stop.

Yes, this is being an authoritarian parent. This is the kind of parent Cherub needs at this age. Cherub already knows that you are a *benevolent* dictator, generous with your smiles, conversation, toy cars, picture books, visits to the park. What he needs to learn now is that you are a benevolent *dictator*: that parents are in charge. When that happens, the kind of testing behavior that can make life with a Two and older so difficult and unpleasant doesn't have a chance to start in the first place. Or if it does, it can be nipped in the bud. (If you would like a taste of Oppositional Behavior, see the Portraits of the Age for Two and Three.)

What Cherub needs to become convinced of is that it is futile to test or challenge your decision. It's a waste of time and also frustrating. No pay-off comes from it. It's not even worth considering. So really, when you act as an authoritarian parent, you are only secondarily teaching Cherub exactly what it is he is Not to Do: bite, pull the cat's tail, draw on the walls.

Cherub has to be taught this concept again, and again, and again, and again. Children learn by repetition. It takes them months of babbling to say Dada, weeks of cruising before they walk. This means that you need to have an effective technique of discipline, one that works and that you can use, *no matter where you are*, again, and again, and again, and again.

This rules out many popular or traditional techniques of discipline. Distraction, or offering a substitute activity, don't work at this age and do not teach the main lesson. Neither does a rational explanation of why a behavior is not appropriate. Other types of discipline are not portable (time-out in the crib). Fortunately, there is an alternative.

For a discipline to be most effective at this age, it needs to fulfill these characteristics:

- **You always do the same thing each time you discipline.**

    This is crucial. Studies show that the best way to encourage a given behavior, to make it happen more and more, is to be inconsistent in your response. Suppose Cherub whines for a toy while you are shopping: one time you leave the store and put her in a time-out in the car; the next time, you give in and buy the toy; the third time, you explain in an angry whisper why she should settle down; the fourth time, you ignore her completely. Well. If your goal was to get Cherub to whine every single time, with increasing decibels and pitch, you couldn't have done a better job.

- **You give an order, not a choice; you do not ask permission.**

    "No scissors!"

    Not: "Give me the scissors now, OK?" Not: "How about putting the scissors down and working on those Legos?"

- **You act immediately, without giving a warning.**

    Don't allow Cherub to persist in trying to do something again and again—grab the juice, explore a delicate instrument, escape out the door. Instead, take him away from the object immediately, and fix it so he can't go back to the testing behavior. Don't give a verbal warning or count 1, 2, 3 to give him time to obey.

- **You remove your attention from the behavior rather than giving it.**

    A response that is close to wordless, with only the briefest eye contact, is ideal. The more words you use, the more Cherub feels rewarded for his forbidden behavior. Even if the words are scolding words, or explanatory words, or even descriptive words ("Now look at poor kitty! Poor kitty is scared of Kyle! She is hiding under the bed! . . .", etc.). A good rule is to use a maximum of one word for each year of age. In the case of a

One- to Two-Year-Old, that would be two words, as in "No Biting."

- **You put a stop to the behavior immediately, without giving a warning, and in a way that the behavior cannot persist or be repeated.**

    Pick up Cherub, ideally with Cherub facing away from you. No eye contact, no hug, no words. Move him to a boring spot, and continue to remove your attention from him for thirty seconds or so. Say, one time only, very firmly, "No Lipstick," or whatever it was.

- **Personal attacks on others are more serious than any other violation.**

    If the behavior was aggression—biting, hitting, pinching, etc.—Cherub needs an increase in your response. Put Cherub facing a wall. Kneel behind him, and hold his arms firmly into his sides. Do not hug! Look at the floor, so that if Cherub turns around he doesn't get your eye contact. Hold him there and say once, "No Biting." Or whatever it was. Continue to hold him (he may wiggle, laugh, turn around and try to flirt with you, say "sorry mommy," or whatever) until he tries to get away from you and starts to complain about being held. At that point, hold him a bit more firmly for about twenty seconds, saying again, "No Biting." Then let him go, but do not make eye contact or talk to him for a minute or so.

*Tantrums*

Tantrums can be a response to frustration, fatigue, or resentment of discipline. The cause doesn't matter. Cherub needs to learn only three things about tantrums:

1. ***A tantrum belongs only to the person who is having it.***

   Once the tantrum starts, nobody should try to coax, reason, threaten, distract, or otherwise try to get Cherub to stop. Other people should go into another room, if that is safe for Cherub; otherwise, an adult should stay within safe distance, but completely occupied with something else. This is a good time for ear plugs.

   The person having the tantrum is the only one who can make it start and stop. Onlookers should not discuss the tantrum or even recognize it is going on, removing their attention until after it has come to a complete stop.

2. ***A tantrum never gets you anything you want, including attention.***

   The threat of a tantrum should never change an adult decision, such as not getting a toy, or having more chocolate, or being allowed to switch the light on and off thirty times. After the tantrum is over, nobody mentions it, or tries to talk to Cherub about why tantrums are not a good idea. It's like what the etiquette books say about Bathroom Noises that are overheard: the audience must pretend they heard nothing, nothing at all.

3. ***A tantrum in public results in loss of attention and privilege.***

That means being carried out by a parent who is in charge, and silent. It means being put in a safe place to finish the tantrum, while the parent stands by not listening or intervening. If necessary, it may mean driving home and going into the crib to finish the tantrum. With a silent parent.

*A note about whining and screeching:* Get a pair of good ear plugs (try the Mack version) and keep them with you at all times. As soon as the whining or screeching starts, *pointedly* put the plugs in your ears, and start doing an engrossing task. Polish the silver. Make a list. Do not tell One to stop whining or screeching; merely be completely deaf to it. When all is silent, count to ten and then resume normal, but not overly attentive, behavior.

*Trying to run the family show*

Often, a One will suddenly decide that she prefers one parent and cannot tolerate being with the other. It is an absolute toss-up which parent becomes the chosen one: sometimes it's the primary caretaker—usually the mother—and sometimes it's the glamorous out-of-the-home parent.

When One is handed to, or left in the care of, or asked to put up with the rejected parent, the histrionics are often quite astonishing: yelling, kicking, pushing away. Moreover, the other rule that One has established is that Parents must not show any affection towards each other. One will pull and tug and push and tantrum, breaking up the parental embrace.

Here's what will help:

- One-on-one time with the rejected parent. The preferred one needs to leave the house, and not sneaking out, either. Simply offer a cheerful, brief bye-bye, ignoring the tears and wails.
- Once the Chosen One has left, the rejected one should not make overtures, but sit quietly, pretending to do a crossword puzzle. One will eventually stop crying and begin to do something else, at which point the rejected parent can express a mild interest. Take it from there.
- Be oblivious to the fuss. If it is Daddy's turn to put One to bed, and One is outraged at the idea, well, so what? Daddy is who One gets. Don't let One bully you into leaving the room and getting Mommy in there.
- When parents are having an affectionate display with each other, and One tries to push them apart, look down on One as if One were having a bowel movement: you are registering what he is doing, but are not particularly interested in it. You certainly aren't going to let that activity run the show. Finish the hug, and speak only to each other for several minutes. If a tantrum occurs, or hitting, biting, or pinching, treat them just as you would if One had not been trying to interrupt your hug.

Above all, don't take One's preference or behavior personally. It's all too easy to feel a certain smugness if you are the Preferred Parent, and a poignant martyrdom if you are the Rejected One. Resist that impulse.

## PREPARING FOR DISCIPLINING

It's useful to read over the principles and imagine scenarios and your response. Ideally, Cherub should sense from you, the adult, very little anger or embarrassment. Rather, the "rays" coming from you ideally are those of omniscience, omnipotence, and slightly fed-up resignation. You, the Adult, the Mighty Parent, are far above the infantile agitation going on at your feet.

## COPING WITH TEMPERAMENTS

When they handed out intensity genes, Chantel was at the head of the line. Her days are filled with seemingly inexplicable tantrums, starting with the apparent insult of having to wake up in the morning. She clings to her mother's knee, then thrusts her away. She is busy, busy, busy, trotting from the cabinet to the sofa to the coffee table, up the stairs and down, touching, tearing, banging. The simple act of getting the mail may mean escaping into the street or a tantrum at being interrupted or screams at being abandoned for thirty seconds.

The child who, like Chantel, charges into each day in a state of red alert needs very great patience, tact, and consistency. The more she is able to gain mastery over objects, the more she is able to avoid situations that are strange, the fewer times she has to make choices or interrupt an activity, the better.

Otherwise, a vicious cycle is likely to develop. The more tantrums Chantel has, the less she feels in control, and the more desperate and poorly focused grows her need for mastery and control. This anxiety just sets the stage for more tantrums.

All this will make her caretakers more desperate. They are less likely to respect her task of growing up, and less likely to try to help her with it. Her life will become unbalanced. Her intensifying and troubling relationship with her loving adults will overshadow her mastery of skills and her exploration of the world.

Helping Chantel may mean, in a sense, handling her with kid gloves; but that does not mean spoiling her. Chantel's parents work on:

- Keeping her schedule as regular and her environment as consistent as possible. This is no time, they agree, to visit the zoo, join a play-group, start taking showers rather than baths, adopt a puppy, or even get a new maternal hairdo.
- Baby-proofing the house to a degree of fastidiousness their friends think is ridiculous. "You could let a chimp loose in here by himself," one jeered.
- Making transitions as easy as possible. They allow extra time to go from one activity to another. They try to get to Chantel in the morning as soon as they hear her stirring on the baby monitor, to be with her with hugs and a glass of milk before she begins her morning weep.
- Allowing a bit of regression, with several cuddles during the day, an occasional

bottle or nursing, and the unrestricted use of a security blanket.
- Making sure that neither they nor Chantel's nanny are encouraging temper tantrums by being inconsistent with rules; making transitions without adequate time or preparation; or trying to make Chantel socialize with other children (she is having a hard enough time with herself, her world, and adults).
- Making sure that no one is spoiling Chantel by focusing too hard on avoiding temper tantrums. Trying to prevent tantrums altogether can result in: allowing behavior that should be forbidden (hitting, biting); giving her attention for unattractive behavior, such as a whining tone of voice, banging her head on the floor, or trying to buy her off with cookies and other treats.

# Day to Day

## MILESTONES

### Vision

Babies see well enough to accomplish anything they want to at this age. A baby who consistently gets right up close to the picture on the page or to the TV, or who seems afraid to climb, may have visual problems. Be especially alert if there is a family history of "lazy eye" or of needing glasses before puberty.

## Hearing, speech, and language

Young toddlers need excellent hearing not only to learn how to talk but how to react to social cues. Even before a baby talks, his "language" is rich with gestures and vocalizations. A baby this age can wave bye-bye, play peekaboo, invite someone else to roll the ball back and forth, present a body part to be kissed or cleaned, etc.

The average number of words a baby can say at the beginning of this period is three, besides *mama* and *dada*; at the end, ten. So it may not seem as if much language is learned. Not so. The number of words *understood* doubles, from fifty to one hundred. By the end of this period, babies can follow a one-step direction or request (if they feel like it).

Babbling with inflection is a charming attribute of the age, and studies show that the inflection and vowel sounds are language-specific. By fifteen months, nearly all babies (when in the mood) like to point at and hear the labels for pictures rather than just turning the pages and eating them.

There seem to be two styles of learning to talk. Early talkers seem to pick up separate words. "He can say anything he hears." Later talkers tend to make a jargon with inflection. This from-another-planet talk often consolidates into discernable language at around Two.

Ones do imitate, though. "When I say anything, I mean *anything*," Chantel's mother says ominously. Beware. Little pitchers don't only have big ears; they've got mouths to match.

## Large motor: walking

No matter how ravishing the accomplishment, the age at which a baby walks does not predict or correlate with IQ, later athletic skill, or any particular aspect of life whatsoever. If the baby had no perinatal complications, starting to walk late, up to about sixteen months, is in nearly every case just programmed in.

Some babies walk as early as seven months, which is downright comical. Some begin at sixteen or even eighteen months. The average is sixty weeks (twelve-and-a-half months).

New walkers hold their arms up in a fixed balancing position, their legs are spread for balance and to accommodate bulky diapers. Their knees point out on either side. Their toes may either point out, in concert with the knees, or a twist of the lower leg called tibial torsion may turn the toes in. (This is checked at the one-year- and fifteen-month-old visits. Except in severe cases, no treatment is necessary.) By eighteen months, a toddler ought to be off the toes, walking heel-toe fashion.

Unless special ones are prescribed, shoes are necessary only for warmth, protection from sharp objects or slippery surfaces, and decoration. They don't "form the foot." The least expensive shoe that allows the toes to wiggle and that fits the heel well is the best shoe. "My mother says even if we have to live on peas, to get her the very best shoe," says Stacy. Not so. Cash in the shoes and get some artichokes.

If you think the baby's gait in walking is not normal, be sure the pediatrician observes it in action. Most of the time all is well, no matter how comical One appears lurching down the corridor. A limp, a very asymmetrical gait, or a real waddle needs attention, though.

*Fine motor*

By eighteen months, babies can and will remove their own diaper. They can hold a cup and most can feed themselves with a spoon. They can press buttons, work handles, and hold something in one hand while performing a directed action with the other. During this age, they learn how to scribble with, rather than eat the crayons, and by eighteen months can stack four smallish blocks.

*Social*

Ones are hams. They realize what cute behavior is, and can turn it on. They also show pride when they accomplish something. They can anticipate what is going to happen next by what goes on around them.

They often show concern when another child cries or is hurt, and will go and comfort a person in distress. This does not mean that, intellectually, they realize that a person can have a different point of view than their own. Nor does it mean that they connect the bite they just delivered with the bitten one's tears in a very meaningful way. That is, those sympathetic tears don't mean the biter will refrain next time the impulse strikes.

During this six months, toddlers don't need much in the way of peer activity. They mostly either ignore or stare at other children, though they may hug, push, hit, and grab. Most social skills develop in connection with adults: learning to get adult attention and keep it, having "dialogues" in which the adult speaks and the baby makes responses of voice or body.

Children this age don't really solve problems. They mostly experiment with manipulation. Cause-and-effect relationships hold great charm for exploration. Nor do they understand the idea of completing a project. Arts and crafts attempts may look fine on the refrigerator door, but they don't give One much pleasure.

Curiosity leads them to handle everything, stare at everything, and mouth some things. If something can wiggle, tip over, rotate, switch on and off, bounce, crinkle, tear, or shut, One is your man or woman to find out about it.

*Sleep*

Most children this age sleep about thirteen to fourteen hours in a twenty-four-hour period, about ten to eleven in the night and about three in the daytime. Because the number of daytime sleep hours goes down, most children may move to only one nap in the daytime—if not in this six months, then in the next. This move from two to one nap may be rough, with crankiness and irritability. It may take several weeks to juggle the schedule back into a reasonable one.

If the afternoon nap lasts later than four P.M., One will want to stay up later and bedtime will be a chore. If bedtime becomes late, One is likely to wake up late, which then puts the nap even later.

If One is an early awakener, and demands an early morning nap, this too can delay the afternoon nap and bedtime.

It's most helpful, therefore, to try to keep the time of waking up and napping a steady goal to be aimed for. You may have to awaken a very grumpy One if she's gone to sleep late the night before. Or you may have to prevent the early awakening One from taking an early morning nap.

If your One attends daycare, and you start having problems with bedtimes, better check the nap schedule at daycare. A rare daycare adult may be encouraging two long naps a day in her charges, which may be nice for her peace of mind but wreaks havoc on the home front.

For more sleep considerations, see *Separation Issues* in this chapter, and *Night Terrors* under *Health and Illness* in this chapter or in Part II of this book.

## GROWTH

### Chubbiness watch

Starting at a year of age, growth really slows down. Petite babies will only gain about three pounds and taller ones about five pounds over this entire year, less than half a pound a month.

You'll see One's belly start to diminish and the creases disappear from thighs and upper arms. One's legs and arms will start to look longer.

If a baby gains as much as half a pound a month, something is going on in her environment that needs to be changed. Perhaps she's watching lots of TV, or doesn't get much active play, inside or out. Perhaps she's drinking juice or soda between meals, instead of plain old water. Perhaps she's being offered snacks she doesn't need, or large portions at meals.

At any rate, if she is "going up the weight for height curves," talk it over with your pediatrician.

## NUTRITION AND FEEDING

Fish-consumpion guidelines may be found on page 7.

As growth slows, the baby's appetite decreases accordingly. Most babies this age eat only one real meal a day: that is, a big breakfast, a fair lunch, and practically no dinner. Since the breakfast meal may be fed after they go to work, many parents don't see their child eat a real meal all year round. This year, though, One needs only about 1000 calories a day.

### Nursing babies

Most nursing babies are weaned by a year. Babies who continue to nurse beyond the first birthday obtain nutritional and immunological benefits, but there are some cautions to observe.

- Increasingly, One may confuse the need for comfort with appetite, and start to depend on nursing to solve problems such as boredom, frustration, and minor upsets. One needs to learn more independent and grown-up techniques for life's little problems.

- Frequent daytime *and* nighttime nursing habits can lead to a mouthful of cavities in Cherub's baby teeth.
- Nursing may unbalance the equilibrium important at this age between the Big Three Tasks: relationship with mother, exploration of the world of objects, and motor skills. Try to keep the nursing for scheduled mealtimes. Nursing a One and older to sleep is a big invitation to a mouthful of cavities for baby and a very interrupted night for Mommy, as the baby is likely to wake up during her sleep cycle and have to nurse herself back to dreamland.

Weaning an over-one-year-old baby is a multifaceted campaign. One way to start is to be sure that the One does have plenty of opportunities and encouragement for motor activities and exploration of objects. One's mother can monitor herself closely and make sure that she herself isn't the one initiating nursing sessions. She can also wear clothing that requires some manipulation to free up; a One finds it easy to crawl into the lap and pull everything apart. Finally, she can "absent herself from felicity" for awhile, letting One be mostly in the care of his father or daycare mother for a day or two.

*Milk*

Unless One has a suspected or proven allergy to cow's milk, most pediatricians recommend weaning from breast or formula to pasteurized (or reconstituted evaporated) whole milk* until age Two, when the fat content should be decreased. Milk and dairy products provide protein, fat, calcium, and Vitamin D.

Sixteen ounces of milk a day will give the correct amount of calcium required for a One.

If your One loves milk and drinks much more than 16 ounces a day, there are a few pitfalls to watch out for.

- First, that's a lot of fat. At this age, that's OK in terms of cholesterol (we think) but it's still a lot of calories, and a chubby One benefits by having her milk quota reduced to 16 ounces a day.
- Second, milk itself contains no iron; and excess milk can keep Cherub from absorbing the iron in cereals, fruits, vegetables, and non-red meats such as chicken and turkey. Iron deficiency can affect development, mood, energy level, immunity, and growth.

If your One doesn't like to drink milk, but will take cheese and yogurt, these may supply enough calcium. The amount of calcium in one and a half ounces of cheese is about 300 mg, the same as a glass of milk. (Yogurts vary so much these days, you need to read the cal-

---

*The American Academy of Pediatrics recommends whole milk up until the age of Two, when fats ought to be somewhat restricted. The chapter on Two Years discusses this fat issue.

cium content on the container.) However, cheese and yogurt aren't fortified with vitamin D, so you need another source: either judicious sunlight, a vitamin supplement, or other fortified foods. Moreover, cheese can be very constipating, and flavored yogurt adds the problem of sweets.

*Foods*
I have not included the Food Pyramid in these sections. I have never known a parent to take kindly to the thing. Its main purpose is to remind us that children need very little in the way of sweet, sugary, greasy foods; that they need the basic amounts of grains (cereals, starches, and breads); that they need more fresh vegetables and fresh fruits than Americans are used to eating.

Indeed, it's crucial to keep this in mind with every meal, every snack, and every beverage. Remember, you are struggling against the mighty forces of fast food, soda, juice, and snack producers, and their media accomplices!

The second crucial thing to keep in mind is serving size. When you serve flat food, such as a piece of meat or poultry, the serving size is the size of your child's palm. When you serve a bulky food, the serving size is the size of your child's fist. Offer three or four different foods at each meal, in these serving sizes. If you keep the nutritional guidelines, the serving sizes, and the milk requirements in mind, your child will most likely be very well nourished.

If your One is taking sixteen ounces or so of milk, that leaves about 600 to 700 calories a day to take as solid foods. If you've ever counted calories for a diet, you know that's not a lot of calories. So it's good to make them all count.

Since One gets protein and fat from milk, it makes sense to concentrate on solid foods that provide what milk doesn't: Fiber, variation of textures and taste, vitamins A, B complex, C, E, K, and iron. (Iron is particularly tricky, because milk intake keeps some iron from being absorbed. Another reason not to overdo milk.)

You don't need to be a nutritionist nor to abandon favorite, cultural, ethnic, or family-tradition foods. The old rule of having foods of bright and varied color is a good one, with white foods limited to dairy products and the occasional rice, potato, or noodle/pasta dish.

Fortified cereals can be an excellent source of iron and B vitamins, and can substitute for red meats in this area. (Read the nutritional content on the cereal box!)

Green leafy vegetables, yellow fruits and vegetables, fruits, whole grain breads, and the occasional (three a week) egg can provide the bulk of One's diet.

In preparing the foods, experimenting with spices and herbs is fine. Hold the salt and sugar, and don't overdo added fats, such as butter.

Ones are likely to develop tastes for foods that are highly addicting and not very nutritious. Bear in mind that it's much easier not to introduce them to One than to have to limit them later on!

## ADDICTING FOODS

- **Beverages:** Juice and soda (including sports drinks)
- **Snacking foods:** Crackers, chips, raisins
- **Condiments:** Catsup, mayonnaise
- **Spreads:** Peanut butter
- **Other:** French-fries

*Overweight*

Like Measles and Oppositional Behavior, extra chubbiness is much easier to prevent than it is to treat. When a baby this age starts gaining more weight than is desirable, it's usually because somebody taking care of her has a misconception about what she needs to be eating.

#### WHERE IT CAN GO ASTRAY

If a baby eats 50 calories above and beyond what she needs for growth, every single day for a year, she'll gain an extra five pounds from that daily oversupply. That would *more than double* the amount she's supposed to gain in this year.

Those fifty calories can come in the form of juice instead of water for thirst. Or in the form of snacks—a handful of crackers, some string cheese, raisins or dates. Or in being coaxed and urged to eat a "good" dinner, when she's not in need of it.

#### WHAT TO DO

Ask your pediatrician whether Cherub is going up on her Weight for Height (make sure it is that one, *not* the Weight for Age curve) curve. If she is above the 75th percentile or has recently gone from a lower percentile up to the 75th percentile, you should intervene.

If your baby is showing an increase in her weight curve, check with your baby's daycare provider about TV-watching, snack habits and the handing out of juice. Review your own dietary patterns, especially the Dinner question. If Cherub spends her weekends helping you do errands, ditch the "handout" snacks and juice boxes that keep her quiet in her car seat.

#### WHAT NOT TO DO

Don't restrict milk below 16 ounces a day. Remember, you decide *what* is offered, *where*, and *when*; Cherub decides *how much* to eat.

## DIAPERING, BATHING, AND MASTURBATING

They go together like a horse and carriage. Or something.

Young toddlers masturbate, without exception. I have never seen a baby this age who has pulled off, worn out, inserted something into, or otherwise damaged his or her genitalia. They masturbate when being bathed or when diapers are changed or whenever they can get their hands on themselves.

Not only should this not be a matter for discipline but it should be a good opportunity for parents to start discussing sexual issues with children. It is the perfect time. The baby is a captive audience, and interested in everything you say. She is nonjudgmental, and will forgive you (and be entertained by) any blushing, stammering, hesitation. It gives shy parents an opportunity to get used to merely saying the words.

"Kyle's penis, that's right. And here are Kyle's testicles: one, two. Two testicles. They live in the scrotum. And here is Kyle's belly button, and his ear, and his nose . . ."

"Chantel is touching her clitoris. Chanty feels good when she touches herself there. Mommy cleans Chanty's vulva with the baby wipe. We don't get the soap inside Chanty's vagina. Hey! Wait! Baby wipes are not for eating! Baby wipes are for cleaning!"

Using the accurate terms now will help in being comfortable teaching them later. (There will be a good time at about age Three to teach more acceptable public nicknames for the genitalia.)

When I said captive audience, by the way, I meant just that. By fifteen months, as early as nine months with many babies, you can't just pick the baby up, put him on the changing table, and change the diaper. You have to chase him, squealing, around the house, tackle him, bring him to the floor, and keep one leg on top of him as you do the job. This is not an indication that the baby has been abused and is frightened of having his genitals handled. It is universal behavior protesting at having to pause and get changed.

## TOYS AND ACTIVITIES

Up until about Two, or even Three, there isn't much real difference between toys and the real world. If One were totally deprived of toy stores, even of homemade toys, One would still find plenty to play with.

The most successful toys for One come in these categories:

Things that can be made to go away and come back again. *Balls, wheeled objects like big safe cars and trucks. (Beware little parts that can come off and be swallowed or choked on, and anything sharp.)*

Things to stack. *Blocks, big Legos, boxes, cushions.*

Anything with hinges. *Books, things with doors and lids. (Beware toy chests with lids. Severe head injuries have been reported.)*

Safe things to ride. *With no pedals and with a low center of gravity.*

Things with which to imitate adult activities. *Anything around the house. Cooking utensils. Pretend vacuums and lawn mowers. Baby dolls, bottles, doll carriages, blankets. Toy telephones.*

Things that go into things. *Containers with smallish but safe items to put inside them.*

*If nothing else, a kettle with kitchen objects is fine.*

Things that fit on things. *Ones love to fit the speculum on the otoscope under close supervision. They sigh with satisfaction at the accomplishment. Anything that replicates the experience but isn't small enough to be choked on or swallowed is a good toy.*

Things that change shape and configuration. *The wire toys with bright wooden balls that can be collapsed and stretched, and on which the balls can be moved, are great hits.*

Playthings that feature different textures. *Wading pools and sandboxes are big hits, and so is the ensuing mud. However, any water source, even a four-foot-diameter wading pool with two inches of water, can be a hazard for a One, who can fall face downward, aspirate, and have his airway clamp shut in a spasm. Water sources like toilet bowels, ornamental fountains, barrels with a few inches of water, even scrub buckets, can also be sources of tragedy. Sandboxes need clean sand without asbestos, and protection from pets and wild animals that regard them as litter boxes.*

Anything to bang. *If you don't want to buy her one, don't worry; she'll make her own. Whatever it was you had in mind.*

### Classes and play groups

Circle time, arts and crafts, listening to a story told to a group, entering into group play such as ring-around-a-rosy: these are inappropriate for children at this age, no matter how bright and verbal. Social development depends on emotional, not intellectual, readiness, and One has important and necessary tasks to accomplish as an individual before she is ready to enter into group activities.

Watching and bumping into and smelling other babies, hugging them and kissing them, staring and staring and staring, pushing and biting: that's what One likes to do in a group. Grabbing is fun, too, and throwing things. But nearly every *reciprocal* activity One is able to perform (with balls, trucks, books, etc.) requires a more grown-up playmate, not another One. Or even a Two. Young Ones may parallel play, watching and imitating another child, but they don't have the resources to play together.

This is not to say that One can't form a poignantly strong attachment to another toddler. Very young children have been observed to comfort, seek out, and even protect each other. But this kind of relationship and commitment isn't necessary to insure future abilities in these areas.

In any group activity, One will inevitably catch more illnesses than she would if she stayed at home and at her regular daycare. The problem is that the immune system is immature still. Later on, she'll be able to "catch" a virus or bacterial illness and make immunity without much in the way of symptoms, and that immunity will be strong and long-lasting. But now she's likely to produce symptoms every time she catches an infection.

And since language development is so important for One, if she tends to get ear infections every time she gets a cold, she can actually be held back rather than enriched by group activities.

"Superbaby" lessons strike me, and most pediatricians and child psychologists, as unnecessary at best and damaging at worst. When you consider all the learning tasks that engage One in her ordinary, everyday world, you have to wonder why anyone would want to distract her with, say, trying to teach her the alphabet or how to hold a violin.

## SAFETY ISSUES

If you haven't baby-proofed your home and stocked your medicine chest as described in previous chapters, please do so now. The safety concerns for babies this age are the same as those described in the last chapter.

# Health and Illnesses

*Part II of this book covers illnesses in babies this age and older: how to assess symptoms, how to tell whether a child is sick enough to require urgent attention, and what kind of home treatment may be appropriate. It also examines problems by body part. The Glossary is an alphabetical listing of medical terms with pronunciations and, when appropriate, detailed descriptions.*

## EXPECTATIONS

It would be delightful if One were always perfectly healthy, with a dry nose and quiet breathing and normal stools.

Unlikely.

It is no doubt the wisdom of Nature that makes the one-year-old birthday the begin-ning of so many thrilling developmental achievements and, at the same time, the start of a series of pretty much inevitable minor illnesses and injuries. Perhaps the achievements are designed to charm and distract parents from the little catastrophes.

However charmed and distracted, it's still nice to be prepared for the most likely con-tretemps of this age.

## COMMON MINOR SYMPTOMS, ILLNESSES, AND CONCERNS

### Colds and ear infections

Once again, the problem with frequent upper respiratory viruses is that many Ones will turn them into middle-ear infections. This is a particular problem now because Ones are busily engaged in learning recep-tive and expressive speech. Recurrent ear infections may temporarily muffle what they hear, thus dulling their responses to, and pleasure in, language. See the essay in Part III, as well as the *Body Parts: Ears* section of Part II.

### Conjunctivitis (pink eye)

A baby this age with tearing or goopy eyes is more likely to have an infection or injury or foreign body than just a blocked tear duct.

### Croup

A cough that sounds like a goose honking or a seal barking; a frightened child who goes "croooop! crooop!" when she inhales; and all of this in the middle of the night. Croup is a respiratory illness usually caused by a virus.

It is discussed fully in Part II in *Breathing Disorders*.

## COMMON VIRAL DISEASES WITH UNFAMILIAR NAMES

There are a number of very common but unpublicized viral diseases of young children, and babies this age can catch all of them. They are discussed in Part II in *Common Illnesses with (More or Less) Unfamiliar Names*.

## VIRAL DISEASES WITH FAMILIAR NAMES THAT ARE BECOMING, OR ARE ABOUT TO BECOME, RARER AND RARER

Measles, mumps, rubella, and chicken pox are all relegated to the Glossary.

## COMMON SCARY BUT USUALLY INNOCENT PROBLEMS

Of all of these, fever convulsions, breath-holding-spells, croup, and night terrors are the most frightening, and are most likely to make their appearance this year. Even if your child doesn't have one, you may open the door one day to your frantic friend desperately seeking help for a seizing toddler. I strongly suggest you review these conditions, which are thoroughly discussed in Part II. Nursemaid's elbow, which presents as a child who (after a yank on the hand) won't use her arm, is less frightening but very guilt-inducing in adults. Don't panic; read Part II.

The new entrant in this category, toxic synovitis, can occur at a later age. But it's more likely to cause parental fear now because it presents as a toddler who suddenly won't walk or who limps.

### Toxic synovitis: one cause of toddler limping

*Toxic synovitis* sounds truly dire, but it is one of the most common causes for a young child to limp and is quite benign. It occurs when an ordinary virus makes the hip joint hurt. This is a common cause of limping in toddlerhood. Any one of a number of common viruses can cause the lining of the joint (usually the hip) to become inflamed and the child limps. Unfortunately, this is what is called a "diagnosis of exclusion," which means you have to exclude more ominous causes for limping. Often this means a blood test and maybe X rays and sometimes even more sophisticated tests. If the child does have toxic synovitis, the condition goes away with rest and time. How you keep a toddler resting is, fortunately, your problem and not mine.

## SERIOUS OR POTENTIALLY SERIOUS ILLNESSES AND SYMPTOMS

During this year, toddlers who are really seriously ill usually look seriously ill to any observer. Earlier, the signs of serious illness were more subtle. Starting now, however, parents are likely to be able to tell when a child is mildly sick as opposed to being in real danger. The discussions in Part II on fever, *Trouble Breathing*, *Not Acting Right/Looking Right/Smelling Right* will be most helpful.

As always, the first main worry for parents

and for pediatricians is serious overwhelming infections, such as sepsis (infection in the bloodstream) and meningitis (infection in the fluid surrounding the spinal cord). And, as with younger babies, many toddlers have to undergo tests to determine whether their symptoms are early signs of such potentially fatal diseases. Most will turn out to have viruses. This doesn't mean the tests were unnecessary.

The second main worry about truly serious diseases is cancer. Cancer in One- to Two-Year-Olds is rare. When it does occur, many forms are truly curable. Statistically, the kinds of cancers parents worry about most— leukemia and brain tumors—are very rare in children this young. And the one rare form of eye tumor, retinoblastoma, can usually be cured by surgery when it is diagnosed at an early stage.

Ironically, two serious diseases that pediatricians worry about in children this age are both preventable and not usually taken very seriously by the public. These are iron deficiency, which can cause impaired intellect and a sluggish immune system, and lead poisoning, which can cause mental retardation and seizures, as well as delayed growth. If you have any worries about your child's iron intake (too much milk, not enough iron-rich foods) or exposure to lead (in chipping paint, water sources, lead fumes, chemicals) ask your pediatrician for specific blood tests, rather than for screening tastes. A normal hemoglobin or hematocrit does not rule out iron deficiency. And the usual screening test for lead, called the FEP test, does not rule out low levels of lead poisoning.

## Minor Injuries

Do review *Head Bonks* and *Head Bonks That Bleed* from the previous chapter. Aren't we having fun!

### Cuts inside the mouth

Biting the tongue/lip; splitting open the frenulum (the little piece of tissue that holds the upper lip to the gum):

These rarely need any attention other than sympathy. The first thing to do is to calm the child. The bleeding is likely to stop on its own after a minute or two. As soon as the baby is calmed down to the hiccuping stage, try to examine the damage. If there are any loose or absent or out-of-place teeth, or if the bleeding won't stop, or if the injury looks really deep (you can see bone, or at least lots of torn tissue), he will have to be seen. But most often, no stitches are needed and everything heals up in less than two days.

### Fingers in the door

It isn't as common as you might think for this to result in a fracture or significant injury. If there is a torn nail, bleeding, blood under the nail, or the child can't be calmed, then he must be seen right away. Sometimes there is a fracture, or damage to the nail bed, which, if not repaired expertly, can cause the nail to grow weirdly or not at all. But if it is a matter of a very slight swelling, calm the child and observe him for half an hour or so before calling the pediatrician (unless it is very close to the end of office hours).

*Bee stings (and wasps, yellow jackets, etc.)*

Toddlers attract, chase, and step on stinging insects. Be prepared: see the discussion in the chapter on *First Aid* in Part II of this book.

## Major Injuries

*These should be avoided at all cost. Nevertheless, this is a good time to be sure all the child's caretakers are up-to-date on CPR training and speak enough English to summon help effectively.*

### Choking

This is perhaps the most serious and dreaded accident. Prevention is crucial. All caretakers should take an up-to-date CPR course, and rehearse its lessons mentally every so often.

### Poisoning

Most babies this age can't unscrew any container tops of medications, much less the child-proof variety. The danger comes rather from plants, household chemicals, and tablets or capsules or open bottles lying around loose. Most pediatricians recommend every household have, locked away, a bottle of the vomit-inducing medicine syrup of ipecac. Call for advice before using, as it is itself a poison. The local Poison Control phone number should be posted by each phone.

### Water immersion

The danger in this age is not limited to pools and hot tubs, though these are a major problem. Pools with soft pool covers are particularly treacherous, as the baby can become trapped, invisible, beneath the cover. Baths, toilet bowls, buckets of water, and fountains are all irresistible and potentially lethal.

### Burns

Hot liquids, stoves, irons, and fireplaces are what parents usually, and rightfully, worry about. Consider also: curling irons, frayed electric cords, exhaust pipes.

### Animals

The German shepherd who loved the sedentary baby may become quite annoyed at the tail-grabbing one-year-old. Even cats can inflict major damage.

### Falls

Serious falls are quite rare in this age group, as babies don't usually climb higher than a couch. Any fall from a height of more than four feet should result in a medical assessment promptly, even if the baby seems fine. See *Head Bonks* in Part II.

### Car accidents

The main problem with One to Twos is that they can be terrible back seat distractions. No matter what the temptation though, they should stay in the backseat; it's much safer there. Any air bag close to Cherub's car seat must be disconnected. If your toddler learns how to get out of the car seat, you face a parenting emergency. Prevention, of course, consists of car seat toys and tapes. If you sense One is getting restless but has not tried to escape, pull over and play a little. But if an escape attempt occurs, take action. The first time it happens, pull over to the side of the road. With your firmest authoritative voice say, "No getting out of the car seat." Place One in it firmly. Proceed slowly in the slow lane. If One tries it again, pull over; do so at the very first sign that One is making an escape try. Make your way home in this fashion and immediately put One into time out in the crib for a whole minute or until the crying stops, whichever is longer. When you take One out, don't mention the car seat incident; do something jolly and loving. But shortly thereafter, make a practice trip, slowly, close to home. If One makes a move to escape, go home immediately and do time out again. Do this as many times as needed before venturing out on a real trip.

*Older children*

Older siblings very frequently become extremely jealous and annoyed by mobile, curious, attention-getting babies this age. Even young children can inflict serious injuries in a jealous or frustrated rage.

# Window of Opportunity

## TACKLE TANTRUMS

If One is having major frequent tantrums, don't assume that he will simply grow out of them. Often a complete physical examination and a cool-headed discussion with the pediatrician can reveal a pattern or even a cause that can be treated.

Frequent tantrums can reinforce themselves. Parents tend to become worn down and less able to set a limit and follow through; and One becomes more and more passionate about pushing that limit. Or parents become so exasperated that they let One know they don't find her pleasant, cute, delightful; and One reacts with tantrums to get the parental attention she isn't getting from endearing behavior.

## CONFRONT CONSTIPATION

When a child starts eating more adult foods, the outcome is often too-firm stools. Nathanson's Second Law states that any food that children really like produces either diarrhea (apple juice) or constipation (cheese, bread, bananas, chocolate, pizza, pasta . . . the list goes on).

A firm stool is difficult to pass; it may even be painful. So One holds back. You may see One grunting and turning red: he isn't trying to push it out, he's trying to keep it in.

If he succeeds, the firm stool sits there, stretching the rectal muscle so that it is less effective in pushing out the stool. And more firm stool builds up behind.

Such a vicious cycle becomes harder to break the longer it lasts. Ask your pediatrician about mineral oil to soften the stool and/or suppositories or other aids to help it exit.

## DIALOGUE AND BOOKS

All of One's caregivers should engage her in dialogue where the adult does the talking but responds to cues from One as if One were speaking also.

This is a good time to consciously suppress the urge to ask the child to say a word or to perform. Because this age is so gung-ho for autonomy, such requests or commands can actually delay speech acquisition.

This is also a good time to vow to keep the TV off and to have regular time with books every day. Interrupting a child's activity to "read" is self-defeating; it's much more effective to wait until she or he is done with an activity and a bit at loose ends. And then to say, "Which book do you want me to read?" (not "Do you want to read a book?" or "We're going to read now," or "Get a book to read"). But don't expect to "read." Point to and talk about the pictures.

# What If?

## STARTING DAYCARE

One is likely to have separation difficulties starting a new daycare or daycare for the first time. It is unusual for a One to go right up and join a group of children, or even one other child. One needs time to stare at and touch everything. He needs to tour the place gradually, giving frequent glances to his parent to see whether all is safe, whether he should continue to explore, whether this is a good place or not.

This behavior is not shyness, it is normal for the stage of development.

- Since One can't maintain real play with another child yet, make sure that the daycare Nanny doesn't expect her to do so and leave her to "stand up for herself" in these encounters. Ones need kind supervision.

- This is very difficult in daycares that are large (twelve at most; eight is better; four or fewer ideal) and that have more than four children for each adult caretaker. Caregivers shouldn't be spending their whole time changing diapers.

- It also can be difficult if the ages of the children don't mix well, especially if the under-threes are constantly harassed by, or harassing, the under-twos. This close spacing, even in families, is difficult. A daycare that has two under-twos and two four-year-olds or older may work better.

- Health and safety concerns include:
  *No secondhand smoke.*
  *Diapering area separate from food preparation area; good hand-washing maintained.*
  *Special precautions about water hazards, including fountains, wading pools, scrub buckets, and toilets.*
  *Avoidance of lead-containing temptations, like chipping paint.*
  *No pets. A One who hasn't yet gotten the hang of playing with other children often seeks out the cat or dog. The cat or dog may not appreciate the attention.*

## EXTENDED SEPARATION FROM THE PARENTS

One of the major goals for a baby this age is to maintain a balance among three important areas: exploring the world, mastering skills, and concentrating on his relationship with loving adults. An extended separation between baby and the closest, most attached parent may make this balance very difficult.

If such a separation must occur, preparation by transferring some of the close attachment to another adult will help, painful as this is to the parent who must leave. And the adults remaining with the baby should do everything they can to preserve this balance, by being as loving, patient, and consistent as possible. A transitional object can be a big help in this situation, too.

When the absent parent returns, it may take the baby several days to "forgive" him or her. Expect hurt, angry looks, and biting; most of all, expect to be totally ignored. But don't despair; babies are truly resilient and forgiving, and in a week or two all will be well, except that One will need a bit more cuddling and be a bit more clinging than before.

## TRAVELING WITH THE BABY

"We met everyone on the plane. Some of them three or four times. Not that I'd ever recognize them again; we mostly examined their feet. We patted many knees."

- Don't try to sedate the baby, even for long trips. Sedatives may put the baby to sleep nicely, but they may also get him just drowsy enough so that he is irritable and alarmed at all the new sights, sounds, and smells. Trying the sedative at home first is not reliable. She can sleep like a lamb after a dose of Benadryl at home, and then on the same dose make everybody's Hawaiian vacation truly memorable.

- Never give alcohol to try to sedate the baby, no matter how desperate you are. It can be dangerous, and at the least is likely to produce vomiting.

- Soap bubbles are great for traveling. Clean, silent, and when you chase them you don't stray very far. (And if they spill they generally don't do too much damage. Fellow passengers who wear silk in Coach are living dangerously anyhow.)

## CONCEIVING OR PLANNING TO ADOPT A SECOND CHILD NOW

If a new baby arrives in nine months, our Cherub will be Two, or just under Two.

A Two who is difficult and demanding is not going to become magically easy and cooperative upon the birth of Baby Number Two. If this is in the cards, it is really a good idea to do everything you can to prepare right now for the event.

This does not mean merely talking about the new baby, taking Cherub along to doctor's visits, or having her help pick out things for the new baby. In fact, these activities are likely to be pretty meaningless to One.

What it does mean is helping One to be as grown up and self-controlled as possible.

Here's what I mean:

- Put emphasis into helping her learn to talk. That means rewarding all her attempts with your interested responses. It also means "showering" her with your own speech: descriptions, in short simple sentences, of what One is interested in. It also means limiting TV or cutting it out entirely: the whole point of language, for One, is that it is reciprocal. TV is never reciprocal.
- Be consistent about setting limits, and try to follow the guidelines in this chapter. If you have to yell or spank to "get her to behave" now, this will become much much more frequent and intense after the new baby arrives.
- Plan ahead for Two to have regular time in a daycare, preschool, or frequent playgroup, starting well before the new baby arrives. This is partly because all Twos need peers to develop social skills and to feel challenged and competent. But it's also true that studies show that mothers of new babies give Twos much less, and more negative, attention than before the baby was born. To protect everyone's ego, Two needs some time with adults not thus preoccupied.

Be aware that when Baby Number Two is born, your first child will suddenly look much older to you—no longer a baby. "Shouldn't you be out earning a living?" is a common and rapidly suppressed inner question.

Some babies sometimes seem to know, magically, that their mothers are pregnant. They will choose the moment that she is most vulnerable, about to throw up, exhausted, or highly emotional, to engage in testing behavior. See *Limits*, above. Since the second trimester tends to become easier, now might be a good time to splurge on a babysitter to give you both some relief. Later, when you're feeling better, you'll be glad you did.

## ARRIVAL OF A NEW SIBLING

Astonishingly, most Ones under eighteen months behave as if they couldn't care less about the new arrival. They'll even pat and kiss him and chortle. And then they're off on their usual rounds about the house, slamming cupboards, eating crayons, and staring at the toilet.

And this honeymoon may last for quite a while, even until the then-Nearly-Two begins to be harassed by his nine-month-old sibling.

The main problem is not One's relationship with the new baby, but One's own developmental needs. It is tough to care for a newborn and also to engage One in dialogue, prevent him from electrocuting himself, and keep your cool during tantrums. Getting help is the best solution, and if possible have the help take care of the newborn while you spend time with One.

Many Ones regress when the new baby comes. Some of this may be jealousy and "wanting to be a baby again," but there are probably other motives as well. One loves to imitate, and if he sees Baby nursing or drinking a bottle, he's just got to try it too. He'd want to even if it weren't his own sibling.

And One's tendency to ask for extra cuddling, throw tantrums, and suck his thumb more may well have to do with the fact that his mother is acting differently towards him, not with the fact that she is also caring for a new baby. It is doubtful that One can make the connection: "Mommy is acting cranky and won't play with me because she has to take care of the new baby." Ones are very egocentric. He is just as likely to think, "I don't matter; I'm not loved."

Here are some hints:

- Spend ten minutes a day with Cherub, down on the floor, focusing only on Cherub. Nobody else in the room. Bedtime doesn't count, watching TV together doesn't count, doing errands or chores doesn't count.
- Let Cherub hear you say, frequently, "Baby, you are going to have to wait until I see what Cherub is doing." She'll be hearing it the *other* way around many, many times.
- Touch Cherub often, without saying anything. A pat, a hug, a ruffle of the hair.

## DIVORCE AND CUSTODY ISSUES

Now that One is an actor in the world, One tends to assume that whatever occurs in his world is his doing. Parental anger, whether expressed in words, tone of voice, expression, or by physical violence can produce a lot of reaction in One, from silence and withdrawal to clinging and thumb-sucking to tantrums and aggression.

Clearly, for parents already stressed maritally, such an acting-out baby makes their relationship even more difficult, and can fuel divorce issues. If parents see this, they may be able to understand how important intervention with a counselor and therapist is— not necessarily to save a marriage, if it is unsavable, but to prevent lasting wounds to their child.

Certainly, One needs loving contact with each parent one on one, and consistent attention from a loving adult who is outside the fray and whose behavior with the baby really does suffer no change because of personal problems.

As to custody decisions, a baby's concept of time at this age is mostly just Today. Yesterday isn't really remembered as Yesterday, but as part of a blurred past that sums up the baby's idea of Normal. Tomorrow isn't comprehended at all. It is kindest to try to keep the schedule for a baby this age as regular and consistent as possible, and to avoid spectacular moves. Spectacular moves to a baby this age mean:

- a change in who it is that cares for him intimately,
- the absence of adults whose love and adoration he is accustomed to, or
- a diminution in that love and adoration.

Thus, a move to another house is less traumatic than is a change in daycare. The absence of a beloved parent can be emotional as well as physical. Even if the parent who has custody has been the primary caretaker all along, if that parent is engulfed in bitterness, rage, and worry, the toddler will grieve as if the parent had died. Worse, the toddler will feel the target and cause of the bitterness, rage, and worry.

If at all possible, mediation, rather than litigation may help parents feel less adversarial and more in control. Therapy for the custodial adult, to help that parent maintain appropriate bonds and boundaries with the child, is usually essential.

## HOSPITALIZATION AND SURGERY

Since One is still so dependent on intimate relationships for security, hospitalization is not likely to be exceedingly traumatic if a parent or loved adult is present all the time. Surgery, too, may be less traumatic now than later, as One doesn't have yet a grown-up image of himself.

While genital surgery may be traumatic now, it is probably less so now than it will be between the ages of two and five. Reassurance, cuddling, and distraction are more likely to ease the doubts and worries of a baby this age.

## MOVING

As long as One's loving adults and security objects stay in place, moving isn't a big deal. But there is one very important warning: the day of the move, One should be absolutely elsewhere, in the care of someone loving and highly reliable. From now until mid-childhood, moving days are very dangerous. These are the days the child discovers the open unscreened window, the bucket of water, the jar of turpentine, the swallowable Ping-Pong ball; the day he decides to play with the garage door opener while standing in the path of the door; the day he stands in the driveway below the level of the rear-view mirror.

# The Fifteen-Month-Old and Eighteen-Month-Old Well-Child Visits

These are visits fraught with challenge. Frequently everyone, afterwards, is ready for a nice long nap.

Here are some suggestions for making it all easier.

- Prepare ahead of time. Dress Cherub in clothing easily removed. Particularly beware the shirt with the tight neck hole.
- Bring your own entertainment. Presented with a selection of toys intended for children from birth to adolescence, the child this age will unerringly find the crayons and eat them.
- Don't bring snacks. If you do give them to the baby, he will smash them into the carpet and throw them up during the throat exam. If you don't give them to

the baby, he will find them in your tote bag just as the examination is about to start, and howl furiously for the rest of the visit.

- Communicate as best you can that you like and trust your pediatrician. Children this age are very sensitive to their parents' feelings about other adults.

- Do not use the words *hurt, cry,* or *hate,* or any synonyms thereof. Babies are quick to understand these words and apply them to their current situations. I have seen babies perfectly happy with the pediatrician, and interested in the otoscope, until the parent says, "Don't cry. It won't hurt." Then, guess what happens. Or the parent announces, "She hates to have her ears looked at." Oh dear, oh dear. She does now, all right.

- Keep your tone of voice cheerful, not worried, apologetic, or even sympathetic, during the exam. When you are asked to place the baby on the exam table and/or to restrain the baby's hands or whatever, try to be brisk and confident and upbeat. Trying to prepare or to negotiate with an already-screaming toddler makes matters worse, as it inspires the child with the idea that you, the parent, are ambivalent about the whole thing. Which is a very scary thought.

- If there is to be a shot or blood test or whatever, keep your tone confident and positive, and explain briefly that it will hurt but that you want the pediatrician to do this because it is good and important. Afterwards, hold and hug the baby and comfort and kiss the hurt place, explaining again that it was important. Saying that you are sorry it hurts is fine; saying that the person who gave the shot was a meanie is not a good idea. Not only does it turn the child against the medical personnel, but it gives your toddler a poor idea of your ability to protect him against such meanies.

There should be a well visit at age eighteen months, and most pediatricians suggest one at fifteen months as well. There may be immunizations at both visits; the types and schedules for vaccines change rapidly. The main point of these visits isn't really just to get shots (you could do that at a public clinic) but to evaluate growth, development, general health, hearing and vision, and discuss what to anticipate between that visit and the next.

By eighteen months, only a few toddlers cling and scream during the entire exam. If yours is one of these, ask your pediatrician for suggestions on preparing the child for future exams. It often helps to make social visits to the office, with nurse, medical assistant, or pediatrician giving the baby a present she can eat afterwards, outdoors. Three or four such visits can often make the next real office visit much more pleasant for everyone.

## Looking Ahead

You are about to make a new friend: your child.

The next six months, from Eighteen Months to Two, transform the young toddler into a social person, ready to deal with adults and other toddlers. Where the one-and-a-half-year-old can understand and act on what he or she hears, Two can actually have a conversation. Two can think out an action before deciding to embark on it. Two can talk out angry feelings rather than just act on them; Two can ask for help when frustrated; Two understands the concept "broken" and how things get that way. Two starts to take turns, and understands that a possession is something that one not only holds on to but something that one can give up.

Getting from Eighteen Months to Two is thus a huge adventure in forming a friendship. As with all developing friendships, it helps to have patience and a sense of humor and to pay close, respectful attention to the interests and feelings of the other person.

And boy, is it worth it.

# Eighteen Months to Two Years

## Taking the Terribleness out of the Twos

*". . . she is the centre and the driving force of her little world. As she is a profoundly irritating person, bossy, horribly energetic and pushing, the others groan beneath her yoke and occasionally try to shake it off: but in their heart of hearts they know that it is she who keeps them going and that life without her would be drab indeed."*

*Forward to E. F. Benson's* Lucia *books by Nancy Mitford, Harper and Row, 1977*

## To-Do List

- Introduce simple plots when you "read" picture books.
- Chit-chat frequently.
- When you give an order, don't ask permission or phrase it as a choice.
- Consider using a harness or leash when out of the house.
- Limit TV to half an hour a day.
- Even better: no TV at all!

## Not-To-Do List

- Avoid overusing addicting foods.
- Don't buy beverage juices.
- Don't give "handout" snacks.
- Don't move Cherub from crib to bed.

## Portrait of the Age: On the Cusp of Thoughtfulness

Chantel, lifted from the crib after imperious shrieks, lands running to the remote control for the TV set. Has she been dreaming about it all night? She turns it on, off, on, off, clicks to one channel after another, her rage and puzzlement growing. Her mother, Kathy, grins to herself: unplugging the set was a master coup.

In fury, Chantel hits the unresponsive TV, hurting her hand, sits down hard to cry, and then is up again, this time to her pile of books. Her father Brad stops at every garage sale to replenish this pile, for Chantel is particular about her books. They must have intact pages, and must be made of paper—not sturdy cardboard or cloth or synthetics. There is a mystery, a mind-boggling secret about the spine of such a book, and how the pages attach, and Chantel is determined to get to the bottom of it.

"Look, Chanty, here's a big book called a magazine. See how shiny. The shiny pages are hard to tear." Kathy donates an ancient copy of a fashion magazine. "Those pages don't crunch up like the others, do they?" She crouches next to Chantel, crunching a few herself.

After satisfyingly trashing the 1999 issue of *Vogue* magazine and then *All About Beavers, Big Day for Boris,* and *The Life of a Tooth,* Chantel turns her attention to the rocking chair, climbs into it in a flash, and begins to rock, so furiously that that item of furniture begins to creep across the floor.

"No! That's no no!" Kathy grabs and brings the chair to a halt. "Hard rocking can hurt the chair. Hard rocking can make Chantel fall."

Could this admonishment have been what Chantel was searching for in her mad pursuits? A look of fulfillment flashes across her face for just an instant before she lashes out, screaming "Mine! More chair!" She clenches the arms of the chair. As Kathy detaches one little hand, the other re-attaches. Finally Chantel gets another idea. She grabs Kathy's hand and drags her into the kitchen pointing and grunting. There, the toddler makes her daily attempt to open the refrigerator. Still sobbing a little, she eyes the handle. As she does every day, she first tries to reach it by standing on tip-toe. Then she picks up her rubber cow and bangs it against the handle. Then she finds the stool and pushes it to the refrigerator and

climbs up: now she can reach and even move the handle, but she is standing so close to it that she can't get any leverage and the door won't open. She turns on her mother, eyes filled with blame, and howls.

Her mother lifts her down, trying not to laugh, and opens the refrigerator. She pours Chantel a cup of juice, and hands it to her.

"No!" Chantel says, reaching for the cup. Holding it in trembly hands, she drinks. "No!" she shouts again. Oddly enough, when she's finished drinking, she sets the cup down very carefully on the table, and dashes back to the family room, Kathy in close pursuit. She aims for the rocker. Kathy grabs her just as she reaches for it, picks her up, and returns her to the kitchen.

"The rocker is no. No rocking. No!" she says firmly. It is time to get out of the house, breakfast or no, Kathy decides. A special treat: they'll picnic breakfast on the beach.

Chantel seems wildly excited. Out of the car, Kathy has a hard time gripping the eighteen-month-old with one hand and her backpack with the other as they race over the grass verge, barefoot. Chantel loves the beach.

They approach the sand. Chantel steps onto the cool, grainy surface—and stops dead. She turns pale, grabs her mother's legs, and lets forth a horrified moan, which then turns into a shriek. She tries to climb up on Kathy.

Kathy looks around in vain for the snake, the shark, the jagged piece of glass. She sits down, *plop*, with Chantel on her lap. She tries to point out enticements—the gentle waves, the children playing. She picks up a tiny handful of sand to show her; Chantel slaps her hand away and buries her face in her mother's shoulder.

Forty minutes later, Kathy steaming in defeat, Chantel hiccuping exhaustedly, they troop to the car.

Once home, Chantel consents to sit on Kathy's lap and "read" the book about the animals. She points to the animals Kathy names, now and then says one: "Panda" or "Kangaroo." Suddenly, she's down on the floor and off again, back to the rocking chair. Kathy intervenes. Chantel attacks her mother, hitting and kicking.

"NO! No hitting!" Kathy turns the baby facing away from her and holds her hands firmly into her body. "No hitting! We don't hit or kick people!" Chantel writhes, frantic. Kathy's hold slips.

Chantel arches back and hurls herself onto the tile floor, landing full force on the back of her head. She holds her breath so long her mother fears she is knocked out, but one look at her contorted face tells her that's not so. The scream, when it comes, rivals the loudest of noises.

Kathy picks up her flailing daughter and, dodging fists and teeth, carries her back to her crib. "No tantrums! Chantel is having a tantrum, and she must stop it, now!" Kathy says in a firm voice. "No screaming. No banging head on the floor!" She wrestles the toddler into the crib, removes the stuffed animals, and leaves her to it. She returns to the kitchen and takes three aspirin. Her ears are ringing anyhow; a little overdose won't hurt.

A war of wills; phobias; tantrums—isn't this supposed to be the province of the Terrible Twos?

Many parents, girded for that second birthday, are shocked when Mid-One blossoms forth with such behavior "precociously." Are they in for a whole extra six months of Terribleness?

Actually, many don't find it so terrible. Though she is demanding, and loud, and she wears them out, Mid-One is not really, well, not terrible. For one thing, the baby is almost comically at the mercy of the universe. Things attack her, elude her, mystify her, frustrate her. The concepts "accident" or "mishap" or "chance" are meaningless to her.

Moreover, this older toddler acts without forethought. She is totally without guile. She doesn't really plan how to open the refrigerator door. She madly tries one thing after another, no matter how unlikely it is to succeed. (What magic, Kathy wonders, has her child ascribed to that rubber cow? She has seen Chantel hurl it at the refrigerator, throw it after an unreachable item under the sofa, and attempt to make it go down the bathtub drain.)

Most of all, many Mid-Ones seem to have no control over their own negativism. "NO!" Chantel shouts, as she greedily slurps down juice. "No!" she wails, at the sand that she so loved only yesterday, staring at the beach with stricken, covetous, terrified, and enamored mien.

A baby this age seems to need to establish some kind of selfhood by hurling herself against boundaries. Her compulsion to reject seems almost programmed. While she can be utterly infuriating, it is still possible not to take all this behavior personally. Even when she bites.

Mid-Ones also seem programmed to try out every new skill, such as climbing, and to practice it again and again. They have a deep urge to master their universe, to unlock the secrets of book bindings, TV consoles, cabinets, toilets, stairs, drawers, and their mother's shoes.

Over the next few months, an enormous change will occur. Mid-One will develop thoughtfulness. She will start to be able to try out something in her mind before she acts. She will be able to think about how to get a block from under the sofa, and use a tool (not her rubber cow) to try to reach it. Words start to enable a child to think about an action before performing it, and to remind herself about its appropriateness as she engages in it.

The first signs of thoughtfulness can be very funny. Chantel at twenty months climbs into the rocker and, as she builds up a head of steam, admonishes herself, "No, no, no, no, no, Chanty, no hard rocking!"

Hand in hand with thoughtfulness comes a more complicated idea of self. At eighteen months, Chantel needs to define herself by bumping against everything and everyone, by the words *mine*, and *no*, and by demanding and demanding and demanding. By Two, both language and skills are so much more sophisticated that she can define herself positively, by setting a goal and accomplishing it. "Do it self!" she can say, removing her diaper or sock, pushing her own stroller, turning the faucets.

When such learning does not go well, Terribleness can certainly result. Such a Ter-

rible Two-Year-Old may have learned, instead, that tantrums and aggression are condoned or even encouraged, or that clinging and whining are considered appropriate behavior. She may be very confused about her own power over adults. She may have no self-control, nor see any reason to develop any. She may have received enormous proof that not only is she special and important, but that she is more special and important than anyone else, and her wants and whims take priority over everyone else's, especially her parents'.

Developmentalists consider this six months crucial. Many parents instinctively feel the same. Fortunately, the payoff for making special efforts during this time is enormous. When a baby does learn these crucial lessons, they really do stick.

What a Mid-One or a Nearly-Two needs most are:

- As safe and varied an environment as possible, so that she isn't continually confronted with the forbidden, and so that her contests can be with the real world and not just, repeatedly, with the adults around her.
- Adults who set limits consistently and firmly and stick to them, who don't take her tantrums and aggression personally, and who can find humor in the predicaments of this age.
- Repeated messages that she is very much loved and respected, but that other people also require love and respect.
- Many, many encouragements to learn language. This means twenty to thirty

interactions an hour, most days, that feature pleasant "baby chats."

By Two, not only is the baby not Terrible, but she is on the way to being a real companion. She is getting to be a person who can negotiate and take turns; a person who can tell the difference, most of the time, between wants and needs, and who can wait a bit to get something she most desperately wants, whether it is attention or a desired object.

Between eighteen months and Two, children also become more aware of, and more able to interact with, other children. While group activities are way in the future, children in this six months enjoy being together. Rather than just watching each other, they will try to engage in the same activity, seeing what each one can contribute.

Even so, babies this age need close adult attention to get the most out of social experiences. They don't have adequate language in which to express feelings, much less ideas and intentions. They can easily run out of ideas (making the cars go round the sandbox) and resort to the next obvious action (throwing handfuls of sand).

Babies this age cannot be expected to share. Most are barely able to get the idea of taking turns, and then only when the other children are older and enforce it. Toys that can't be used by all simultaneously are best used when a baby is by herself.

## Separation Issues: Here We Go Again!

Eighteen months is a second peak of separation anxiety. Here is someone who is starting to be able to anticipate events, to try things out in her mind. She has a rudimentary concept of "what if." Here is someone who is preoccupied with the subject of permanence and disappearance. Here is someone who is just starting to understand that words can "stand in" for people and objects, so that beloved people and objects can, in a sense, be held on to even when they aren't physically present. So here is someone for whom separation is a very big deal.

Separation at this age needs to be handled tactfully. Babysitters need time to warm up with the child before parents leave. If both parents must be away from the child overnight or longer, it's crucial to be sure the baby has had recent, prolonged care from the person staying with her. A visit with Grandpa six weeks ago isn't recent enough. An hour with the new sitter isn't long enough. If the trip isn't really necessary, parents should consider postponing it.

*Bedtime separation*
Many babies who started off falling asleep in their cribs without aids at an early age continue to do so, and get through their normal night awakenings by soothing themselves.

But some don't.

Early on, a baby who was just discovering the fact of separation became anxious when a beloved parent left her sight. Now, that same baby not only knows about separation but has discovered that her parents, her beloved parents, sometimes—oh, horror—get ANGRY at her! Parental anger is a new kind of separation, a very scary one. Going to bed, losing the assurance of enduring parental love just when one is most vulnerable and stressed, can be just too much to bear.

Much bedtime refusal at this age seems to stem from daytime problems of limits and separations. Inconsistent approaches to bedtime make matters worse.

Before honing in on what to do at bedtime itself, consider whether Cherub's daytime experiences might be contributing to the problem:

- **Lack of practice in physical separation.** Separating at night is tougher when daytime separation hasn't been mastered. Remedy: a regular playtime at somebody else's house, with Mommy saying a cheerful brisk good bye when she leaves. Ideally, at least three times a week: daily would be even better.

- **Bedtime is the only one-on-one time with a parent.** Bedtime is supposed to be a separation ritual, not quality time. When that's the only time Cherub gets full, focused, attention from Mommy, especially, it's a guarantee of bedtime rebellion. From Mid-One's point of view, gee, she was just getting started having enough of Mommy, and now Mommy is going to make her go into that crib and then disappear. Remedy: make time for ten or fifteen minutes of one-on-one time earlier in the day. Then have regular life events—dinner, bath, books—intervene before the bedtime ritual.

- **Parents give mixed signals about their desire to have Cherub go to bed.** They say that's what they want, while at the same time prolonging the bedtime ritual, or just not being consistent about it. Sometimes this can be because of marital stress, sometimes because the parents are guilty about not having enough time with Cherub, and sometimes simply because the baby is so cute and so much fun. Remedy: face up to one's own parental motives.

- **Taking too many naps at daycare.** Most Mid-Ones need only one nap a day. When they are given two naps, or the one nap is longer than an hour and a half, or ends later than about three P.M., they aren't ready for bedtime at a decent hour. Remedy: discuss this with the caregiver.

- **Sleeping late the next morning, after a sleep-deprived night.** This just gets One's whole schedule out of whack. Remedy: get her up at her regular time, and have her nap at her regular time, no matter what.

### CRIB TO BED

If Cherub is not yet in a real bed, I urge you to wait until Two or later. If you can't, you need to be sure that Cherub cannot get out of the room in the middle of the night, and that the room itself is baby-proofed thoroughly and fastidiously.

I shall always remember Noah, who was discovered at one A.M. by his parents as they returned from a party. Noah, stuffed lamb in hand, was just making his way down the front walk. The sitter was sound asleep in the living room.

If Cherub must transfer to a bed, one way to keep Cherub confined is to erect a bedroom barrier, either a gate or a Dutch (half) door. Another solution is a lock on the outside that can easily be opened by an older child or adult. If you have to erect a barrier for the first time now, make sure you spend time in the room with the baby with the barrier erected, so she comes to regard it as a normal condition of life.

If there is room, it may help the emotions of the transition to have bed and crib coexist for awhile, allowing the toddler to play and nap on the bed.

## Limits

*Please go back and read the section on* LIMITS *in the last chapter. This is a sequel to that one.*

Over the next six months, Mid-One needs to become convinced that when you tell her to do something or not to do something, that is exactly what will happen— whether she agrees with you or not. She needs to understand that there is a realm of behavior in which she has no choice, and no voice.

Because Cherub is older now, and may be talking some, you are likely to discover real oppositional behavior starting now. That is: you give Cherub a direct order, Cherub says "No," you give the order again, louder, and add "Right now!" and Cherub says "NO!" and on it goes.

Whining gets harder to deal with, too, especially if it started months ago and is getting worse. Tantrums may be longer and louder. For the first time, Interrupting (parents on the phone, or in the midst of a task) can become a problem.

How do you make a Mid-One understand that "arguing is not an option?" That when a parent gives a direct order, it must be obeyed? Here's how. If you follow this guide consistently, by the time Mid-One turns into a Two, you'll be well on the way to avoiding the dreaded Oppositional Behavior.

### Only give an order if you can enforce it

If you keep giving orders that you cannot enforce, Cherub will come to think of orders as optional. Worse, Cherub will come to think of you as ineffectual. That's a real problem, because at this age Cherub needs to learn that parents are completely in charge.

Fortunately, in this situation the adult holds all the good cards. What you need to do is to make sure that each and every time you give an order, *you* can see to it that that order is obeyed. If you tell Chantel to get down off the coffee table, and she hesitates for even a second, or says "No," you simply pick her up and move her. Not with a smile, or an explanation, or an apology; you simply say "No" and place her somewhere else. And then you don't give her direct attention for a minute or so.

### How to give an order
- **Use as few words as possible**—ideally, only two or three. Don't apologize, describe, or explain.

- **Don't give a warning.** As in, "If you don't get down off there . . ." or "I'm going to count to three and then . . ."
- **Don't ask permission.** As in, "Hey, Chantel, get down off there, OK?"
- **Don't present it as a choice.** As in, "Wouldn't you rather come down here on the floor and play Legos?"

*If you can't see to it that Cherub obeys the order, don't phrase it as an order.*

- Say Chantel is whining and whining and whining. You can't make her stop; it's *her* voice, after all, and she's in charge of it. What you can do, though, is completely ignore the whining. Get those ear plugs in your ears. But here's the trick: *Just do it.* Don't say, "I can't hear you when you whine." Don't say, "That's not Chanty's nice grown-up voice."
- Or suppose Chantel is pressing the buttons of her busy book, and that cow noise that she's pushed about two hundred times is driving you wild. But you don't want to confiscate the book. After all, you got it for her to play with.
  Some other options:
- Join her in the game, then segue to something different.
- Or get down beside her, push a few buttons yourself, then grab one of her plastic animals and pretend to make its noise. Get her started with the plastic animals, and rise from the floor, the busy book unobtrusively in your hand.

- Or turn on some music and start to dance, without saying a word to Chantel, and she'll probably join in.
- Or produce from your Secret Closet a toy that she hasn't played with for weeks. Don't try to lure her away from the busy book. Just start playing with the new toy yourself. She'll probably join you in a minute or so.

*Don't let Mid-One challenge you repeatedly.*

Peter likes to stand on my exam room table and play with the expensive instruments. His mother stands there, watching him, and every time he reaches for them, she takes his hand away and says "No." They play this little game (as I watch from the door, amazed) for about two minutes.

Peter's mother is giving him a tutorial in Oppositional Behavior. Look, she says to Peter. You can say No to me and the most I do is let you do it again. What she should do, of course, is get Peter off the exam table even before he reaches the first time. She needs to anticipate when a challenge is going to come. Then she has a choice: prevent the challenge, or immediately remove the baby or the object.

If Cherub is already highly oppositional, and life is very difficult, please see *Limits* section of the next chapter, "*Age Two to Three.*"

# Day By Day

## MILESTONES

### Vision

Babies should see well enough to accomplish anything they want to at this age. They should spot an airplane or hot air balloon you point out, recognize a picture in a book from normal adult reading distance, and recognize Barney from four feet away from the TV.

For signs of problems, see previous chapters. Below are some new signs, appearing at this age:

- Having to get right up close to the picture on the page or to the TV.
- Seeming afraid to climb, or often misjudging distances. For instance, if you hold something for her to reach and grab, she reaches past it or to one side.
- Apparent difficulty judging distance: not wanting to tackle stairs, or climb the ladder of the slide.

### Hearing and language

Ideally, Cherub ought to have 20 to 30 encounters an hour of chatting with an adult. By chatting, I mean that the adult talks to Cherub about whatever Cherub is doing or is interested in. "What a big box! Many toys in that big box!" "Here is your spoon. You can eat the pudding with the spoon." Cherub, in return, can initiate and respond, and then needs to be rewarded with adult attention and words. Reprimands ("No!" "Stop that!" "Dirty!") and judgments ("Nasty dog food!" "Good girl!") don't count as chatting; even if they are not negative in tone, they are not reciprocal.

By eighteen months, a baby should be able to follow a simple, one-step direction like "bring me the book" (if she's in the right mood), and point to a body part that you name. If she knows the name of a picture in a book, she should be able to point to the picture and say it. She should be saying at least 5 words besides Mama and Dada. Mostly, these are commands: "Up!" "Down!" "Juice!" "Out!" "Bye-bye!" (You may think that Bye-Bye is a polite recognition that someone is leaving. Not at this age, it's not. It means, "Get me out of here" or "Go away!") At least some of the time, you should have the feeling that "She understands everything we say!" (for better or worse). She can repeat words that you say (also for better or worse).

At the beginning of this six-month period, many babies jabber in a beautifully inflected language of their own, with only a few understandable words. They show that they have language "inside their heads," however, by their play. Making the cars and trucks zoom along with their zoom noise. Playing hide and seek with Mama and Daddy. Taking care of a baby doll or stuffed animals.

**If at Eighteen Months a baby says fewer than five words, or doesn't seem to understand simple directions, or doesn't engage in pretend play, or *consistently* plays with toys in a strange way—not recognizing that a toy truck moves zoomingly, but just spinning the wheels and staring at them,**

**for instance—hearing and development should be checked by the pediatrician.**

By Two, Cherub should be able to show that she understands more than a hundred words, and make telegraphic sentences, as in "More Juice" or "Daddy down boom" or "Bye-Bye Doctor." A Two who can't put two words together, or who says few enough words so that you can easily list them, needs a hearing test. A Two who isn't engaged in pretend play, who consistently plays with toys inappropriately (for instance: just lining up all the cars, or all the stuffed animals, rather than playing with them; still eating the crayons, rather than scribbling with them) needs a developmental evaluation.

It is never useful or successful to ask a baby to "say thus and so" or to "count to ten." I have never in three decades of practice seen this work once on a child this age. The only, and best, way to evoke speech is to make every utterance of the toddler into a dialogue you participate in.

Many early talkers make substitutions for consonants. "Yellow Truck" can come out "yehyo twuck" until kindergarten. But if a baby seems to be very unclear, or simply not saying some consonants at all, a hearing test may be in order. Check with your pediatrician.

### Books

At the beginning of this period, Cherub starts to understand that a book has pictures of familiar items. This is the time to learn to love what books offer, rather than just eating them or investigating how pages turn. A few minutes each day of turning the pages and talking about the pictures ("I see a cow. Do you see a cow? There's that nice happy cow.") is ideal. As the age of Two approaches, you can start developing plots. "Does Clifford want that bone? Oooh, look, he's going to take that bone. What will he do with it?"

### Large motor

By eighteen months, a baby who is not walking needs full evaluation. The gait should be heel-toe, not on tiptoe. Most can run, walk upstairs holding on, and all can climb. Rarely, a youngster may be able to jump, but lots of children can't get both feet off the ground simultaneously until age Two to Three, and a good thing, too. During this period, Cherub should be starting to throw (a ball, a shoe), and kick a big beach ball.

**When Cherub is stooping and tries to rise to a stand, he shouldn't need to "climb up himself" to do so—pushing his hands against his thighs. This can be a sign of serious muscle weakness; a visit to the pediatrician is in order.**

### Fine motor

For months now, toddlers have been able to pick up a tiny object, like silverfish or other disgusting minutiae, with thumb and forefinger. They can hold a crayon in a fist and scribble with it. They can feed themselves with a spoon held in a fist, and can hold a cup. Some will be showing a handedness preference, but it is a preference, not a necessity; they can use both hands equally well. Most don't automatically put things in their mouths, the way they did only six months ago (unless it is crayons in the pediatric office, which clearly taste better than other crayons).

## SLEEP

Most children this age sleep about thirteen hours out of twenty-four, with about two of these being during the daytime.

It's not unusual for a baby this age to seem ready to give up the daytime nap, but almost always that turns out to be a mistake. Such a baby usually is one full of intensity and determination, and it's easy to think that he really doesn't need as much sleep as other children his age. What happens is that he gets more and more frantic, dashing from one activity to another, easily frustrated, easily enraged. Sometimes he even pushes his bedtime back, so that he's going to bed at the same time his parents do. Often, but not always, he's hard to wake up in the morning after only eight hours of sleep. The funny thing is, go for a ride in the car, at almost any time of the day, and he falls asleep.

Do not be deceived: this is a child who is absolutely sleep deprived but reacting to this paradoxically.

Solution: start a four- to seven-step bedtime routine about twenty minutes before the time he usually conks out from sheer exhaustion. Make sure he actually falls asleep in his crib: not on your lap, on the floor, or in a big chair. Get him used to the routine over a few days. Then start that same routine, exactly, right after lunch. Don't let him up to run around right after lunch. The chances are great that he'll fall asleep, and I mean deeply asleep.

Now, the trick here is to awaken him after only two hours of napping at most. He won't like it, he'll be in a terrible mood for the rest of the day. Then that evening, start the bedtime ritual at around seven-thirty or eight o'clock.

Most of the time, he'll sleep through for about eleven hours, awaken at the decent hour of about seven A.M., and be ready the next day for his after-lunch nap.

One more nap comment: as in the previous six months, beware the daycare provider who still tries to keep children this age taking two long naps a day for her own convenience. If your baby is sleeping four or five hours in the day at her house, he'll only need about eight hours sleep in the night at yours.

## GROWTH

Growth continues at its comparatively slow rate. Babies gain about two-and-a-half pounds and grow about two-and-a-half inches over this six-month period (five pounds and five inches over the whole year from one to two). By eighteen months, the soft spot (anterior fontanel) in the skull has usually filled in, though sometimes a shallow firm depression remains. At around the age of two, the second-year molars come in, usually without a great deal of fuss.

### Chubbiness watch

If Cherub's weight for height is above the 75th percentile, there is a chance that she is overweight. If it is at or above the 90th percentile, she is very likely to be overweight. Ask your pediatrician, or see the essay in Section III: *Chubby or Not, Here We Come.*

### THE NORMAL PATTERN

Her total caloric needs are about 1100 to 1300 calories per day. If she's drinking 16 ounces of milk a day, she's got about 800 to 1000 calories left for food. That's not a lot. No wonder many normal-weight babies eat only one big meal a day and hardly any dinner.

### WHERE IT CAN GO ASTRAY

When a baby this age starts gaining more weight than is desirable, it's usually because somebody taking care of her has a misconception about what she needs to be eating

If a baby eats 50 calories above and beyond what she needs for growth, every single day for a year, she'll gain an extra five pounds from that daily oversupply. That could *more than double* the amount she's supposed to gain in this year.

Those fifty calories can come in the form of juice instead of water for thirst. Or in the form of snacks—a handful of crackers, some string cheese, raisins or dates. Or in being coaxed and urged to eat a "good" dinner, when she's not in need of it.

### WHAT TO DO

If your baby is showing an increase in her weight curve, check with your baby's daycare provider about TV watching, napping, snack habits and the handing out of juice. Review your own dietary patterns, especially the din-

ner question. Make sure Cherub is not nibbling from "handout" snack baggies filled with Cheerios, crackers, raisins, and so on.

Like Measles and Oppositional Behavior, extra chubbiness is much easier to prevent than it is to treat.

## TEETH

Here is a big surprise.

Believe it or not, tooth cavities are contagious. I am perfectly serious. To get a cavity, you need acid-forming bacteria which go by the name of *Streptococcus mutans*. And how do babies catch this bug? Pretty much from their parents! Sharing a spoon, mouth kissing (probably not such a hot idea for reasons having nothing to do with teeth), holding the baby's pacifier (which should have been discarded by now anyway) in the parent's mouth—that's how. How to prevent it? Try not to indulge in these behaviors, but, most importantly, get your own cavities fixed and brush and floss yourself.

And why do I mention this for the first time at this age? Because, even more incredibly, there is a window of time, from nineteen to twenty-eight months of age, during which babies tend to "catch" these bacteria from their parents.

Since you're going to go to the dentist yourself, this is a good time for your baby to have a checkup also. Some general dentists are very comfortable with babies, and there are more pediatric dentists in practice every year. You could call 800-544-2174 outside Illinois or 312-337-2169 inside Illinois for a referral from the American Academy of

Pediatric Dentistry, or ask your own dentist or pediatrician.

## NUTRITION AND FEEDING

Nutritional guidelines are exactly the same as in the previous chapter, One Year to Eighteen Months.

Toddlers should be feeding themselves by now, and be fairly adept with spoon and maybe fork. A bottle that is used for mess prevention now and then won't do any harm, nor will one that is taken in a parent's lap as a ritual cuddle.

A baby who is using a bottle more often, and especially one who carries the bottle about with her, has a problem:

- If the bottle contains juice or milk, she runs a serious risk of tooth decay, as her teeth are constantly coated with sugar (even the "natural" sugars of juice and milk).
- The constant presence of a bottle (or pacifier) occupies a mouth that is now supposed to be uttering jargon, words, and, eventually, sentences.

## DIAPERING, BATHING, AND MASTURBATION

See the section on these interrelated activities in the previous chapter.

## TOILETING

Once in a while, a baby this age will learn how to use the toilet. This usually happens when there are other, rather older children in the family, and the baby undertakes the task in the spirit of curiosity, imitativeness, use of tools, etc., that are the keynotes of intellectual development now. Or an only child, usually easygoing, eager to please, and verbal, will simply start on his own.

Most babies demonstrate clearly that they are in no mood for such a task now.

- Intellectually, they are just working on developing thoughtfulness—that is, thinking through an activity before trying it out.
- Physically, they are more interested in conquering the environment than in performing a task.
- Emotionally, they are dealing with the aspects of control and bodily integrity and loss that are aroused with the act of pooping.

Babies this age seem to have a very special feeling about pooping. It is as if the effort, the control aspects, the feeling of ejecting something or losing something or relinquishing something; the highly sensuous and variable result, in color, texture, and odor; the response to that result—well, it's just too much. As if to contemplate the whole subject (there doesn't seem to be a privacy issue involved), many will retire to a quiet, hidden place to perform. Popular retreats include closets, behind sofas, and under large pieces of furniture, such as tables and pianos.

Readiness to use the toilet for a baby this age is shown by:

- Being able to recognize that pee or poop is about to happen. Pointing and saying the words or grunting.
- Being able to go for several hours with a clean, dry diaper.
- Preferring to have a clean dry diaper rather than a dirty or wet one.
- Being over the most intense of the normal oppositional, negative, and possessive behavior.

For an Under-Two, learning how to go to the potty is a complicated task. Over-Twos often can learn just by watching, but the younger toddlers often do better going step by step.

- Carry the potty chair around with him and let him sit on it with his clothes on.
- Show him after he's gone in the diaper that the poop can go in the potty.
- Let him watch other people going to the bathroom.
- Let him go around (outside) naked with the potty chair nearby. When he goes in the grass, remind him casually that he could have put it in the potty.

Being too intense about potty skills, whether in a positive or a negative direction, can backfire. Most Nearly-Twos ready to use the potty want to do so because it is an interesting, grown-up challenge, and not just to please their parents. If Nearly-Two thinks of it in that way, he is less likely to reward a parent by going in the potty—but also less likely to punish her by refusing to do so.

## ACTIVITIES AND TOYS

Parents want a Nearly-Two to learn skills that will equip her, eventually, for the social and intellectual challenges of preschool and beyond. They want her to make friends. They want her to trust adults as teachers and guides. They want her to develop curiosity, to enjoy language, and, in a word, *think*.

Many parents of Not-Yet-Twos feel obliged to enroll toddlers in an array of classes: gymnastics, swimming, art, music, dance. They attend playgroups and even prenursery schools where the children make craft projects and have "circle and sharing" time.

These are really better aimed at children Two and older.

When the crafts project is approached, the babies seem to be shaking sprinkles on glued colored paper as if this were an incomprehensible endeavor they are doing to please a doting parent, not as if they were creating a product. Circle time finds most of the children looking up at their mothers for clues rather than at the teacher. Many Nearly-Twos just get up and wander.

Pride in performing a task can appear at this age, but it is the task, not the construction of a product, that gives Twos a sense of achievement (e.g., scribbling on the paper and making a mark rather than producing a collage for the refrigerator). Imitating adult behavior is fun for Nearly-Twos, but actually playing a role in a fantasy comes after Two. Waiting while an idolized older child takes a turn is just possible now; understanding the concept of taking turns, and allowing an age mate to go first, is not.

For all these reasons, organized classes for babies this age are mostly social fun for parents. They can only do harm if they are medically or physically risky (e.g., swimming or trampolines) or if they take time away from other important activities. They may also harm parents, who may feel disappointed that their babies don't "appreciate" the classes, aren't "social," or don't learn what the classes are billed to teach.

Activities that *are* useful to children this age are ones that stimulate:

**Curiosity:** *A Nearly-Two is interested in everything novel, but at her own pace. In the yard one day, for instance, Chantel is occupied in mashing a little stone with a big stone. Kathy, a little bit away, sights a praying mantis. Thrilled, she calls to Chantel, who ignores her. Kathy nonetheless picks her up and brings her to the praying mantis. Chantel gives one bored glance, wiggles away, and resumes stone-mashing.*

**Fantasy play:** *A child this age engages in rather uncomplicated fantasies. She can make her fire engine go, with sound effects; she can cuddle and nurse a baby doll; she can make her toy cow moo and her kangaroo hop, though she has never seen a kangaroo. But she can't make the fire engine put out a pretend fire, or make the baby doll cry for her bottle, or milk the toy cow. That is, she can pretend an object is a real object, but she can't really role-play yet. She can't actually think of herself as a fire-fighter, mommy, or farmer.*

**Making sense of the world:**

Noticing: *A Nearly-Two is able now to notice some small details in her toys. She is fascinated with the steering wheel in her fire engine and the belly button on her anatomically correct doll. She also can tell when something is "wrong" with a toy or object.*

*Noticing discrepancies is the first step to a sense of humor. When she puts on her father's sunglasses and looks in the mirror she thinks she's hilarious.*

Anticipating and relating: *She can also anticipate consequences and associate two or more unlike items. In the bath, for instance, she knows that when she pours the water in the top of the big bath toy it will, after turning the waterwheel, come out the bottom. Adults can cultivate this activity greatly.*

Dealing with abstractions: *Most children at twenty months already have figured out at least one abstraction: animals, which they call Doggy. In Nearly-Two's mind, Doggies or animals have tails; also fur and four legs, probably. When an animal doesn't fit this category, she may become upset. She is also aware of quantity and direction. When she wants to climb higher, for instance, she tells her mother, "more up."*

Many Nearly-Twos are able to say the numbers from one to ten, but the idea of actually counting comes much later.

## SAFETY ISSUES

See the *Safety Issues* section in the chapter *"Nine Months to One Year."* If you haven't baby-proofed your home and stocked your medicine chest as suggested in the previous chapters, please do so now. Just three reminders:

### *Getting out of the car seat*

Just as they escape from the crib, so do toddlers this age learn to undo the belt of the car seat. The first rule, of course, is to try to make staying in the car seat fun, with simple toys, and sing-along tapes. Try to make eye contact, smile, and start the playing before you turn on the ignition, and then keep your presence for the toddler an active one with conversation and frequent turns of the head, but not eyes of course, towards him.

If he still escapes, try practice runs. Get him into the car as if you were going someplace, but in reality drive very slowly around safe streets close to home (I hope you have some). At the first sign of rebellion, drive home with no eye contact but just a firm voice saying "No getting out of the car seat" over and over again. At home, immediately take the toddler to the crib and institute two minutes of time out or until the crying stops, whichever takes longer. You will only need to do this two or three times.

This may seem extreme, but an escaping or escaped toddler on the freeway is a fatal accident about to occur.

Also, keep doors and windows locked from the inside at all times. And of course never ever leave the baby in the car alone, even for a minute.

One other car warning, for those with a new baby plus a toddler: this is the age when the new baby gets set down in the car seat while the toddler gets settled inside, and then the harried parent drives away with the baby still ensconced on the roof or hood of the car. A nightmare? You bet. It happens? You bet.

### *Swimming*

The American Academy of Pediatrics, once again, discourages water-immersion activities until age Three for these reasons:

1. Being able to swim in a supervised environment does not make a baby water-safe. If she falls into the water unexpectedly, she is just as likely to panic and drown as a baby without any swimming instruction. And being "friendly" with the water may make her even more likely to plunge in on her own, unsupervised.

2. Infection is still a problem. Water immersion is one of the predisposing conditions for ear infections in many babies. Babies in diapers also generate bacteria in stools, and it is a rare pool that is large enough and well treated enough to preserve hygienic conditions.

3. Most important, babies can suffer from water intoxication by swallowing, unnoticed, large quantities. This can happen without any sign of distress at the time. Water intoxication can appear after the baby is out of the pool, with intestinal symptoms (distended abdomen, vomiting, diarrhea), signs of shock (pale or

blue, cold or mottled skin) and, devastatingly, lethargy, convulsions, and coma.

### Gymnastics

Children this age are still very vulnerable to shaking and whiplash. The head is still disproportionately large and the neck weak.

Another caution has to do with trampolines. Most babies are just learning how to jump and are fascinated with any item that allows them to practice the activity. Trampolines, even the mini ones, are unfortunately dangerous. The American Academy of Pediatrics has issued a statement against use of this equipment. The concern is with spinal cord injuries, and it is a concern even if the child is supervised.

## Health and Illness

Health issues of illness and injury for the second half of the year are the same as for the first half. Please see the previous chapter.

## Window of Opportunity

### SECOND LANGUAGES

Some evidence indicates that children who are exposed to, and imitate the sounds of, a second (or third, etc.) language before the age of two will be able to learn that language with no accent later on; and the more exotic the language, the more difference this might make. That is, this may not be so important

if the first language is English and the second German, or even Spanish, but might be more important if the second were Japanese or Swahili or Persian. There is no sense trying to "teach" a second language at this age, of course. Mere exposure in a natural setting is all that is required.

### SQUELCHING THE TV HABIT BEFORE IT STARTS

Many babies this age show a distinct propensity to becoming addicted to TV, especially to videos like *The Little Mermaid* and to action cartoons. By age Two and Three, the "good" children's shows like "Sesame Street" and "Mr. Rogers' Neighborhood" may, in fact, seem too tame for children who cut their video teeth, so to speak, on the snazzier productions.

There is every good reason to quench this tendency immediately. Of course, everyone knows that TV is too passive, sets bad examples for violence and language, advertises toys and sugar-loaded foods, and so on. But at this age it is particularly devastating to turn on the tube.

First, it is too tempting to use TV as a babysitter, especially at this demanding age. To an exhausted working parent, if half an hour of peace and quiet with Toddler watching Ariel cavort on the tube is heaven, an hour is even better: time for Parent to take a bath, cook dinner, read a report, grade papers, call a friend. It is all too easy for an hour now and then to become an hour a day, two hours a day, every single day.

Second, once started, the habit is hard to

stop. A child who has watched Ariel every day for three weeks is not going to take it calmly when, in week four, her parents suddenly realize that she is hooked and turn off the set. A very young child depends on the constancy of her environment.

Third, the baby who watches TV will see things that his parents do not intend him to see. Do you really think that the twenty-month-old who sees that monster in the toilet on TV will realize that it's just an ad for bowl cleanser, and feel calm about sitting on that horrific piece of furniture himself in a few months?

Also, never underestimate the ability of a young child to understand what is going on in the news, commercials, previews. When little Jessica McClure fell down the well, even the under-two set was distraught.

Fourth, whatever annoying noise or utterance is produced on the TV set, a child this age will reproduce, loudly and in public. Enough said.

## What If?

The sections for *Starting Daycare, Extended Separation from the Parents, Traveling with the Baby,* and *Moving* are the same for this six months as for the first six months of this year, and are covered in the previous chapter.

### ADDING TO YOUR FAMILY

By two-and-a-half, a toddler can anticipate and get excited about a new baby coming. He has his own social life, independent of parents and siblings. He can talk about negative feelings rather than just acting them out, and is thoughtful enough to consider an action and its consequences before he undertakes to perform it. He can be of some help, fetching and carrying; he can be very gentle and protective when he feels like it. On the other hand, Mid-Two's veneer of civilization is fragile indeed.

For many children, this seems to be the very hardest time for a new baby to arrive. "Just as I was getting the hang of it!" they seem to say. "Just when I was getting the idea of being a separate, precious, independent person, who can be powerful and in control and still be loved and cherished by my parents—along comes THIS!"

The sight of parents enthralled by the dependent, loud, demanding new baby just sets off Mid-Two. Tantrums and aggression resurface. It's worse if the Mid-Two is home most of the time, and is constantly having to compete with the baby. Mid-Two doesn't understand how to compete effectively, how to win back the doting attention of his parents, especially his mother. He believes he can achieve this with force, and rarely tries charm.

Three is simply more his own person than Two-and-a-half, more confident that he has a life separate from his parents. Three has more tools for getting and keeping adult attention in a civilized and acceptable way. He also is more likely to have a niche in a group of other children, whether at preschool or daycare or in the neighborhood, a niche undisturbed or even enhanced by the new arrival.

Most important, Three is less ambivalent about wanting to be a baby herself. She is also able to role-play, trying on a behavior without losing herself in it. She can *pretend* to be a baby, *pretend* to be the mommy. On the practical side, Three is likely to have learned to use the potty, and to stay in her bed reliably.

## ARRIVAL OF A NEW SIBLING

Many toddlers this age are "all over" the new baby, kissing and hugging (once in a while a little too intensely, of course) and take out their negative feelings on their parents, especially on their mother. And these negative feelings can be intense.

Such a young toddler is just out of babyhood herself. Her self-control and new strategies for getting and holding attention, for feeling successful and good about herself, are fairly precarious. The sight of a new baby, and of the kind of unconditional nurturing the baby receives, can arouse in the older toddler an intense ambivalence. The call of babyhood is a strong siren song.

### Helping Almost-Two Adjust to the New Baby

- Keep schedule and activities as uninterrupted as possible, avoiding new challenges. Try not to start or change care, move Cherub into a big bed from the crib, or bring up toilet training, Defer elective surgery for six months, if possible.
- Make sure each parent spends ten to fifteen minutes down on the floor with Cherub, one on one, every day. Bedtime, classes or going to the park, running errands or doing chores don't count.
- Give Cherub many touches, pats, hugs, without saying anything at all.
- Keep discipline as consistent as possible.
- Control behavior rather than feelings. It's one thing, and quite appropriate, to let the toddler know that she can't poke the baby's eyes or shake him. It's quite another, and like speaking Martian, to tell a child this age that she should love the baby.
- Put a sign on the front door reminding guests to greet the toddler before asking to see the new baby.
- Make eye and voice contact with the toddler before tending to the new baby, and continue your interested "conversation" as you complete the baby's care.
- Talk to the *toddler* about the *baby*. "Look, Max, what a funny sticky-outy belly button she has, and what little toes. YOU have big-boy toes, but she just has baby toes. Let's count them" seems to go down a lot better than "Sweetheart Milly, what a lovely little belly button you have, look at your little tongue, look, this is your big brother Max."
- Tell the baby to wait while you tend to the toddler, rather than always the other way round.

## MATURING OF A YOUNGER SIBLING

The toddler this age with a younger sibling who is just starting to crawl and to test limits usually finds life hard, and so do her parents. The younger one grabs; the older one flails out and yells and hits.

The problem is, it's very easy to think of the older child as much older than he really is. It's one of those normal human phenomena. So it's natural to tell Big Brother to "share with the baby." And to rebuke him when he doesn't. And to blame and punish him when he lashes out.

This won't work.

First, you're not really asking Almost-Two to share. What you're asking him to do is give up his toy when somebody grabs it.

Second, Almost-Two is still working out the rules of "possessionhood." He is just figuring out that a possession is something that belongs to someone. "Mommy's lipstick." "Daddy's sunglasses."

Now you've changed the rule on him, before he's come to the part about Giving, Taking Turns, and Sharing. Not fair.

This is a set-up for turning Almost-Two into an angry, oppositional person.

The truth is, you can't expect the two of them to play together unsupervised. You have to be there every second and intervene before, or immediately after, such an encounter occurs.

In fact, the only fair thing to do is to keep the two separate and in the hands of different, competent, loving adults most of the time.

## DIVORCE AND CUSTODY ISSUES

Toddlers at this age require so much attention that parents may not have much energy

left for each other—or themselves. The turmoil created by an intense baby, like Chantel, can leave caretakers unhinged. This may be the final straw in a relationship already in trouble.

But this stage can also cause problems in a relationship that is basically solid. It's important for parents to realize that this is an extremely stressful time in the development of many children, and to seek counseling if the toddler's demands are causing a great deal of strain between the parents. It's also important to realize that this isn't a stage that will get unimaginably and inevitably worse when the child reaches the Terrible Twos. Instead, it's the stage in which parents can help a child grow into a delightful Two.

For a child this age, divorce can threaten the most important factors in constructing a mature social identity by around age Two. Parents can be very helpful if they recognize these needs and make sacrifices to ensure that the toddler gets what is required.

First, the emotions surrounding a divorce can deprive her of the support she needs from the adults closest to her.

- A child this age needs to be sure of the boundary between herself and others, especially her parents. A distraught parent who looks to the toddler for comfort and reassurance can be very confusing and upsetting.
- Toddlers learn behavior by how it is modeled by their parents. Complaining, criticizing, shouting, cursing, all can be

seen by the toddler as appropriate ways of behaving.

- To get to Two in good shape, toddlers need a great deal of input from parents who aren't heavily preoccupied. They need to learn how to get and keep adult attention in civilized ways. They need to learn that language is reciprocal, and not confined to orders, criticism, demands, rewards, and emergencies. They need to be free of worries about the adult world in order to practice their mastery of the physical world. And they need adult support when they make their first social contacts with other children. Parents who are engrossed in a divorce may not be able to satisfy these needs.

If this is happening, parents do well to find another adult who loves the child and who is out of the fray to stand in for them during these months.

Second, custody decisions may make the toddler's life so unpredictable that it deprives her of the stability she needs to establish her own autonomy.

The ideal custody arrangement for most toddlers is "bird-nesting," where the baby stays in one home and the parents alternate who lives there with her. This almost never happens, however. If this can't be done, staying with one parent and having the other parent make daytime visits without overnights is probably the best arrangement for most toddlers.

## HOSPITALIZATION AND ELECTIVE SURGERY

Despite the increase in separation anxiety and the often tempestuous emotions of this age, it may be better to perform some elective surgical procedures now rather than wait until after age Two. This is because from Two to Five or so children become intensely preoccupied with concepts of bodily integrity and differences in appearance. The Under-Two is still establishing a baseline of what normal is; the Over-Two has that idea implanted already, and with it the clear idea of Abnormal.

If elective surgery is to be done, it's crucial that a parent be with the toddler all the way, up until anesthesia takes over and from the moment it starts to wear off. Painful recuperation will be best tolerated if the parent behaves as if everything that happens is expected and normal and not frightening, while at the same time recognizing and soothing the discomfort. "I know it hurts, oh, it's too bad that it hurts so much. But it is very important that we have this done now. Mommy will help it feel better," is much better than "Oh, that must hurt terribly. Mommy's so sorry. Please forgive Mommy. Mommy didn't mean to let anybody hurt you."

# The Two-Year-Old Well-Child Visit

Many Twos are active, happy participants in this visit. So happy and active that most women pediatricians have learned not to wear wrap-around skirts. (For a while, we were considering making "able to unwrap the doctor" a chartable developmental milestone.)

What helps most are arriving ahead of time, so that Two can become used to the office and exam room; bringing Two's own favorite toys, plus one or two inexpensive new ones, to make Two feel at home; projecting a cheerful and matter-of-fact attitude, voice, and countenance; and allowing Two to stay on the parent's lap and "help with" the examination. Most Twos will fit the speculum (little cap) on the otoscope (though you have to be quick and sure Two won't put it in the mouth) and look through the "little TV screen" at his parent's ear. A rare Two will "blow" a pediatrician's or parent's finger so as to take big breaths for the stethoscope. No rational Two, however, will open his mouth big and wide for the tongue depressor.

Even if Two is potty-proficient most of the time, consider pull-ups or a diaper for the visit. Stress and puddles (or worse) tend to go together.

Try hard not to bring snacks to the visit. They coat the mouth and throat, are likely to come up when Two is gagged with the throat stick, and make crumbs that irritate the office staff. They also will probably get you a discussion of snacking and handouts from the pediatrician, when you could be discussing something more dear to your heart.

## Looking Ahead

Fear not! Two is a wonderful age, full of conversation and friendship. Stock up on the wonderful picture books of early childhood—make friends with the children's librarian, with the owner of the local bookstores. Clear the refrigerator door for the first artwork. Find a tricycle with real pedals. Get ready for a fun year.

# {TEN}

# Two Years to Three Years

## Two's Company!

## To-Do List

- Address Oppositional Behavior.
- Find Two Year Olds for Cherub to play with regularly.
- Teach Manners.
- Introduce daily chores.
- Consider introducing a second language.
- Get a potty.
- Get a bed.
- Make a dentist appointment.
- Change the smoke-detector batteries (each year on Cherub's birthday).
- Help Two to love books.

## Not-To-Do List

- No TV or computer in Cherub's bedroom.
- No TV while eating; no eating while watching TV.
- Better yet, unplug the TV entirely.
- Avoid beverage juices and sodas.
- Avoid "handout" snacks that you carry with you.
- Don't believe that swim lessons make Cherub water-safe.

## Portrait of the Age: The First Adolescence

Steven likes to sit on his potty—with all his clothes on. He likes to put things in the potty, too: his bear, his trains, and yesterday his mother's credit card case, which she has reported lost.

At two and a half, his mother Miranda thinks, Steven ought to be getting the picture. According to her pediatrician's checklist, he's ready for potty training. He wakes up dry from his nap. He knows what peepee and poopoo are. He has watched many interesting demonstrations of other people, children and adults, using potties and toilets. He says he would like Big Boy Pants, with Thomas the

Train Engine on them. They are sitting in his bottom drawer right now, ready to go.

*They* may be ready to go. Steven is not.

He is not, even though he doesn't like going around with a wet or soiled diaper. When he's wet, he'll come to Miranda right away and tell her. He doesn't need to tell her when he's had a poop: it's obvious. Steven gets a funny look on his face and goes behind the sofa, into the corner between the coffee table and the TV. He squats and grunts and turns red. There he goes now. That is Miranda's signal to get the diaper equipment and room spray.

As she cleans his tushy, she points out the Facts of Life to him. "If you went in the potty like a big boy, you wouldn't have to stop playing to get your diaper changed. And you could wear your Thomas the Train Engine pants. And Mommy and Daddy and Grandma and Grandpa would be so proud." She glares at him. "And David, too. David was in Big Boy Pants when he was *much younger* than you are now." David is Steven's hero, nemesis, and brother. David is now Six.

It doesn't work. Other things that have not worked include rewarding any positive Potty-Behavior steps with M&Ms, a star chart (Steven merely stared at it), and a trial on the big grown up toilet. David has been enlisted. Every time David "goes," Steven is corralled and brought in to watch. Any luck?

Nope. Nada.

Why? Miranda wonders. Steven has such a nice life. He goes to a lovely daycare three times a week, which he adores. Three of the six children are already potty trained. There

is no particular stress in Steven's life. He is very healthy, has never been constipated.

He's a bright little boy, too. He can tell you anything he wants to in perfectly grown-up talk. This morning at breakfast, he even said, "Actually, I'm ravenous." He was copying David, of course, but even so. But stubborn? Oh my.

You should see Steven and David when the two of them get into a fight. It is always Steven who attacks David, of course, punching him, crayoning on his pictures, grabbing his toys and running and hiding. Most of the time, Miranda and Jim try to let the boys work it out by themselves, but when somebody (David, almost always) screams in pain, they have to go in. But putting Steven in Time Out is futile. He won't stay in a corner, and now he can get out of his crib. So Jim has decided to give him a swat on the bottom, instead. It brings an end to Steven's yelling, and it shows David that his little brother can't hurt him, but so far the swat has not diminished the fighting.

The more stubborn Steven is, Miranda thinks darkly, the more stubborn he gets. What will become of him?

Right now, for instance, it is time to take him to daycare, but Steven won't get dressed. That is, he won't let Miranda dress him. (Steven's fine motor skills are excellent—he is even allowed, under supervision and with David's permission, to play with David's little Legos—but he can't seem to get the hang of putting on his own clothes. Even teaching him which holes in the pants his legs go through was a failure.)

Two Years to Three Years

"Steven, we have to leave now. Miss Patricia will miss you. You won't get to sing the Morning Song. Besides, Mommy has a dentist appointment, she can't be late." No. No no no no no. Steven doesn't stamp his foot or shout; he just hides behind the chair, dodging around it as she chases him. But he's grinning at her angelically, what a charmer! At one point he even blows her a kiss. But No. No clothes.

Miss Patricia doesn't like children coming late, and lets you know about it at length. Time is marching on.

"Steven! Listen, we have to get dressed right now!" she shouts. "Right now! OK?"

"No!"

"Now, I said!"

"No!" and on it goes.

Finally, she wrestles him to the floor and puts on his simplest outfit, shorts, shirt, shoes and socks. And manhandles him into his car seat. By the time they arrive at Miss Patricia's, Steven has removed everything except his shirt (it buttons in back) and, thank heavens, his diaper.

What, Miranda wonders, is wrong with Steven. Or with her. David never acted like this.

*This, my friends, is Oppositional Behavior.*

Two is all about feeling powerful and in control. It's all about discovering where you fit in your family. Two is always figuring out:

- How much power do I have? As much as my big brother? As much as Mommy or Daddy? More power than anyone? No power at all?

- What's permitted? What's *not* permitted, and what happens when you try it?
- Who's really in charge? Daddy? Mommy? Me? Nobody?
- Who does Mommy and Daddy love more? Me or Him?

Steven is making fine headway on these questions this morning. So far, he is by far the most powerful one, and everything is permitted, and Steven is in charge. Miranda, to be honest, feels pretty much the same way.

Oppositional Behavior starts out with ordinary defiance by a small child. When that defiance escalates to an argument with the parent in charge, the stakes for both parties go way up. For Cherub, it's all very exciting. It's also astonishing (You mean Mommy really *isn't* in charge here?) and thrilling (If I really push, will something terrible happen?) and scary (I know deep in my heart I'm not ready to be the one in charge). And it's deeply sad (Mommy is angry at me. I can tell I am not the apple of Mommy's eye) and angry (*David* is the apple of Mommy's eye).

Once a child-parent pair start in with oppositional arguments, the battles tend to escalate in both intensity and frequency. In worst-case scenarios, parents may become so outraged that they spank harder and harder. More often, the oppositional behavior merely poisons the atmosphere and makes everybody feel on edge and anxious.

When Oppositional Behavior never gets started, or gets fixed soon after it starts, life with Two can be a pleasure. When Steven is not in his oppositional mode, he's a perfect charmer, full of amazing two-year-old observations. On seeing a new baby with an umbilical cord stump, Steven remarked "Somebody picked her flower."

So what is wrong with Steven? Why didn't his older brother David engage in such obnoxiousness?

What Steven needs is for Miranda to rethink her disciplinary approach. Right now, they are in a war that Steven is winning. Here is what Commander-in-Chief Miranda needs to do:

- She needs to redraw the battle lines, and only engage in a fight when she is capable of winning it. Instantly.
- When there is a situation in which she cannot win, she finds a way to overcome Steven's objections without engaging in argument, explanation, or discussion.
- She needs to restructure important aspects of Steven's lifestyle—especially potty, daycare, and sibling battles—so that Steven *wants* to mature, rather than fighting it every step of the way.
- She needs to pay special attention to letting Steven know she adores him, because Steven spends a lot of time aware that—not to put too fine a point on it—she could wring his little neck.

These strategies of Miranda's are discussed in the section below on *Limits*.

Once freed from oppositional behavior, Steven will be able to dress himself, use the potty, sleep in his own bed. When his mother tells him to do something, or to stop doing something, he will be very likely to do what she says. In fact, he will remind Miranda, happily, of David at Two—David, who had the advantages of a very easygoing temperament and of being an Only Child for four years. Yes, it can happen.

That is why getting Limits right is one of the three most important focuses of this age. The other two important developmental areas are Language and Social Skills.

## CRUCIAL NEW TASKS FOR PARENTS AND CARETAKERS

Being the parent of a Two is like being top secretary to a very demanding CEO of a major firm. You need tact, organization, patience; you need to think on your feet, you need intuitive skills. But what helps most is to be absolutely convinced of the importance of the firm's undertaking.

The firm, the Two-Year-Old, is engaged in a vital effort: to consolidate control over his body, to express his feelings, needs, and objections in words rather than only in deeds, and to develop a style of relating to other children and adults beyond the family.

This isn't just any CEO. This is someone who is sophisticated enough to issue demands, make comparisons, and deal with basic abstractions and symbols. This is someone who can dress himself, wash himself, and help with chores. This is someone who can make a friend and take turns (but not really share yet).

But this is also someone who believes that he can become invisible and inaudible by going behind the playhouse to poop. He wants to be completely in charge of his body, his schedule, and his possessions. Except sometimes. And then he wants assistance, and he wants it now.

**Until about age Two, nearly any environment featuring loving consistent adults willing to set reasonable limits will foster development without special effort. After age Two, though, the skills and understanding of adults, and the play environment, play a big role in helping children to achieve optimal development.**

This is especially true in three areas:

- Language development and intellectual concepts;
- Social interaction, including fantasy play, peer relationships, and leadership and competition;
- Self-reliance, including self-care and the use of the potty, self esteem, and perfectionism.

## Language development and intellectual concepts

Two-year-old language starts out as "egocentric" and "field-specific." That is, Twos use language to express themselves rather than to communicate, and they talk primarily about things that are here and now.

It usually doesn't work to ask a young Two to discuss something that isn't right there. Asking her what she got for her birthday, or where you just went to lunch, or what she thinks Daddy is doing right now, is not going to advance the conversation much.

Catering to, nurturing, and deciphering a Two's speech can be fairly wearing, but not doing so can be even more exhausting.

Many Twos also stutter. This usually starts after speech has been fairly well established, and occurs at first when the child is very excited or stressed, and then seems to become a habit. Signs that this is "innocent" stuttering are:

- The stuttering occurs at the beginning of a word, or is one word at the beginning of a sentence. "Why why why why why is a cir, why is a cir, why is a circus" is a typical normal Two sentence. Stuttering does not occur in the middle of a word, such as "moth-th-th-th-ther."
- The child doesn't show any facial contortion or distortion during the stuttering.
- The child usually doesn't seem upset by the phenomenon.

If this is not the profile of the stuttering, or if it lasts longer than four weeks, or if parents or others are very concerned about it (perhaps because of family history of stuttering), then a speech evaluation is in order.

Otherwise, the best way to deal with it is to tone down the tempo or stress level of the child. Calling attention to the stuttering by asking the child to slow down, saying the word for her, etc., is not a good idea.

Another thing: Twos don't know that some words are considered harder or more sophisticated than others.

Once a Two has simple conversation down pat, parents can feel free to introduce sophisticated terms that are both concrete and easy to explain. Examples would be titles such as "construction worker" or "astronaut," verbs such as "demonstrate," "lounge," "plead."

Each word, if offered in a context that is interesting, or better, exciting to the child is easy to define. Not one is any more difficult to pronounce or has any more syllables than "Snuffalupagus," the Muppet mammoth.

Is this different from urging a child to read early, to memorize the state capitals, to do simple math? You bet it is. When parents introduce their normal speaking vocabulary to a child, they aren't force-feeding specialized skills. They are establishing a ground of mutual discourse and respect.

Skilled adults can also encourage perceptions of concepts. Since two-year-olds are rooted in the Here and Now, and the concrete, they aren't likely to understand references to past events ("This toy is like the one Aunt Margaret gave you for your birthday last year") or to talk about something not concretely present ("Tell Daddy about all the airplanes we saw this morning, Petey").

Instead, it is useful to comment about the item the child is already interested in. "The rabbit in the book is driving a steam shovel. I guess he wants to dig a big hole with that shovel. I wonder what he could do with that hole" (anticipating consequences of actions). "There is the handle that makes the shovel work" (noticing a detail). "It looks like the gear shift on Mommy's car, but Mommy's car doesn't have a big shovel" (pointing out a discrepancy). "Wouldn't it be fun if all cars had big shovels?" (bringing up an abstraction).

## Social interaction

If the last year has gone well, Twos have already made enormous progress in their interaction with adults. They have learned to get and keep adult attention much of the time in acceptable ways (not by tantrumming or whining or clinging). They can express feelings, both positive and negative, without physical attack. They know when they are being "cute" or "adorable" or "a character," and can ham it up.

Now, two-year-olds are ready to venture into the world of peers. Over this year, parents can help a two-year-old develop socially in several realms. Two areas in which we often just leave children to work things out on their own, areas in which that may not be the best tactic, are fantasy play and peer interactions.

### FANTASY

A few months ago, Two could feed a doll with a bottle or make a fire-engine noise. Now he can say, "I'm the daddy, and this baby is bad and goes in time out," and "Here comes the firemen to put out the fire." He can copy a huge range of TV figures, from Ninja Turtles to Peter Pan. Little girls, perhaps because TV features so few female heroes, often play complicated domestic roles.

Some parents worry that Two is already too locked into gender stereotypes. However, the role-playing that Twos do mostly reflects healthy identification with the same-sex parent. It also reflects the still developing ability of Twos: they can act out a role by repeating, endlessly, its component parts, but they can't yet make up a story. So Two may change her clothes seven times a day, and apply makeup, and paint her nails. Or he may endlessly brandish a sword and insist on his Batman cape.

Many parents I know try hard to enlarge and form the fantasy life of little girls. At home, two-year-old girls may have mothers who work as attorneys, artists, accountants; they may know or see women who drive trucks, run dangerous and powerful machines, go up in spacecraft. They may be surrounded by books featuring female trains, steam shovels, horses, bears, as active and capable heroines. But still they dress up and comb their hair by night light.

I don't know of any studies that culminate in scientifically validated advice to parents who wish their little girls to be less fettered by stereotyped gender roles. What I have seen work in my practice, however, are these suggestions:

- Make sure the little girl has lots of real-life challenges as well as much pretend activity.
- Make a conscious effort to recognize and encourage risk-taking, boldness, curiosity, and originality.
- Allow as much loud, active behavior as you would if she were a boy; and as little whining, tantrumming, and clinging.
- Try to make sure that she believes her mother, and mother substitutes, enjoy being female, and that their definition of female is not limited to feminine stereotypical behavior.
- Try to make sure that her father, or the man who takes the place of father, prizes her human virtues (courage, strength, originality, humor).
- See the section on Potty Skills, this chapter, about naming the external genitals.

A little girl who knows that a boy has a penis (and maybe testicles or balls) but only that she has, at best, a mysterious area called "gina" may wonder what in the world is wrong with her and all other females. She may believe that a penis, and maleness, can be acquired or removed.

- Make sure that the child feels physically attractive, enchanting, fun to be with.

GUNS AND SWORDS

Then what about the little boys, with their weapon play and hair-raisingly cheerful enactment of decapitation and disembowelment?

From seeing so many children, I feel certain that during this year such play is almost entirely imitative.

At about age Three, so many little boys will invent a gun, using their fingers, a stick, whatever, if they don't have a toy gun, that I have to believe there is something preprogrammed here by the Y chromosome. By Five, most are fully armed and can be considered warriors. But during this year from Two to Three, such play preoccupies a child to the extent of being his main fantasy *only* if the child sees such action on TV or has older children as role models.

If Two does have such exposure, however, gun and sword play can take the place of all other fantasy. And no wonder. The focus of a two-year-old is on self-control and power, and what an easy and thrilling way to gain the illusion of both. You don't need to bother with language or negotiation or taking turns. And the less practice you get at these skills, the less you are able to play with anything except guns.

There are three problems with such play at this time, no matter what parents think of gun play in general.

First, a two-year-old doesn't yet have all that much self-control and judgment, and plastic swords can cause damage to eyes and other body parts (besides knocking things off tables).

Second, gun and sword play can often interfere with acquiring other skills. The child who uses his gun or sword to attract adult attention doesn't learn how to do so with words. If he is always playing Shoot'em, he doesn't get to play firefighter, astronaut, Daddy, doctor, lion tamer, locomotive engineer, etc.

Finally, two-year-olds caught up in gun play find action-oriented TV absolutely mesmerizing. It may require more moxie than a parent can muster to withdraw the Saturday morning cartoons from an addicted two-and-a-half-year-old. Then, of course, the programs reinforce the gun play, and the gun play reinforces watching the programs, and you have a powerful cycle running all by itself. Such TV watching is totally noninteractive, and may in fact turn the child off from interactive TV watching, reading, and story time.

When all this starts so early in life, it doesn't seem to wear itself out as the child grows; rather, it becomes more entrenched.

Better not to start in the first place.

### PEER PLAY

Peer play is about power relationships, fantasy, and common interests. Since a two-year-old is just learning about all of these, peer play requires a great deal of informed, tactful, good-humored adult intervention.

The easiest form of peer interaction at any age is a power relationship. The hardest form is cooperation. With adult help, two-year-olds can learn to take turns, produce a project (dig a hole to China), play simple games in which everyone wins, role play with imagination (play doctor or circus). Without adult supervision, a high-energy child may resort to pushing, grabbing, and hitting; and a quieter child may simply retire to the side, giving up. This sets an unproductive pattern.

As the year goes on, each child can, with adult help, learn to lead, follow, and share in the accomplishment of a task.

### COMPETITION

Twos bent on power are hot to compete. Left to their own devices, in fact, Twos will turn everything into a competition.

An adult supervising two-year-olds can be very helpful in nipping in the bud overly competitive tendencies which can dampen a Two's joy in life, as well as cause problems later on. Children with such tendencies include:

- The child who engages in an activity **only** in order to win;
- The child who becomes perfectionistic;
- The child who never does anything "best";
- The compassionate child who opts out in order not to win.

One of the most helpful ploys such an adult can use for all of these children is to put competition on hold until the child is more self-confident. Of course, you can't do this merely by telling children not to compete. Moreover, adults are so used to overseeing behavior by the use of praise and criticism that they may not realize they are fostering more and more intense competition.

One tactic that rarely fails is simply to converse about what is going on rather than giving praise or making comparisons. A discussion of the bugs, worms, rocks, etc., in the holes being dug redirects interest from who is digging the deepest hole.

Finally, praising the group rather than individuals often works: "Jamie and Petey and Melissa gave that ball a good game today. They all ran so fast that ball didn't stop for a second. I'll bet that's a really tired-out ball."

**SELF-RELIANCE**

Since being two is all about self-control, autonomy, and empowerment, the more success a Two has in these areas the nicer a person she is likely to be. Parents can help by:

- Offering choices instead of issuing demands (unless setting limits!). But if a parent does offer a choice, it ought to be a true choice: no criticism, verbal or unspoken, of whichever option is chosen.
- Teaching skills in ways that make the child feel empowered rather than merely obedient: see *Potty Skills*, in Window of Opportunity section in this chapter.
- Bearing in mind that Twos are perceptive about the picture their parents have of them, and being careful to convey respect for autonomy in both direct conversation and in anything the Two is likely to eavesdrop on. For instance, in talking on the phone in Two's hearing, it's helpful to describe some new achievement rather than to complain about some failure or recalcitrance. It's even more effective if you do so while pretending that you don't know Two is eavesdropping.

## Separation Issues

Separating at daycare or preschool and separating at bedtime are the two most common problems for Twos.

### GOING TO DAYCARE OR PRESCHOOL

What seems to bother Twos most about separation at daycare or preschool is not so much anxiety (Mommy's left; how do I know she'll be back?) but the fact that they are not in charge of the separation.

Some teachers tell parents to make the good-bye quick and cheerful, explaining to the child carefully that Mommy or Daddy will be back and when, and then leaving. The child may cry for a bit, but usually can be coaxed into playing, and often doesn't even want to leave when the time comes. Indeed, this often works.

Unfortunately, this doesn't make things easier on Mommy or Daddy. Even though Two clearly loves his group once he has adjusted to being there, the crying and protest may continue to be repeated day after day.

One wonderful aspect of Twos that can be used here and elsewhere is their utter transparency and lack of suspicion, coupled with their will to power. The trick is to construct a scenario in which the course that Two believes gives her power is the same course that you want her to follow. Two is unlikely to see through this.

What works in the daycare situation is for the parent to go with the child and plan to stay for one or two hours the first few days. Usually only three days are needed. For the first day or two, let the child set her own pace and enjoy the activities. Pick an opportune time between activities to leave *with the child*.

When Two has clearly started to enjoy herself and look forward to the experience, you can then embark on the next phase. Stay with her as she becomes absolutely engrossed in an activity. Try to pick the height of her involvement, then go to her and say, "I am sorry, we have to leave now." Usually, tears and kicking ensue. Stand firm. Do not be swayed. Pick her up, if necessary, and leave.

The next day, take her to the daycare, say a cheerful and firm good bye, telling her when you will return. Nearly always this works. The child often practically pushes the parent out the door, saying, "You go now. I have to stay here and play."

This can work even after a child has become habituated to crying when the parent leaves in the morning.

For it to succeed, however, it is important that:

- The daycare or preschool really is designed to be a happy and fascinating one. Activities are age-appropriate, peer relations are well and tactfully overseen, and the atmosphere is unpressured.
- The parent is not too ambivalent (nor shows ambivalence) about separation, and the child is not worried that something awful could happen if she is not right on the scene at home. Marital stress, an impending move or parental illness, etc., can make a child just determined not to go anywhere.
- The child doesn't feel deprived of one-to-one parent/child time outside the daycare situation, doesn't feel uncertain that she is liked and enjoyed and thought cute, bright, funny, etc.
- The time gap between daycare visits is not too long. A once-a-week daycare, even a twice-a-week one may simply not work for a Two with a short memory. Every visit seems like a brand-new experience. This may even happen in the daily daycare if the daycare caters to drop-ins or if the turnover in children and caregivers is rapid.

For further discussion, see Window of Opportunity, below.

## BEDTIME AND SLEEP

The second big separation issue is bedtime.

The three big problems I hear about for this year are: prolonging bedtime with demands, waking up scared in the night, and transferring from crib to bed and not staying there.

As always, night problems are often a reflection of problems that don't get solved in the daytime. For a Two to go to bed without a lot of fuss, put herself to sleep at the onset and again when she awakens during the night, and stay in her bed to do so, she's got to have a lot of things down pat.

- She has to have had enough satisfying time with her parents in the daytime so that she doesn't feel empty and deprived. For time to be satisfying, she has to feel liked and enjoyed and respected, not just loved and cuddled.
- She has to feel a degree of independence and self-reliance and power. She has to have experienced the joys of separating.
- She has to be sure that her parents wholeheartedly want her to go to bed, and that neither is secretly or not so secretly yearning for her presence, for whatever reason.
- She has to be sure that she isn't missing something by going to bed, whether that something is scary (a marital fight) or wonderful (popcorn[1] and a video).

If all these are fulfilled, you can deal with the mechanics of bedtime. (See also the *Sleep* section in the previous chapter, "*Eighteen Months to Two Years*.")

### Prolonging bedtime

See the *Sleep* section in the previous chapter.

Sometimes a Two who is just fine with separation in the daytime will still have anxiety at night. A couple of things seem to help.

- Make sure that she is spending some happy time in her bedroom during the daytime so that it isn't just That Place To Be Exiled To.
- Try giving her a "piece of Mommy" or "piece of Daddy" to hold on to. This is an item that "speaks" to her of the parent, that is closely identified with parental security and warmth. An article of maternal or paternal clothing can go on her teddy bear, who is her guardian for the night.

### Waking up scared at night

What kind of waking is it? There are two kinds, night terrors and bad dreams.

**Night terrors** are really terrifying for onlookers. In the grip of a night terror, the child acts as if possessed. He doesn't seem to recognize his parents, and may stare with hate or rage as well as with terror. He pushes them away. He trembles and shakes and screams and screams. It is just awful. Nothing you do can comfort him.

What has happened is that something has roused Two from the very deepest stage of

---

1. Still very dangerous and chokable for Twos.

sleep, and he is truly not aware of his surroundings. The primitive parts of his brain are awake, but not the thinking and remembering parts. At Two, the cause of waking often is physical, such as pinworms. (See *Night Terrors* in *Frightening Behaviors* in Part II of this book.)

**Bad dreams** are upsetting but not terrifying to parents. The child recognizes you. He demands comforting and closeness and insists that you stay there while he falls asleep.

A Two with a bad dream usually can't tell you what it was about. At Three, you may hear about monsters and bears, about rockets and guns, but at two you are likely to just get a despairing wail and outstretched arms.

Most often, bad dreams are associated with stress and change in daytime life. The stress needn't be "bad" stress like sickness or divorce or punishment. It can be a happy change of schedule to a wonderful preschool; or the arrival of welcome and loved visitors; or even the acquiring of a new skill, like pedaling a tricycle.

Calming down daily activities is helpful. During especially exciting or stressful times, it often helps to spend one-on-one time down on the floor with the Two and play repetitive and boring (to an adult) games, such as roll the ball or stack the blocks. These seem to lull the demons.

If bad dreams persist, more analysis is needed, and pediatricians are usually good at it.

### Getting out of bed; changing from crib to bed

This isn't just a behavioral problem; it's downright dangerous. Two is sometimes even able to get out of the house, as did Noah, the one A.M. explorer, mentioned in the last chapter. Even inside the house, there are all kinds of dangers, from toilets to not-quite-empty glasses of alcoholic beverages.

Thus, if your Two is still ensconced in a sleeping bag at night, keep her there. It will deter night traveling. Moreover, if Two still fits in his crib or has a crib extender and can't climb out, I would not change him to a real bed at this age. Closer to Three, he'll have a greater ability to exercise caution and restraint. Three-year-olds don't usually explore the toilet, leave home without saying good-bye, or decide that two A.M. is a good time to get into the broom closet and eat Drano.

But if the bed is an inevitability, there are several ways to approach the big change.

First, many Twos are very attached to the crib and don't want to give it up. This can be a problem particularly for parents with a second, younger baby who is outgrowing the bassinet. One way to help the Two get used to a bigger bed is to move the crib mattress to the floor for him to sleep on, leaving the crib in the room. After he gets used to the lack of containment in the crib, the transfer to a real bed seems easier.

Second, before he transfers from crib to bed, get him used to a barrier at the door at night. A Dutch door, a gate, even a closed door (with a night light in the room) are all fine. This is important even for the Two who sleeps like a log and has never awakened a parent for night-time consultation.

Third, the child who awakens at night and calls you to his crib to be picked up is not going to change his pattern when you put him in a bed. He will merely be harder to deal with. If possible, end this pattern first before trying out the bed.

## Limits

Day after day, Steven and Miranda engage each other on the domestic battlefield. She gives an order, he refuses to obey; she becomes more firm, his refusal escalates: Up and up the decibels go, until one of them—often Miranda—collapses. She either gives up, acts ineffectually (getting him dressed, knowing that he'll just take off his clothes again) or, like Jim, gives Steven a swat on the behind. That may jump-start Steven into obeying, but it doesn't help the continuing cycle of fights. In fact, they may even be getting worse; and she doesn't know where to escalate to from here.

Here's what else Miranda has discovered:

- Giving bribes and rewards for ordinary good behavior makes it worse. That's what happened with giving M&Ms for using the potty. After three days, Steven said, "Don't like candy," and that was that.
- Time Out doesn't work; Steven escapes and defies her even more.
- Sympathizing with what she believes to be Steven's feelings doesn't work. ("*I can see that you're angry. That's why you're hiding under the table. It's OK to be angry but it's time to come out now. Come on now, Steven. It's really time to come out. I know you don't want to but sometimes we have to do things we don't want . . ." and on and on.*)
- Having a long talk with Steven and asking for cooperation doesn't work. Steven starts to fidget, and after she's all done says something like, "I want my red truck."
- Telling Steven firmly that she expects him to listen to, or obey, or respect her doesn't work. He cocks his head at her and blinks, as if she were speaking Martian.

In fact, neither punishments nor rewards are effective for truly Oppositional behavior. That's because of the toxic effects of back-and-forth arguing. Once this becomes a pattern, it becomes irrelevant whether a particular battle is won or lost by Two. What Two is seeking is self-definition, and where he's finding that is in a power contest with parents and siblings.

Why does Time Out often work at a good Daycare or Preschool? Because the way a Two gets self-definition there is by interacting with other Twos. The teacher is not the focus of Cherub's drive to power and mastery, and Cherub feels little impulse to contest her rule. This is what I mean when I say that when a Two doesn't have other Twos around, she'll turn the nearest adult into a Two.

This is what Steven has done with Miranda. He's pulled her right down into the Field of Combat with him.

## FIXING OPPOSITIONAL BEHAVIOR

- Only engage in a contest of wills with Two if you can win the battle by immediate action.
- Recognize the times when giving an order is asking for trouble, because it's not a battle you can win.
- Prevent "oppositional" fighting between siblings.
- See to it that Two has a satisfying way in which to test and define his powers—for instance, a good, daily preschool or daycare.
- Restore Two's certainty that Two is the Apple of your Eye.

However, there is something that will indeed work. Here are five strategies that, if Miranda uses them together, will prevent or very much reduce oppositional behavior in this age group.

What Miranda needs to do is to *prevent* any successful challenges to her authority, and stop the toxic back-and-forth battles. Here's how she needs to do it. She should:

- **Only engage in a contest of wills with Steven if she can win the battle by immediate action.** She can't make Steven go on the potty. She can't make him stay dressed, not once he's out of her reach in the backseat. She can, however, pick him up and put him in the car seat and take him to Miss Patricia's with his clothes in a bag, to put on when he gets there. And she can do this without engaging in any back-and-forth arguments with him.
- **Recognize the times when giving an order is asking for trouble, because it's**

**not a battle you can win.** Getting and staying dressed or using the potty, for instance. In this case, have other options available. Putting his clothes in the bag, for instance. For his potty refusal, she could use the techniques in the essay in Section III on *Potty Resistance*.

- **Prevent oppositional fighting between siblings.** Miranda doesn't suspect it, but it is David who is starting most of the fights, and Steven who is getting blamed for attacking David. Steven is now beginning to try the same tactics on David. They are fighting not to get a toy, or to take revenge, but to determine which is the dominant sibling. It's even better when Miranda or Jim comes and administers "justice:" that way, the boys get an absolute decision from the referee. Fighting is much more exciting than playing. They are not ever going to learn how to negotiate their way out of a fight. The brothers are learning only two things from their fights:

- How to get each other in trouble
- How to get parental attention by fighting

What they need to learn, of course, is how to play together *without* fighting. See the essay in Section III on *Sibling Battles*.

- **Restructure Steven's daycare.** If Steven could attend his very good daycare every day, three advantages would be gained. First, he'd calm down about going because it would be a daily ritual. Second, he would benefit from the maturing experiences at preschool. Third, simply being out of the house allows Miranda some much needed R and R, and cuts down on the toxic back-and-forth fighting.
- **Make sure that Steven *does* know that he's the Apple of her Eye.** At the moment, Miranda can't do that. She is too angry, frustrated, and exhausted. But once she sees the importance of this factor in her new strategy, she'll be able to take this on with a willing heart.

For a complete discussion, see the essay in Part III on *Oppositional Behavior*.

# Day to Day

## MILESTONES

### *Vision*

Two can see a plane in the sky, a bird on the wing. Two may not be able to find Waldo, but can identify him when pointed to. Two can follow plots on TV, identifying characters as they pass from one situation to another.

Again, a clue to vision problems is the Two who squints, gets up close to, covers one eye, or tilts the head while watching TV.

### *Speech and language*

Lots happens this year.

Two can understand nearly everything addressed to her, and respond to one- and sometimes two-step commands, if she feels like it. Two can make two- and three-word telegraphic sentences ("Daddy down boom") and repeat nearly every word she feels like. Two may still refer to people in terms of proper nouns instead of as I, me, you, he, she.

Three can follow three step or more commands if he chooses, and can express in words anything he feels needs expressing. Three stutters and stammers often, but when he does say things, a stranger can understand nearly all the words. Many Threes still put on their "thocks," which may be the color "lello," before "wunning" outside. Threes understand On, In, and Under, but often not yet Behind. (Why is that? Is it because they are so egocentric they only understand directions they can concretely see?)

By Three, many children can name most of the primary colors, understand the concept of two, and may be able to "count" to ten but not manipulate the numbers. Most have memorized at least one short jingle or song.

Two likes to look at pictures in a book, and find items. Three wants to hear the story, with no words missing, over and over and over again, until both of you know it by heart.

### Large motor

By Two, toddlers can walk heel-toe, not just on their toes. They can throw overhand, kick a big ball forward, and walk upstairs holding on to somebody's hand. By Three, they can jump, getting both feet off the floor at the same time, and stand on one foot for several seconds. (But if you just ask a literal-minded Three to stand on one foot, without demonstrating, she will give you an odd look and carefully place one foot directly on the other, then topple over.) Some Twos and all Threes can pedal a tricycle. When getting up from the floor, Twos should be able to stand up without climbing up their own bodies.

### Fine motor

Two can hold a crayon in her fist and scribble. Three can hold a pencil with the "tripod grasp" and draw a sort-of-circle, that is, a shape made by a continuous line that connects on itself. ("It's a baked potato," many Threes tell me.) Threes can use a spoon and fork, but usually not yet chopsticks. Two can turn a lever-type door handle, and long before Three can manage a door knob. Twos are learning, and Threes are adept at, unscrewing caps, especially those on medicine bottles. Over this year, many will start to demonstrate a hand preference; though some boys don't show such a preference until about age five.

### SELF-CARE

Two can be taught to put on her own clothes, wash and dry her hands. Most impressively, most Twos learn to use the potty this year—to pee, if not to poop. See the section on *Potty Skills* in Day to Day Care.

### CHORES

Two is a fine helper, when he feels like it. Good chores for Two include "making" his bed (extricating himself gently, then pulling up the covers); putting dirty clothes in the hamper, hanging up clothes on low hooks. Two can put toys into the toy box and put away books.

This is a good time to start referring to these things as "Two's Jobs," rather than as "helping" anybody. Helping, after all, is optional—a generous bestowal of one's time to assist someone else at their job. Owning a job yourself is quite a different kettle of fish.

### MANNERS

What one learns at Two, for good or evil, tends to stick there for years, so it's a great investment to start Two on Good Manners.

If adults say "Please," "Thank you" "Excuse me" and "*Gesundheidt*" to Two and in Two's hearing at every single opportunity, Two is almost inevitably going to do so as well.

Two is old enough to "write" thank you notes for gifts after her Two Birthday party—a scribble with adult writing.

Two is old enough to "give me five," and to be taught that in some situations it is more polite to shake hands (either hand will do).

## SLEEP

Most Twos still sleep about thirteen hours out of twenty-four: eleven hours at night, the rest as a single daytime nap.

Problems with napping and bedtime are often very much the same as those described in this section of the last chapter.

However, a new dimension has been added: bad dreams that the child can partially describe afterwards. If Two is having this kind of bad dream more than rarely, it is likely that one of three problems is going on:

- Two has a pronounced fear of, or awareness of, being considered "bad" or "naughty" or failing his parents in important areas. Such a Two often engages in aggression, tantrums, and other disturbing daytime behaviors. This needs to be discussed with the pediatrician if the "home remedies" in this chapter don't help.
- Two is feeling insecure because of stress in the family: marital discord, financial problems, illness, a new sibling on the way or just born. The solution here is helping the adults deal with the stress.
- Two may be watching violent TV programs, with the biggest culprits being cartoons and children's shows meant (I hope) for children who are much older. Even violent ads, such as the one with the monster in the toilet bowl, can upset Two.

## GROWTH

There is a persistent rumor that you can double a child's height at age two and arrive at his adult height. Well, sometimes. If a child stays on the height curve he has attained at age two all the way to adulthood, then, yes, this will be roughly accurate. But that "if" is a big one, dependent on inborn patterns of hormonal maturity (and, of course, on health and nutrition). If the child is more mature hormonally than her age in years, she'll have puberty earlier and stop growing earlier and be shorter than predicted. If less mature, she'll have puberty later and stop growing later and be taller.

A family history can be useful. A boy whose father shot up fourteen inches his senior year at high school may well do the same. A girl whose mother towered over her buddies in fourth grade but now shops in the petite department is likely to follow in the same footsteps.

Most children will gain about four-and-a-half pounds during this whole year, or less than half a pound a month. And most will grow about two-and-a-half to three-and-a-half inches during this whole year. Over this year Two will lose that pot belly and start looking like a lean and muscular little kid.

See the essay in Part III, *Growing in All Directions*, for more information.

### Chubbiness watch

It's hard for most adults—even doctors and medical students—to identify which Twos are overly chubby. That's because we have a picture fixed in our mind's eye of what "normal" is. Because so many children in real life and on the media are overweight, our idea of "normal" tends to be overly chubby, too. So to many of us, a chubby Two looks normal, a normal weight Two looks thin, and a slender normal Two looks emaciated.

For that reason, ask your pediatrician for your child's growth percentiles or BMI, or figure them out from the instructions in the essay, *Chubby or Not, Here We Come.*

When a Two starts gaining more weight than is desirable, it's usually due to one of three reasons:

- Somebody taking care of her has a misconception about what she needs to be eating, and is offering her portion sizes that are too large; unnecessary juice or snacks; or unwarranted second helpings. Maybe she is drinking too much milk.
- Her daily life features too much TV and too little activity. She may be getting in more napping than she needs.
- She has a predisposition to overweight. This may be inherited, or it may be partly inborn and partly learned. Such a Two is "always hungry," seeks out food, never turns down a snack or a second helping.

While helping such a Two is a bit trickier than helping one without such a predisposition, it's even more important to do so. Left unattended to, such a Two may have a serious weight problem early in life. A very overweight little girl may be triggered into puberty at the age of Seven or Eight. All very overweight children are prone to sleep prob-

lems, a predisposition to type II diabetes and high blood pressure, and teasing from their peers.

If your doctor agrees that you should be concerned, please read this section in the last chapter, as well as the essay on *Chubby or Not, Here We Come.* All the information in the last chapter applies to Twos. The only difference is that Twos can safely have their fat intake reduced to adult recommendations of 30% of their daily calories. That means that it's OK to give nonfat milk and dairy products.

## TEETH

All the previous caveats about brushing and snacks still apply. The new wrinkle has to do with the invisible, unerupted teeth.

Between two and four years of age, the upper front teeth are calcifying. If there is too much fluoride going into the child, these very cosmetically important teeth may display "unsightly" white spots. They aren't dangerous, but at least at this time in history they are not stylish, and may cause unhappiness in the teenage or adult owner of the teeth. You can prevent them now by being specially vigilant about your child's fluoride intake. Is the water in your community fluoridated? Is the child on the right dose of fluoride supplements? The dosage of supplements may increase this year, and again next year. Are you sure he or she is not ingesting an unregulated amount of fluoride by taking an older child's fluoride chewables or drinking fluoridated mouthwash or swallowing considerable amounts of fluoridated toothpaste?

If you haven't yet taken Two to the dentist, it's time. Some Twos may need sealant on the primary molars to prevent cavities and to protect the permanent teeth. And some may benefit from a topical fluoride treatment, an exciting and challenging enterprise.

## NUTRITION AND FEEDING

Fish consumption guidelines may be found on page 7.

**Hold to the Mantra:** You decide *what* foods you offer, and *when*. Two decides *how much* of which one to eat.

**Choking is still a potential problem, so avoid high risk foods:** Anything round (like a slice of hot dog), spherical (like a whole grape or a cherry with a pit), pellet-like (peanuts, popcorn, hard candy), chunky (raw carrots or celery) and very sticky (peanut butter).

### Foods

I have not included the Food Pyramid in these sections. I have never known a parent to take kindly to the thing. Its main purpose is to remind us that children need very little in the way of sweet, sugary, greasy foods; that they need the basic amounts of grains (cereals, starches, and breads); that they need more fresh vegetables and fresh fruits than Americans are used to eating.

Indeed, it's crucial to keep this in mind with every meal, every snack, and every beverage. Remember, you are struggling against the mighty forces of fast food, soda, juice, and snack producers, and their media accomplices!

The second crucial thing to keep in mind is serving size. When you serve flat food, such as a piece of meat or poultry, the serving size is the size of your child's palm. When you serve a bulky food, the serving size is the size of your child's fist. Offer three or four different foods at each meal, in these serving sizes. If you keep the nutritional guidelines, the serving sizes, and the milk requirements in mind, your child will most likely be very well nourished.

Normal, healthy Twos are often "Finicky Eaters," refusing to taste new foods, demanding favorites such as macaroni and cheese, or scorning vegetables. If you have one of those, don't be bullied! Try treating Two as a real person. Put out the foods you want Two to choose from. That is, don't put out Oreos if you really want Two to choose to eat broccoli. Do be sure that at each meal there is one portion of one food you know Two likes.

Then, sit down at the table and talk with Two about cars, dinosaurs, whatever. Don't talk about the food, or what Two is or isn't eating. Don't even reach for a spoon, or dab Two with the napkin. Don't appear to notice that Two is eating.

A few normal, healthy Twos may still be fed by someone else. *This is not a good idea.* First, Two is likely to become very confused as to body signals of hunger and fullness—signals that need to be in good working order to attain appropriate weight and to avoid eating disorders. Second, being fed is likely to increase Two's temptation to return to babyhood. It may exacerbate oppositional behavior. Two needs every opportunity to feel competent and independent. Try the above suggestion for treating Two as a real person. No normal healthy Two has ever starved when there was sufficient food within reach.

*Two's food guidelines in a nutshell*
- Use 1% or nonfat milk, averaging out to 16 ounces a day.
- Make sure Two gets at least one excellent or two good iron sources each day—red meat, iron-fortified cereal, iron-rich vegetables.
- Try to serve two or three different food groups at each meal.
- Don't expect Two to eat big quantities of anything. Average servings are very small: 2 ounces of meat, ½ a sandwich, 1 egg, ½ cup cereal, 1 ounce of a vegetable.

*Supplements*

- If your Two isn't drinking water with adequate fluoride, either natural or added, Two needs a fluoride supplement.
- If Two won't or can't take dairy products, Two needs another rich calcium source.
- If Two takes a great deal of dairy products, or if your family doesn't eat red meats, Two probably needs an iron supplement. Calcium in dairy products can reduce the iron we absorb from nonmeat sources.
- If your family is truly vegetarian, without dairy, meat, fish, poultry, or eggs, Two

needs a nutritional consultation to make sure Two is getting enough protein, calcium, and iron. Two definitely needs a supplement with B and D vitamins.

## BATHING

Again, bathing, diapering, and masturbation all go together: see this section of the last two chapters.

Bathing is now primarily a safety issue. A Two left alone in the tub is guaranteed to be able to turn on the hot-water faucet, grab the plugged-in hair dryer, or bump his head and sink beneath the surface of the water.

At the very least, she'll drink the shampoo (our daughter Sara did, and smelled of Herbal Essence for days).

On the hygiene front, keep bubble baths and sitting in soapy water to a minimum for little girls, to avoid vaginitis. Make sure that you pull back the foreskin of uncircumcised boys, **if it retracts easily,** and that after you wash and rinse, you pull the foreskin back up again so that it doesn't become a constricting tourniquet.

As you teach your little girl potty skills, teach her to wipe first in front, then in back, to avoid carrying germs from rectum to vagina. At two and three, her arms won't be long enough to do it in one swipe, so teach her to wipe front, drop tissue, get new one, then reach behind and wipe in back. Good luck.

## DRESSING
Two is old enough to dress himself or herself first with supervision and teaching and then

without. It's most fun to learn this one article of clothing at a time, breaking down the task to smaller ones. For instance, lay the pants and shirt out with tag in back. Then get them on.

Avoid pants with zippers and anything with small buttons or buttons in back. Shoes that have a design on the lateral (outside) border make it easier not to reverse feet. Shoes with Velcro fastenings are great, and this is the only age when children will be interested in doing and undoing the Velcro rather than just wrenching the shoes off whole, as it were.

## ACTIVITIES

Other toddlers are vital for Twos, whether as neighborhood friends or relatives, in play-groups, or at preschool. It's helpful if the other toddlers are slightly older, to illustrate the virtues of talking, taking turns, and going to the potty. If the group is an informal one, just a few friends, it still needs close adult supervision and direction. See the extended discussion in Portrait of the Age, under *Social Skills*, in this chapter. And the section on *Daycare* in Window of Opportunity, this chapter.

Swim lessons can be started close to Three, but need close supervision.

## TOYS AND EQUIPMENT

Recommended toys:

*Intellectual:* Picture books of several kinds: find the object ones (*Where's Waldo?* and Richard Scarry's *Goldbug*), books with repetitive rhymes and phrases (*The Cat in the Hat*, etc.), books with pictures for the adult and child to figure out the story, and books with simple plots.

*Fine motor:* Simple jigsaw puzzles in which each piece is a recognizable shape; crayons, fingerpaints, play dough (store-bought or homemade).

*Fantasy:* "Guys" and dolls and all their accoutrements, cars, trucks, things that go, clothing that can be used for dress-up.

*Gross motor:* Tricycles, very small bicycles with training wheels, balls of all sizes and uses, things to hit the ball with, climbing toys with a safe surface to land on when you fall, closely supervised swings.

Cautions about toys:

- Beware toy boxes with lids, as they can fall and crack young heads (and older ones, too).
- Tricycles that have a low center of gravity and big wheels are safe, but remember your child won't be seen in a car's rear-view mirror. I urge you to start helmet use with the first "wheels" so that vehicle and helmet seem to always go together.
- Beware dangerous, grown-up toys: trampolines (condemned by the American Academy of Pediatrics), darts, swords, BB guns, skateboards. And dangerous baby toys: especially rubber balloons and plastic gloves blown up like balloons. If inhaled, a balloon is stuck, and a Heimlich maneuver won't pop it free.

## SAFETY ISSUES

Take a CPR course. It's been a long time since the last one, and your child is bigger and the instructions are different. Besides, new studies update guidelines every two years or so. Ideally, take a course every year to review and update. Remember, you might save your child's friend, as well as your own child.

Every age has the defects of its virtues. Twos, fearless and social, get a lot of bumps, bites, scrapes, cuts, and bone and joint accidents. They are vulnerable to drowning, even in small collections of water in containers like toilets, fountains, and barrels. They can get out of their car seats when so inclined. They pull out dresser drawers and climb up them and pull the dresser down on top of themselves.

Particular problems include:

- Nursemaid's elbow. This occurs when the tissue band at the elbow slips out of place and traps the elbow so that the child can't turn the hand palm up without excruciating pain. It can happen after anything that jerks or pulls hard on the arm. Nursemaids yank their charges out of danger: that's where the name comes from. But a Two who grabs on to something and then lunges or falls can dislocate her own elbow. She won't point to the elbow as the source of pain, but to the wrist. Putting the elbow back in place is a simple office procedure. Usually the hardest part is soothing the adult "yanker's" guilt feelings.

- Burns from curling irons seem to peak at this age, perhaps because children are so imitative and so gender-involved.

- Since this is the age of the first tricycle or bicycle, if possible fit out Two with a helmet, so that it seems like part of the equipment for riding any wheeled vehicle. If the habit is engrained this year, next year Three will delight in enforcing it as a Rule of Life.

- Beware the driveways of safe neighborhoods, where people backing out in a hurry can't see a little person on a Big Wheel.

- Beware water immersion. Even the best swimmer this age is still not drown-proof. Especially beware thinking that somebody else is watching the child at the pool. Beware taking a nap together with the child: she may wake up first, and test the water on her own.

- Beware Jacuzzis. The suction of the drain can be very powerful when the system is turned on. Sitting on, getting a hand or foot into, or the hair caught in the suction of the drain can be fatal. Be sure everyone knows how to turn off the Jacuzzi.

- Beware escalators, and little feet getting caught. The emergency off button for the escalator is right by the side, at the bottom. Learn where it is.

- Beware real guns. Many of your neighbors have real ones, loaded ones in their homes. (I know; I ask.) A Three, and even a Two, can point and pull a trigger. Teaching safety is not an effective protection. Ask what the house gun rules are before Two goes visiting.

- Beware burns that result from curiosity as well as accident. Hot coals at the beach, handling the hot hair-curler, stepping on the floor furnace.
- Make sure the car seat is a deeply engrained habit. Put your child in charge of making sure everyone is buckled. If possible, install the car seat facing backwards in the backseat. No matter where you install it, deactivate any air bags that could impact it, as this could cause fatal injuries to the baby.

## MEDICINE CHEST
Add:

- A new bottle of syrup of ipecac, to induce vomiting, as your old one from last year may have gone beyond the expiration date. Review the rules for using syrup of ipecac (never use if the child is unconscious or sleepy or has taken something caustic that will burn as much coming up as it did going down). Always call poison control before giving syrup of ipecac.
- Real adhesive strips (bandages) for invisible as well as real booboos.
- Over-the-counter cortisone cream for contact dermatitis.
- If you live in a stinging-insect area, ask your pediatrician for a prescription for "bee-sting kits" containing easy-to-use injections of adrenalin, just in case somebody gets a bad reaction.
- "Artificial tears" to flush out foreign bodies from eyes.

- After-swim eardrops to prevent swimmer's ears. You can buy them over the counter or make your own: one part vinegar to one part rubbing alcohol. Put a few drops in each ear after a day in the water.
- Antibacterial ointment for cuts and nicks (ointments containing neomycin can cause allergic reactions).
- Don't include sore throat and cough lozenges, because the child with a cough is at particular risk for inhaling one.
- Don't include throat spray containing phenol, as some children are very sensitive to it. It may temporarily numb the area, and then the reaction may make the throat hurt even more, so you give more spray.

# Health and Illness

*Part II of this book covers illnesses in babies this age and older: how to assess symptoms, how to tell whether a child is sick enough to require urgent attention, and what kind of home treatment may be appropriate. It also examines problems by body part. The Glossary is an alphabetical listing of medical terms with pronunciations and, when appropriate, detailed descriptions.*

## EXPECTATIONS
It is the task of Twos to catch minor illnesses and make immunity to them. Nature is very clever in this regard. The immune system matures at the same time that social

behavior does, so by the time a child is ready to go out and make friends, she's also ready to handle the sicknesses she catches from her buddies.

Most of these illnesses are caused by viruses, and most of these viruses cause respiratory symptoms. Many Twos will spend much of this year with a runny nose and cough: the average number of respiratory illnesses is about eight. That is, one a month except for summer.

So most Twos will spend all the days of the months from October through April coming down with, coping with, and recovering from, upper respiratory infections. Think of this as a good investment for the future. Because Two's immune system is fairly mature, she may "catch" many viruses without getting any symptoms at all: you've escaped having to deal with those. When she does show symptoms with a virus, they are likely to be milder than when she was younger. And she is much more likely than earlier to make substantial, long-term immunity to the ones she does catch.

This is also true of bacterial illnesses. In particular, Two is somewhat less likely to have overwhelming infections such as sepsis, bacterial pneumonia, and meningitis (see Glossary) than younger children.

There are two enduring health questions about young children that are discussed in other sections:

- Which infectious diseases mean that he has to stay home from daycare?
- What if Two's infections aren't just minor, but interfere with eating and

sleeping and socializing and learning? What if they are constant, and we keep recycling them through the family?

## COMMON MINOR SYMPTOMS, ILLNESSES, AND CONCERNS

### Colds and ear infections

Most of Two's colds are harmless and don't bother Two nearly as much as they do the surrounding adults. Twos never blow; they always sniff, or just drip, or, worse, explore with interest. Sometimes they will wipe, using one tissue per wipe. My suggestions for parents of Twos with one cold after another are few:

- Attitude adjustment is paramount. Try to regard colds as your friends.
- Have a parent/daycare-provider discussion that clarifies the rules for staying home from daycare.
- Force yourself to get into the habit of washing your hands before touching your eyes and nose. Do this whether you have touched Two or not: viruses cling to inanimate objects. Remember, touching your face is your primary way of catching these viruses.

Finally, many Twos will turn colds into middle-ear infections. If Two already has very good receptive and expressive language, this may not be so much of a problem. But if Two isn't talking in sentences and understanding pretty much everything you say, the ear infections may be blunting hearing.

Also, a Two with an ear infection or sinus infection acts very crabby, and this crabbiness may be confused with the Terribleness associated with being Two. So the underlying illness may not be suspected.

See the Essay in Part III if you are concerned about otitis media.

### Conjunctivitis (pink eye)

See the discussion on *Eyes* in *Body Parts, Bodily Functions, and What Ails Them* in Part II.

### Croup

A cough that sounds like a goose honking or a seal barking; a frightened child who goes "croooop! crooop!" when she inhales; and all of this in the middle of the night. See the discussion in Part II.

### Common viral diseases with unfamiliar names

There are a number of very common, but unpublicized, viral diseases of young children, and babies this age can catch all of them. They are discussed in Part II.

### Viral diseases with familiar names that are becoming, or are about to become, rarer and rarer

Measles, mumps, rubella (German or three-day measles), and chicken pox (*varicella*) are preventable by immunization and are also discussed in Part II.

## COMMON SCARY BUT USUALLY INNOCENT PROBLEMS

Twos may frighten parents with the same problems mentioned earlier: fever convulsions, night terrors, breath-holding spells, not using one arm due to nursemaid's elbow, limping because of toxic synovitis, puffing up with hives, waking up honking in the middle of the night with croup. These are all covered in previous chapters, and also in Part II.

## SERIOUS OR POTENTIALLY SERIOUS ILLNESSES AND SYMPTOMS

### Infections

Two is beyond the age where fever, all by itself, is a worrisome sign. Many Twos with temperatures of 104° and 105° gambol cheerfully about, spreading their viruses as they go. When a Two has a serious infection, the major indications are acting or looking sick, having trouble breathing, becoming dehydrated, having complex vomiting or diarrhea. These are all addressed in the *Frightening Behaviors* or *Body Parts* and *Body Functions* discussions in Part II.

### Noninfectious diseases

Of all the worries, it is leukemia that seems to haunt the parents of young children. This is particularly a problem from now until kindergarten, because young children have frequent illnesses. They are supposed to. But somehow the idea has spread that frequent illnesses are a presenting symptom of childhood leukemia. Actually, this is not the case.

Children with leukemia present with other symptoms: acting sick, looking sick, bruising and rashes. (These are all covered in the *Frightening Behaviors* and *Body Parts* discussions in Part II.) Remember that leukemia is a rare disease, attacking only 2 to 4 children out of 100,000. If you find that you are still worrying, be honest with your pediatrician.

## MINOR INJURIES

Two still sustains bumps (particularly on the head) and bruises and minor burns, and these are discussed in Part II in *First Aid*. Two is particularly prone, however, to sustain:

*Curling-iron burns*: Twos are magnetically attracted to curling irons. Burns on the palm of the hand always need to be seen even if blisters are tiny. Don't break the blisters, ever. Run cold water for as long as you can over the burn, then get in touch with the pediatrician or emergency room.

*Unwitnessed injuries that cause a limp*: Most often this is caused by a splinter, blister, bruise, sprain, bee sting, or toxic synovitis. But any limp that persists for more than two hours, without a very obvious minor traumatic cause, needs pediatric attention. The worry is that it could be due to a tiny fracture, the beginning of a serious bone or joint infection, or some other potentially serious problem.

## MAJOR INJURIES

*These should be avoided at all cost. Nevertheless, this is a good time to be sure all the child's caregivers are up-to-date on CPR training and speak enough English to summon help effectively.*

### Choking and poisoning

You'd think that Two would be less likely to suffer these, since the intense mouthing phase is past. This is partly true. But Two mouths things now not because Nature impels it instinctively but because Two thinks the object looks attractive, or food-like, or because Two can't figure out what it is by looking at it or throwing it or whatever. So Two sees what it does in the mouth. Big problems for choking include items like Ping-Pong balls, marshmallows, and coins; and for ingestions, adult medicines and delicious-smelling liquids. Beware adult iron tablets that look and taste like candy. These can produce life-threatening poisonings. Especially beware alcohol. Many a Two and older has awakened the night after an adult party and finished off all the dregs left in the glasses Two's parents were too tired to empty. Alcohol can lower the blood sugar in a young child to the point of convulsions and coma.

### Falling from heights

A fall from more than four feet high is statistically associated with serious injury, and Two should be seen after any such event, even if Two seems fine. Such falls are more likely because Two can get to high places and have no realistic idea of the consequences of leaning over and out, or jumping, or pushing someone.

Two Years to Three Years

*Water immersion*

The danger in this age is not limited to pools and hot tubs, though these are a major problem. Pools with soft pool covers are particularly treacherous, as the baby can become trapped, invisible beneath the cover. Baths, toilet bowls, buckets of water, fountains are all irresistible and potentially lethal.

*Burns*

Hot liquids, stoves, and fireplaces are what parents usually, and rightfully, worry about. Consider also: curling irons, frayed electric cords, exhaust pipes.

*Animals*

The German shepherd who loved the sedentary baby may become quite annoyed at the tail-grabbing two-year-old. Even cats, when teased, can inflict major damage.

*Car accidents*

See this section in the previous chapter.

*Older children*

Older siblings very frequently become extremely jealous and annoyed by mobile, curious, attention-getting babies this age. Even young children can inflict serious injuries in a jealous or frustrated rage.

# Window of Opportunity

## SOCIALIZING WITH PEERS

Twos need other Twos. If they don't have playmates, they will try to turn everybody else into a Two.

Most Twos do best when their attendance at daycare, preschool, or play group is more frequent than once or twice a week. Social memories fade after two or three days, and the child who attends infrequently may have to spend most of each session reacquainting himself with the other children and calling up his social skills. This is a big cause of separation anxiety and school refusal.

Twos are much more at the mercy of the social atmosphere than are Threes and older. They need much more adult protection from their own impulses and can't be counted on to work things out on their own with their age-mates. They also depend on all their adults to encourage autonomy and exploration.

*Health and safety at group activities*

It is the job of Twos to catch viruses to build up immunity to diseases. Parents and teachers have to try to cope with the philosophy that getting repeated minor illnesses is a benefit of group activity at this age. Though, of course, the child who is too sick to enjoy herself (as well as the child who has a potentially dangerous illness or whose symptoms require a great deal of teacher attention) should stay away.

Having said this, there is no sense in actively encouraging nonstop illness. The diapering area should be separate from food preparing area. Excellent hand-washing will help keep both staff and children from catching every single thing going around. And a few diseases are either so dangerous (like serious H. flu infections and measles) or so exceedingly contagious and demanding (like some kinds of diarrhea) that a child with that illness ought to be kept away.

Safety is a major concern for Twos. Ideally, the area ought to be so safe that few activities are forbidden. This can be a special problem if older children use the same facilities, if dangerous areas are enclosed with gates an ingenious Two can open, or if there are pets to be annoyed.

Since many Twos are just consolidating potty skills, this area ought to be very user-friendly and clean.

Knowing first aid is important. This is the age of bites, falls, and fingers jammed in the door. Ear infections, gastroenteritis, and other minor diseases are very common. And since 5 to 7 percent of young children will have a fever convulsion, teachers need to know how to handle both a sick child and the frightened onlookers without panicking.

### Between teacher and child

Twos are particularly vulnerable to turnover of teachers. It may take days to weeks for a Two to become close to a new teacher. Or, just as bad, if turnover is frequent, the Two may choose either not to bond with any teacher or to be overly receptive to any adult.

Language learning is an essential task for Twos. Teachers are models and facilitators. A withdrawn teacher or a driven chatterbox can each be a problem. The teacher who engages her Twos in dialogue frequently, expanding and drawing out their ideas, is a teacher indeed.

Twos need frequent adult contact, both one-to-one and in small groups. Children rarely can be left on their own in groups to play for very long. The intervention of the teacher needs to be tactful and kind. Blame and justice have no place in a classroom of Twos. Shouting, labeling ("That was mean!" "Don't be such a baby!") spanking, slapping, or any physical punishment should be non-existent.

Nearly all interactions between teacher and children should be positive and exceedingly tactful. A teacher who praises Jessica's prowess throwing balls and Nicki's ability to climb the slide, but who is silent in the face of Jordan's block tower, has just criticized Jordan.

Putting Twos into a large group is unproductive for everybody. A brief circle time might be attempted, but lots of Twos will get up and wander, as is their perfect right.

The style of many Twos is still to watch closely while assessing a new activity, with frequent glances to their special adult to see what she thinks of it: is it safe, fun, gender appropriate, etc. Teachers who understand this important role are invaluable. They don't hurry a child into participation, or project discomfort if the child's activity seems to be unconventional for his or her sex, for instance.

### Activities in the group

The primary activity of Two should be play, not formal learning. Story reading is best when it is interactive. Pointing out the letters of the alphabet or how words look is an annoying distraction, unless the book is an alphabet book. Even then, Twos regard letters as pictures, not as symbols. A triangle shape may be recognized as looking like an A. That doesn't mean that Two has any idea of the function of A in helping to spell a word.

Twos don't do well being marshalled from one activity to another. Two is about choices and control and power, about autonomy. Twos are also not yet geared to appreciate the construction of a completed project; they are more involved in process, exploration, and fantasy. Art projects and such are not a big part of the program.

Twos need one nap a day. It's best for parents if the afternoon nap doesn't last longer than an hour and a half and is over before three o'clock. Otherwise, Twos may stay up much later than they should and be hard to awaken in the morning.

Twos need planned nutritious snacks, not ad-lib handouts of crackers and juice.

## Preventing TV Addiction
Twos can fall in love with the TV, and a habit formed at Two is very hard to break later.

In particular, many Twos fall in love with programs that later on can lead to behavioral problems: nightmares, separation anxiety, and intensely gender-oriented play. These programs are cartoons, series like *Pokemon*, and fantasy videos: *E.T., The Wizard of Oz, The Little Mermaid, Beauty and the Beast*, and so on.

Alas, Dumbo's mother and Bambi's mother die; E.T. and friend nearly do; Dorothy and Toto nearly can't get home; Ariel's father gets transmogrified to a horrid wizened little creature. Meanwhile, on the same TV, real little girls fall down wells and are rescued (or not), real people starve, drown, shoot and get shot; cartoon monsters emerge from toilet bowls.

But it isn't the TV-addicted Two who gets upset. Two picks up the action in bits and pieces and doesn't appear to take it to heart.

Rather, it's the Three who has become "addicted" to these videos who has the bad dreams and refuses to sit on the potty. Three tends to be convinced that boys fight with swords and girls comb their hair in a mirror; that guns are fun, that old women with warty noses are evil witches. They may even be certain that if you have a cape you can fly.

If Two becomes addicted to TV and videos, stopping or changing at Three is almost impossible. Threes have wonderful memories, incredible brand loyalty, and exhausting persistence. Do yourself a favor and don't start the videos in the first place. Save them for the Over-Fives for whom they are intended.

One of the difficulties with TV is that for tired working parents, watching TV with a toddler on your lap can be charming, undemanding, and intimate. But there may be a heavy price next year, and an even heavier one later on, with regrets about missed opportunities to play and read aloud and explore.

## Potty Skills
(If it were really "toilet training," you'd train the toilet.)

The major task of Two, of course, is learning to use the potty.

Many advisers suggest breaking down this task into parts and teaching it one part at a time: sitting on the potty dressed, then with just a diaper, then diaper off. Or showing the toddler where the poop goes by putting it into the bowl.

For most Twos, however, this approach isn't necessary. The potty task is not complicated in terms of its component skills. It is difficult because it has to do with control and genital sensations.

Indeed, many Twos can learn how to use the potty mostly by watching. *How* is not the problem. The trick is to make the Two want to learn to use the potty.

For a Two to want to do so, she must feel fairly unambivalent about growing up. It's hard to feel that way if there is an attention-getting younger sibling in diapers. Two also needs to feel that her previous self-care accomplishments have been respected by her parents as achievements in autonomy, rather than valued just because they make life easier for them. She also needs to feel relatively unthreatened by the act of peeing and pooping. If she thinks of the act as shameful and dirty, or if she has had infections or constipation, which make it painful, she's going to need a great deal of recuperation before she can tackle the potty.

Given that these conditions are met, the easiest way to teach potty skills is usually to have potty use demonstrated in a cheerful and proud way by someone: a parent, sibling, friend, or peer. Often, the Two, after watching and thinking for a while, will ask or demand to use the potty himself.

But be careful how you intervene in the process. Don't moralize. "See what a big boy Gary is. Wouldn't *you* like to have big-boy pants like Gary?" is likely to backfire. It is just too tempting to take the easy road to self-assertion: "No! I don't want to be a big boy like Gary!" Far more likely to succeed is, "I wonder what kind of big-boy pants you will choose, the Turtles or the Snoopies?"

Another good trick is to profit from Two's "magical" thinking. Twos assume that every item in their world has a will of its own. They also assume that the entire focus of that will is to please, hamper, or attack the Two himself.

So try telling Two that his poopoo wants to go in the potty. Use a little high or low growly voice and pretend you are the poopoo, saying, "Oh, I want to go in that potty. I wish Jason would put me in the potty."

### Balky refusal

Sometimes a Two reacts with balky refusal.

A frequent scenario for balky refusal: Jessica has to use the potty or she can't start preschool and her mother's job starts in three weeks and there is no refund on the tuition and she'll lose her place in *Highly Desirable Beginnings* and she just digs in her heels and won't do it.

There is one ploy worth trying in this situation: that is, rewrite the scenario so that choosing to use the potty is in the child's self-interest. Jessica at the moment controls the whole situation. She WON'T use the potty. She determines when she will pee and poop in her diaper: it is when she feels like it. She determines when her diaper will be changed: it is when she informs her mother that it is dirty or wet. She determines where she will pee or poop: the former, any old time or place she feels the urge; the latter, in her parents' closet with the door slightly ajar.

To turn the tables, the adult simply has to regain control. Every half hour (when she sees that Jessica is involved in something very interesting) she calmly sticks her hand into Jessica's diaper, explaining "Just seeing if that diaper needs changing."

When it does, she must interrupt Jessica from whatever Jessica is doing to change that dirty diaper (and the interruption "just happens" to occur when Jessica is thoroughly engrossed in her play).

If Jessica announces that the diaper needs changing, Jessica's adult is much too busy right then to tend to the problem. It must wait. In fact, it must wait until Jessica has gotten involved with something else highly interesting. Then, she must be interrupted for a brisk, not entertaining, diaper change, in which her adult seems not annoyed, exasperated, etc., but merely very, very preoccupied with her own thoughts.

I have not seen this method take longer than three days to work, given that the problem was really balky refusal in the first place.

### Developmental refusal

Besides balky refusal, however, there can be true emotional unreadiness.

This is usually a child who pees in the potty just fine, but can't bring himself (it is more often a boy) to poop in the potty. He asks for a diaper and goes off under the table or wherever, like Steven.

Frankly, I sympathize with Steven.

From his point of view, what, pray tell, is his poop?

Is it something separate from him, that ought to be on the outside? Or a part vital to him, that ought to be retained? If it is something like a body part, the idea of losing it is appalling. Is it a possession, like the Batcar?

On the other hand, the act of pooping has qualities of an artistic construction. Producing it is an effortful accomplishment, garnishing mixed reviews from an interested audience.

But here is the kicker: all the effort is concentrated Down There, where one's genitals so vulnerably and excitingly live. . . . And afterwards, being cleaned up is both reassuringly babylike and rather, unavoidably, stimulating. Well! The whole thing is a bit overwhelming, and best done with concentration, in private, into a safe, enwrapping diaper, behind the playhouse.

My feeling is that this is a child working out important issues of control and worry about loss and giving things up. I would not rush him. One diaper a day for a few weeks is not going to break the bank, use up a landfill, disqualify him from any reasonable preschool, or exhaust a caring parent.

### Panicky refusal

"Dani won't have anything to do with the potty chair. She won't even sit on it with her

## HINTS FOR POTTY USE

Don't get a potty with a penis guard. They inevitably bump and scrape and hurt.

Once the basic task has been learned, make a game of peeing in other receptacles, such as a bucket or specimen cup, or outdoors, behind a tree. Children can become very fixated on the idea that the potty is the one and only right place.

Beware constipation; it can make peeing as well as pooping difficult, because the hard stool presses on the bladder.

Make sure that when the adult toilet is used the child has a smaller inset seat so he doesn't feel at risk of submersion. And make sure that there is a place for him to rest his feet so that they don't dangle. He's got to have them braced so that he can push out the stool.

Fear of pooping can arise from the splash the stool makes in the adult toilet. A solution is to start with or return to the potty.

Teach a little girl to wipe in front first so as not to carry germs from rectum to vagina and urethra. However, be sympathetic: until about age five or six, her arms are too short to accomplish this feat. You'll have to help out and be philosophical about hygiene.

Beware toilet cleanser ads on TV. Would YOU sit down bare bottomed on something that had a monster inside it?

---

clothes on. When I show it to her, or show her her big-girl panties, she pushes me away and uses her bad words, dammit and dummy. I've tried everything they suggest in the books, star charts, and M&M's, and I've tried to read her all the little potty books. I have friends with older children and they tell me horror stories about constipation and so on. I don't want to blow it. But sometimes I just break down and cry."

I have this image of the potty in Dani's mind. It is sort of like that inscrutable, Power of the Universe, mind-boggling slab in Kubrick's movie *2001*. Potty knows her

every thought and wish, it follows her from room to room, gets into her books, causes deprivation of candy, makes her mother (her mother!) cry. If I were Dani, I'd panic too.

Dani's mother made an enormous effort to lighten up. "I can see that Dani must be mixed up about whose achievement this is," Michelle says. "But I just feel as if I have to be in control. Not just about the potty, but about everything. It's a constant battle."

Michelle started to concentrate on sharing interests with Dani and cutting down on her directive language. "Every statement I make to the poor child has to do with something she's supposed to do that she has or hasn't done. I see that. But it's my nature."

Dear Michelle. Her friends and I gave her suggestions: Loosen up, stop and smell the roses, enjoy making a mess with Dani, pick a bouquet of weeds, learn to enjoy her weird clothing combinations. Put her in diapers until she goes to kindergarten, if you have to.

What Michelle finally did, however, was perfect for her. She wore a rubber band around her wrist and snapped it, hard, whenever she caught herself overcontrolling Dani. And Dani learned to use the potty when she was just turning Three. She insists on having the door closed, however, and always flushes twice.

## Naming Private Parts of the Body and Their Products

I feel that body parts deserve to be accurately named. In one study* only 30 percent of the boys could name their penis and testicles correctly. Only 21 percent of girls were able to say they had a vagina. Many used nicknames, but even then, only 65 percent of girls (but 85 percent of boys) used any name at all!

Oddly (or not so oddly), only 58 percent of girls who were told the right names for penis or testicles were told the correct name for their vagina. And only one girl of sixty-three in the study had a word to describe her external genitals, the vulva and clitoris.

Penises get called *whizzer, dingdong, dinkus, peepee, dingadoo,* and, since Robin Williams, *Mr. Happy.* Vaginas are usually either ignored or called *gina* or *bagina* (rhyming with Carolina). One rather endearing title I have heard is *her little peach.* Clitorises, if mentioned at all, get called *button* or *special place.* Anal areas are generally called *bottom.* The whole region is mostly referred to as *down there.* Little girls are often taught to refer to their nipples as *boobs* or *boobies.*

This makes for two problems.

First, diagnosing illnesses can turn into a real wild-goose chase. Hundreds of pediatrician-hours have been spent searching a child's body for the "Ouchie in my whistler," and determining if the statement "It hurts Down There when I go potty" means painful urination, constipation, vaginitis, pinworms, self-inserted foreign body, or, alas, molestation.

Second, and much more important, the world of the Two is all about naming. Naming doesn't merely mean assigning a label; it means possessing the thing that is named. When a most important entity, which the child knows is exciting and interesting, is not given a name, this has got to be puzzling and worrisome. If the name is one that gets an adult reaction, like *boobies,* the child is both mystified and titillated.

One solution is to teach the real names to the young Two, and then, at around Three, to teach a public "nickname" for them. But

*(Fraley, M. C. et al.: *J. Dev. Behav. Pediatr.* 1991 12:301)

at home, or in the doctor's office, the real terms can be used.

Finally, there is the issue of urine and stool. Many families use the euphemism "going to the bathroom" for both, which confuses medical personnel. (What are we to make of the problem when a parent reports that a newborn hasn't "gone to the bathroom?")

Here it isn't so important that "stool" and "urine" be used. In fact, these terms may make the whole subject seem a bit stark. Also, it is difficult to see what verbs go with them. "To stool" and "to urinate" sound humorless. If you teach a young child to use the terms "defecate" and "micturate" you are setting him or her up for terminal teasing.

I myself use *poop* or *caca* and *pee*. Steven "makes a wee" and "does a job." The terms *shit* and *piss* are regarded by most middle-class people as nasty, and their use can prejudice medical personnel.

Air emerging from the mouth is no problem; it is a burp. Air from the rectum is a problem. *Passing gas* or *passing wind* seem to be popular; the term *fart* is often used but usually by parents who look embarrassed and at a loss for another term.

Throwing up and vomiting are clear terms. The word *barf* is felt by many middle-class pediatricians to be a kiddie-slang word, and they are startled when a parent uses it. I don't care, myself; I'm just pleased to be warned that it could happen, before it does, on me.

# What If?

## STARTING DAYCARE

At Two, this isn't a question of "what if?"; it is a suggested option for finding social opportunities, and thus is discussed in Window of Opportunity, above.

## EXTENDED SEPARATION FROM THE PARENTS

Two, and even more so Nearly-Three, can be prepared to a good extent for such a separation. It's important to go over repeatedly and in detail who will care for Two, and where the parents will be and why, and for how long, in terms Two can understand. Since future time, even tomorrow, is a hard concept for Two, some help with the idea is indicated. It often works well to have lined up small, inexpensive, wrapped presents in a row, one to be opened the morning of each day the parents will be away. When all the presents are gone, the parents return.

It's kindest to keep schedule and bedtime routines as consistent as possible, and crucial for the person caring for Two to be someone Two knows and loves well. It is important that this person has had experience putting Two to bed, bathing Two, and taking care of all potty issues, ahead of time—before the parents depart.

Finally, the daily play group or friends in the neighborhood or preschool will help Two to keep in mind that she has an independent and joyful identity not totally dependent on the presence of her parents.

## TRAVELING WITH A TWO

The main problem of traveling with Two is that if Two doesn't have access to other two-year-olds, he tends to attempt to turn the nearest adult into one. And often succeeds, merely by refusing to accept that this is truly an adult.

However, Two can also be a joy to travel with if you have a good, friendly relationship beforehand. This means that your interchanges are mostly friendly discussions about the world, not harangues of monitoring narrative. (See Portrait of the Age, this chapter.) Even so, I'd try to keep schedules and meals regular and familiar, and keep the bedtime ritual inviolate.

Taking Two's security object plus at least one "double" in case it gets lost is vital. A portable potty seat to put on a big toilet is dandy for Twos who have learned potty skills, but only if they've previously been introduced to it at home and on local outings before the big trip.

## CONCEIVING OR PLANNING TO ADOPT A SECOND CHILD

If Two will be over Three before the second child comes along, the whole event is likely to go much more smoothly than earlier. Three is a person with a sense of himself as an independent actor in the world, a reputation to uphold, and a place in his group of peers. Three is pretty adept at self-care, and has come to terms with the awe-inspiring act of letting go, whether it be letting go of a possession, a parent who must go elsewhere, or a poop into the potty. Three can express anything he feels or thinks in words, if encouraged to do so. Three can role-play, and can be a mommy or daddy to a "guy" or doll or stuffed animal, emulating the real thing. Three can be aware of the status that goes with being a big brother or sister.

## ARRIVAL OF A NEW SIBLING

I am moved by the consistent reaction most Twos have to a new baby sibling. They "love" the baby, showering her with attention and kisses and sometimes-just-a-little-too-forceful hugs. Never once do they say a hostile thing. They become upset when she cries, and become agitated if a parent doesn't tend to the baby right away.

And they punish their mothers.

They regress, hit and bite, tantrum, reject and whine.

Not surprisingly, studies show that a Two in this position does experience a change in maternal behavior. Mothers don't look at or talk to a Two as much after the newborn arrives as they did before. And when they do address Two, it is frequently critically. Even if they don't remonstrate, command, or scold, they spend a lot of energy redirecting Two's activities, requesting help or quiet, explaining to Two that the baby's or the mother's needs take precedence over Two's.

Naturally, this holds a danger of becoming a vicious cycle. The more Two behaves obnoxiously, the more time and energy her mother has to put into disciplining her and the less time she has for making Two feel like the beloved center of the universe. So in response Two behaves even more horridly.

What will help:

- Get off to a good start. Don't make any major changes or demands two months before or two to four months after the baby's birth. This is not a good time to get Two out of her crib to a big bed or start potty training. It isn't even a great time to take her *out* of daycare group while Mom is home on maternity leave. After the baby's birth, either Mom will have to cut down on Two's attention dramatically or Two will have to return to a barely remembered daycare group. Or both.

- Prepare for the hospital stay. Whatever Two's arrangements for care might be, practice them ahead of time. Talk about it in a calm and positive way beforehand.

- When Two comes to the hospital after the birth, don't be holding the new baby. Twos are great symbolists; the place in Mommy's arms is more than an anatomical location. It's where a Two's soul lives.

- When Two first goes to touch the baby, guide her hand and praise her. If she is unguided, she may go for the eyes or fontanel, and her first experience will be scary and humiliating.

- When you bring the baby home, put a sign on the front door to remind visitors: "Please greet and make a fuss over Two first, before you ask to see the new baby."

- Don't expect that calling her The Big Sister or him The Big Brother is going to cut much ice. Twos are not yet into role-playing; they're just establishing their own identities, and taking on an identity that only exists as a relationship is meaningless to them. Two may pretend he has a baby in his belly, but that's different from pretending he's the mother or the daddy. Moreover, as Two grows into Three and does begin to role-play, the roles are egocentric ones: firefighter, Wonder Woman, circus performer, ballet dancer. The role of Big Sister is hardly egocentric; it exists only in relationship to Little Sibling. Not pleasing to most Twos.

- Crucial: No matter what else doesn't get done, each parent should spend fifteen minutes to half an hour a day ALONE with Two, just playing or vegging out. Get down on the floor with her. If you have no energy for anything else, disregard the TV warnings above and turn on the tube, but cuddle and chat while you're watching. ALONE means alone. No phone, no other relatives (even the other parent), no baby.

- Many times Two will hear, "Just a minute, Two. You'll have to wait while I change Baby's diaper." So many times let her hear, "Just a minute, Baby. You'll have to cry and wait until I help Two with her truck."

- Twos have great receptive language. Let her overhear you talking to other adults, on the phone and elsewhere, praising her. Not about her role as Big Sister, but as cute, funny, original, brave, smart on her own terms.

- Twos need lots and lots of physical attention. Hugs and cuddles and pats galore.

- A Two who regresses needs more, not less, physical affection. And fewer demands or requests to help.

## DEALING WITH A GROWING UP YOUNGER SIBLING

A Two whose younger sibling is between nine and twelve months of age may spend most of her time in the attack mode. A Two whose younger sibling is older than a year may spend half her time attacking and the other half being attacked. This can be a major problem.

Unfortunately, the nature of the problem isn't something parents can easily fix. This is because there really isn't anything much that two children with this age difference can actually play together. Both are still in the watch-and-imitate mode of social encounters. Neither is yet able to do much in the way of fantasy role-playing. The kinds of reciprocal activities they do like usually require an older playmate: roll the ball, swing on the swing, read the book. Two has just been consolidating her idea of what a possession is and is not yet ready to share, and the younger sibling hasn't even figured out what it means to possess something and so grabs without inhibition.

Half the battle for parents, I think, is simply facing this fact.

Once expectations are changed, the fighting between the two often lessens. Or perhaps it just seems more appropriate and expected.

The most helpful advice I have been able to give is to get Two into a regular play group, and treat each sibling as much as possible as if he or she were an only child.

## DIVORCE AND CUSTODY ISSUES

Twos have spent their whole lives establishing a base-line feeling of "This is the way things are around here, and here are the ways in which I can be in control." A divorce, especially one that is bitter, or results in the total or near-total disappearance of one parent or in a big change in lifestyle, can be devastating. Of course, this needs to be balanced against the bad effects of an awful marriage. If "the way things are around here" features drugs and alcohol, abuse, emotional battering, or consistent fear and anger, divorce may be the only healthy option.

If divorce is going to happen, the special needs (as opposed to the general needs of all age groups) of Two are:

- Preservation of a sense of autonomy. No matter how unhappy or preoccupied, a parent needs to protect this for Two by continuing to stress self-reliance, pride, and control.
- Consistent setting of limits. A Two in the midst of chaos will out-chaos everybody. Parents owe it to Two and to themselves to control the adult chaos and set limits on the toddler form. This may be extremely difficult in the face of feeling guilty, apologetic, and protective.
- Predictability. The basis of Two's power lies in a predictable environment over which she can exercise control. Custody decisions should reflect this need. There

isn't any one best answer (unless it's the rare one of bird-nesting, where the child stays in the "nest" and the parents stay there alternately) but certainly the predictability factor ought to be monitored. A profusion of daycares is especially to be avoided.

- Gender issues. Twos are deep into genderness. A divorce may confuse Two about what is appropriate, necessary, and valued gender-related behavior. Everything, and I mean everything, is imitated at this age. Twos who see the same-sex parent physically or verbally attacking the other or *being* repeatedly attacked by the other, or freezing with silence, or whining, or talking bitterly about the other parent, may deduce that this is appropriate gender behavior. Once learned, such a concept is hard to undo.

## HOSPITALIZATION AND ELECTIVE SURGERY

Many psychologists feel that, if possible, elective surgery shouldn't be undertaken during the years from two to five. This stems in part from Freudian psychology, which postulated that children this age were going through an Oedipal stage with enormous fear of genital mutilation. Any damage to the body, it was thought, could serve as a symbol for such mutilation, and could produce disproportional fears in a child this age.

Another consideration is that children this age seem to be forming a "social" physical identity. They are more susceptible to public, peer, and cultural cues in deciding whether they are cute, tough, pretty, plain, graceful, clumsy, and so on.

Threes, and many older Twos, spend a lot of time studying physical differences: they are consolidating gender identity, they are aware of racial differences, they notice oddities (like birthmarks, glasses, French braids, a new hat) and comment on them. They earnestly desire power in shaping their appearance before their public. Threes are constructing a social identity. Any procedure that endangers this fairly precarious identity can be threatening.

This is not to say that any child who has surgery or a procedure now will be devastated. But it is extremely important for parents to be with the child as constantly as possible, and to expect fears and regression afterward.

## MOVING

A young Two is not likely to have much trouble in moving if her parents stay sane, she has a backpack securely strapped on that contains her most precious toys and security items, and her schedule and especially her bedtime routine can be preserved. If possible, her crib and room should be set up early on, and in a way as close as possible to the previous arrangement.

If Two temporarily will be sharing a room with parents, make sure Two has a sleeping area of his or her own. Preserve that special separateness and the bedtime ritual, and try to have a physical separation of beds such as the blanket-curtain in the great old movie *It Happened One Night*.

An older Two can be prepared for, and help plan, the move. There are good picture books about moving. Two can be shown photographs of her new home and neighborhood. "Here is where you can ride your tricycle." "Here is where we will plant flowers." "Here is a good place to dig."

It will help the whole family if each parent spends some relaxed time alone with Two every day, one on one, even during the throes of the move.

Most crucially: Moving day is so disturbing and exciting for a Two that he inevitably gets into trouble. Tragically, this can mean trouble like getting through an upper-story window or drinking the turpentine or choking on a faucet washer. Moving day is more dangerous for a Two than perhaps for any other age group. Two should be absolutely, positively elsewhere on moving day itself, with a trusted and unpreoccupied caregiver Two knows well.

## THE THREE-YEAR-OLD WELL-CHILD VISIT

This can be a delightful visit, but not in the way many parents expect.

The Three who talks nonstop at home is usually as mute as Marcel Marceau in the office of her dear pediatrician, a person she has known and visited frequently since her birth.

The Three who parades about the neighborhood nude will refuse to pull down his jeans.

The Threes who know all their colors, can count to twenty, sing all the verses of

"The ants come marching one by one," and draw a recognizable person with a head and arms and legs coming out of it, will sit under the exam table until lured out.

All this refusal is generally performed in the most charming way possible, with Three glancing out of the corners of his eyes, ducking his head, and flirting.

Needless to say, it's most helpful to prepare Three for the visit ahead of time. A picture book about going to the doctor may be helpful, but even more so is a visit to the office a day or so beforehand, just as a reminder. A particularly good idea is to invite Three to make something special for the doctor and for the nurse or assistant. Another good one is to have Three bring her own doctor kit along, so that there's a sense of equality about the whole thing.

I suggest not telling Three that there either will or won't be a "shot" or a "poke." At this age, I wouldn't even bring up the possibility. In the unlikely event that Three asks, just say you're not sure, but if there is one, it will be explained first, and it's OK to cry if it hurts, and then a pleasant reward will come afterwards. Don't be drawn into a long conversation about the subject, as that only heightens Three's anxiety.

If there is a test or shot that hurts, parents can help Three best by giving a brief, simple, confident explanation that allows for no argument, and then getting the procedure over as briskly as possible. When Three cries, it's nice to say that crying when something hurts is always OK. Then review that the procedure was done not because Three was bad, or sick, or the doctor or parent was angry at Three. It was done because it is the rule. The test is to make sure Three is healthy. It is the doctor's job to do the test, and the parent's job to pay the doctor to do the test, and Three's job to have the test, to cry when it hurts, and to get a treat afterwards.

Three really likes it when everybody does their job.

## Looking Ahead

Having Three around is like having a very endearing person from another planet. You can talk with Three about almost anything if you define your terms and simplify words and concepts. Three's part of such a conversation is likely to be highly original. It isn't so much that Three believes in magic; the term "magic" has no meaning for Three. Rather, Three assumes that adults can do

anything, and that if they don't, it's not because they can't but because they don't want to. Three assumes that inanimate objects are alive and are either friends or enemies. Three is impressed with the power of language, and learns only slowly that saying something doesn't make it so.

Threes require challenges and safety. They need to trust the world is a friendly place, but know not to go off with a stranger. They need to feel that adults are in charge, but that Three has a say in most matters that Three cares about. Threes can become avid readers or avid TV aficionados, but not both.

Threes reward your periods of undivided attention, your respect for Three's feelings, and your ability to set limits in a dignified manner. The reward Three gives is friendship between child and parent, which lays a foundation for such a friendship throughout childhood and adolescence.

# Three Years to Four Years

## The (Mostly) Benign Dictator

*"A Person's a Person, No Matter How Small."*
—Horton Hears a Who, *by Dr. Seuss*

## To-Do List

- Read real books.
- Restrict the Security Blanket a bit.
- Make a dentist appointment.
- Find a good preschool, or substitute thereof.
- Anticipate sex play and questions about sex.
- Keep an eye on weight issues.
- Teach compassion; it doesn't come naturally.

## Not-To-Do List

- Don't let Three get used to TV, including videos.
- Don't introduce fast food, sodas, and juice.
- Don't expect Three to confess when accused.

## Portrait of the Age: Power Through Predictability and Language

"Remember, Petey, it's a school day today. Time to go see Miss Kate and play in the sandbox and finger paint and swing on the tire swing!" Linda carols.

"I'm not going to that old place today." Petey scowls at the trucks he is lining up along the doorway of his room, a veritable phalanx. "That stupid old place."

"But Petey, you always love it when you get there. Besides, this is Mommy's work day, too. We can't stay home and play every day, and today is Monday and it is a school day and a work day."

"Vroom, vroom." A race car gathers speed on the floor.

"We're going to be late. I want you to get that shirt on right now. We have to leave *right now*. I told you five minutes ago to get dressed."

Linda steps over the trucks, and Petey rolls so as to get under her descending foot.

"Petey!" As Linda contorts herself, she puts her hand out to balance and gasps.

"Peter! Look at this! There are big red circles all over your wallpaper. Why, that's lipstick! That won't come off, ever!" These aren't just circles. They are huge and jagged with spikes and dots inside them. Linda kneels before Petey and fixes him with a gimlet eye. She is so angry her hands are shaking. "Now, Petey, I want to know how that happened. How did big red lipstick circles get on your wallpaper?"

"I don't know."

"Peter. Tell me the truth. Did you write on your wallpaper with my lipstick?"

"Miss Kate did that." Peter looks straight into Linda's eyes. With one red-smeared hand he cranks his dump truck, which flings its load, a silver monogrammed cylinder filled with red cosmetic, on the floor. "Vroom, whuwhuwhuwhu Vrooom."

In a most unpleasant three minutes, Linda delivers a sermon on telling the truth with a corollary on writing on the walls, gets Petey's shirt over his head but gives up on getting the arms in the sleeves, wrenches her shoulder hauling him, kicking, from the room, and finishes, ". . . So because you wrote on the wall, and you didn't tell mommy the truth, there will be NO cupcake in Petey's lunch and NO video after preschool!"

Trembling with rage, she hurls both of them into the car, fastens the belts, and bangs the doors shut. They drive in silence for five minutes, during which Linda covertly makes rude hand gestures and mouths obscenities at her fellow drivers. I could get us killed, she thinks grimly. One of these guys could get out a gun and shoot us. So what.

At a stop sign, she can't help herself. She turns to Petey. "Now Mommy feels sick. I don't like getting so angry and upset and now my stomach is sick."

Petey considers this. "Your body will make antibodies," he tells her reassuringly. "When Jules gets sick, he makes antibodies."

Linda can't decide which is winning within her: fury at Petey's behavior, despair at his lying, amazement that he knows the word antibodies—much less what they do; or annoyance that Jules has reappeared. Jules is Petey's imaginary friend—a sixteen-year-old male babysitter, says Petey, despite the fact that Petey's three babysitters are all female college students. She glances at Petey, who is holding his crotch, a recent habit. Last night in the bathtub, he sang a little song: "I love my penis, I love my penis."

Before she can introduce any coherence into her thoughts, they have arrived at The Children's Place, and Miss Kate is smiling a welcome. "Oh, Petey, we're so glad you're here." She turns to Linda. "Petey is such a leader, and he follows our rules so well. Petey never has to sit in the time-out chair."

Linda growls something polite and dashes back to the car. Time-out chair! Time out for Petey at home means dragging him into the "safe, dull place," the laundry room, and standing on the other side of the door clinging with dear life to the handle until he finishes his three minutes of "time out."

Two hours later, she is regaling her colleagues with the antibody story. What a memory Petey has! It's been at least a week since they read the picture book about germs.

And suddenly, in mid-sentence, she stops. Those germs, in the book: they were red and fierce. With spikes. And dots.

When she picks up Petey that afternoon, he looks uncharacteristically sad. "I didn't have a cupcake in my lunch," he tells his mother, as if he had never heard of such a thing.

"That's right, and no video either today." Linda finally feels as if she has a handle on the situation. "We don't draw big germs on the wallpaper. The wallpaper is ruined. And Mommy wants Petey to learn to tell the truth. It is wrong to lie. So no video."

Petey stares as if his mother were speaking Martian.

"Besides, remember, the book says that germs are tiny, so tiny you can't even see them." Linda realizes she's gotten off track. "But anyhow, no video. So you will really remember about no crayoning on the wall and about telling the truth."

Petey, after one incredulous look, turns an ominous purple. By the time they are home, the tantrum is fully developed, and Linda, clinging for dear life to the laundry-room door as Petey shrieks and kicks on the other side, feels her already-sore shoulder begin to ache like fury.

They don't ask for much, really, these Threes. They merely want to be in control of every situation, and to be Perfect. Living with somebody in the grip of these needs isn't always easy, but once you identify just what Three is aiming at, and why, Three can be wonderful fun.

Linda and Petey aren't having as much fun as they might be. Linda thinks that Petey's schedule, Monday at preschool, Tuesday at home, Wednesday at play group, Thursday at home, and Friday at preschool, should be perfect for him—a nice mix of home, play group, and school: Petey thinks his life is chaos. Linda thinks that Petey should tell the truth: Petey thinks he can change the reality of an unpleasant event by denying it. Linda thinks that if Petey does something wrong, then he should lose a privilege. Petey can't make such a sophisticated connection, and thinks that his mother is speaking Martian.

Here's how far apart they are. Linda thinks that Petey understands germs and antibodies. Petey has forgotten all about germs, but he does believe in "Ant-y Bodies," which *our* bodies make by—you don't want to know.

Oh, the mind of a Three-Year-Old.

Right now, Petey and Linda both need a bit of remedial work. See the section on *Limits* in the last chapter, from Two to Three.

Linda can help Petey be his best self, and help them to have a better time together, if she can see the world more from his point of view.

- Petey would be much calmer and more self-controlled, less oppositional in general, if his schedule was the same every day of the week. His preschool is a good one (see Window of Opportunity) and the challenges there help him to grow. If he attended each day, it would become a Life Habit, and would go more smoothly.

- Linda needs to recognize that only Petey owns his Potty Behavior. She needs to change the rules so that it is in his interest, as an autonomous, powerful person, to use the potty. See the essay in Section III on *Potty Refusal*.

- The more Linda asks Petey to confess to a naughty action, the more oppositional he'll become in general. The whole focus of his misbehavior will be on denying it. This is not the age to demand the truth about wrongdoing; that just complicates matters beyond belief. Much more effective, at this age, to discipline the behavior without asking who-done-it.

- Linda needs to let Petey be more accountable, and let the consequences of his behavior demonstrate that accountability. Instead of depriving Petey of a cupcake because he wrote on the wall, she might have him help her try to clean the wall instead of playing outside, having a story, or watching his video after preschool.

- She needs to ditch Time Out; it's only making matters worse. When Linda has to hold the Time Out door shut, Petey is not really in Time Out: he is in a War Zone. And he eventually wins, because after three minutes Linda lets him out. Petey doesn't know that it was the timer, rather than his own energetic attacks on the door, that got him released. This makes Petey's oppositional behavior

appear to pay off, so he behaves even more obnoxiously.

- She needs to pay special attention to letting Petey know that he is the Apple of her Eye, so that he continues to want to please her.

As Linda puts these operations into effect, Petey will become more and more amenable to Authoritative Parenting. In a few months she'll be able to say, "Time to get dressed for preschool," and Lo! Petey will get dressed. Most of the time, anyway.

A Three who *is* potty trained and cooperative about preschool and getting dressed is more likely to indulge in more sophisticated forms of oppositional behavior, such as whining, interrupting, and procrastinating. See the section on *Limits*, below.

## POWER THROUGH LANGUAGE

Three wants control not only of herself, but of her world. Not an easy task, when you're only a yard tall and weigh about thirty pounds. Fortunately, Three has a secret weapon: language.

Three uses language not to describe reality but to create it.

It's called Magical Thinking. If I say it, it Will Be. This is extremely useful. It allows you to pretend that a wish is reality. If you do something wrong, you can make it Not So: Miss Kate wrote on the walls, not Petey.

Consider Jules. A Three with an imaginary friend has one person in his life over whom he has total power. Jules must be sixteen, he must be a babysitter. When something goes wrong, Jules must take the blame. When something is nasty but required, Jules can be the one who doesn't like it. (Jules doesn't like pork chops, the white medicine, or being kissed by Aunt Amanda.) You can order grown-ups around on Jules' behalf ("You can't sit there. That's Jules' place!").

If you are being pressured to do something scary, you can make up a story and use Jules to get yourself off the hook. (Jules blew bubbles under water and ended up in the hospital.)

This isn't lying. Three is not out to deceive, but rather to call into being a preferable reality. Getting Three to "tell the truth" doesn't work, because Three thinks he or she can create the truth.

So if you know that Three committed the forbidden act, don't ask, deal with the act itself!

"Petey, you made big red circles on the wallpaper," Linda might have said. "That ruined the wallpaper and also Mommy's lipstick. Mommy is VERY angry. Mommy is going to be angry and not smile and not play or make jokes on the way to school. When we get home, we'll have to try to wash off the lipstick, instead of playing or watching a video."

Just as adults may misunderstand what Three says, so Three may misunderstand grownups.

Metaphors and figures of speech are beyond Three, who is still involved in matching up words and things. I once told a Three that his pink antibiotic was our "big gun" for his ear infection. NEVER do that. If you ask a Three to stand on one foot, she will; she will look at you as if you are insane,

and then carefully, precariously, plant one foot on top of the other.

Telling a Three to "calm down, now" won't work either. You have to be specific. "Please climb down off the table and give me back my pocketbook and sit on the floor with the crayons and paper and draw me a rainbow to put on the refrigerator." Telling him to "be gentle with the baby" may bring no results, but you might have luck if you ask him to "touch the baby's feet with your finger and look how tiny those toes are. Does each one have a toenail?"

Threes can comprehend "if, then" relationships to a degree. If you drop an egg, then it will break. If I open my mouth wide, the doctor won't have to use the tongue stick. But Linda uses "if, then" in a way Petey can't understand. Linda thinks she is stating a rule: "If you write on the walls with lipstick, then you lose a privilege." But to Petey, the IF part is not an IF at all. He has already written on the walls. Moreover, he does not regard cupcake and video as privileges but as part of his daily life and pleasures. Even if he grasped these concepts, the discipline isn't very effective, because there's no "reality" connection between writing on the walls, a cupcake, and a video.

So Threes and adults really speak two different languages, and it's the adults who have to be "bilingual," taking into account what Three can understand and what he or she is likely to get confused and upset about.

As a reward for all this confusion, though, Three has insights that are unconventional and sometimes incisive. Jenna goes around asking people, "Are you naked under your clothes?" Henry, on hearing an angry political speech by a candidate, says, "That's a bad man. He doesn't like children." Our Sara, at three, was cautioned about falling from heights and was impressed by how much easier it was to hike down the mountain than up. Why? she asked. We told her about the force of gravity, and she became very solemn. A few days later, she told us about Grabbity, the big white lobster who could eat little girls who fell down.

## POWER THROUGH PREDICTABILITY

When you're a Three bent on power, but speaking and thinking in a language foreign to adults, your best friend is a predictable universe.

Many years ago, they say, most Threes who lived in middle-class families stayed home. No preschool, gym glasses, formal play groups. At most, there was a bit of formal radio instruction (*Charming Children* in the 1940s, if you can believe it) or TV for toddlers (anybody remember *Romper Room*?).

Back then, Three had breakfast with a stay-at-home Mommy, played with neighborhood friends, had lunch, a nap, more play, then time with Daddy, and bed.

Did those days ever exist? If they did, they don't now. Threes whose mothers stay home are in a minority. And even these Threes, if they stay home with Mommy, may have no one to play with, if many potential friends are at preschool or daycare.

So Three doesn't have an automatically predictable, controllable world. It has to be created for Three.

A world that reasssures Three means that the important adults in Three's world stay constant, and aren't emotionally labile. It means that playmates are not a different bunch every time, or a mob of forty in a playground, but the same old friends day after day, playing with the same old things in the same old surroundings most of the time. It means that Three doesn't go bopping from one parent's household to the other's every few days.

The time frame of Three is about three days: Yesterday, today, and tomorrow. If the schedule is the same for two out of three days, at least, life seems to go more easily. One of the problems Petey is having is that his schedule, which seems so regular and reassuring to adults, is chaos to him. Never is Today the same as either Yesterday or Tomorrow. When he goes to Preschool on Monday, he has a dim memory of Preschool on Friday, enough to make him grumble, rather than scream, at the idea of attending. But by Friday, Preschool is as far away in his memory as would be an adult's weekend in the Berkshires three years ago. Play group once a week is even more difficult.

To top things off, most of the other children in preschool attend on a daily basis. They are pros. They know that Circle Time is followed by the Sing, and then Playground. Petey gropes from one activity to the next, and doesn't know the words to "Itsy-Bitsy Spider." Moreover, even Threes make friendships, have social pecking orders, and fads and fashions in play. Petey may have a wonderful friend in Georgie on Monday. They play for "hours" in the sandbox.

But by Friday, five days later, Georgie has bonded with Meg, and Petey doesn't understand the fantasy they are playing that involves the hole in the dirt, three trucks, and all those blocks.

And if Miss Kate gets sick and has a substitute on a Friday or a Monday, Petey will asssume that Miss Kate has gone forever, and his universe has crumbled.

All this anxiety and unpredictability makes Petey's days at home with Linda less fun. He isn't sure what he is "supposed" to do, what comes next, even though she tells him. No wonder he insists on watching the same video (Petey prefers *E.T.*) over and over again when he's home, and no wonder Linda, against her better judgment, lets him. (See Window of Opportunity: *Television, Videos, and Movies,* this chapter.)

So most Threes do better with a schedule that replicates the same activities at least three, and preferably five, days a week.

Another way to make the world predictable is to put everything into a category. And keep it there.

Threes sort everything: laundry, toy vehicles, colored ponies, fruit cocktail.

A Three told to clean up her room becomes immobilized, but when invited to impose categories on the mess—"Let's put away all the toys with wheels, then all the red toys"—may well enter into the project with pleasure.

Some categories are more important than others, especially those having to do with personal identity, danger, and authority. Never tease a Three about such holy categories. I know a misguided pediatrician who

tried to tease a Three, "You're not a little girl, you're a little boy!" The parents of that Three changed doctors and took the whole neighborhood with them.

I know other Threes who become frightened and angry when an adult, parent, or teacher appears to lose control and act "like a baby," either on purpose (throwing sand, like a Three, and squealing) or by accident (throwing up).

This is another reason why Three is so interested in sex and gender, so aware of personal differences, and so untactful in announcing discoveries about these categories. It is also why discussing good touch/bad touch and good strangers/bad strangers with a Three has to be done very delicately (see Window of Opportunity, this chapter).

## Separation Skills

Since the age three brings the first experience with public identity, and an enormous will to control private and public worlds, it isn't uncommon for change and separation to be overwhelming.

Every bit of energy spent anticipating big changes is well spent for Three. If he is starting or changing preschool, going to the dentist or doctor, starting swimming lessons, about to acquire a sibling, flying to visit Grandpa, even going to the circus, talk about it, show pictures, act it out ahead of time. Three's sense of control, sense of himself, and ability to separate will all be enhanced. (As mentioned in the last chapter, though, I would not go over ahead of time the possibility of a shot or blood test at the doctor's. See The Three-Year-Old Well-Child Visit, previous chapter.)

Indeed, helping Three have a sense of control over situations in which he is really quite helpless is one of the rewarding aspects of being parent to a Three. If Three is going to the doctor, bring along Three's doctor kit or a picture Three has drawn to show the doctor. If Three is starting a new preschool, visit ahead of time and take pictures of the classroom for Three to look at again and again.

Many Threes still have a security object. By this time, it has usually been reduced to a shadow of its former self. Miranda's "Blanky," having evolved to "Stringy," is now thought of by her mother and father as "The Shred."

Most Threes do well with having the use of their security object a bit restricted.

This is a good idea for many children for several reasons. Three can postpone satisfying her needs, and postponing Blanky is good practice. Moreover, many Threes suck their thumbs when they comfort themselves with Blanky, and constant thumb-sucking can distort teeth and jaw.

A good way to help a Three is to restrict the security object to Three's bed. Say, "I know that when you need to suck your thumb and use Blanky, you are very tired. So when I see you need to do that, I want you to go and lie down in your bed. When you are rested, you can get up and play/watch TV/help me cook dinner."

## Bedtime Monsters

Three tends to find monsters under the bed and in the closet.

Monsters for Three seem to be handy ways of disowning scary and forbidden feelings. If you are feeling frightened of separation, angry at a beloved person, conscious of having behaved in a way your parents disapprove of, or worried about having too much or too little control over events, those feelings may be too overwhelming to put into words, even to yourself. Far better to turn them into Something Awful Under the Bed.

Some popular monster-makers include:

- Recent or projected major changes and challenges: starting preschool, spending the night away from home, mother returning to work, separation or divorce, moving, hospitalization, or, the Tyrannosaurus of under-the-bed monsters, a New Sibling (see *What If ?*, below).
- Feelings that Three is not liked, enjoyed, thought wonderful and special (see *Limits*, this chapter).
- Loss of control in important areas. Diarrhea that produces potty accidents, a sickness that reduces ability to perform (and this can be an ear infection that affects hearing and balance), or a care provider who discourages or forbids autonomy are effective Monster-Makers.
- Exposure to scary television, whether program, ad, or news coverage (see *Window of Opportunity*, this chapter).

A monster-control program first addresses any fixable underlying problem. It also makes sure that Three is getting satisfying time alone with each parent every day. Time with both parents together, with a sibling, or while the parent is preoccupied with worry or interrupted by phone calls, doesn't count.

Of course, there will always be some monsters that can't be eliminated by changing Three's lifestyle. These are monsters that, like vitamins, are necessary to growth. Every Three will have some scary feelings and fantasies based firmly in reality and the nature of growing up. Rivalry with a sibling, possessive feelings about one or the other parent, anger at obstructions to power and autonomy—all these are important and inevitable parts of life.

Some Threes feel so magically powerful that they think their thoughts, angry or otherwise, can make things happen. A Three who loves his mother possessively and wishes his father would vanish can become terrified when his dad goes to an evening meeting or if he hears his parents quarreling. Parents who display affection and regard for each other openly can help allay these fears. They can also help by giving Three a clear idea that they, not Three, are firmly in control of the adult aspects of life.

For these normal, developmental monsters, parents also can help by providing:

- Security objects, or "a piece of Mommy or Daddy" to watch over Three in the night. I know one Three who formed an attachment to her mother's silk slip;

another who preferred to take her soccer ball to bed with her.
- A consistent good-night ritual. The ritual search of closet and bed doesn't "prove" that no monsters are lurking, but it does insure a few minutes of close parental attention. A cuddle and hug, a quiet book (like *Goodnight Moon* by Margaret Wise Brown) or a story about "when Three was a tiny baby" can be very reassuring.

The main thing to avoid is a vicious cycle, in which parents become so distraught about Three's fear of monsters that their distress creates more monsters in Three's mind.

## Limits

Setting limits for Threes needs to be a bit more sophisticated than setting limits for Twos. Not much, but a bit.

Threes don't usually indulge much in tantrums or aggression, such as hitting, biting, kicking. If Three is still producing this kind of behavior on a daily basis, check to see if Three is engaging in these behaviors at Daycare or Preschool. If so, Three needs an evaluation by the pediatrician, to sort out what might be going on. If not, then readjusting parenting techniques is likely to help.

If Three is *not* attending a regular daily setting, this may in fact be a large part of the problem. Threes need a structured, cheerful daily routine, full of age-appropriate challenges, to feel good about themselves A Three whose days are spent with much younger toddlers is pretty much guaranteed to behave obnoxiously.

Two's oppositional behavior was likely to be pretty simple: aggression, tantrums, refusal. The obnoxious behavior of Threes, however, is more varied and sophisticated. It comes in a variety of flavors: whining, interrupting, procrastinating, making irritating noises, raising havoc in the backseat of the car, mouthing off, digging in their heels and refusing to budge, tearing about regardless of the rights of others, refusing to confess a wrongdoing.

## WHINING, INTERRUPTING, AND PROCRASTINATING

### Whining

*Whining* can drive a parent mad. And that's a problem. A vicious cycle lurks here:

- Three feels out of sorts or "neglected" and whines.
- So parents react with exasperation and spend more and more energy condemning Three's behavior;
- So Three feels more out of sorts and more neglected and steps up the whining.

Whenever you confront a vicious cycle, the most effective response is to intervene at each spin of the wheel. Here's what helps:

- Analyze Three's lifestyle to see how it can make him feel more competent, grown-up, and independent. If he's going to Preschool only two days a week, but he behaves beautifully there, is it possible to make it more frequent? If not, what playmates and activities could fill his day satisfyingly?
- No matter how annoyed and exasperated you are, give Three positive attention whenever he is *not* whining.
- Instead of showing exasperation, put in your earplugs, just the way it's suggested back in the eighteen months chapter. Don't say, "I can't hear you if you whine." Don't say, "Use your big-boy voice." Show, don't tell, that whining is a noncommunicating noise.

### Interrupting

*Interrupting* is also a great pastime of Three. Moreover, Three will be scrupulous in saying, "Excuse me!" each and every time. I know a Three who managed to interrupt a single one of her mother's sentences fourteen times, in the most polite manner imaginable.

On occasion, Three interrupts because she really *does* need your attention. If you are dealing with a phone call or some other uninterruptable occupation, a good way to deal with this is to make eye contact with Three and put your hand, gently but firmly, on her, massaging her head and shoulders, drawing her in for a hug until you can give her your full attention.

But when interrupting becomes a way of life, not a need but a manipulative habit, you need a different technique: a brief, intense tutorial in Not Interrupting. What Three needs to learn is that you are not going to give in and stop what you are doing to attend to his nagging.

- Tell Three right beforehand that you must make a call, and you don't want any interruptions. Then make a phone call that you have already, in secret, set up.
- Call your friend, who is expecting the call. Then let your friend help keep you from responding to Three's imprecations. Ignore the interrupting.
- Don't even recognize that Three is interrupting. Don't make eye contact, don't make a rebuking face, don't show any response. Just keep talking with your friend, until Three is worn out and goes on to something else. (If the thing he goes on to is something forbidden, such as a tantrum or destructive behavior, end the conversation and address that behavior immediately. But don't mention the interrupting problem. Let Three believe that you really couldn't "hear" his interruptions.)

You may need to do this three or four times.

*Procrastinating*
Ah, this is a favorite.

What can you do with Jacques, who won't put on his clothes in time to leave for preschool? With Melody, who can't tidy her room before visitors come?

Threes procrastinate for several reasons. Jacques could be having separation anxiety, delaying the time of parting from his mother, perhaps because the rush and annoyance of morning has made her seem inaccessible. He could be very involved in another, addicting activity (watching the morning cartoons is popular). Or he could find putting on his clothes difficult and no fun.

If one of these is the cause, the way to fix it often seems obvious. (Start the morning earlier with better planning and no cartoons. Make time for a pre-preschool cuddle. Make sure he has a sense of control in other parts of his life. Show him step-by-step ways of putting on his clothes; let him pick them out; avoid buttons and zippers and snaps.)

On the other hand, he could be using dressing as a control issue, to say "I'm in charge here; I'm big and powerful."

When that happens, you have very limited options.

- You can't physically force him to get dressed. He's big, wiggly, coordinated, and able to kick.
- Even if you did get him dressed, he could get everything off by the time you got him to preschool.
- Yelling at him, scolding, bribing, spanking: none of these gets him dressed. You are, indeed, teaching him that all of these behaviors of *yours* are ineffective. He, in fact, has won. When a Three wins a battle with his parents, his behavior just becomes worse.

You could, of course, punish him by keeping him home from preschool. But then what? A long, unpleasant, conflict-filled, resentful day lies ahead. Whereas if Jacques did go to preschool, he's likely to come home a tiny bit more mature and willing to cooperate. (If it's a good preschool.)

So. Don't *order* him to get dressed. Tell him that it's time to be dressed for going to preschool, and that you will be leaving in five minutes.

If he doesn't respond, put his clothes into a bag. Tell him that he will get dressed at preschool, since he didn't get dressed at home. Period. Don't draw a moral or make negative comments. Stay very quiet indeed.

If the weather demands it, get him into his outerwear (coat, etc.) over his pajamas or nakedness. Or just heat up the car. When he gets to school, he will have to spend the first few minutes dressing rather than playing. The other children will be curious and may tease. That's OK.

He is likely to figure out that getting dressed on time is more fun. But at any rate the mornings will be easier on all concerned.

*Making irritating noises, mouthing off, raising havoc in the backseat of the car, digging in their heels and refusing to budge, tearing about regardless of the rights of others.*

These are all forms of oppositional behavior. Oppositional behavior means that the child is behaving obnoxiously *on purpose*, in order to provoke a reaction from her parents. Very often, the child who is most oppositional at home behaves just fine elsewhere, and may even be well-behaved with the other parent.

Usually it starts when Three produces an obnoxious behavior, just because Three feels like it, and parents react ineffectively. You tell Three to stop: Three refuses to obey. You escalate the demand by yelling or

threatening; Three digs in his or her heels and does it again. This nastiness continues until one or both parties explode or until one gives in to the other—often the parent being the one who knuckles under.

Once such conflicts begin, they tend to spiral out of control, becoming more and more frequent and upsetting. When this happens, the whole family suffers. This is addressed in the essay in Part III, on *Oppositional Behavior.*

## TANTRUMS AND AGGRESSIVE BEHAVIOR

If Three does start having frequent tantrums or behaving aggressively, it's important to decide whether the behavior is truly beyond Three's control because of factors that render him at the mercy of his emotions; or whether they are contagious—caught from another child or children; or whether they are manipulative.

### Uncontrollable tantrums

Tantrums that are beyond a Three's control are often a sign of discombobulation in some important area of Three's life and the behaviors won't get better by usual methods until that something is fixed. At home, marital discord, financial problems, or ill health of a parent may make a Three feel totally helpless and panicked. Or perhaps there is something wrong at preschool, with too rigid or too lax discipline, a teacher who doesn't like the Three, a bully, or scary physical surroundings (a dirty toilet area, for instance).

Health problems that can make Three grumpy, frazzled, and short-fused to the point of frequent tantrums include hearing loss, sinusitis and low-grade ear infections, allergies, and constipation.

Finally, a combination of factors—innocent in themselves—that make a Three feel powerless can trigger uncontrollable tantrums. Petey is feeling desperate because his life seems intolerably unpredictable. His daily schedule is disjointed. He didn't realize writing on the walls was forbidden: after all, the wallpaper has pictures on it. He spends all day doing his best at preschool, following rules and procedures, and then his mother refuses him cupcake and video. It is the video deprivation that does him in. Videos are highly predictable, highly emotionally rewarding, and watching *E.T.* on a daily basis helps Petey feel, vicariously, in control.

The only cure for uncontrollable tantrums is finding and fixing the cause.

### Contagious tantrums and aggression

A Three who is in near-constant contact with younger children, especially from eighteen months to two-and-a-half, can almost be counted on to have tantrums, bite, kick, and in general drop all semblance of maturity. If you really tune in to such a Three, you will detect a false note about it all, a kind of glint in Three's eyes that shows that Three knows quite well that she is behaving "like a baby." The pull on Three to grow up is strong, but so is the pull to regress. Moreover, Threes have an enormous drive to imitate and emulate behavior.

This scenario can happen when Three is home all day with a younger sibling, or when her daycare group has shifted so that she is the oldest, or when her preschool either has no more room in the three-year-old group and leaves her in the younger one until space appears or when the preschool, seeing her behavior, judges her not yet ready for the older group.

In every one of these situations, the only adult action that really works, and it works promptly, is to move Three into an environment in which she is one of the younger children. For this "treatment" to be most effective, Three needs to be in this group regularly, daily if possible.

Usually nothing else is needed. And usually, nothing else works.

### Manipulative tantrums

These are tantrums "designed" to get attention, or to get one's way, or to make life unpleasant for others. If a Three (who is developing and talking normally) is still having manipulative tantrums, I urge you to take action. When manipulative tantrums persist, there is something out of kilter in Three's life.

- Three's days are not filled with challenging age-appropriate activities with Three's age-mates. This gives Three little way in which to feel competent and independent. Instead, Three devotes all his or her energy and skills to winning a power-struggle with Mommy or Daddy.
- Three feels "invisible" when behaving well at home. Perhaps Three's adults are

preoccupied with work at home or with other siblings; perhaps it has not seemed necessary to them to design a challenging, satisfying environment for Three; perhaps there is family stress.

- Three is getting rewarded for having tantrums. Perhaps a parent is so stressed that giving in is the only option. Perhaps a parent is deeply angry at the world, and Three's tantrums give secret satisfaction—a kind of outlet. Perhaps a parent hasn't thought through the answer to the question, "What am I teaching Three when I respond to a tantrum by giving in?"
- Perhaps parents have not found effective ways of dealing with the tantrums. They may have been advised that the best treatment is to hug Three until the tantrum is over, for instance. Such responses increase the frequency and intensity of tantrums.

If any of these fit the bill, the solution is to fix the underlying problem. If none of these seem to be likely, or the tantrums don't improve within two weeks of fixing the problem, it's time to ask the pediatrician for a consultation or for a referral to an expert in children's behavior.

## GIVING PRAISE AND ATTENTION

Sometimes praise backfires with Threes. A Three may feel that some kinds of praise encroach on her independence. She may even resent categories such as "Good girl" or "Fine big sister" if she suspects they are being used to manipulate her. Moreover, some kinds of praise may distract her from what she is doing, and give her less joy in her work.

Another kind of praise that can backfire is telling a child repeatedly that he or she is "wonderful," or using other superlatives about the child's existence. I suspect that Three's reaction to this (usually a confused, slightly haunted smile) may reflect Three's suspicion that a parent who has to keep saying these things is really suggesting the opposite. Most Threes and older respond better when their wonderfulness is assumed and reflected in attentive, respectful conversation and physical affection, not stated explicitly.

Indeed, the kind of praise that almost never backfires is the unstated praise that comes from a parent's focused interest in what Three is doing. Unlike conventional praise, this kind of attention focuses on the "work" itself rather than on the behavior of the child. While conventional praise may elicit all kinds of unfortunate responses, focused attention increases Three's interest and pride.

*Not recommended:*

Conventional praise: "That's a beautiful picture you drew."

Likely Response: "It is not. I made the rainbow crooked. I hate it."

*Recommended:*

Focused attention: "You blended the yellow and red there so you have an orange stripe in that rainbow."

Likely Response: "And there is a yuck color where I put the green on the brown."

Very often a Three will chatter on about her accomplishments and interests. Her train of thought is sometimes hard to follow, and what she wants to tell us is heavily in doubt. The adult impulse is to try to figure out what she is saying, and then to respond with Judgment, either to praise or to blame. Unfortunately, this bores and annoys the child and bores and exhausts the parent. Both people get the idea that the adult isn't really listening.

What can work better is simply reflecting back what the child has just said.

"And we had the big blocks out and Gary made a tunnel and we put all the cars in and Jessica took the chair and Miss Sarita said we had too many cars and Gary got the truck and we went outside."

*Not recommended:*

Judgmental comment: "Yes, it's important to share the trucks and cars."

Likely response: A blank stare, or more incomprehensible narration.

*Recommended:*

Reflective comment: "So you left the tunnel and took the truck outside."

Likely response: "And then Miss Sarita said that was good and we all had milk."

Sometimes the Three who has a reputation for endless chatter is really going on and on in search of appropriate parental response. Using reflective responses rather than judgmental ones may help give her a more satisfied feeling about communicating.

# Day to Day

In this chapter, there is a section on Toileting after Activities. The actual learning of potty skills, however, is discussed in the previous chapter, *Two Years to Three Years* under *Window of Opportunity*.

## MILESTONES

### Vision

Focus should be perfect, and when tested vision should be at least 20/30 in both eyes. Three shouldn't squint, even with one eye, or tear in bright sunlight; these may be signs of, among other things, nearsightedness. Head tilt or closing one eye to attend to something visually may be signs that focusing, as well as vision, needs to be checked. Three should be able to see pictures in a book when it is held at arm's length and should watch TV from four feet away. If there is any suspicion of less-than-perfect vision, or a family history with a parent or sibling needing glasses or patching before kindergarten, or having a "lazy eye" at any time, a formal vision test is needed.

### Hearing

Some Threes can cooperate with a screening hearing test, in which they raise their hands when they hear a tone, and this is fine for children with no suggestion of hearing problems. A Three with signs of hearing impairment, however, needs a more formal test. This is usually done by a children's audiologist, and makes use of a sound-proof room and play responses to different pitches; it's called "play audiometry."

Signs of hearing loss include:

- Delayed speech and impoverished vocabulary.
- Unclear speech. A stranger should understand Three nearly all the time. If in doubt, ask your pediatrician to listen to a tape you make at home with Three talking conversationally. (Three often won't say a word in the pediatric office.)
- Odd voice quality: nasal, hoarse, or unpleasant.
- A very loud voice.
- Inattentiveness to voices and music and sounds.
- A family history of hearing loss.
- A complicated medical history: Recurrent otitis media (stubborn or complicated) is the most common, but some children who had other medical problems or who have required special procedures may be at risk. Such problems include bacterial meningitis, a severe case of true measles, and ear trauma. Complicated births can also cause subtle hearing problems. Premature babies, especially those with complicated neurological courses, babies who were given high-frequency ventilation as a newborn, or those who underwent more than a very brief course of antibiotics called *aminoglycosides* (such as gentamycin or kanamycin) may also need a hearing test. So may a child diagnosed as having CMV (*cytomegalovirus*) as a fetus or newborn.

### Speech

Three should speak clearly enough to be understood by strangers and should be able to express anything he or she needs or wants to say. Threes routinely speak in sentences that are at least three or four words in length. They have a vocabulary of about five hundred words. They should be using pronouns such as *I* and *me* correctly.

Some lisping and mispronunciation is normal: "My lello thockth," a Three says, getting dressed. "I weally like my lello thockth." Grammar can be strange. "I bringed it." Narration is highly idiosyncratic: Three tells you what he remembers in the way he remembers it, not in the way that makes for a coherent plot. "The big dog came up and he bited the ball and Wendy had the ice cream and then we all played on the swings."

Most Threes go through another "dysfluency" phase, as did many Twos. "I-I-I-I-I want to get that g-g-g-g-guy." They may say "uh" or "um" as fillers in sentences. Perhaps it's because so much language learning has produced temporary gridlock. Sometimes the stuttering is repetition of a phrase, rather than a sound: "Iwant Iwant Iwant Iwant Iwant to get that guy."

Adults can help Three by waiting patiently, with no show of anxiety or concern, for Three to get the sentence out. Asking Three to slow down, or to think about what she wants to say, or telling Three not to stutter are guaranteed to make the dysfluency worse. Adults can help most by making the environment easy for Three to talk, slowing their own speech, reducing stress, paying

## NORMAL STUTTERING (BENIGN DYSFLUENCY)

- The child doesn't become anxious, self-conscious, or show any tension about it. You shouldn't see her tensing up, making a face, staring to one side, or holding her hand to her mouth.
- Repetitions occur at the beginning of words, not the middle.
- There aren't any "prolongations," in which a child gets stuck on a sound and makes it go on and on: "Whaaaaaat are you doooooooooiiiing?"
- It only happens occasionally, not more than one sentence in ten, and tends to come and go. It happens especially when the child is excited or tired or learning something new.
- The stuttering doesn't last longer than three or four weeks without improving.

friendly attention to what Three has to say and not asking many questions.

Signs that Three needs help with fluency problems, then, would be:

- Looking tense, blinking or having tense lips, staring to the side as she tries to get the word out.
- Making long (up to a whole second), stutters, and doing it frequently: more than about one sentence in ten.
- Stuttering most of the time, rather than when having an emotional reason for doing so.
- Looking embarrassed or saying that she is worried about the stuttering or asking an adult why she stutters.
- Continuing to have a problem for as long as six to eight weeks.

*Cognitive*

Over this year, Three learns about categories and explores them. He is fascinated with same and different. He doesn't reject picture books, but becomes ever more interested in stories with plots and suspense. He is very vulnerable to that which he hears and sees, and will act out again and again anything he finds upsetting.

*Large motor*

Three can stand on one foot for several seconds and can jump in place. She can walk up and down stairs on her own. Over this year, she'll learn to pedal a tricycle and catch a ball, most of the time, if you aim it at her fairly gently.

*Fine motor*

Over this year, Three perfects the adult grasp on a crayon or pencil, and can draw a person that usually consists of a head with legs and arms coming out of it. Three can use a spoon and fork with adult grasp, and over this year will start to be able to use chopsticks.

## SLEEP

Most Threes sleep twelve hours out of the twenty-four, with, yay, hurray, still one daytime nap of about an hour. Sometime over this year, many Threes will drop that nap: some letting it drift off gently like a balloon; some giving it up with ambivalence and grumpiness. The nap should not, however, be torn from a Three who still needs it.

A daycare provider who insists on more than a one-hour nap for Threes, and who insists on a nap for each and every Three, is likely to be using the nap for her purpose and not for Three's well-being.

## GROWTH

### Chubbiness watch

Over this year, Three will gain about four and a half pounds, less than half a pound a month, and grow on the average two and a half to three inches in height.

From Three to Four, Cherub should start to look longer and leaner. The tummy gradually becomes a pleasant slope rather than a cute little bulge; the arms and legs are creaseless. If Cherub gains more than one-half pound a month during this year, she is probably gaining more than she should, unless she is very tall indeed.

When a Three starts to gain too much, something in her environment and lifestyle needs fixing. Maybe she's watching TV or videos every day. Maybe she prefers quiet, imaginative games to running around. Maybe her daycare enforces a daily nap and offers a couple of routine snacks.

It's important to look into it promptly, for two reasons. First, extra chubbiness in itself seems to promote even *more* chubbiness. A Three who is just a bit chubby by mid-year can grow into a Four who is seriously "off the chart." Second, the next two years are the last ones in which parents have so much control over children's activities, exercise, and diet.

So if you have concerns in this area, please

see the Essay in Section III: *Chubby or Not, Here We Come.*

Indeed, it's crucial to keep this in mind with every meal, every snack, and every beverage. Remember, you are struggling against the mighty forces of fast food, soda, juice, and snack producers, and their media accomplices!

The second crucial thing to keep in mind is serving size. When you serve flat food, such as a piece of meat or poultry, the serving size is the size of your child's palm. When you serve a bulky food, the serving size is the size of your child's fist. Offer three or four different foods at each meal, in these serving sizes. If you keep the nutritional guidelines, the serving sizes, and the milk requirements in mind, your child will most likely be very well nourished.

Threes still are likely to eat only one "solid" big meal a day, with the other meals being smaller. Usually this is breakfast. If a Three eats her big meal elsewhere than home, some parents may never see their child eat a "normal" meal. Most Threes still eat practically no dinner, by adult standards.

A child this age requires on a daily basis about forty calories for every pound he or she weighs. The average girl weighs about 30 pounds, the average boy 32; but the range of normal weights is large. A tiny Three may weigh only 25 pounds, and a very tall one nearly 38. So the average Three needs about 1200 calories a day, but needs can vary from 900 to over 1600. Of those calories, only about 30% should come from fat.

Like Two, Three should drink low fat or even nonfat milk. The main difference between Two and Three is that Three's portions of food are slightly larger, and Three can start cautiously to eat stickier, chewier, crunchier foods safely. The operative word is Cautiously.

Eating social meals with Three can be a bit tricky. Three is still restless sitting for long periods, and Three doesn't eat much, but Three is clearly ready to be included in regular family dinners. Serving tiny portions, in courses, and making sure the conversation includes and can be understood by Three (and that it isn't about what Three is or is not eating) are both helpful.

When Three goes to a restaurant, preparation should focus on the rituals of eating out, not on the food. In fact, it's a good idea to make sure Three understands that the food won't come right away, and to take along some quiet entertainments and quiet, crumbless snacks to while away the time.

## TEETH

Three won't lose or get any more new teeth until age six, unless he injures them, a not terribly unlikely event, or unless tooth decay warrants major intervention. This is a good time to read the section on *Mouths and Teeth* in *Body Parts and First Aid* in Part II. Most Threes can attempt to brush their teeth on their own, but benefit from parental help on the molars.

As mentioned in the last chapter, the years from two to four are when the permanent front teeth are developing their enamel. During this time it behooves parents to be specially careful that Three gets enough calcium (see *Nutrition and Eating* in the chapter *Two Years to Three Years* and also in the present chapter) and just the right amount of fluoride. Too little fluoride may allow the development of cavities. Too much fluoride can produce white flecks that show up when they develop on these very obvious teeth: so be careful Three isn't getting too much fluoride. If you are using a supplement, the water Three drinks should not be fluoridated. And either way, don't let Three ingest fluoridated toothpaste. Restrict the amount on the toothbrush to a tiny streak.

The dose of supplemental fluoride changes in some areas at age Three; ask your pediatrician or pediatric dentist. Speaking of dentists, Three should be seeing one twice a year.

## NUTRITION AND EATING

### A Good Eater

- Eats when she is hungry, and stops when she is full. Thus she eats erratically meal to meal and day to day.
- Is not allowed to demand special meals, but is given a choice of healthy foods— she decides how much she does or doesn't eat of each.
- Eats a variety of foods because she doesn't regard some foods as "punishment" and some as "rewards."
- Is not habituated to juices, sodas, or "handout" snacks.
- Drinks milk for nutrition and water for thirst.

### Foods

I have not included the Food Pyramid in these sections. I have never known a parent to take kindly to the thing. Its main purpose is to remind us that children need very little in the way of sweet, sugary, greasy foods; that they need the basic amounts of grains (cereals, starches, and breads); that they need more fresh vegetables and fresh fruits than Americans are used to eating.

## BATHING AND HYGIENE

The main problem with bubbles in the bath water is vaginitis. ("That's why they call it Mister Bubble," a Six told me confidently, "not Missus Bubble.") However, boys can show an allergic reaction to bubbles, with puffy swelling of the penis. Uncircumcised boys can get into difficulty with soap being trapped under the foreskin.

An occasional bubble bath is a necessity of life, but making it a nightly ritual is something else again. Bathtub toys are a good distraction.

Threes should not be left alone in the tub. Hot-water burns and submersion accidents are still too strong a possibility, as is drinking the delicious-smelling shampoo.

## ACTIVITIES

### Preschool

Does a Three have to go to preschool, or a daycare with preschool program, to develop optimally?

No. But it is a challenge for parents to

give a Three all the factors present in an excellent early child education program. This is a program that:

- Reinforces skills in getting and holding the attention of adults in socially acceptable ways. This means that the class size is small enough and teachers well trained.
- Encourages children in cooperative play, making sure that each develops abilities to lead and to follow, and that aggressive behavior is managed appropriately.
- Stimulates curiosity and imagination by having a variety of materials and adventures available.
- Develops joy in active dialogue and play. TV has no place, unless teachers have a special, well-defined reason for using a specific program for enrichment.
- Recognizes differences among children and helps children value such differences.
- Employs teachers who understand the subtleties of this age, with expectations that are neither too low (putting up with tantrums and aggression) nor too high (expecting consistent sharing, absence of potty accidents, and "truth-telling").

*Appropriate activities for threes*
- Crafts. Threes, unlike younger children, can go beyond exploring the properties of materials to actually making constructions.
- Fantasy. Dressing up, role-playing, fantasy playing with dolls, cars and trucks, and cooking utensils, in play environ-

ments for dolls such as circus, farm, garage, etc.
- Group activities. Threes can sit for fifteen minutes or so in a circle, singing, sharing, and listening to a story read interactively.
- Playground. Besides swings, slides, and climbing structures, a Three playground can include tunnels, a large hammock, and sandpile with a water source. The ground should be soft, not cement.

Threes generally should come home pretty dirty. A pristine Three is probably not in the right preschool program.

Any preschool or daycare should fulfill the basic requirements established by the National Association for the Education of Young Children.

Threes at home can have just as rich an experience, but it's harder in some ways. One of the challenges for many parents of home care Threes is to find other children for them to play with in neighborhoods where everybody's in preschool or daycare.

*Swimming*
When Threes begin swimming lessons, they benefit from:

- Being prepared ahead of time. Going to watch a class before you even put on a bathing suit can inspire confidence and yearning IF it's a good class.
- Confident and relaxed parents. A parent who is so eager to have Three water-comfortable that he is far more excited than the child can expect problems. A

parent who recounts his own fear of the water and the ways he overcame them, before the child even participates, may plant seeds of worry. A parent who introduces the pool as a nice thing to do, which one can take on a step at the time, has the best chance.

- A friend, an established buddy, someone with whom he already has a give-and-take relationship to learn with. But only if competition and comparison is not encouraged. Holding up for emulation the ability of Ollie to jump from the low board is not going to cut much ice with Miranda, who is still clinging to her instructor and won't put her head under the water.
- An instructor who makes friends with the child poolside before going in the water, and who is attuned to the child's temperament and style of learning.
- A clean, warm pool without a lot of splashing going on.

**A Three who can swim is not water-safe; he is water-comfortable. He is more likely to jump into a body of water than a Three who is afraid. He is not likely to wait for an adult to check out dangers such as rocks and depth, and just as likely as a non-swimmer to panic when unexpectedly dunked. Drowning of children one to four years of age is strongly correlated with the presence of backyard swimming pools.**

### Prereading

Threes learn three things about literature this year. First, they get the idea of a story, with characters, and a beginning, middle, and end, often with a lesson or a moral. (Yes, even today.) Second, they get the idea that printed words have a relationship to spoken words. Third, they get the idea that the act of reading is a valued skill, a grown-up ability.

If a Three can learn all these things with pleasure, she or he is well on the way to wanting to learn to read, and barring specific disabilities or teachers that get in the way, reading is likely to come naturally.

Reading out loud now goes both ways: Three will want to do some of the reading. If parents recognize that this "reading" is really consolidating all the age-important insights above, all goes well. Three may make up a story that has nothing much to do with the story in the book, and the lesson drawn at the end may be highly idiosyncratic. The parent who wants to encourage reading will play right along, asking interested questions and admiring the story and how well Three "puts so much expression in his or her voice."

Three may also discover that she or he can actually recognize certain words: Exit, names of some cars, advertising labels. This isn't actually reading; it's symbol recognition. The parent who wants to link this with reading can give Three additional information. "You're exactly right, that's the Exit sign, and the sign right above says Emergency. That means that we can only go out that door if there is an emergency. For instance, if the electricity went off and the store wanted all the people to please leave."

These are quite different responses from the parent who wants to *teach reading*. In this case, a parent may correct the Three's story, or point out that the picture on this page doesn't have any relation to the picture on that page. Or a parent may stop Three to point at and sound out an easy word, such as *cat*. Or the parent may excitedly tell Three, who has found the exit, that he can read; and then Three may be very disillusioned when it is obvious that he can't, not really. The problem with this approach is that it may make reading seem such a chore, or such a magical and unattainable skill, that Three doesn't want to do it anymore.

*True early reading*

Once in a very great while, a Three really does learn to read.

The main short-term advantage for young children is that if one is already reading by the start of school, one doesn't have to learn reading in kindergarten. This may make life easier for some children who don't learn well in groups, who are very shy and don't like to read out loud, who are easily bored, or who have the misfortune to encounter a poor kindergarten environment.

The main disadvantage to early reading is that a Three who reads is regarded as something of an intellectual oddity. Adult comments may persuade a reading Three that he has achieved the pinnacle of learning and nothing more will be required of him; that reading and godliness are identical and that he is virtuous because he reads; or that reading skills, power, and maturity go together. Another disadvantage is the tendency to drop reading out loud when a child can read on his own.

It won't do to actively prevent a Three from learning to read, but it is helpful to prevent unwanted side effects.

If early reading seems like a good idea, the best way to promote it is to respond to a Three's questions about letters and sounds and how they are related, and to read aloud often.

But never push. If pushed, Three is likely to dig in her heels and refuse to have anything to do with the printed word. A child who develops an adversarial relationship to books early on is a child in academic trouble. Worse, she may develop the same attitude towards academic learning in general.

## TOILETING

Some Threes, mostly boys, may just be completing toilet skills now. Many Threes will be dry at night, but a significant number won't be; 15 percent of children are still wetting at night by age five.

A good number of Threes manage the pee just fine but insist on pooping into a diaper, in private. For these reluctant-to-poop-in-the-toilet Threes, I suggest reviewing the *Potty Skills* section of the last chapter to diagnose the reason for the behavior, and the corresponding appropriate teaching response.

It is a rare Three who doesn't have an occasional wetting or pooping accident during the day, and preschools should be supplied with an extra clothing change for each child just in case. Such accidents should never be punished.

There are two styles of toileting that can get a child into trouble medically, though.

*The wait-until-the-last-minute pee-postponer*

This is a Three who holds her genitals, sits on her foot, dances around, and otherwise delays getting to the potty. If this becomes a real engrained behavior, she can stretch her bladder to hold amazing amounts of urine. The bladder, thus stretched, produces a weakened signal to urinate, and its muscle is more lax, thus making it less likely that all the urine is voided at a given time. This produces a vicious cycle, with even more urine being held in an increasingly stretched and less responsive bladder. This can contribute to bed wetting and, alas, to urinary tract infections.

The trick is to get the Three to empty her bladder promptly when she first feels the impulse, but to make her think that it is her idea to do so. One way to do this is to make use of Three's magical thinking. Tell her that the peepee wants, yearns, needs to go into the potty, and that it is her job to put it there. When Three does her job, she can stick a star on the special daily calendar hung in the bathroom. (This seems to work better than if a parent gives Three a star for being a big girl or a good girl.) Another trick is to have a timer to remind Three that this is when the peepee wants to come out. If the reminder is couched in these terms, the parent isn't in the position of a power struggle with Three. A timer set for two hours is probably adequate.

*The constipated nonpooper*

The Three who thinks it will hurt or be difficult to have a stool, or who is afraid of any aspect of the potty or toilet, or who doesn't want to take the time to go, will hold back stool. You may see him grunting away, but all that effort may be directed to holding back rather than to pushing out. If he succeeds in holding back, the stool retained will become bigger and firmer, and Three's reluctance to go will become bigger and firmer as well.

This can become an appalling problem later in childhood. Some children develop encopresis, in which they habitually have bowel movements (either the real thing or significant staining) in their underwear.

As the muscle of the rectum stretches to accommodate the held-back stool, it weakens. It can't signal to the brain that a stool is ready to be ejected. So then the child "just goes." If the backed-up stool is bulky enough, looser stool may leak out around it, soiling the underpants. If it is soft, out comes the whole poop, into the pants.

This is much more easily prevented than treated. A Three needs to have a diet that produces soft stools, easy to pass. He needs a regular time to sit on the toilet for at least five minutes. It should be a pleasant, unrushed time. Privacy or company, a tape to listen to or a book to read, are individual options. What is very important is that:

- The toilet area is clean and user friendly.
- Three feels secure about not falling in or about the poop making a big scary splash.
- Three has a place for his feet to rest so that his legs don't dangle. This gives him the needed oomph to push out the poop.

- Parents are relaxed and encouraging, without making bathroom use a power contest or a shameful action. Pooping or not pooping should not be the main way in which Three gets parental attention. The strategy above for pee is helpful here: "The poopoo wants to go into the toilet, and it is your job to put it there."

### Wiping

Three's arms aren't long enough for him to reach to wipe very easily or thoroughly. This isn't much of a problem for boys, but girls need to be watched so that the perineal area (the opening of vagina and urethra) doesn't get contaminated with germs from the poop and cause infection. She can be helped to learn to wipe in front first, drop the tissue in the toilet, get a new one, and then take care of the rectal area.

## SAFETY ISSUES

**If you haven't child-proofed the environment and updated the medicine chest as suggested in previous chapters, please do so now**.

Every Three will have a few accidents during the year.

Three doesn't get into danger the same way Two did out of impulse, fearlessness, and ignorance. Three is more likely to be tempted into danger by overconfidence and magical thinking. ("Have you flown recently?" I asked a father of a Three with otitis media, looking for risk factors. "Yes!" Three said excitedly. "Off the garage roof when I'm Batman.") Three lacks experience of the world, and can't think ahead to consequences very clearly yet.

When you watch over a Three, you have to cherish Three's curiosity and self-confidence, but you also need to predict the most likely cause of bad accidents. You need to preserve trust, in the world and in people, but protect against abuse and molestation.

It's tricky, and parents may find the task complicated by their own childhood fears and traumas. A frank discussion with the pediatrician can help to confront these and to see whether further counseling is in order.

### Drowning

It's the leading cause of accidental death at home in this age group. Beware the exploring Three who has had swimming lessons and is lured to water with confidence; beware the Three who hasn't but who thinks she is Ariel and wants to try out her tail; beware the Three who is pedaling along furiously on his trike and simply doesn't notice the pool is there. Beware the Three who naps with you at poolside and wakes up before you do and, politely, allows you to sleep while she tries out the Jacuzzi on her own.

### Choking

Threes are no longer mouthing objects indiscriminately, and their coordination of chewing and swallowing makes choking less likely. However, foreign-body choking is the second-leading cause of accidental death at home for children younger than six years. At this age, one of the common causes is food, round or spherical, given as a snack when the child is particularly likely to aspirate when she is excited or active or has a cough. The worst foods are nuts and popcorn, though celery and carrots are close runners-up.

### Mouth and throat injuries

Many Threes will sustain a tooth injury or a bitten tongue or other minor but bloody mouth accidents. It's a good idea to prepare for this by reading the section in First Aid in Part II. The most dangerous kind of accident occurs when the child is running with something, like a lollypop, in the mouth and falls. This kind of injury always needs pediatric or emergency room attention. A firm rule against such activity is much happier for everybody.

### Poisoning

Again, Threes are less likely than younger children to get into something and eat it. Seventy-five percent of poisonings occur in children younger than three. However, it's still a problem: a finicky Three who won't even touch kiwi fruit may decide that Grandma's cough medicine is just delicious. Sugar-coated iron pills can be a big temptation: they look like M&M's and taste sweet initially. Iron is a very dangerous overdose, and can be fatal.

### Burns

If curling irons were a problem at Two, they are even more so at Three, the age of gender imitation. Bathing alone can inspire turning on the hot-water faucet. Helping cook can produce burns with boiling water and hot oils.

### Being very short

Threes on a Big Wheel cannot be seen in a car's rear-view mirror. This tragic type of accident seems only to happen in the best of neighborhoods, where parents allow Threes out on their trikes. The only solution I know of is to make sure everybody is excessively careful about backing out their cars, perhaps having an older child make sure the way is clear. Threes are also at eye level for adults holding lit cigarettes and cigars.

### Being very curious

Escalators are the biggie here. No Three can resist an escalator. Make sure that Three has shoes on and tied, that you hold hands, and that you know where the button is that stops the escalator in an emergency.

### Toys

Threes like bows and arrows, swords, water pistols. These need to be safety-tested and their play supervised. Bear in mind that if you are so "cruel" as to deprive Three of such toys, he or she will make imaginary ones, which aren't nearly so dangerous, never break, don't cost anything, and don't have to be stored somewhere.

### Sexual abuse and kidnapping

These are every parent's nightmare, of course. Protecting a child as he or she grows older means informing the child of danger and teaching him or her appropriate responses. This is different from making the environment safe, and is discussed for children this age in the *Window of Opportunity* section of this chapter, as is sex education for Threes.

# Health and Illness

*Part II of this book covers illnesses in babies this age and older: how to assess symptoms, how to tell whether a child is sick enough to require urgent attention, and what kind of home treatment may be appropriate. It also examines problems by body part. The Glossary is an alphabetical listing of medical terms with pronunciations and, when appropriate, detailed descriptions.*

## EXPECTATIONS

Since Three focuses on mastery of the world, it's a good thing that hearing, vision, speech, and coordination are well advanced. Since Three is such a social being, it's a good thing that the immune system is now well developed to handle inevitable, often frequent, infections.[1]

Three has a good chance of contracting at least one of several specific viral illnesses, such as fifth disease, oral herpes, or herpangina. See *Common Illnesses with (More or Less) Unfamiliar Names*, in Part II.

As to accidents, Three is more likely to have serious ones because Three seems so grown up that adult supervision becomes a little lax. Three is more likely to have minor ones because of overconfidence and lack of experience: fortunately, Three's increased coordination and ability to understand concepts helps protect him or her.

The social life of Threes also makes them candidates for such contagious skin conditions as lice and scabies. (See *Skin* in Part II; also, the Glossary.) Three may well catch pinworms (which really are tiny little worms) and ringworm, which is not a worm at all but a skin fungus. All of these are treatable annoyances rather than serious problems. (See *Body Parts, Bodily Functions, and What Ails Them* in Part II.)

The recent acquisition of toileting skills makes Threes vulnerable to constipation (see under *Toileting*, above) and to vaginitis from wiping back to front.

Limping is not uncommon for Three. The Three with a limp is most likely to have had trauma (sprain, toddler's fracture, insect sting, foreign body stepped on) or the viral illness called *toxic synovitis*. But take a persisting limp seriously. Every Three who refuses to bear weight or who limps without apparent innocent cause, like a splinter or tight shoes, needs prompt pediatric attention, whether or not there has been a known accident, whether or not Three has fever or other illness.

Foreign bodies, in the nose, ear, and vagina, are also part of being Three. The problem may present with discharge, pain, or confession. But if an object stays in nose or vagina and festers silently, what you may first notice is a really awful, awful smell that doesn't respond to bathing. A foreign body in the ear may produce pain, but more often muffled hearing.

---

1. From the months of "October through April, a preschooler is likely to be starting, having, or getting over a cold about half the time" (*Report on Pediatric Infectious Diseases*, Oct. 1992 21111 [9]: 35.)

## COMMON MINOR ILLNESSES

### Upper Respiratory Illnesses

Colds, ear infections, croup, and intestinal upsets with vomiting and diarrhea are all likely menaces for Three, but they are generally milder and less frequent than they were when Three was younger.

A Three who gets recurrent ear infections is a bit more likely than a younger child to have allergies or enlarged adenoids as the culprit. Three may be less in danger of having language skills impaired by recurrent otitis media, but social activities and balance may still be affected.

Three may also contract a true sinus infection. Children younger than Three can have sinus infections as well, but Three is more likely to have this suggested as a diagnosis. Three's maxillary sinuses, the ones under the eyes, are developed well now, and Three is more able to show and complain of the symptoms of sinusitis: a painful nose, prolonged thick nasal discharge, puffy areas over the cheekbones, and a lousy disposition. A morning tummy ache and cough go along with sinusitis too.

## COMMON SCARY BUT USUALLY INNOCENT PROBLEMS

### Night terrors

While night terrors are not unheard of in younger children, they can seem to parents more frightening in children three and older than they did in younger children because Three's daytime behavior is so comparatively grown-up and in control. In a night terror the child partially awakens from a very deep sleep. He doesn't seem to recognize his parents, and may react to their approaches with terror or rage. He can't be soothed. However frightening, these are benign spells. (See *Frightening Behaviors* in Part II.)

### Delirium with high fever

Again, because Three seems like such a grown-up, it is disconcerting to have Three "act weird" with a high fever. And of course, since fevers tend to go up at night, this usually happens at night. Three, flushed and with a pounding heart, sees things that aren't there and talks about fantasies as if they were real. He makes no sense at all. Certainly, such a reaction calls for an immediate phone call to the pediatrician, but don't be surprised or offended if, after asking all the usual questions, the pediatrician advises you to treat Three's fever and make an appointment in the morning.

Acting weird and delirious without fever, or with signs (such as headache and vomiting) that something else is going on, demands an urgent evaluation. So does any suspicion Three could have ingested a drug or medication or poison.

### Nursemaid's elbow

Threes are still prone to this common accident, which parents almost always think is a sprained or fractured wrist. See the discussion under *First Aid* in Part II.

### Reddened vaginas and anuses

These strike fear into parental hearts because of the worry about molestation and abuse. However, it is common for Three to have these symptoms because of hygiene mishaps and infection.

Three loves bubble bath, polyester tights, and masturbation. Threes can't wipe very efficiently, and often make a swipe from back to front, carrying germs from rectum to vagina. Strep, as in strep throat, can cause reddened vaginas as well as reddened anuses of both sexes. ("How do they get this?" parents ask. Well, generally by picking their noses and then, er, scratching.) Often there is gooey pus, and on the surrounding skin, impetigo, looking like pimples or honey-colored crusts.

Pinworms can also play a role in producing anal and vaginal pain and itch.

Of course, the pediatrician needs to evaluate and often take cultures to make sure of the cause of these genital and rectal symptoms. If there are any risk factors for molestation known, or if the culture shows a potentially sexually transmitted disease, then of course further investigation is mandatory. But most of the time, these findings are just part of being Three.

### Nosebleeds

Three just loves orifices. Most nosebleeds in Three stem from nose-picking. Threes pick their noses because the mucous membranes are dry: a coating of petroleum jelly and a humidifier will help. They pick because their noses are itchy from a cold or allergy. They pick because it feels good and/or they are bored or anxious or because picking gets disgusted adult attention, especially if they then eat what they have picked. Threes pick their noses when they sleep too.

Most nosebleeds are innocent and common and do not signify leukemia or a bleeding disorder or high blood pressure or a tumor.

To treat the nosebleed have Three sit up, lean forward, and if you can get him to, blow out any clots. Then pinch the nostrils together, firmly, for five minutes, without peeking. Easy for me to say. In 95 percent of cases, this will stop the bleeding. Coat the inside of the nose with petroleum jelly, and cut Three's fingernails short. If the bleeding won't stop with this, or recurs frequently, or is accompanied by other symptoms (see *Nose* in *Body Parts*, Part II) then of course see your pediatrician.

## SERIOUS AND POTENTIALLY SERIOUS ILLNESSES AND SYMPTOMS

Three is much more protected now from overwhelming infection than was the case earlier. Moreover, Three can usually be very communicative about how she is feeling, can say where and how much it hurts, and can even ask to go to see the doctor.

### Fever

High fevers of between 104° and 106° can occur with innocent infections in Threes, and by themselves do not constitute an emergency. When I say *by themselves*, though, I mean that. It is perfectly possible to have a Three with a virus and a temperature of 105° who is happy, eating well, and asking to go to the zoo. If Three has other symptoms, however, such as vomiting, rash, trouble breathing, pain, or scores in even the slightly worrisome range in terms of "acting sick" or "looking sick" (see *Frightening Behaviors* in Part II) then such a high fever mandates urgent attention.

A fever of any degree that lasts longer than three days, even without any other symptoms, deserves a phone call to the pediatrician, and one that lasts longer than five days requires a visit.

### Asthma

A Three whose previous wheezing episodes were chalked up to recurrent viruses may now be showing clear signs of asthma: wheezing as a response to allergy, exercise, change in temperature, anxiety, or other kinds of stress.

Many children only have mild asthma, a tendency to wheeze now and then, with symptoms easily relieved by oral medications. However, some children will have frequent, severe episodes that incapacitate them, and others will have a constant wheeze that may come to seem "normal," but it's not.

Any pattern of asthma other than the very mild, infrequent wheeze in Three deserves a careful work-up and meticulous environmental control and medications. See the essay on *Allergies* in Part III.

### Epiglottitis

A child this age who has severe trouble breathing and makes a noise when she inhales may have the very rare and very dangerous disease epiglottitis. This is when the little lid covering the airway becomes infected and swollen. Please see the section in Part II, *Body Parts,* under *Airway and Lungs*, for a full description.

## MINOR INJURIES

Threes' social life produces a bit of new sophistication in the type of injuries sustained, though the ones mentioned in the last chapter (curling-iron burns, unwitnessed injuries that cause a limp or nursemaid's elbow) still occur. So do bonks on the head and lacerations inside the mouth. Here are some particularly common injuries Three might suffer:

### Teeth knocked out

This can happen from a fall or a direct blow to the mouth. Primary teeth are *not* reimplanted, but the dentist needs to check that no other damage has occurred.

### Foreign bodies in orifices

As mentioned above, Threes like to put small or even quite large objects into ears, noses, and vaginas. The first sign of nose and vagina FBs is sometimes blood and pus; sometimes an unbelievable odor. If you see Three insert something, or Three confesses, you may be able to get the object out before this complication occurs. (Delay makes it harder to remove the object because the surrounding tissues swell.) Foreign bodies in the nose are worrying because they may get sniffed back into the nasopharynx and then get aspirated into the airway, where they can block breathing.

### Jumping off or out of things with no thought of consequences
Usually such leaps result in minor fractures or sprains, but bear in mind that this is the age when children (especially those who watch TV cartoons) think they can leap from windows or roofs and either fly or get up and walk away afterwards.

### Bee Stings
These are discussed in *First Aid* in Part II.

## MAJOR INJURIES
These certainly overlap with those in the last chapter (choking, poisoning, water immersion, animal attacks, car accidents) but there is a major new one:

### Real guns and dangerous toys
Study after study shows that young children cannot be merely educated to "be careful" with real guns, bows and arrows, darts, or other toys that can be used as weapons. The only real solution is not to have this kind of toy, or any real guns, in the house, or in the houses of friends.

If you do have both a gun and a young child and are determined to keep both, don't even think of simply telling the child about gun safety, or showing him what guns can do to hurt people, or telling him not to touch the gun. At Three, this simply won't take.

Keep the gun unloaded. Keep it locked up. Keep the ammunition locked up, separately, from the gun.

## Window of Opportunity

Three is a very special time, still part baby but already part "kid." Parents can take advantage of this in-between state to inculcate some important values, nip some bad habits in the bud, and have a lot of fun at the same time. The areas that seem to be most important have to do with attitudes about sex and gender, categories and stereotypes, and active versus passive learning, with special emphasis on TV and the much-beloved videotapes.

Of these, the one most worrisome to most parents has to do with nurturing healthy attitudes about sex and about trusting the world at large while, at the same time, protecting Three against sexual molestation and abuse and against kidnapping.

## SEX EDUCATION FOR THREES
In their urge to make the world controllable by assigning categories, Threes are no dummies. They understand that the most potent category around is sex. Everything important in Three's environment is assigned a sex, including quite genderless stuffed animals, security blankets, and enigmatic pets like turtles and bugs. Three is high on the discovery that all human beings have sexual identities, and can be rigid about sexual roles. ("You can't be a lawyer," a Three of my

acquaintance told his mother. "Girls are doctors. Boys are lawyers.")

Many parents are filled with doubts and hesitation about when and how to tell children "the facts of life." If you think it's tough at Three, wait until Six. Three, in fact, is the perfect time to impart a little formal education.

Three is close enough to babyhood so that it is still natural to look at and talk about private parts of Three's body. Three still likes to be cuddled and tucked into bed. Three still thinks parents are wonderful and flawless, and, even more important, that they are authorities on every subject.

Three needs to know that sex is an adult subject. Therefore, Three needs to know that parents understand sex, and that they are not frightened, angry, worried, shocked, or shamed by the subject. A Three who thinks a parent is frightened, angry, worried, shocked or shamed by any subject is very unsettled. Indeed, this is the primary reason to try to rehearse one's attitude toward masturbation and sex play and intrusion into the parental bedroom at the wrong time. Outraged, horrified, or grief-stricken reactions won't alarm Three about sex as much as they will about her parents and their ability to cope in the world.

Three is already fascinated by the mystery of sex: not just in terms of body parts but as a force determining personal identity and destiny. Three is usually eager to learn.

Many books advise that parents wait for children to ask questions. I don't think that this works. Many children, whether they sense reluctance and apprehension in their parents or have already picked up lots of societal cues about the topic, never ask. They are Eight and have seen news programs about rape and videos of *Look Who's Talking* or, heaven forbid, *Fatal Attraction*, and hear God-knows-what at school but they never ask their parents about sex.

Even Six is already modest, embarrassed, and has gotten the idea that parents are pretty uncomfortable with the subject. A Six who hasn't been talking with parents right along about sex is likely to be highly resistant.

But what about preserving innocence?

Innocence really means an attitude, not an ignorance of facts. If Three is taught facts at home, and has parents who are loving and respectful towards each other, Three has protection against what parents really worry about when they talk about "loss of innocence." This worry is really about premature exposure to the erotic, violent, or prurient aspects of sex. It is about what Three might see and hear on TV and at the playground and the homes of friends. If parents don't give Three facts and security and role modeling, all these other exposures can, indeed, rob Three of innocence.

Three needs to know the names and basic functions of body parts that show. Three needs to be told explicitly that boys are born with penises and girls are born with vaginas. Otherwise, all kinds of confusions may occur.

My friend Heather thought that all babies were born "the same" and acquired either penises or vaginas, depending on character traits and clothes preferences. When she had

## FACTS OF LIFE FOR THREES

"Lily's mommy is going to have a baby. Did you see how big her tummy is? How do you suppose that baby got in there? And how will that baby get out?"

"Here is a picture of you when you were a baby. My, what a lot of hair you had! You had just come out of Mommy's tummy. I bet you wonder how you got in there and how you got out."

"There is a special place in Mommies for babies to grow in. It's called the womb (or uterus). It's a special place just for babies, not for food or peepee or poopoo."

"Your belly button is the place where you used to get fed when you were inside Mommy. There used to be a special tube there, called an umbilical cord. You weren't big enough to eat food, so Mommy fed you through that special tube."

"After you were born and you could eat, your umbilical cord came off. It didn't hurt."

kidney surgery, she was convinced that the surgeon had removed the penis that was inside her waiting to come out, so that she had lost her choice to become a boy. She was relieved and astonished when shown brand-new babies equipped at birth with penis and testicles or vulva and vagina.

My friend Greg is still confused. His pregnant mother had an ultrasound that showed the potential new sibling to be a boy. Well, that was an error; the umbilical cord got in the way and looked like male genitals. But Greg was told that he would have a baby brother. When the little girl was delivered (surprise, surprise!) Greg was undisturbed: all babies, he announced, are born girls, and get their penises later on. He is still waiting.

Three needs to know that babies are started by sperm and egg, and grow in the mother's womb. Three needs very much to know that where the baby grows and exits is not the same place that food goes and poop exits! (If you want to see real potty refusal, find a Three who thinks that he or she may just possibly give birth with the next push.)

Three can also be introduced to the idea that the sperm and the egg meet during a special hug called "making love" or "having sex." She might ask how they meet, in which case the answer is "the daddy's penis gets big and hard and goes in the mommy's vagina."

When Three asks, "Can I watch?" the answer, of course, is "No, that's private."

There are some good books about the facts of life for very young children. But I don't advise them for the initial discussions. Learning something from a book, even though parents read it out loud and discuss the pictures, puts a barrier between child and

## ANSWERING THREE'S QUESTIONS

"A mommy and a daddy make the baby together. The daddy makes a sperm and the mommy makes an egg. The sperm and the egg are smaller than a dot. They are so small you can't see them at all. So the baby is very, very tiny at first. It takes a long time for the baby to grow big enough to be born."

"They make the baby when they are alone together and very private in their bedroom. Nobody else can watch because it is very private."

"The daddy's penis makes sperm and Mommy makes the egg way in the inside of her body. They make the sperm and the egg come together in a special hug."

"The mommy has a special place in her body for the daddy's penis to come inside, so the sperm can come to the egg."

"When Mommys and Daddy's do that special hug, they call it making love."

subject, and child and parent, that I think isn't helpful. Looking at a book after the topic has been discussed in person a few times is different.

Parents have a fine opportunity to teach Three about adoption when they teach the facts of life. If parents *all* presented adoption as "one way to have a baby," adoptive families would have a much easier road to hoe. As it is, only adoptive families tend to teach this as a matter of course.

### SEX PLAY

When Threes indulge in sex play with other children, it's as much about exploration of difference as it is about genital excitement. It tends to be very much about the mystery of sex in its relation to power and roles. ("I am the doctor, giving you a shot in your penis.") Sex play also has a lot to do with the recent scary and empowering act of toilet control. ("Here comes the big enema suppository.")

When Threes engage in sex play with each other, they mostly look and touch for reassurance of identity. Their role-playing, their questions, and their fights during this activity have to do with Three's excitement about the defining quality of sex. They may look, and sometimes touch, but they don't usually carry out the "realer" aspects of the role playing. Nobody really gives the shot or the big enema suppository.

Coming upon Threes engaged in such play can be a shock. It's wise to have a prepared response. I suggest the following:

- Name the behavior. Threes are very sensitive to the implications of not giving something a name. Not naming makes the behavior fraught with forbidden excitement. "I see you are playing show-and-touch with your clothes off." Or even, "I see you are interested in each other's private parts."

- Give a good reason for telling them to change activities. "I know you are very interested in private parts and what they do, and that is fine. But I want you to ask me those questions instead of playing with each other. Those are very important questions and I want you to have the right answers."

- If you are discernibly upset, tell them why. "I am upset to find you playing this game because it is about such important and private things, like making babies. Mommies and Daddies need to be in charge of telling you about these things. Not other boys and girls."

- Make it clear that you will, indeed, satisfy Three's curiosity. "After Bobby goes home, we can talk about private parts and how babies are made. Now let's go outside with the bucket of water and paintbrushes and paint the back porch."

Besides engaging in sex play with other children, Threes often touch themselves and may even masturbate in public. They don't do so (normally) to be exhibitionistic; they do so because it feels good, and because they have, for the moment, lost a sense of themselves as performers with an audience. Thus, they may unselfconsciously touch themselves both in public or at home. They may sit with crotch exposed or scratch themselves. They are more likely to engage in any of these behaviors when tired, bored, or anxious.

It usually works well simply to tell Threes that touching their private parts is a private activity. This avoids suggesting that it is dirty, shameful, or dangerous. To appeal to Three, try a rousing discussion of *public* and *private*, making it a category game:

- Private rooms mean rooms whose doors you knock on before entering. This can be demonstrated by always knocking on the door of Three's bedroom or the bathroom, even if the door is open and even if you are absolutely going to enter that room. It's the gesture that counts.

- Private parts of the body are those covered by a bathing suit, and we touch (and maybe uncover) those private parts only in the private rooms of the house.
- Private names for those parts are the real medical names like penis and clitoris, and we use those private names only at home (or at the doctor's).
- Public names for those parts are nicknames, like whizzer and gina.

   If your Three is found engaging in sex play that strikes you as unusual, see *Beyond Show and Touch* in the *Window of Opportunity* section in the next chapter, *Four Years to Five Years.*

The payoff for helping Three handle the potent categories of sex and gender, public and private, is enormous.

- You set the stage for future conversations about heavy topics in an atmosphere of trust and calm.
- You protect Three from keeping molestation or inappropriate behavior a shameful secret.
- You protect yourself from terminal embarrassment. Jenna, at Three, got up on her chair and announced to an entire restaurant, "My mommy has hair on her vulva." Jenna had mastered an important fact, yes, but not the private-public category.

## GOOD TOUCH-BAD TOUCH

Most Threes are still so close to babyhood and diaper changes and assisted wiping and bathing that teaching about good touch-bad touch is fraught with difficulty.

I have given up instructing parents to look for pinworms (examining the vaginal and anal area with a flashlight after bedtime) and doing cellophane tape tests for the same diagnosis (placing a piece of cellophane tape over the anus to catch the eggs for microscopic examination) because Threes have described these activities to other adults, and parents have been accused of molestation. I have also seen Threes so terrified of Bad Touch that they won't let a doctor examine them, even after rehearsal and reassurances from both parents that it's OK.

I have come to the conclusion that it is not a good idea to sit down with Three and have a serious talk about good touch-bad touch. Instead, it's best to mention briefly and casually to Three that genitals are private, and that grown-ups only touch Three's private parts to clean them, or when the doctor has to look at and touch them. There are opportunities to do this frequently, while helping Three with baths and bathrooms, or when looking at TV or magazine ads featuring naked or near-naked children.

It is inviting trouble to call a Three's nipples her "boobies" or to treat or tease her in any way as an erotic person.

When going to the pediatrician, you can make a point that the doctor only undresses, looks at, and touches Three's body because Mommy or Daddy is right there and says that it is important and the doctor's job.

If you make all this clear, you shouldn't have to tell Three explicitly to tell you if something disturbing happens. You will be communicating that expectation and gaining Three's trust merely by your positive behavior. If you warn Three explicitly, Three is likely to worry and show separation anxiety.

## SEXUAL MOLESTATION AND ABUSE

Nearly every parent worries at one time or another about the possibility of sexual abuse, in both boys and girls. Leaving one's child even with a good friend or respected teacher, much less a comparative stranger or an estranged spouse or the estranged spouse's new romantic interest, is an act of trust.

But there are other times when a parent may actually suspect sexual molestation or abuse. Warning signs include:

- Marked change in behavior, such as becoming very withdrawn or very hyper and aggressive.
- Refusing to go to or stay with a previously accepted adult.
- Truly indiscriminate kissing and fondling of strangers (NOT a hug for the teacher and or the pediatrician).
- Increase in frequency of or preoccupation with masturbation.
- Intensity or frequency of sex play that is unusual. If Three is engaging in sex play that strikes you as unusual, see *Beyond Show and Touch* in the *Window of*

*Opportunity* section in the next chapter, *Four Years to Five Years*.

- Physical findings such as an inflamed vaginal or rectal area. Even if a mundane cause is found (see below), abuse is not ruled out if other behavioral problems coexist.
- Recurrent urinary tract infections or severe constipation despite investigation and treatment.

While once in a great while, such inflammation or infection may signal sexual abuse, more often the cause is a mundane one—hygiene problems or pinworms. Only a visit to the pediatrician, with an objective examination and, if necessary, cultures, can determine what is going on. Again, a mundane cause does not rule out abuse. Once an innocent cause is ascertained and the problem cleared up, the child should appear well and back to normal. If not, be suspicious.

## STRANGERS AND KIDNAPPING

Many Threes are very friendly with strangers, and people certainly are drawn to Threes. How do you warn a Three away from danger without producing paranoia? This is especially tricky because of Three's tendency to categorize. Trying to tell Three the difference between good strangers and bad strangers is futile. Three is likely to regard all strangers as Bad.

One solution is to reinforce the idea that Three is never to go anywhere with another adult unless you say it is all right, and then practice it and stick to it. Mention this casually, briefly, on appropriate occasions. For instance, "I bet if that lady's little dog ran away, she'd ask you to go with her to find it. But you would say, 'Nonono, I have to ask my Daddy first.'" Do this often enough and "Nonono, I have to ask my Daddy first" becomes second nature.

This is a particularly useful ploy if you are worried about abduction by a noncustodial parent. However, it is best for the child's mental health if you stress that the rule applies in all situations. This makes it easier for the child to apply it with the other parent without feeling a burden of guilt and woe.

Of course, if someone else is going to pick him up from preschool you have to tell Three ahead of time. And remember that most child sexual abuse is performed not by strangers but by someone known to the family.

## STEREOTYPES, TACT, AND CURIOSITY

Children also become aware of racial, ethnic, and personal differences at Three. Unfortunately, this great interest in announcing categories and labeling in terms of categories occurs a little bit before most children really get the hang of compassion. Compassion relies to an extent on the intellectual ability to see something from another's point of view. Many children can't really do this until Four.

(Sit across a small table from a Three. Between the two of you place, from your angle, a carrot, a key ring, and a pencil. Ask Three to name what she sees. She will say, "A pencil, a key ring, a carrot." Now ask her,

"And what do I see from My point of view?" She will say, "A pencil, a key ring, a carrot." Ask a Four, and she will, after consideration, say, "A carrot, a key ring, a pencil.")

So when Three labels someone as a fairy, or calls a friend a racial epithet, or asks loudly "Why is that man so fat?" or "What's wrong with her skin?" or "Why does he smell so bad?" Three isn't really being callous. He just wants to know in what category to place this novelty, and it doesn't occur to him that feelings could be hurt along the way.

To keep this from happening, it is useful to explain to Three that certain things about people are private: their appearance and smells and abilities. When Three spies a person with a difference in these categories, no matter what it is, the prudent parent can then catch Three's eye and say softly, "I see what you want to ask, and we can talk about it in private later."

In that discussion, it can't hurt and might help to point out how that person would feel if loud questions were asked. Sometime over the next year or so, light will dawn.

But what about the child who is being labeled and categorized offensively? Certainly, if the teasing is happening at preschool, the teachers ought to intervene. But how this is done is crucial.

Sometimes teachers will secretly or unconsciously, or even overtly, agree with the teaser. Usually this comes in the form of, "Don't tease such and so; he (or she) can't help it if. . . ." This is devastating. This can even happen in families, with a parent "chastising" one sibling for teasing another.

Sometimes all that is needed is to simply

correct the categories. Threes want to get it right.

"Everybody has different color skin, Marcia. Here are some crayons. Look, I think my color skin is apricot, and yours is peach beige. Trudy is burnt umber. I am Irish-American and Trudy is African-American."

At other times, it helps to make a rule for everybody. "In our classroom, nobody can call somebody else a hurting name. Not the teacher and not the children. If you ever hear me call somebody a hurting name, I will have to go and sit in time out."

This is particularly effective when there are two teachers, and one can call another a hurting name such as Poop Face and then be made to sit in time out.

## Gender and Assertiveness

One of the trickier aspects of being parent to a Three is helping a very timid Three to be assertive and a very macho Three to curb obnoxiousness.

It is tempting for parents to counsel children what not to do to assert themselves: don't hit, bite, punch, kick, call names, etc. When confronted with the question, What should Three do when attacked?, we often pass the buck: "Just don't pay any attention," we say. Or, "Tell the teacher."

So when Hubert punches out Melissa, she runs and tells the teacher, who puts Hubert in time out.

By kindergarten, however, Teacher will very likely be too busy to attend to Melissa's complaints and by third grade Teacher will start thinking of Melissa as a goody-goody. By kindergarten, Hubert may be well on the way to a reputation for bullyhood, and by third grade he will be too big to put in time out.

Both Melissa and Hubert need some ideas and some behavior modeled and practiced to learn ways to assert themselves.

- Use *I* statements, not *you* statements to say what you don't like. "I am mad that you took my truck, I need it back now!" instead of "You shouldn't take my truck, you're a bad boy." *I* statements make the victim feel better and don't diminish the offender.
- Use body language. Stamp that foot, talk big and loud. Make faces.
- If you call names, make them public names not private names. Names that have to do with how you look and smell are private names. You can make up names, like "Boogy-doogy" or "Fifflestamper."
- Don't forget the old standbys, "Sticks and stones may break my bones"; "I'm rubber, you're glue, whatever you say bounces off me and sticks on you." Three doesn't understand what they mean, but they feel great to shout, and to each generation are new and fresh.

A very demure little girl or shy boy may need lots of practice before trying an assertive behavior. Playacting with toy animals is helpful, especially if Three gets to be the bully and the parent gets to be Three.

Little boys who would prefer using fists and feet and weapons may be swayed when told that using words is harder and stronger. Rather than praising Three for merely not hitting or kicking, try praising him for the force and devastating effect of his words.

Once in a while, a parent may be concerned that a boy is not "masculine" enough, a girl not "feminine" enough. A thorough discussion with the pediatrician, or with a counselor specializing in gender issues, is most helpful.

### TELEVISION, VIDEOS, AND MOVIES

Threes love the predictability, the action, the emotions of these. They are hypnotic, addicting, and can be very detrimental to Three's emotional, intellectual, and social wellbeing.

A Three who, at Two or younger, became addicted to fantasy videos may show a change in that addiction, a more anxious and dependent behavior. Three may gaze raptly while biting the nails or twiddling the hair, and may search for a security object. I believe that this is because Three is newly able to understand the implications of the plots of these popular videos: *Beauty and the Beast, The Little Mermaid, Aladdin, Dumbo, Bambi.*

In all of these, and in most of the other very popular videos, a parent dies, or is seriously endangered, or is transformed into something hideous, *because* of the existence or actions of a beloved child. A Three who grasps this concept gloms onto it. It feeds all of Three's developmental anxieties: ambivalence about independence, normal hostility towards parents, normal fantasies of displacing one parent and marrying the other. No wonder that Three "has to" watch over and over again, to see the magical resolution.

Threes can't differentiate between something bad happening to someone else and something bad happening to them. When they hear a story being told or read, they can stop the adult, ask questions, halt the story in its steps. But TV is designed to be all-engrossing, and Three is at its mercy. If the heroines are all beautiful and flawless, and the heroes all muscular or preternaturally agile, Three is going to take that at its face value and be upset that real life doesn't live up to these standards. A Three who loves the morning cartoons can start imitating their frenetic energy and violence, and can get off to a bad start every day.

Television isn't interactive. It gives a child feelings he hasn't earned, and poses examples of behavior he hasn't experienced in life. Jackie, a mother I know, tells her children: "TV doesn't help children grow."

A Three who is addicted to TV or to videos needs rescuing just as much as a teenager on drugs. A preschool or daycare that incorporates TV (other than a half hour of *Sesame Street* or *Mr. Rogers' Neighborhood* or other age-appropriate fare) or videos in its daily program should be ashamed of itself.

One good way to discover whether a Three is too enthralled by videos is to have the TV set "break" for a few days and see what happens. If everybody starts sleeping better and laughing and talking more, you have your answer.

## LITERATURE

If a Three learns to love books, the passion has a good chance of lasting a lifetime, no matter when Three learns to read. Being in a home where adults read for pleasure is a great boon to a Three who loves to imitate grown-up behavior.

Reading out loud, both on impulse and as part of a ritual, is one of the biggest gifts parents can give children. Reading aloud doesn't mean just announcing the words on the page, of course; it means talking about the pictures and plot and wondering what's going to happen next.

Books along the lines of *Where's Waldo?* or ones that make noises or tunes are novelties, not books.

## SELF-ESTEEM AND PRAISE

Self-esteem grows as a child conquers age-appropriate challenges. Certainly, a healthy family shows a child unconditional love (which also implies setting limits) and gives appropriate praise. But nothing substitutes, in my experience, for the child's own appreciation of his achievements.

A Three who is praised for nonachievements, for endowments that Three has nothing to do with (what a pretty dress, what a handsome fellow, what a tall girl), doesn't get much out of the praise, and it can contaminate meaningful praise. Even praising accomplishments can detract from rather than enhance self-esteem. "Oh what a beautiful picture" may focus Three on the flaws in her drawing or may be such a frequent comment that it is meaningless. The most sincere form of praise is to make sure she has plenty of sharp crayons and clean paper, a place to draw, and your silent interested attention.

Watching out how you praise can be very hard with a precocious Three. It is difficult not to *kvell* over a Three who has learned to read, hit a golf ball, complete a difficult puzzle, jump in the pool. But praise can actually rob the child of Three's own achievement, putting it into the context of "being good" and pleasing Three's parents. After all, Three didn't whack that ball or unlock the key to literature in order to please Mom and Dad. Three did it because he or she saw the challenge and liked it and accepted it and achieved.

## MANNERS AND CIVILIZED BEHAVIOR

By now, Please, Thank You, Excuse Me, and *Gesundheidt* should all be very familiar terms to Three. Three's thank you notes now might include laboriously printed phrases ("THANX") and first names spelled with the help of an adult.

Threes can shake hands and smile and say "How do you do." It will get them so much attention it should be easy to teach.

Three can learn to eat neatly. By Four, most should be able to hold a fork and spoon in a tripod (pencil) grasp rather than a fist. Chopsticks are harder, and Four may be a better time to start.

Sometimes a shy Three benefits by learning to bow or curtsy a greeting. (You don't have to say a word or make eye contact, all you have to do is move; and the attention you get is heartwarming.)

*Helping at home*
Three can still "make his bed," and put toys away. Three can set the table with plate and utensils. Three can dust low furniture. If these are presented as Three's family chores, rather than as helping Mommy, they are less likely to be the focus of oppositional behavior.

Threes are very interested in roles and imitation and emulation, and they are pretty unambivalent about wanting to please their parents. Once a habit of doing one's family chores is established, it's easier to keep it going. By Four, some children acquire enough defiance and independence that this becomes more difficult.

# What If?

## EXTENDED SEPARATION FROM PARENTS
Threes often tolerate such separation well if they are prepared ahead of time and if they have a regular life not intimately and totally dependent on the parent's presence. See the suggestions in the previous chapter, *Two Years to Three Years*, in this same section.

## TRAVELING WITH THE CHILD
Traveling with a Three can be wonderful, especially if you prepare Three ahead of time. Be sure to include in your preparation not just what will happen when you get to where you are going, but how you will arrive there. Try to keep schedule, bedtime ritual, and security objects as consistent as possible. It's a good idea to have regularly scheduled, one-on-one time alone with Three. This time is often best spent doing something that from the parent's point of view is boring, such as repetitive games or reading aloud, as an antidote to the strangeness and stimulation of the journey.

## CONCEIVING OR PLANNING TO ADOPT A SECOND CHILD
When the next child arrives, Three will be well on the way to being able to enjoy and help with the new baby. See the next chapter for usual reactions of Four to a new sibling.

## ARRIVAL OF A NEW SIBLING
Three's powerful use of language can be an excellent aid in helping her learn about, anticipate, and cope with a new sibling. This is one of the big advantages of waiting until Three to embark on such an undertaking.

*When do you tell Three?*
Some Threes know that something is up incredibly soon into their mother's pregnancy. Certainly, Three ought to be told at least at the same time you tell adult friends and relatives. Waiting will ensure that Three overhears mysterious and anxiety-producing

whispers, and then he may not believe you when you finally do tell him.

If you are adopting the next sibling, the same thing goes. Whether or not Three was adopted, your covering both possibilities in your sex education chats ought to pave the way for Three being excited without regarding adoption as unnatural.

*What do you tell Three?*
- Even if you are sure of the baby's sex, consider holding off. Ultrasound examinations have been known to be wrong. Remember my young friend who is still waiting for his little sister to turn into a brother.
- Nine, or even five, months is a long time to wait for a new baby. Give Three a reference for when the baby is due: right before Halloween, or after all the snow melts, or after it is warm enough to go swimming.
- Be realistic but jolly about what having a new baby will be like. "The baby will be red in the face and he or she will make funny faces. The baby will cry a lot and hiccup and burp." Don't promise that the baby will be a playmate to Three, or he'll anticipate a trip to the sandbox and swings as soon as you arrive home from the hospital.
- Allow Three to figure out her own feelings. "You will love the baby," "You will be such good friends," can strike a Three as encroaching on her autonomy. "If you want to, you can help me put on diapers and feed the baby," is as good an enticement as any.

- To state a reassurance is to suggest its opposite. "We will love you just as much as ever when the new baby comes" makes Three think, ominously, "Whatever makes them think they wouldn't??" Assume that your behavior, body language, and track record assure Three of your unconditional love and your pleasure in him.

*How can you prepare Three for the big changes in your life?*
- Point out babies that you see, and tell Three how old those babies are and describe how they are behaving. Don't use ages in months, but make statements like, "That baby was born a pretty long time ago. Already he knows how to hold his rattle. He doesn't just cry and eat and sleep all the time."
- Use Three's own baby pictures to illustrate what a newborn is like.
- Try to accomplish any big changes well before the new baby arrives: new preschool, change of sleeping arrangements, new babysitter.
- Let Three help pick out things for the new baby, but don't focus too strongly on preparation. Three still has his own engrossing life to lead.
- Let him talk to the new baby through your belly button. You can find out a lot of Three's worries and hopes in this way. *You* can talk to the new baby too: "Now, Cherub inside, we want you to be very strong and cry real loud and make giant peepees and poopoos." A Three hearing

you give such instructions is likely to repeat them back in a gratified manner when the baby comes home and obligingly carries them out. I'd avoid requests that depend more on chance: "Please, Cherub inside, have lots of black curly hair, and be a girl, and don't have colic."

- Before the delivery, rehearse what will happen when you go to the hospital. If Three might be awakened in the middle of the night and taken to neighbors, make sure he is reminded about this frequently, and has a suitcase with a reminder sticker on it to take along his security object. Three also appreciates spending a night there beforehand under calm conditions, just so he knows what to expect.

- Presents do help and are not a bribe. A present delivered to the adults with whom Three is to stay, for them to give to him in the morning, is a concrete reminder of parents' love. If Three is to stay home with a sitter, a nice ploy is to hide the present in a fairly obvious place for sitter and Three to hunt for together. Should the present be from the new baby? Or is that a deception? One way around this is to say "the new baby is too little to give presents. But if he could, he would give you this."

- A hospital tour, books, videos, and a sibling class are all useful, but be sure they don't paint too rosy OR too bleak a picture of life with a newborn. It is helpful to find the ones that emphasize the older sibling's leading a life away from the baby, at preschool or with friends, as well as the older one's ambivalent feelings about the newcomer. And be attuned to when Three has simply had it with discussion about new babies.

- After you come home, put a sign on the front door that asks visitors to greet Three (be sure you spell out Three's name; some of your visitors may have forgotten it) first and ask Three to tell about or show the new baby.

- When Three wants to take the newborn to Show and Tell, provide a photograph of the baby and carefully wrap it in a receiving blanket, for Three to carry gingerly. A few photos of Three holding the baby are a nice addition.

- Anticipate that Three will want frequent hugs, cuddles, and physical attention. Floor play, when you get down there and let Three lounge on you and just chat or play board games, is great for tired parents.

- Assume that Three will be ambivalent and may need to act out as well as to talk about negative feelings. "We could take a ride in the car and throw him out the window," a Three I know suggested in the politest possible way. "I think it's time for that baby to go back to the hospital now," sighed a man-of-the-world Three, who had been carefully prepared. Clinging to and hitting out at a parent, even for a Three who long ago gave up such behaviors, is pretty common.

- Most crucially, keep time alone with Three after the baby's birth. Put it at the top of the priority list, ahead of thank-you notes, dish washing, phone calls, finishing the draft of your thesis, paying the bills. Alone means really alone, without the other parent or visiting grandparent. Let Three know that you are interested in him not just as an extension of yourself but as an independent personality. Talk with him about things other than his behavior, the new baby, or your own feelings. Let him overhear you telling flattering stories about him that have nothing to do with the baby.

*What if Three is sick right when the baby is born, and has to be kept away from the new baby?*
This is such a common event that I have come to feel that the possibility ought to be addressed *before* the birth. There is something about oncoming labor, I am convinced, that triggers upper respiratory infections in the older sibling. Until someone writes a picture book on the subject, a good way to introduce it is, "When I was a little boy and my brother Uncle Fred was born, I got the measles. So I couldn't see the new baby until he was old enough to smile! And I was the first person he smiled at!"

## DEALING WITH A YOUNGER SIBLING GROWING UP
If Three has a well-established life with peers, Three is likely to tolerate the invasions of a nine-to eighteen-month-old sibling fairly well. However, to do so, Three needs to have private time alone with each parent, a private place (even a dresser drawer will do) to keep objects safe from the younger child,

and a sense of self as an older, wiser, and more privileged person. All the suggestions in previous chapters for helping Older deal with Younger continue to be helpful with this spacing.

A more difficult situation occurs when the younger child is so close in age that Three feels pulled back into the stage Three just matured out of. See the discussion in this chapter on *Contagious Tantrums and Aggression* in the section on *Limits*. The best way to cope with this, as described there, is to supply Three with a true peer group in which the other children are Three's age and slightly older. Also: See *Sibling Babies* in Part III.

## DIVORCE AND CUSTODY ISSUES
Three's focus on controlling the world, and his drive toward autonomy and achievement, make it easy to see how discombobulating divorce can be.

A Three who has heard parents fighting often is convinced (and sometimes correctly) that Three is at least one major cause of the anger. Three simply can't imagine that he or she is not the center of the parents' universe.

Moreover, Three has probably harbored angry or jealous wishes ("I wish Daddy would go away and leave Mommy and me alone!"). When Daddy does go away, Three is certain that he caused it.

It is inconceivable to Three that his parents could divorce for reasons of their own. So the departure of a parent is likely to be seen as abandonment. Three is going to need special understanding, patience, physical affection, and comforting. He is going to act out, either with aggression or withdrawal or both, alternating.

### Custody

While your therapist and/or pediatrician will help guide your decisions, it may be useful to know about two large studies (Wallerstein and Johnston: "Children of Divorce: Recent Findings Regarding Long-Term Effects" and "Recent Studies of Joint and Sole Custody," *Pediatrics in Review*, Jan. 1990 II [7]).

One study showed that whether joint or maternal custody was decided on, the children who had the fewest problems were those whose parents didn't exhibit a great deal of hostility and conflict, and whose mothers were less anxious and depressed. The amount of time and frequency of visits with fathers seemed not to make much difference. Children without siblings, and whose baby temperaments were difficult, had the hardest time; and boys had a harder time than girls.

A second study looked at hostile divorces. In this special situation, joint custody and sole custody with frequent visitation appeared to be harder on the children than did living with one parent with only rare visits from the other. "Children who shared more days each month with both parents were significantly more depressed, withdrawn, noncommunicative, had more somatic symptoms, and tended to be more aggressive."

These are difficult studies to perform and interpret, but their outcomes are not unexpected. Joint custody, meaning frequent access to both parents, most likely involves more continuing contact of hostile parents with each other and more preoccupation with the divorce and previous marriage than sole custody with fewer visitations. The temptation to emote, verbalize, and act out hostility in the presence of the child, and to use the child as a spy, go-between, or weapon, is increased.

Unfortunately, what helps Three most are the things that seem the very hardest for parents to ensure:

- Avoid regaling Three with the sins of the absent spouse. Do not express with body language, actions, or words that Three is "contaminated" after returning from a visit from the Other Side.
- Forge a strong mother-child relationship, in which the mother doesn't turn to Three for adult companionship, support, or romantic involvement. A mother who stays a mother—cuddling, protecting, setting limits, enlarging horizons—is a godsend to Three.
- Be a father who doesn't withhold attention to the child or child-support payments as a weapon against the mother; a father who doesn't use visitation time as a holiday from discipline, everyday activities, and chores.
- Strive to preserve ongoing relationships with both parents, who act as parents rather than as pals or lovers of the child.
- Plan for Three to have regular activity that is separate from the disturbed family situation, activity that is engrossing, rewarding, consistent, age-appropriate.

- Make a schedule that is predictable from day to day.
- Make custody transfers on neutral territory, like preschool or daycare, not from the arms, home, or car of one parent to the arms, home, or car of the other.

If all these are in place, books about divorce may help a child voice his feelings and come to terms with his situation. Even if all goes as well as possible, however, parents are well advised to consider therapy for the child as soon as a hint of problems appears.

## SURGERY AND HOSPITALIZATION

Three has both strengths and weaknesses in dealing with this kind of trauma. Three's main strength is the new ability to use language to frame experience. Preparing Three when possible is of the utmost assistance. Many children's hospitals and wards have tours and classes for children undergoing planned procedures. At the very least, picture books such as *Madeline* (who gets her appendix out) and the comforting books by Sesame Street and others can make a child feel more in control of the situation.

Three's major weaknesses have to do with magical thinking, and with Three's new sense of self as a person with a social identity in which physical attractiveness and strength are important.

As for magical thinking, it is good to anticipate Three's worries. I would assure Three as if you took for granted Three was worrying about it that:

- Going to the hospital is not punishment for being bad, and that no one is angry at Three.
- Being sick does not occur because Three did or thought anything bad, mean, or vengeful.
- Three's body is strong and will heal itself. Three makes new blood right away to replace any blood that was removed.
- A parent will be with Three all the time that Three wants a parent there. And, of course, make sure this is the case. A parent should be with Three up until the anesthesia takes over or until Three happily goes off with the anesthesiologist to breathe magic air. A parent should be there when the anesthesia wears off. A parent should be with Three during the day and the night.

As for bodily intactness, it may be necessary to go over again and again with Three the reason for any surgery that removes or adds or makes a mark on the body that Three can see. Sometimes play therapy is required, especially for anything involving the face and/or genitals. Be sure to discuss these issues ahead of time with your pediatrician, as well as with the surgeon.

If elective surgery can be postponed until after age Five, your pediatrician may recommend this be done.

## MOVING

As with any big change, the more Three can be prepared and participate and feel in control, the better. Photos of the new home and of Three's room will help. So will photos of any neighborhood children, even videos if possible.

Three will need a special small carrying case for precious objects. A backpack is excellent, but don't depend on Three to keep track of it. And if it gets lost, this is a disaster, so watch out.

Three will want to have photos of the friends and home Three is leaving, as well. There are many good picture books about moving and making new friends, but don't forget to include stories about the pain of leaving old friends and about staying in touch with them.

### DEATH OF A FRIEND, RELATIVE, OR PET

It is a good idea for parents to think about presenting Three with a concept of death before any such event occurs. Three doesn't think of death as inevitable or as permenent as yet, but still has major fears and concerns. Most of these have to do with fears of abandonment or fears that Three has caused an event by wishing for it, thinking about it, or misbehaving.

Parents can introduce the topic of death within the context of religion or as a natural phenomenon. A classic book for children this age is *The Life and Death of Freddy the Leaf.*

If a close relative or friend is very ill and expected to die, I strongly recommend you talk over this crisis with your pediatrician.

## The Four-Year-Old Well-Child Visit

Where Three was quiet and coy, Four usually likes to take over the visit and be in charge. Parents can help Four feel powerful during this visit by teaching Four the name of each instrument and what it does, and how to help the pediatrician examine each part of the body. Opening the mouth wide and saying "aaa" as in *cat*, not "ahhh" as in *father* may help the pediatrician to examine the throat without having to use the dreaded tongue-depressor. Taking big smooth breaths without spitting can be learned by "blowing" a parent's finger. Parents can assist in the belly exam by not mentioning the word "Tickle!"

Finally, parents can help Four to practice the genital exam. Girls can be shown how to sit frog-legged, perhaps on a parent's lap, and spread their own labia; boys, how to let penis and testicles be examined. This is a good time to briefly mention that this is part of the doctor's job, and that Four is helping the doctor do that job. The examination is sanctioned because the parent is right there, reassuring Four and cooperating with the doctor.

Four is likely to ask if a shot or blood test will be part of the visit. I suggest telling Four that you are not sure.

## Looking Ahead

Three wants power and predictability, so as to be in control. Four wants to be perfect, and wants to be the judge of what is perfect. Parents find themselves in the position of being judged and challenged in a new way. Three says, "I didn't do it." Four says, "You're mean, that's not fair." Where Three is obstinate, Four is bossy. It's a subtle difference, maybe, but it's definitely present.

When parents are prepared for this development, prepared to negotiate and reason, prepared to set limits without feeling personally threatened, to have a sense of humor at being challenged by such a small and relatively helpless person, they can enjoy the great things about Four.

For the first time, Four can be compassionate not just from emotional empathy, but because Four can see another person's point of view intellectually. Three could take turns, but Four can really share. Three thought reality could be created by words; Four starts to get the idea of the difference between real and imaginary, telling the truth and lying. Three can understand the difference between public and private; Four can keep a secret. Three is just getting the hang of the concept if-then; Four can make a promise, and keep it, and hold you to one as well.

Dealing with Three is often a lot like dealing with a charming person from another planet. When you're dealing with Four, you're usually pretty certain that this is an Earthling.

## {TWELVE}

# Four Years to Five Years

## Here Comes The Judge!

### To-Do List

- Laugh at Four's jokes, even if you don't understand why they're funny.
- Use sophisticated words repeatedly. Four will catch on.
- Time out can be added to the list of useful limit-setting techniques.
- Seize teachable moments to talk about compassion, honesty, fairness.
- Pre-reading skills start now.
- Start early to plan for kindergarten.
- More sex education this year.
- Have Four memorize the home phone, cell phone, and 911.
- Change the smoke alarm batteries every birthday.
- At Four Years and 40 pounds, get a booster car seat!

### Not-To-Do List

- No TV or computer in Cherub's bedroom.
- No eating or drinking (except water) while in the car or watching TV.
- Regard TV as a potential threat to Four's intellect and character development.

### Portrait of the Age: Forming a Conscience

"There is a witch that lives under the front yard," Amanda whispers solemnly. "Here is her door." She pats an embedded rock. "If we talk too loud she'll hear us. If you step on her rock, she'll come out and cook you and eat your toes and throw the rest in the garbage."

Petey stamps on the rock. "Ho! Hello, Witch! I can beat you up!"

Amanda's face contorts with absolute terror. "Help! Help! I hear her coming! Here she comes!" She clasps her hands over her ears and squints her eyes shut.

This is contagious. Petey turns pale and dashes away, trips over a tree root, crashes, scrambles up, and weeping flees inside.

Amanda's mother emerges holding Petey's hand. "There isn't any witch under the front lawn," she says firmly. Yesterday, Amanda had decided that the laundry basket could be made to fly if she and Petey sat in it, closed their eyes tightly, held on and rocked gently, and concentrated. The basket had been filled with clean folded sheets at the time.

"There is! She's in there!" Amanda seems close to panic.

"Look. I'll move the rock and you'll see. There isn't any witch's hole down there." Amanda's mother bends to lift the rock, ignoring the cold chill down her spine. Amanda can be very convincing.

"No! No! No!" Amanda jumps up and down, clasping her hands between her legs for comfort. "I hear her coming! I hear her coming!"

"Amanda Jennings, you stop that right now. You are getting into a real tizzy fit!" Amanda's mother abandons the rock and frowns at her daughter.

"You made her mad! She's going to come out and eat us! You poophead!"

"To your room, Amanda!"

"No, I won't, and you can't make me!" Amanda is in a furor. "You *are* the witch! You're the bad mean nasty old witch! I hate you!"

Welcome to Four.

Four is the entering year to real childhood. All entering years are a bit painful, and Four has its share of tumult.

In Four, you can see a conscience struggling to be born. Three had an almost comical need to control the universe, and believed he could create a reality by stating it. Four has the same need and the same tendency to make up his own version of the truth, but with a new impetus: Four seems to be getting the idea, ever so dimly, that at some point he will have to judge his actions himself, and may have to find himself in error. Over this year, Four will start to learn about telling the truth and being responsible for mistakes.

Three lived in the here and now and was vulnerable to being scared, while Four has a grasp of "what if?" and knows how to worry. Four understands what a promise is, not only when someone else makes one, but when she does. Four can keep a secret—even from her parents.

But it's a great year. Four is a real little person, parents say. Four can negotiate and solve problems. Four has a sense of humor that is more than a sense of incongruity; it includes jokes and riddles. Where Three could take turns, Four can share. Three could take on a pretend identity, like Peter Pan or a monkey; Four can act out a scenario with plot and characterization and a moral.

Making the most of Four takes an enormous amount of energy. You have to help Four develop a good conscience, one that neither terrifies and inhibits nor condones too easily. You have to take all the fantasy seriously enough that Four isn't insulted, but make sure that Four learns the difference between real and pretend. You have to protect Four from harm, and teach Four how to protect herself, but without making her timid or offensive.

And then there are the Outbursts of Four. These should be rare by now. Oppositional behavior—the toxic back-and-forth fights between parent and child—should be a thing of the past.

Amanda, for instance, would not think of hitting, or of refusing to stop a behavior her parents order her not to do. She takes for granted going to the potty, going to preschool, putting away the toys on the floor. She and her seven-year-old brother play together a lot, without fighting.

But Four is not perfect.

**There are likely to be defiant outbursts.** Very often, they happen because something has disturbed her idea of herself. Amanda was "in character" as the Scared Narrator of a Spooky Encounter, and her mother spoiled it.

**It is extremely difficult for Four to admit to obvious guilt**. Four is unlikely to blame an imaginary friend, but she is perfectly capable of digging in her heels and saying "I didn't do it and it's not fair!"

**Four is prone to the more subtle forms of oppositional behavior:** Four is unlikely to hit, bite, kick, or pinch. Four should have already learned that a parental order must be obeyed, and that there is no use in engaging in toxic back-and-forth battles. Four should have only rare tantrums, and these are usually not oppositional tantrums directed at parents, but frustrational tantrums directed at Life.

On the other hand, besides the Blow Ups and the Refusal to Confess, Four is likely to mouth off (bad words; defiant statements), nag, whine, interrupt, procrastinate, and tattle.

Four still needs help with manners, too. The Big Words (please, thank you, excuse

> If Four is stuck in a true oppositional style of behavior, with frequent toxic back-and-forth battles with parents, she needs to be checked by her pediatrician. If all is well, go back to the sections on *Limits* in the previous chapters, find the age that Four is stuck at, and start there.

me, I'm sorry, *gesundheidt*) ought to be second nature by now. What is new is the concept of not hurting the feelings of other people, whether family, friends or strangers.

These are all addressed in the section on *Limits*, below.

And on top of all that, you have to figure out what Four means by "Why?"

"Why?" asks Four, as did Three and even Two on occasion. But now there is a difference. Two and Three asked why as a means of expressing a feeling; the *why* was attached to a statement in order to get someone's response. Four asks why in a more sophisticated manner.

Sometimes he really means *Why*. Why are you separating the egg yolk from the white? Why do you have to shift the gears before the car goes? Or, harder: Why do blue and yellow make green? Why is that horse getting up on top of the other one? Why is the moon so much bigger when it's low in the sky? Good luck.

At other times, Four seems to be checking out whether you and he are on the same wave-length. He is just discovering that other people can see things differently from the way he does. Why are you wearing that hat? can mean Why are you wearing that hat when it isn't raining, or Do you like that hat,

Does it feel good, Is that what grown-ups wear? Or, "why" masks a worry. Some conversations with Four can go on forever if you don't take into consideration this use of *why*:

"Now I need to look in your ears."

"Why?"

"To see if they are healthy."

"Why?"

"Because I want you to be strong and feel good."

"Why?"

"Because it is my job."

"Why?"

"Because I am a doctor."

"Why?"

"Because . . ."

The correct answer is: "You have great ears, and they are clean too, so we won't have to take the wax out with the little instrument the way we did last time."

Sometimes Four will use *why* as if he has a tic, almost singing it as a refrain. There is a literature report of a normal Four who asked more than four hundred *why* questions in one day. Or maybe it was one hour; I forget.

## Separation Issues

Most Fours have had considerable experience in separating from parents. So you wouldn't think there would be many separation problems at this age. But often there's a resurgence of separation anxiety.

This stems from the very special new aspect of Four. In developing his own conscience, Four is starting to judge other people's behavior. This didn't happen at Three. Even though this is an inevitable, important step in developing a conscience, it makes both Four and Four's parents newly aware of the painful side of being independent, separate, and autonomous.

In fact, Four can be so taken up with this discovery, she may convince herself that she is in control of her parents. This conviction is made intense by Four's fairly frequently expressed hostility and independence and subsequent guilt, and by Four's continuing belief in magical reality. ("If I believe it strongly enough, it will be so.")

"I hate you! You can go away forever!" Four screams. And then Mommy leaves for work. Four is in a panic. She has committed the ultimate act of separation. She has declared herself the judge of her own mother. More: she has uttered words that, if words were reality, would banish or kill.

This kind of separation fear can easily be transformed into problems at bedtime. At Four, there tend to be all kinds of monsters under the bed, in the closet. They're there even when Four isn't under any special kind of stress, such as a move, new sibling, illness, or family financial or marital problems, though such stress can make the monsters more numerous and powerful.

These are developmental monsters, and can only be banished by Four taking another step in growing up. Four has to learn not only that she can be independent of, and can judge, her parents. She also has to be sure that they are strong and mature enough to welcome her new maturity rather than take it as a personal attack from an equal. Parents must neither overreact, nor crumble in the face of Four's new behaviors.

She needs to learn that though she must judge her own actions to be less than perfect, she is still lovable. She needs to learn that mistakes and accidents and errors, broken promises and spilled secrets, are all part of being a person. Finally, she needs to learn how to express all these discoveries in ways that assert her independence without being arrogant, offensive, or manipulative.

This is a good reason to treat the hostility of Four with care. You don't want her hurting your feelings, and you don't want her guilty about doing so.

What Four really wants when she screams "I hate you! Go away forever!" is:

- To have her angry feelings recognized and dealt with as normal.
- To be told that her words are hurtful and unacceptable but not dangerous.
- To be given an appropriate discipline for the words, which she can control, but not punishment for the feelings, which she can't.

- To be assured that her hostility hasn't destroyed your love for her.
- To have her fear recognized, and assuaged, that what she wishes and says in a moment of passion can cast a spell and vanish, hurt, or kill someone.
- To be taught, when everyone has cooled down, appropriate ways of expressing anger, resentment, jealousy, and other negative feelings.

These developmental monsters feed on anything that threatens Four's confidence in her own conscience. Harsh discipline and no discipline can both shake up a Four, no question.

As with Three, videotapes can be upsetting. All the most popular ones focus on a threat of death or evil transfiguration of a parent, a threat engendered by the actions or immaturity of a child (*Beauty and the Beast, Dumbo, Bambi, The Little Mermaid,* and on and on).

To Four, the idea that a parent could be demolished or threatened in any way by Four's actions or existence is particularly devastating. It makes the developmental work of forming a conscience very hard, because that task demands that a child separate from parents in the most intimate way. That is: the child starts to replace her parents as the judge of her own and others' actions.

It is normal for a Four to be ambivalent about this developmental task. It has to be seen as risky. The more ambivalent Four is, the more these "classic" videotapes feed into these normal fears. Each time Four watches, the fear is aroused and dramatized, and then the whole issue is magically resolved. Watching these videos allows Four to "live through" the most anxious work of development without Four having to do or say anything. No wonder they're addicting!

Unfortunately, this can turn into a vicious cycle. The more Four relies on videos to do this "work" passively, the less she gets her own real developmental work done. And the less she's able to accomplish this work of Four, the more she relies on videos to soothe and reassure. She may become hysterical if deprived of them. In the face of this, parents often give in. But the developmental monsters aren't fooled by the videos; they feed on them. So Four wants even more.

These videos are really for children who have already worked through the initial tasks of conscience formation, and who aren't so ambivalent and threatened. What Four needs is real contact with other children and with grown-ups, contacts that teach Four that independence is not so scary. Four may need the same monster searches and night-lights and hugs and reassurance already recommended for Three.

And Four needs limits, and needs them to be set and followed through appropriately.

## Limits: Finally, the Authoritative Parent Appears!

Up until now, you were setting a foundation for behavior. You were teaching Cherub that if you made a real demand, gave a real order, *that thing was going to happen. Period.* Up until now, there were few and brief explanations of Why you were giving that order: what One and Two and Three needed to know was that you were In Charge.

Four needs explanations for the limits set on Four's behavior. When Four trespasses on a limit, she needs to hear not only that this May Not Be Done, but *why*. It's a good idea to think about how you do so, because Four takes what you say to heart, and may bounce back to you things you'd rather not have said in the first place.

She needs praise for good behavior, too. But some kinds of praise can backfire.

Here are the most frequent foibles of Four: • Mouthing off (bad words; defiant statements) • Nagging • Tattling • Whining • Interrupting • Procrastinating.

*The last three, Whining, Interrupting, and Procrastinating, are dealt with in the* Limits *section of the last chapter.*

## Mouthing Off with Epithets

Jane Jennings is lucky: Amanda only called her a mean old witch, and said she hated her mother. It can get worse than that. Much, much worse.

When it does, it's important to realize that Four is still basically guileless. Four has not had enough experience to know that certain words are so shocking, coming from Four's mouth, as to make adults turn pale and shake. Kenny, for instance, called his beloved grandmother a GD, MF SOB.

No matter what type of Mouthing Off occurs, what Four needs to know is *What* the infraction was, *Why* it is not allowed, and *What* to do about it. Making a bigger deal of it when the words are really, really awful is a mistake. It is the calling of names, not the degree of insult, that needs to be addressed. So tell Four the following:

- She has yelled hurtful words. Hurtful words always must be followed by a cooling-down period and an apology. "No hurtful words," her mother says in a serious, disappointed voice. "You hurt Mommy's feelings. You need to sit on your bed and think about not hurting other people."

- It doesn't matter why she did so. It doesn't matter what the other person did. Calling people hurtful names is simply not allowed. There are other ways to express anger.

- Instruction in how to apologize. "Then, you need to tell Mommy that you are very sorry that you hurt her feelings. You need to say, 'Mommy, I am sorry I called you bad names. I am sorry I said 'Hate.' "

- A suggestion of what to do instead. "I think what you meant to say was, 'Mommy, we are playing a big game of pretend, and I don't want you to spoil it.' "

- To have her fear recognized, that what she wishes and says in a moment of passion can cast a spell and vanish, hurt, or kill someone. And to be reassured that that can't happen. Amanda's outburst is not dangerous, and she has no power to eradicate Mommy. In fact, she is a powerless little kid who has Blown Her Top.

Period. "I know you didn't mean what you said about hating. Just because we think something or say something bad and mean doesn't make it come true. But we still need to say we are sorry."

After going through the full version a couple of times, shorthand is fine. "I heard bad words, Amanda. Bad words need an apology. We can't go to the park until I hear an apology." Then make sure you stick to your word.

## DEFIANT MOUTHING OFF

There are two versions of this: The specific: "You can't make me!" and the general: "You are not the boss of me!"

Ideally, when Four makes such statements, the adult should keep cool and count to ten before responding. That's because only the following three replies are unanswerable on the part of Four, and you need to decide which one to use.

**"You are not the boss of me! You can't make me!"**

Choose one of the following:

- "Oh yes I can."

   Just make sure you *can*. And then make Four do it. Right away.
- "No, I can't, and I don't care if you do it or not."

   This to be used only when you have phrased your words as a choice, not as an order.
- "No, I can't, but you need to know what will happen if you continue to disobey."

   A loss of video privileges? No visit to the park? No story at bedtime? Tell Four. Then stick to it.

## RUDE BACK TALK

When a Two shouts "Shut Up," you know he's just imitating what he's heard. When a Three does the same, it usually just sounds funny and pathetic—such a little person making such a big protest. But when Four does it—it's rude. Four has a much better idea that such an outburst is insulting: he knows because he's been on the receiving end of it himself, from peers (I hope not from family). Any such back talk needs to be squelched, partly because it makes a terrible impression on other people; partly because it will just get worse; but mostly because it lowers—no, it devastates, the tone of parent-child discourse.

The first thing to do, I'm afraid, is never ever model such behavior yourself, nor to allow it from older siblings. (And isn't that a whole other book.)

The second thing to do is to immediately make a big deal of how inappropriate it is. The removal of a highly prized privilege, toy, or activity is ideal, as long as Four feels *immediately* deprived. If it's only noon, canceling an evening's entertainment won't be likely to work. If your Four will stay in time out, and truly regards time out as a reproof that matters, that would be fine. You need to show your firm displeasure by your facial expression, your disappointed reproachful but firm tone of voice; and your removal of positive attention for at least five or ten minutes. Better face it now, my friends. It's easier with a Four than with a Fourteen.

## NAGGING

Nagging is one of those behaviors that is a true judgment call. What happens if Four really wants something (a toy, a privilege, your attention) and after saying No you realize that you didn't have a good reason to do so? If you then give in, Four will think the nagging worked. If you don't, you feel guilty. So think before you say "No."

### Rules About Nagging

- Asking for something is OK. Asking after the adult has given a definite refusal is nagging. Nagging automatically means that the answer is No.
- Nagging means that you think you can change somebody's mind by being unpleasant. The only way you can change somebody's mind is by showing them a reason, and being polite.

## TATTLING

Tattling is a behavior that you don't want to quench too successfully. If Petey decides to climb up the rose trellis, you definitely want Amanda to tell you this.

### Rules About Tattling

If someone might get hurt, or something might get broken, come and tell us right away. But telling just to get somebody else in trouble is Tattling. And give her some examples. When Four does tattle, ask Four to remember the rule: Tell you only if someone might get hurt or something might get broken. Be prepared to hear, "Well, Timmy took my truck, and I might get really mad at him and maybe I'll punch his nose!"

Most tattling is done between siblings to show who is dominant and to get parental attention. This is most likely to happen if the siblings fight a lot. Take a look at the essay in Section III on *Sibling Battles*.

## MODELING APPROPRIATE ASSERTIVE OR ANGRY BEHAVIOR

"You were really angry at me," Amanda's mother observes later, in a calmer moment. "And that's OK. What do I say when I get mad at Daddy?"

"I don't know," Amanda says shame-facedly.

"Well, I can say: "Dan, I am so angry at you. I wish you wouldn't leave my car without any gas in it. I had to get Bernice to drive me to the gas station to buy a gallon. You wouldn't like it if you found your car without any gas. PLEASE make sure you don't let it go below a quarter of a tank."

Amanda shrugs and looks away but she is listening.

"See, I didn't tell him he was a poophead. I didn't tell him I hated him. I told him why I was angry and what he should do about it."

Of course, that has to be what Amanda's mother really does say under those circumstances. Because whatever Amanda really does hear Mommy say will be reflected accurately in Amanda's anger vocabulary. ("Where did he ever learn that!" is one of the most frequent remarks of parents of Fours.)

## LYING

This is such a big part of Four. Four is going to lie when confronted with evidence of naughtiness or error. It's not just to "get out of trouble." It's because Four cannot conceive of not being perfect. There is a difference here.

But Four needs to be held accountable for her behavior, even more than Three did. If Tiny Baby Brother got lipstick all over his face, or the cookie jar is empty and there is a trail of crumbs, then you need to confront the situation.

But don't ask "Who did this?" or "Did you did this?"

It's too much to ask even an adult (!) for such a bald confession. Instead, start from what you know. "Baby Brother got lipstick all over his face. That was not a good thing to do, scribbling on Baby Brother. Lipstick isn't good for babies, and he might get a rash. I would like to hear you say, 'I am sorry, Baby Brother.' We cannot play (have a story, go to the store, have a friend over) until I see you telling Baby Brother you are sorry."

Then be prepared to accept the most subtle apology, such as a kiss on Baby Brother's toe, or the gift to Baby Brother of some not-very-cherished toy. Say, "I see that you are saying 'I'm Sorry' to Baby Brother. That is exactly right. That is called Apologizing."

Then be sure to tell another adult in Four's hearing, proudly, that "Four was very grown-up today. She apologized to Baby Brother." And watch Four look self-consciously virtuous and smug.

As with assertive and angry behavior, modeling apology is important. Be sure to apologize to Four, to Baby Brother, and to anybody else at the drop of a hat. (So be sure to drop a few hats.)

## BE SURE TO MAKE PLENTY OF MISTAKES

Four really confronts a parent with dilemmas. Was his taking a small bite out of each and every chocolate in your anniversary present a crime of passion to be forgiven, or a dreadful violation of limits? Is the crayoned-in library book Four's problem, *my* problem, or shared? Is her refusal to kiss Grandma disobedience, appropriate assertiveness, or cruelty?

Parents can't always give an accurate answer or mete out the appropriate response, and it would be too bad if they could. Part of the important lesson of Four for both parents and children is that parents are not infallible, do not have to be, and can negotiate, discuss, and when necessary apologize.

Most parents have been doing some of this all along, but it feels different when you make a mistake (and realize it) with Four. It feels as if Four *won*. There is a great temptation to ignore, to contest, to deny that you made a mistake.

One of the major achievements of this year is for both parents and Four to overcome this interpretation. Four needs many moral activities to be modeled, and when parents show enough self-esteem, dignity, and respect for righteousness to apologize to Four, they have taught an important lesson.

## Use consequences or time out, not spanking

First, spanking confuses Four about her own role in being the judge of her own behavior. Many Fours would prefer a spanking—a judgment and punishment from outside—to having to judge and deal with their own misdeeds. Spanking doesn't help a conscience to grow; it gives it an excuse to abdicate.

Second, Fours can be so annoying and challenging that a spanking can easily turn into a beating.

Third, Four can recognize that spanking is morally illogical; that you can't enforce a rule of No Hitting by whacking somebody. This can make Four lose respect for parents.

Finally, Fours can be made depressed and intimidated, angry and rebellious, or calloused and uncaring by frequent spankings. This is the age of compassion, and Fours now understand that others can have a different point of view and feel different emotions than they do. If a parent spanks, Four regards the parent as deficient in compassion, or in caring, or in self-control. Time out can be a minute on a quiet chair for a low-key child, four minutes in the (nonscary but very boring) laundry room for an intense one.

Prepare Four for time out before you need to use it. She can be told that it is cooldown time for hot tempers, that it is time for her to think about other ways to behave, that it is time for her parents to get hold of their tempers also. Show her where time out will be, and the timer that will be used. Run the timer so she can see how long time out will be: four minutes, one for every year of age. Long enough to read *Hop on Pop*, listen to three songs on her tape, run around the house several times.

Let her know that if she leaves the time out place, the timer will start again. It will also start again if she yells or uses forbidden words.

And list some of the actions that will always mean time out: hitting, calling hurtful names, giving way to tantrums, waving the sword indoors, throwing the ball at the cat.

## Manners and Civilized Behavior

For the first time, limits begin to mean not only restrictions on doing that which should not be done, but doing that which should be done. Fours are expected to say please, thank you, you're welcome, excuse me; to have reasonable table manners, help out at home, take care of themselves, and curb rude impulses such as pointing out that Aunt Mabel smells really peculiar.

Quite a list. No wonder Four needs some aid and assistance. What helps most with getting Four to act in a civilized fashion is to treat her with respect and dignity, assuming that Four will want to live up to standards.

Fours can feed themselves without much spilling, and can start to be taught the finer points of table manners, such as holding utensils, including forks and chopsticks, the adult way. They can be expected to use please and thank you. Many Fours can be pleasant company at dinners that are not too formal, if they are prepared with an hors d'oeuvre from home so they don't become too impatient for food right away, and with a quiet activity to assuage boredom.

Fours who monopolize the conversation can be a trial. Four can have an intense personal style, and chattering may be just an expression of this. The goal is to help Four become an intense listener as well. Try at home using a three-minute egg timer, the kind with an hourglass, and pass it around the table. Whoever holds the egg timer "has the floor" for those three minutes, while everybody else listens. Try to make the conversation interesting to Four, and when the next turn is Four's, end your turn with a specific question for Four to answer.

Four also can chatter as a way of dealing with performance anxiety and excitement, for instance in the pediatric or dental office, or out to dinner, or at a party. Rehearsing what's going to happen ahead of time, acting it through, can dampen the torrent a bit.

Once in a while, Four chatters "on automatic" because nobody really and truly listens. Try to make special efforts to listen to Four, making eye contact and appropriate listening noises. This may take enormous patience at first, but when Four really gets the hang of a dialogue, the chattering turns into conversation.

In general, when teaching Four manners:

- Make sure he understands what is expected of him, and whenever possible give him a choice. "When Uncle Irving comes, would you like to give him a hug and kiss or would you rather shake hands and say how glad you are to see him?"
- Give a reason; and if there isn't any practical, logical reason, say so. Appeal to his desire to excel. "When you hold your chopsticks like this, you can pick up the tiniest little bits of mushroom." "I know it's easier to hold your fork like that and it works just as well, but we are a sophisticated family, and at the table we all hold our forks like this, the way kings and queens do."
- Be specific and try to make it fun: "After you make your bed each morning, you could arrange your animals (which are all over the floor) to tell a story. And then you could tell me the story when you're done." "Here are the knives and forks and spoons. After you set the table, you could make a place card with a flower on it for each person and I'll write their names on them."
- Anticipate problems and enlist the help of Four in thinking up strategies to fix them. Four can help think of other ways to behave. Not at the time, maybe, but in anticipation and retrospect. "When we go to the store we always have a fight about buying a new Matchbox car. I hate to fight in the store and you never get to buy a car. What could we do about that?"
- Make use of Four's new ability to see

another's point of view. "Amanda doesn't like it when you push her off the swing. She skins her knees and it hurts her feelings."

- Let logical consequences occur rather than impose punishments. "When you push her off the swing, her mother gets angry and sends you home. Then you can't swing at all."
- If Four can't think of alternatives, suggest some. "You could take turns or you could find something else to do until Amanda gets tired of swinging."
- Follow through, be consistent, or Four will remember and quote you back to yourself. "But when I pushed Amanda off at the park you said she was swinging long enough and it was my turn."

### Startling Questions

As Four spends more and more time at other children's houses and in the world in general, you are more likely to hear startling questions or remarks. These are not forms of oppositional behavior; they are not accusations, or efforts to shock. To remain an "askable and tellable" parent, treat this kind of question and remark with respect:

- "Are you and Mommy getting divorced?"
- "What does Gay mean?" "What is French Kissing?"
- "Nathan's daddy is having an affair. What's a Naffair?"
- "What is a sizarian?"
- "I'm glad I'm little and I don't have to have sex so I won't get AIDS."

These are all verbatim quotes of Fours.

Four needs to know that you can be told and asked anything. But Four does not need, and is not entitled to, an accurate and complete answer to every question.

Some questions need to have an underlying, urgent worry addressed: "Mommy and I are not getting divorced. I think you are worried because you heard that Nathan's mommy and daddy are getting divorced. That is sad, but Nathan will be fine. And your mommy and daddy are not getting divorced."

Others can be answered very simply: "Caesarian is one way for a doctor to help a baby come out of the mommy's belly."

Some question, at some point, is going to catch you off guard. When it does, I have a strong suggestion: no matter what else you do, follow through on the question. After you have reacted, no matter with what outburst, spend a few moments gaining control and then explain to Four.

Explain why you reacted: "I'm sorry I shouted at you. I didn't expect you to ask that kind of question. It is a very grown-up question. In fact, I need to think about how to answer it for somebody your age. I will tell you later" but give a specific time, and then do so.

If you really can't think of a truthful answer you can stomach giving to Four, you can say, "Well, I thought about it, and I decided that an answer to this question needs to wait until you are older. But I will be sure to tell you all about it when you are (whatever: in kindergarten, ten years old, about to get married, running for political office—I'm just joking)."

Then reassure Four about any consequences for Four, and for Four's friends if necessary. "We don't know if it is true about Nathan's daddy. Even if it were true, his daddy still loves Nathan just the same. And *your* daddy isn't having an affair."

Finally, tell Four what is expected of her. "I am glad you told me what you heard, but we mustn't tell anybody else. It could hurt Nathan's feelings and his mommy and daddy's feelings very much. Let's promise each other we won't tell anybody else." And if you mean not telling Daddy, say so. If you intend to tell Daddy, say that.

## Day to Day

There is a discussion of toileting problems common to Four in this section. Learning to use the potty is a task usually taken up at around age Two, and is discussed in the chapter *Two Years to Three Years*.

### MILESTONES

#### Vision
Every Four needs a good reliable hearing and vision examination before entering kindergarten. This can be performed just before the child turns Five. If there is any concern that either sense isn't up to par, have it checked out with a specialist right away. If either eye is 20/40 or worse, Four needs to see an eye doctor—an ophthalmologist. By five, vision ought to test at 20/20.

Four will give you no clue if he can't see well. He thinks it's normal. He has no idea that he is supposed to see any differently. "I used to think the trees were big green blobs," a Four in corrective lenses tells me. "And the leaves were little chips coming off."

For the first time, color vision is tested in many Fours, and some boys who know their colors pretty well won't be able to tell red from green. This kind of color blindness doesn't cause much trouble until Four tries to get his pilot's license or pick out ties. Red-green confusion is inherited as a "sex-linked recessive" by boys through their mothers, who have normal vision. Rarely, a *change* in color vision can signal true eye disease, however.

#### Hearing
Four should be able to hear all pitches of sound at fifteen decibels or softer; twenty or over needs careful follow-up.

As with impaired vision, Four also won't recognize it if her hearing is diminished. She will simply say "What?" as often and as cheerfully as she says "Why?" Or she may be accused of not listening or not obeying. She may be fidgety, distractible, and hyperactive. Her voice may get louder and louder.

#### Speech
Four is such a highly emotional age, and Four is so interested in telling about everything, and I mean everything, on Four's mind, that adults can become dazed. When Four sees the listening adult glazing over, Four is likely to become dysfluent and stammer and say "Uh, er, um" a number of

times. This is normal. But a Four with real fluency problems needs help.

As with Three, *normal* stammering:

- Is not accompanied by a tense, ticlike facial expression.
- Occurs at the beginning, not the middle of words, most of the time, and with repetitions or *uhs* that only last for a fraction of a second.
- Only happens occasionally, not more than once in every ten seconds, and tends to come and go. It happens especially when the child is excited or tired or learning something new. Or when the listening adult glazes over.
- Doesn't last longer than three or four weeks.

Adults can help Four by waiting patiently, with no show of anxiety or concern, for Four to get the sentence out. Asking Four to slow down, or to think about what she wants to say, or telling Four not to stutter are guaranteed to make the dysfluency worse. Adults can help most by making the environment easy for Four to talk, slowing their own speech, reducing stress, and paying friendly attention to what Four has to say. They can ask Four questions that are easy to answer.

Signs that Four needs help with fluency problems, then, would be:

- Looking tense, blinking, or having tense lips, staring to the side as she tries to get the word out.
- Making long, up to a whole second, stut-

ters, and doing it frequently—more than about one sentence in ten.

- Stuttering most of the time, not only when having an emotional reason for doing so.
- Looking embarrassed or saying that she is worried about the stuttering or asking an adult why she stutters.
- Continuing to have a problem for as long as six to eight weeks.

As with Three, a Four with these signs needs a referral to an excellent speech therapist experienced with this problem. The Stuttering Foundation of America can provide referrals. Their toll-free number is 800-992-9392.

*Motor*

During this year, Four learns to somersault, hop, and pump the swing. Some Fours can skip, and a fair number who live where there's good weather and sidewalks learn to ride a two-wheeler without training wheels, but this is as much a cultural as a motor achievement.

*Fine Motor*

The wonderful person Three used to draw, with arms and legs growing from its head, now becomes a person with a body and legs, although the arms may still come out of the head. Most Fours will be able to print some letters by the end of this year, get dressed (even managing buttons in front) and use utensils well, including a knife and chopsticks, if so taught.

## Cognitive

Fours have a firm idea of what is normal in the world, and now want to experiment with alternatives to "normal." They ask *why* a million times a day, and can handle future events such as "what if?" and "if then" and "promise." Four starts to have a more grown-up sense of humor and starts to like riddles and jokes. Fantasy and reality still get confused, but Four is more likely to resent the fact when Four discovers it. For instance, magic tricks don't dumbfound Four as much as they did Three, but it's still important to make it clear that when it looked as if you took your thumb apart, it was just a trick, not real. Scary movies are real and mesmerizing to Four, and not recommended. Four is likely to become angry at the person who allowed or invited such a viewing.

## Sleep

Well, here we are. Most Fours sleep eleven to twelve hours a night, with *no daytime naps.* That doesn't mean no nap ever; any Four, or Forty for that matter, needs one occasionally. Many Fours still like and need a quiet time in early to mid-afternoon.

Night sleep for Four is commonly postponed and interrupted by monsters and bad dreams. These are generally related to what goes on in the daytime, and are discussed in the earlier sections of this chapter.

An important key to good sleep in Four is exercise. Many parents are shocked, when they sit down and think about it, how little sweaty, aerobic exercise Four gets. A Four in a preschool setting where playground games are monitored or organized, who doesn't walk to school, who likes fantasy play and books and, above all else, TV and computer games, gets very little aerobic exercise, and has to have it introduced as a priority into Four's schedule.

The absolute minimum is twenty consecutive minutes of sweaty exercise three times a week, but that's for cardiovascular fitness. For good sleep and feeling good, Four should get at least an hour every single day.

## Growth

Four will again gain about four to five pounds in the coming year, less than half a pound a month, and grow two-and-a-half to three-and-a-half inches.

### Chubbiness Watch

Fours definitely look like little kids, without "baby fat." They have only one chin, no arm or thigh creases, hardly any belly curve, and arches in their feet rather than pads. When they sit, their bellies barely, if at all, overlap their thighs.

Ask your pediatrician for Four's percentile curves, or plot them yourself—see the essay in Section III, *"Chubby or Not, Here We Come."* If Four is above the 75th percentile of weight for height, or at or above the 85th percentile for BMI, do some detective work and find where you can intervene.

## Teeth

The permanent front teeth are finishing their job of forming enamel, so it is still important

- A TV or video habit of more than an hour a day, total.
- Preferring quiet, nonconfrontational, nonaggressive activities to a rowdy playground.
- Not having an appropriate place to play.
- Drinking more than 16 oz of milk a day.
- Juice or soda instead of water to drink.
- Caloric lunches with cheese, sausage, fries, etc.
- Large helpings; unnecessary seconds and thirds.
- Addiction to many helpings of a rich food, such as macaroni and cheese.

to make sure that these highly visible teeth are protected from cavities by getting the right amount of calcium (see *Nutrition*, below) and fluoride. Too little fluoride won't protect against cavities, and too much may stain the teeth with white flecks. Ask your pediatrician or dentist for advice about supplements. Whether Four is getting fluoride as a supplement or in the fluoridated water supply, make sure he or she isn't also getting an unregulated amount from swallowing fluoridated toothpaste. Use only a tiny smear.

Many Fours can learn to floss now, and all Fours should be seeing the dentist regularly, twice a year for routine care.

## NUTRITION AND EATING

### *Foods*

I have not included the food pyramid in these sections. I have never known a parent to take kindly to the thing. Its main purpose is to remind us that children need very little in the way of sweet, sugary, greasy foods; that they need the basic amounts of grains (cereals, starches, and breads); that they need more fresh vegetables and fresh fruits than Americans are used to eating.

Indeed, it's crucial to keep this in mind with every meal, every snack, and every beverage. Remember, you are struggling against the mighty forces of fast food, soda, juice, and snack producers, and their media accomplices!

The second crucial thing to keep in mind is serving size. When you serve flat food, such as a piece of meat or poultry, the serving size is the size of your child's palm. When you serve a bulky food, the serving size is the size of your child's fist. Offer three of four different foods at each meal, in these serving sizes. If you keep the nutritional guidelines, the serving sizes, and the milk requirements in mind, your child will most likely be very well nourished.

The guidelines for Four are exactly the same as those for Three, in the last chapter. Be aware, though, that a Four is much more ingenious about getting addictive treats by trading for them at preschool, or finding them quietly in cupboard and refrigerator, for example.

## BATHING

Four still needs supervision. Three got into trouble at bath time inadvertently, accidentally sliding under the water, exploring the taste of the shampoo, turning on the hot-water faucet with no idea that the water would burn. Four is more likely to get into trouble against his better judgment and from a very poor tolerance of boredom. How would it feel if the water came all the way up to the top of the tub? If a little shampoo makes a few bubbles, how much will the whole bottle make? Will this fit up my nose? In my vagina? Up my penis?

A report of thirty-one boys with urinary tract infections, ages five to fourteen, says that a third of them admitted they injected water into their urethra while taking a bath a few days before the infection. They used a syringe, a rubber bulb, a plastic bottle, and a hand-held shower massager. There isn't any mention of possible previous sexual abuse, but any such activity or attempt should be reported to the pediatrician.*

Speaking of penises, most uncircumcised Fours will now have retractable foreskins. Be sure to teach him to pull it back, wash, rinse, and pull it up again. A foreskin that is left retracted can form a kind of tourniquet when he has an erection. This can be a true surgical emergency, and at best is a dreadful experience.

Neither Four nor anyone else should use cotton swabs to clean the ears. Even if you don't manage to injure the eardrum, you can push cerumen (wax) back so far that the drum can't be seen unless the wax is removed first. (There's a warning on the Q-tips box that nobody will admit to reading.)

A big issue has to do with showering or bathing with a parent or with a sibling of the opposite sex. Many respected pediatric authorities suggest that this not be done after age five. My own feeling is that if you have asked the question, you've already answered it: somebody is uncomfortable with this situation, so better not do it. Certainly any suggestion of eroticism in the bathing activities ought to bring an end to cobathing. It's confusing enough learning about sexuality in the modern world; why add to the chaos?

## DRESSING

Most Fours can dress themselves completely but can't yet tie shoes. (This doesn't matter because it is not cool for anyone under age 21 to tie their shoes any more.) They can button buttons in front.

*(Labbe, J. Pediat. 8 Nov. 1990 86:703–706.)

My advice on Fours and clothing is the following:

Fights about what Four will wear can be diminished by having in Four's closet and drawers only the clothes appropriate for the activities and season, and then gritting your teeth and really not caring which combination is donned.

Male Fours should not have pants with zippers, as they can zip their penises.

Of course, most children's clothing now is unisex, and Fours do like to dress up, often without regard to the gender issues of clothes. However, sometimes parents are concerned that a boy wants deeply and consistently to wear dresses, or that a girl refuses to. I would be somewhat concerned about gender identity in these cases. I would also be concerned about the girl who for a prolonged period refuses to wear anything but dresses, if dresses are inappropriate for her activities and not mandated by her culture. I would be concerned that she is anxious about her worth as a person, and worried that female attributes are of more value than generally human attributes. A conversation with the pediatrician about any of these concerns is worthwhile.

## ACTIVITIES

Teams and rules can be inflicted upon Four, but I don't know of any early childhood authority who thinks this is a good idea. Precocious exposure to the world of organized competition can cause bad side effects. It can rob children of the opportunity to discover the world of rules on their own. Left to their own devices, children up until about seven or eight love to make their own rules, shifting them, dropping them, finding new ones. "Calvinball," in the comic strip "Calvin and Hobbes," is an excellent example.

Competition also can lead to premature labeling and self-labeling. A Four who already "knows" that he is bad at soccer is a poignant sight; but so is one who "knows" that she is a champ at making goals. There is plenty of time for such self-definition later on.

And it can lead parents to excesses of team spirit, elitism, and vicarious thrills. Not you, of course; and certainly not me. But other children's parents. And you don't want your Four to be around that, do you?

## TOYS

A new and special problem with toys arises at four. Four sees a toy advertised on TV and demands it. Over and over and over. With her new skills of persistence, argument, and negotiation, Four can be a real nag.

What to do?

Four is not likely to understand about money, or abstract statements about how TV ads take advantage of children. So here is one parent's solution. One day, Marcus saw what looked like a wonderful toy you could get if you ate a certain meal at a certain fast-food restaurant. Marcus's father, after being nagged to his wit's end, realized that he had a golden opportunity. He talked carefully with Marcus about what the ad showed about the toy. How big would it be? What would it be able to do? Wow, it sounded great! He then took Marcus to the restaurant and purchased him his meal, which came with the toy. Marcus stared at it and moved it and stared at it some more.

Finally, "This isn't what it looked like on the TV," he said, sadly. "They made it up."

A sad, necessary lesson, one that some may feel should not be taught so early in life. But one that has to come someday.

Another problem with toys is safety. Especially:

Toy guns: *There are three main problems with toy guns:*

1. The way you play with a gun (including a squirt gun) is that you point it at people, pull the trigger, and "get" them. Shoot them. Kill them. Whatever. This is not a wholesome activity for Four.

2. If Four is allowed to play with toy guns, toy guns may be the only thing that Four chooses to play with. Play is the work that children do, and Four needs many more kinds of experience than Gun experience.

3. Toy guns may be colorful, so that the Police won't accidentally shoot someone armed with one. But they still are guns, with handles, barrels, and triggers. If Four encounters a real gun, he'll have no hesitation in treating it like a toy. Eliminating toy guns as appropriate toys for Four won't prevent this, of course, but may, just may, cause Four a lifesaving moment to pause and find an adult.

But suppose all the other children in the neighborhood play with guns? You can talk with the other parents (who may feel just as you do, but are too intimidated or worn out to do anything about it). Or you can say, "No guns in our house or our yard. Check them at the gate."

Certainly, Four (especially male Fours) will create their own guns from anything at hand: their own forefinger, a stick, a crayon, even (I've seen it) High Fashion Barbie, using her head for the handle and her thoracic protuberances as a trigger. But that's an impulse of the moment, much more easily given up when an alternative activity comes along.

Trampolines: *The American Academy of Pediatrics has actually issued a statement discouraging trampolines. This means all trampolines: the big backyard ones, the mini trampolines with a bar to hold on to, the ones at family fun centers and gymnastic programs and fitness outfits. The reason: spinal cord injuries. Even with spotters. Even with a coach.*

All-Terrain Vehicles: *Not below age twelve, and never three-wheeled.*

Bikes, skateboards, scooters and whatever new wheeled fad will appear in the twenty-first century: *Every such purchase or gift should have, as part of the vehicle, a helmet and appropriate padding. (Does anyone besides me have a baby picture of herself on ball-bearing roller skates with a pillow belted on to her bottom?)*

Nintendo and other video games: *They sharpen eye-hand coordination, but many people who study children feel that they are too addicting for young children.*

*When the interaction on the video screen is so instantaneous and rewarding, children may come to prefer these rewards to those of real, complex, often difficult human interaction.*

## TOILETING ISSUES

Four is as likely as Three to procrastinate about going to the bathroom. In some families, mornings are so rushed that Four is reluctant to go to the bathroom because a) he might miss something, b) parents are so frazzled he is afraid that they will fight with each other if he isn't there to distract and chaperone, or c) everyone is about to go their separate ways and Four hasn't had a sufficient infusion of parental affection.

- Make sure clothing is easy to remove and put back on. If Four has to remove culottes, tights, and panties, or suspenders, zip-up trousers, and underpants, and then redress afterwards, going to the bathroom becomes an ordeal.
- Avoid zippers. I have watched Fours of both sexes routinely unzip after they have already started to pee, and zip back up before they have finished. More often, they don't zip back up at all. Worse case scenario: the penis caught in the zipper.
- Make the time for having a bowel movement regular, sacrosanct, and pleasant. It usually works best to have it scheduled for after breakfast or before bed, when the call of the wild is not so alluring. The ten minutes or so should

be relaxed; no sounds or intrusions of flustered parents rushing off to work. Again, the toilet should be Four-friendly, with a seat Four can't fall through and a place to rest feet so that they don't just dangle. Provide nearby a pile of books, tape to listen to, even (God help us) a TV for cartoons (making sure no one can touch it when wet or tip it into the bath).

- Be there to cheer Four on and to help wipe (Four's arms are still proportionately short) and redress.

## SAFETY ISSUES

Four is somewhat less likely to choke on food or objects or to ingest something toxic, and much less likely to suffer nursemaid's elbow than earlier. However, some risks increase.

**Drowning** is still a major problem, especially since many Fours are water-comfortable and neither they nor their parents are aware that they are not water-safe. Fours can climb over fences to get to a pool or pond. They may be "nice" and not wake up a dozing parent before heading into pool or Jacuzzi. And while playing for an extended time in the water, they may involuntarily swallow enough that blood dilution dulls their mental processes, making them more vulnerable to drowning.

**Burns** are more likely now to be from accidents while helping than from impulsive grabbing.

**Getting lost** is a particular foible of Four, because of growing independence and self-confidence. Four is old enough for some specific instructions:

- If lost in a store, ask one of the people who runs the cash register for help. Do not leave the store.
- If lost in the outdoors, hug a tree. Stay with the tree and talk to the tree. People will come and find you.
- If lost in an airport or bus terminal, find someone wearing a uniform and a name tag. If you can't find such a person, find the escalator. (Fours are geniuses at finding escalators.) Stay by the escalator and do not move, even if someone tells you your parents are looking for you and they will take you to them. (Make sure searchers know you have taught Four this rule. When they find Four, sobbing by the down escalator, they are to page you, not try to take Four away from the escalator.)

**Beware iron tablets**. Fours, like Threes, can mistake them for candy. Iron in overdose is a potentially fatal ingestion.

## Health and Illness

*Part II of this book covers illnesses in children this age and older: how to assess symptoms, how to tell whether a child is sick enough to require urgent attention, and what kind of home treatment may be appropriate. It also examines problems by body part. The Glossary is an alphabetical listing of medical terms with pronunciations and, when appropriate, detailed descriptions.*

## EXPECTATIONS

As with Three, it isn't uncommon for Four to have at least one upper respiratory illness each month during late fall through early spring. This year also may bring at least one of the named childhood illnesses, like herpangina or hand-foot-and-mouth, and a bout of intestinal upset. By age Five, half of upper socioeconomic and 90 percent of lower socioeconomic children will have had a case of the virus that causes infectious mononucleosis. Most of the time, the illness is so mild or nonspecific that it doesn't get diagnosed as "mono."

The big symptoms for Four, however, are headaches and stomachaches. Most of the time they are related to fatigue of body and soul. Four is a very wearing age: you're thinking all the time and worrying a good part of it.

Once in a while, though, these can be true medical symptoms. The outlines in Part II will help you decide when to call or visit the pediatrician. (It is worth knowing that Fours who have strep throat classically complain of headache and/or stomachache, not of a sore throat.)

In any event, if the onset is not acute and recent, do write down a diary of the headache/stomachache symptoms before you visit or call the doctor. When did they start? How often do they occur? How long do they last? Are they now increasing or decreasing in frequency or intensity, or staying the same? Do they occur at one particular time of the day or week? Are there any associated symptoms at all? Do the headaches or stomachaches keep the child from play, preschool, or sleep? What helps them? Can you tell Four is having a symptom if he doesn't tell you, by how Four looks or acts?

Leg aches are another feature of Four. Usually these are "growing pains" which may have nothing to do with growth at all, but with sore tendons from jumping. These are usually thigh, shin, or calf pain and occur at night, in both legs or alternating. They are not in a joint. They are vague, with no specific point of pain. There isn't any fever, redness, pain, swelling, limp, or any symptom in other parts of the body. It helps to rub them. Jumping on the beds makes them worse.

Yep, sounds like growing pains. Even so, if they are very severe or persistent, see the pediatrician. (If they don't fit all the characteristics above, they may not be "growing pains" at all, and must be brought promptly to the attention of the pediatrician.)

Four is much more inclined than Three to provide a detailed, intense, self-directed, magical explanation for sickness and procedures. Four believes that illness is a punishment for misbehavior, and that necessary medical or surgical procedures are acts of malicious intent. He believes his body is complete as is, and that anything taken out (like blood) is gone for good and won't be replaced. He thinks that if something hurts now, it will hurt forever. He often is sure that the only reason a scratch or puncture wound heals is because there is a Band-Aid on it.

It is just barely possible to overcome these beliefs, but it takes goodwill, energy, and tact.

Four is never to be teased in a medical context. I know a Four who had a small infected boil on his buttock which needed draining. The inexperienced doctor, trying to joke the boy out of what he thought was exaggerated terror, told him: "We'll have to cut it off, here and here," indicating an above-the-tush amputation. Getting the child down from the ceiling was quite a task.

Anticipate what Four is worrying about. "No shots today," if you are positive there won't be. "If you open your mouth big and wide, the doctor won't have to put the stick in," if you are certain there won't have to be a strep test or culture. "He has the tiny washcloth (alcohol swab) to clean his stethoscope, not because you are going to have a needle." Avoid the use of the term *blood pressure*; anything with the word blood in it is assumed by Four to involve a needle. Tell him it is a tight-squeezer cuff that gives his arm a hug, and measures how strong his heart muscle is.

Dress Four for the office visit in loose, unlayered clothing that can be lifted up rather than taken off.

If Four has to have a blood test or injection, most do best with a brief explanation right before the test. Parents seem to help Fours most by expressing support rather than too much sympathy: "Boy, I know that hurts. That's right, make a good big yell. Wow, that looks like terrific blood in the tube. Nice and strong and rich. But only a tiny bit, and right this minute your body is making more blood."

This is more empowering than sympathetic moaning noises and piteous exclamations: "Oh, poor Justin, oh it hurts, oh dear, oh I'm so sorry." (This also tends to unnerve medical personnel.)

Address Four's magical thinking. "When I was little I thought that when I was sick it was because I had been bad. But that's not so. Germs make people sick. It is good for children to get sick so that their bodies learn to fight the germs. Then when you are older and go to school, you don't get sick so much. I'm glad you have a good strong body to fight off those germs."

Anticipate misconceptions. "I know your ear still hurts even though the doctor looked at you. It is the medicine that will help your ear to feel better. The medicine helps your body to get rid of the germs."

The more Four can participate in taking medicine the easier it is likely to be.

*Eye drops, nose drops or spray:* Let Four wash his hands, then put a drop of medicine on a clean finger and touch it to eyes or nose so he sees it doesn't sting and what it smells like. Let him feel a puff of nose spray or drop from the dropper on his cheek so he knows that it is cold. If he is still reluctant about nose medication, put a drop or spray on a cotton swab and have him touch the inside of his nose with it before you use the sprayer or the dropper.

*Oral medicine:* If you are sure it tastes good, including the aftertaste, let Four smell it and try a drop on his finger first. If it doesn't taste good, ask Four if he wants to put it in food (if your doctor says that is all right) or if he wants to try the grown-up chaser method. In the chaser method, the child takes a sip of very icy water before and after the dose. You can demonstrate with something delicious, like chocolate, that the ice water "puts your tasters to sleep."

*Chewable tablets:* Watch it be chewed, give a swallow of water, and check Four's mouth afterwards. A fair amount of medicine is probably still lodged in her molars. And sometimes Four is exceedingly clever: the mystery of one young man's intractable ear infections was revealed when a cache of pink and yellow chewable tablets was found secreted between his mattress and box spring.

*Swallowable tablets:* These aren't a good idea for most Fours. Their swallowing coordination isn't very sophisticated, and if they hold a tablet in their mouth and then are startled they can aspirate it into the airway.

## COMMON MINOR ILLNESSES

### Outgrown Infections

Four has already been exposed to, and become immune to, several common illnesses. This is likely to be the case even though Four has no history, and you have no memory, of any such disease: most of these illnesses produce immunity without symptoms in many children. These include rotavirus, bronchiolitis, and most important, roseola. In a Four, a high fever without other symptoms is almost never roseola (the diagnosis that leaps to mind in the case of younger children).

### Other Viruses

Four is still prone to some common viruses with unfamiliar names such as hand-foot-and-mouth, herpangina, and oral herpes (see Part II).

### Croup vs. Epidglottitis vs. Foreign Body

Croup is less common after age three, but the very serious rare disease sometimes confused with it called epiglottitis (see *Airway and Lungs* in *Body Parts*, Part II) becomes more common. An inhaled foreign body is another possibility. A Four who is making a noise while inhaling and has trouble breathing *could* have croup but don't assume that this is the cause of the problem.

### Eczema

The infantile form of eczema generally has either resolved by now or turns into the childhood kind. This kind can also appear for the first time at four. It consists of itchy patches behind the knees and in elbow creases, and often goes along with very dry skin.

### Asthma

Below the ages of three or four, wheezing most often is triggered by respiratory viruses, both in children with true asthma and children who just tend to wheeze with infection. Now children who have asthma are likely to start wheezing in the absence of infection. See *Breathing Diseases* in Part II, and the essay in Part III on *Allergies*.

### Car Sickness

Four can be prone to this, especially if she tries to do close work to pass the time, such as looking at a book or following the dots.

### Vocal Cord Nodules

When your Four starts sounding like a foghorn, it's probably just laryngitis. But if the hoarse voice persists for longer than a week, two at the most, this is probably because Four has some vocal cord nodules. These are growths on the vocal cords that are produced by—aren't you astonished—voice overuse. Shouting. Screaming. Pretending you are a lawn mower, fighter jet, elephant. Whatever. Since rarely the nodules can abscess, or become so substantial the child turns into a croaking whisperer, most pediatricians refer Four on to a pediatric ear-nose-throat specialist, and often to a speech therapist after that.

## COMMON SCARY BUT USUALLY INNOCENT PROBLEMS

These are identical to the ones for Three, and are covered in the previous chapter: night terrors, delirium with high fevers, nursemaid's elbow, reddened vagina and anus, and nosebleeds. Of course, Fours have a speciality of their own: Whipperdills.

### Whipperdills

As far as I know, this term was coined by Melinda Willett, whose five children are my patients. A Whipperdill is a bizarre, hair-raising, ultimately innocent accident of childhood in which the child is not, or only slightly injured, but the parent's heart stops. Fours are particularly prone to Whipperdills, especially those involving bunk beds, automatic garage doors, swimming pools and other bodies of water, older children's wheeled toys such as skateboards, and pointed objects they hold in their hands or mouths while running.

Moral of the story: A Whipperdill is a potential tragedy, and is only a Whipperdill in retrospect. Some can truly be avoided: never leave a Four and a garage-door opener alone for an instant.

After a child sustains a Whipperdill, it is politic to have a very serious and stern talk with the child, at least two such talks, with ensuing brief reminders. But don't, no matter how tempting, don't tell it to other people in the child's hearing: Four is an avid performer, and the temptation to repeat such a colorful and obviously unique feat may overwhelm Four's better judgment, such as it is.

A Four who has repeated Whipperdills is probably seeking attention and applause better given for other endeavors.

## SERIOUS AND POTENTIALLY SERIOUS ILLNESSES AND SYMPTOMS

The guidelines for Four are identical to those for Three regarding fever (it's not so much the height of the fever as its duration and accompanying symptoms that determine urgency and seriousness), urinary tract infections, and asthma. Please return to the previous chapter.

For parents worried that frequent illnesses may signal the onset of leukemia, please see this section of the *Two Years to Three Years* chapter.

A major worry of parents of Fours is child sexual abuse. Four-year-old sexuality is discussed in this chapter, with guidelines for when to worry about Four's behavior. Vaginal, rectal, and urinary problems are very common at Four and do not usually signal abuse.

## MINOR INJURIES

*Foreign bodies in the eyes*
It is my impression that these start to become much more common at Four, though I cannot document this. Perhaps it is because Four is able to give a coherent history. Perhaps it is because Four likes to stare so hard that she forgets to blink even during a sandstorm. Or perhaps it is because Fours enjoy throwing sand at each other.

The first rule for foreign bodies and chemicals in the eyes is to hold the eye open and wash with clear cool water for at least five minutes; up to fifteen minutes if there is a chemical in the eyes. Do this even before seeking help. Never try to "neutralize" a chemical, say flushing with baking-soda water for an acid splash. Always call the pediatrician after a chemical splash in the eyes.

Trying to flip the upper eyelid of a Four and look for a foreign body stuck on the underside is very difficult but can be done. See Part II.

Even if you think the foreign body is washed out or you have picked it out with a clean twist of hanky or cotton swab, watch the child for signs of a corneal abrasion. The cornea is the clear covering, like a natural contact lens, that covers the iris and pupil. If it is scraped, there is intense pain. The child won't open the eye, or if he will, tears and squints and complains, especially when looking at a light. Any such sign means a prompt evaluation.

If the eye injury involved any force (a direct hit with the fist, rock, ball) the child must be examined for deeper eye injuries such as a hyphema (see Glossary) or detached retina. If there is any penetration by a sharp object, like a pencil point or a thorn or whatever, even just *near* the eye, exquisite care must be taken to evaluate the child. Such penetrating injuries, even though they seem minor, can mean the loss of not just the injured eye but the good one as well.

## Major Injuries

### Real guns and dangerous toys

Fours are even more at risk than Threes from these potential agents of fatal injury. It is perfectly appropriate to canvass the parents of Fours' friends about the presence in their homes of firearms, and to ask how such weapons are stored. They should be unloaded, locked, and ammunition stored in a separate locked place. Same goes for fireworks.

If Four *must* have toy guns, the guns must not resemble real guns in any way except general shape. They should be brightly colored and obviously toys so that no one—police, frightened adult, or the child himself or herself—can make a mistake.

Darts, bows and arrows, slingshots, BB guns, air rifles—I've seen Fours given all these as presents. The adult who thinks these are appropriate for Four suffers from impaired mental and/or moral development, and deserves a thank-you note (written by parent and signed by an unreading Four) that says:

*Dear Aunt Boopsie,*

*Thank you for the darts. They are lots of fun. But now all my friends are in the hospital. I tried to play with Mommy but now she is lying down on the floor and won't play. So please send me some more and come play with me.*

*Love, your niece, Lizzy*

## Window of Opportunity

Four is the year to prepare for entrance into the big world. To be suitably armored, Four needs to know some sophisticated basics about sex, safety, and peer behavior. Parents also need to evaluate the specific prospective kindergarten to make sure there will be a good fit. All this takes time and energy.

### Sex Education for Four

Four lurches wildly between modesty ("No! You can't look!" Four grips his pants white-knuckled at the approach of the pediatrician;

## WHEN FOUR ASKS ABOUT MENSTRUATION, OR STARES, OR ACCIDENTALLY INTRUDES

Yes, that is called a tampon (or a sanitary pad). It does have blood on it, but nobody is hurt. It's a special kind of blood.

It is blood that a mommy makes in her womb for a soft place for a baby to grow.

Every month, if a baby doesn't get made, that blood comes out. That is called a period. Then the next month, the mommy makes more.

Girls: Oh, when you get to be twelve or thirteen. A long time from now. We'll talk about it before it happens so you'll know all about it.

It is private to have a period. We don't talk about it with people outside our family, except at the doctor's.

sometimes he won't even take off his shirt) and exhibitionism. (Here come Amanda and Wendy, giggling hysterically, each holding up the back of the other's skirt to expose the panties.)

Four's attitude towards sex play has changed. Now it is flirting with the Forbidden. Three merely retires to her room with a friend and doesn't even close the door, and looks up unalarmed but startled at intrusion. Fours sneak off to the bathroom with the door closed and are flustered when discovered.

Like Three, Four masturbates, but can usually be persuaded to do so in private. A female Four may be very curious about her father's, and a male Four about his mother's, genitals. Little girls may want to look at themselves in a mirror to see where everything is. Both sexes may be very curious, but may not ask overtly, about the paraphernalia of menstruation.

If you didn't get around to much in the

way of sex education for Three, look at the corresponding section of the previous chapter. If Three is pretty well grounded in the basics, you may want to go ahead to further explanations.

## GOOD TOUCH/BAD TOUCH

Four can be made fearful, even phobic, by serious discussions about strangers, good and bad touch, or death. Sitting down with a book on the topic at hand may make them anxious also, as it may subtly tell them that parents are uncomfortable with, scared by, or uncertain about the subject. Moreover, many books use figures of speech that Fours can't grasp. One book for children about strangers, for instance, says that most strangers are good but there may be an occasional "bad apple." This leaves most Fours utterly bewildered.

One solution is to have a couple of sensible rules that you announce in a brisk and businesslike voice, and then reinforce briefly on several occasions. There are three good rules about touching.

- Your body belongs to you, and that includes your private parts. Private parts of the body are those covered by a bathing suit. Private parts can be fun to touch and play with, but should only be touched that way by the person who owns the private parts. When people grow up, they can decide who else can touch them in that way, but not until they grow up. Be gentle when you discuss this rule in relation to normal sex play.
- Any kind of touch or looking, to the private parts or otherwise, that makes Four uncomfortable is not to be allowed, whether the person doing the touching is a child or grown-up, friend or family member. Four can and should say no and tell someone he trusts about the problem right away.
- If a doctor or medical person has to touch or look at the private parts, it's OK, but only if the parent tells Four that it is OK.

Good times to tell and then to remind Four are during bath or toileting activities, before and during a doctor visit, or when Four is found masturbating or having sex play with a friend. Another good time to discuss such things with Four is before visiting "kissy-huggy" friends and relatives. Remind Four that doting relatives love a hug and kiss. Usually this doesn't feel like "bad

touch." But if Four really objects to such touch, perhaps he or she can shake hands, bow, curtsy, and smile instead. If Four says that such actions do indeed feel like "bad touch" and not just funny or yucky, believe Four. Teach alternative affectionate greetings, and don't leave Four alone with that particular relative or friend.

### STRANGERS

Once again, it's best to communicate information briskly and casually, in small bits rather than in a serious planned discussion. Four needs to know two things:

- Most people are friendly, helpful, interesting, and polite. It is appropriate to smile, be polite, say hello, shake hands or "give five" to strangers when you are with your trusted adults.
- Four is never ever to go anywhere with anyone—stranger, friend, or relative—unless a parent has told him personally and specifically that this is OK.

Then run through some examples. You and Four have a nice chat with an older man in the park. After you have left, you talk about how nice that was. But what if the man had asked Four to come home with him for lunch? No. Not unless Parent was there and explicitly said yes. And what if that nice young woman, who just told Four that she was cute as a button, had asked Four to help her find her lost bunny rabbit? No. Four has to say that her mommy or daddy doesn't allow her to go off without telling them. Not even for a lost bunny. Not even if the young woman says that Mommy or Daddy told her it was OK. Not even if the young woman knows Four's name.

If you feel particularly concerned about strangers, ask a trusted friend or two to "test" Four by encountering him at play or at preschool and offering a ride home or asking for Four's assistance. If Four follows the rule, give much praise and a prize. If Four takes up the offer, have the friend remind him of the rule, and have a brief chat again about the rule.

If you have a concern about an estranged spouse or family member abducting Four, you need to be explicit but not frightening. "I want you to be very sure that you know that this rule includes Daddy (or Mommy or Grandma). Let's just pretend one more time. You pretend to be Daddy (or Mommy or Grandma) and I'll pretend to be you. You try to make me come with you, and I'll show you what to say or do."

Don't discuss what could happen, or give scary visions of kidnapping, torture, abuse. Just make it a rule.

If such a case is being publicized on the media, try to keep it from Four. If Four does hear about it, make sure Four knows that such cases are very rare. Remind Four that the rules have to do with Four always being sure Mommy or Daddy is in charge of where he is, until he is grown up. They do not have to do with a clear and present danger of harm.

## BEYOND SHOW AND TOUCH: WHEN IS SEX PLAY NORMAL, AND WHEN IS IT WORRISOME?

Sex play is as inevitable at Four as it was at Three.

But is it normal for Fours to insert magic markers into each other's orifices? To French kiss? To make Ken and Barbie have intercourse?

A large study looks at different sexual behaviors at ages two through six and six through twelve.* The 880 children studied lived in a community that was predominantly (98 percent) white, with parents who had some college education and earned more than $15,000 a year.

None of these children had a known history of sexual abuse, but since only a portion of abused children disclose such a history, it is likely that a few had been abused.

While the demographics make the study pretty circumscribed in its implications, it gives an interesting idea of the range of sexual behaviors in young children, and flags some areas of possible concern.

Some behaviors were very common. About half of the young boys and girls undressed in front of other people, touched their own sex parts at home, scratched their crotches. About a third of boys and nearly half of girls touched or tried to touch their mother's or other women's breasts. (No little boys, but about 8 percent of little girls, put their mouths on their mother's breasts.)

---

*Freidrich et al., "Normative Sexual Behavior," *Pediatrics* Sept. 1991 88 [3]

About a third of both sexes tried to look at people undressing.

Moreover, about 20 percent of boys and 15 percent of girls were observed to masturbate with their hands, and 35 percent of boys and about 20 percent of girls touched their sexual parts in public.

But there were some behaviors that only a few children engaged in. Fewer than *1 percent* of the young children studied were observed by their parents to:

- Put mouth on another child's or adult's genitals.
- Ask to engage in sex acts.
- Masturbate with an object.
- Insert or try to insert objects in vagina/anus.

And fewer than *10 percent* were observed to:

- Imitate intercourse.
- Make sexual sounds (sighing, heavy breathing).
- French kiss.
- Try to undress other people (children or adults) against their will.
- Ask to watch sexuality explicit TV.
- Imitate sexual behavior with dolls.
- Talk about sexual acts.
- Touch other's sexual parts (though 8.9 percent of little boys did this).
- Use sexual words.
- Draw sex parts in pictures.

The study showed that the more explicit sexual behavior in young children correlated with increased family sexuality, such as family nudity, bathing with adults, seeing pictures or TV with nude adults.

The study doesn't, and can't, say that a child who engages in one or more of the very infrequent behaviors is behaving abnormally. It does suggest, however, that *"a child who is exhibiting several of the least frequent behaviors appears to be unusual. A careful assessment of the family context by the pediatrician may lead to the conclusion that for this child witnessing adult sexuality is quite likely, and in addition greater-than-average levels of family distress may be present."*

This doesn't necessarily mean that the child has been abused. Further studies would need to be conducted to be sure. But, the authors state that *"preliminary research with other samples seems to point to that likelihood."*

A Four who seems to be involved in several of the rare behaviors, or even one behavior on several occasions, deserves a consultation with his or her pediatrician.

Another question about sex play that comes up has to do with homosexuality.

"Rob refuses to go to preschool," a mother told me. "His best friend is a little girl named Jessica, and he likes to play with her a lot, and another little boy, Harvey, started calling him a sissy!"

In the study quoted, about 8 percent of both little boys and little girls had said on occasion that they wanted to be the opposite sex. (In the seven-to twelve-year-olds, only 1.9 percent of boys and 1.1 percent of girls expressed such a wish.) Moreover, 16.9 percent of boys and 20.6 percent of girls pretended to be the opposite sex. Also, 53.9

percent of boys played with "girls" toys, and 63.3 percent of girls played with "boys" toys.

From Rob's mother's anecdote, it is clear that even very young children can pick up some of society's anxiety about homosexuality. Parents can worry too. Rob's mother was appropriately more concerned about Harvey's teasing than about Rob's playing with Jessica.

Once in a while a child shows a strong, definite, explicit preference to dress, play, and present himself or herself as the opposite sex. Since this may reflect a problem with relationships in the family, it's appropriate to discuss this with the pediatrician or with a recommended, trusted counselor.

## PREPARING FOR KINDERGARTEN

One of the most important things parents do for Four is to decide thoughtfully what to do about kindergarten entrance.

Don't be so engrossed with Four that you forget that he or she will soon turn Five. Unlikely that you would, but the earlier this year you start thinking ahead to kindergarten, the better. If you wait until Five is upon you, you may feel pressured to make fast and not well-considered decisions. Moreover, depending on the school, some preparation for kindergarten may need to take place this year.

Most critically, it may be obvious to you even now that *your* Four is not going to be ready to enter kindergarten at Five with her friends. This does NOT reflect on Four's intelligence, character, or potential. Many bright, creative, active Fours can't pay attention unless all distractions are removed, can't

learn unless something is shown them as well as told to them, and have a hard time sharing. If these traits don't develop by kindergarten entrance, Four will have a difficult time of it in kindergarten. Many Fours-into-Fives are in this situation, and do much better if kindergarten is postponed for a year.

The whole point of kindergarten readiness is to make sure that the kindergartener counts school a pleasure and himself a success at it. Nothing much else matters about kindergarten. One final consideration: it isn't quite enough to think ahead to kindergarten. First grade is the real "school entry" at many schools. First grade brings homework, cliques, team games, competition, reading out loud in front of the class. Some children who do well in kindergarten just aren't ready for first grade. It's not a bad idea to talk to Four, and even to other adults in Four's hearing, about kindergarten as if it is *usually* a two-year enterprise. It is much easier to deal with a repeat of kindergarten if that is what is expected.

*Research the schools.*
Even if you don't have a choice among several schools, there are some things to find out about the designated one.

- Is there a choice of classrooms with teachers of differing ability and personality? If one is clearly better than another, is it certain he will be there next year? Or is she pregnant and taking a leave?
- Is there a year-round system with tracks? Which is the most desirable track, and why? If a track seems desirable in one

way (the best vacations) but not in another (poor parent participation), what could you do to make it more desirable in every way?

- What is the average age of entering kindergartners of your child's sex? Are all the little boys closer to six than to five?
- What happens on the playground? Don't limit your observations to the classroom. Does the whole school get out there and play Torture the Little Kids? Do they expect kindergartners to play games with teams and winners and losers?
- What are the bathrooms like? If they are dreadful, what is the PTA doing about it? How would the school like to have parents come in and show them once and for all what a clean bathroom is? (Uh oh. Schools tend to just love it.)
- Is what they say they expect from incoming kindergartners different from what they really do expect? This can go both ways. The letter or brochure may say that all early reading skills are taught in kindergarten, and then you arrive and discover every single child except Robbie knows the alphabet song and can print his name. Or the school may boast that it caters to "gifted" children, but actually takes children without or despite a screening exam, so that Becky spends the first three months bored to tears, "learning" the primary colors.
- How do you get to school? Is there a bus, and if so, is it fairly civilized? Is the route fairly safe, or vulnerable to sharp curves and floods? Has the PTA done anything about seat belts on the bus? If not a bus,

can you carpool? Or, unheard of in this day and age, walk? If so, is there a trustworthy older child to serve as escort?

- What happens when somebody wets his pants? Does everybody have a change of clothes? Is the teacher able to refrain from exasperation? Is the child humiliated in any way?
- What happens when somebody throws up? Is there a school nurse? Does the teacher tend to throw up too, and then hold a grudge against the vomiter?
- And speaking of sickness, what is your plan if the kindergartner gets sick and can't attend school? Or gets sick *at* school, and has to be brought home? Parents who work outside the home can be well and truly trapped if a good solution hasn't been worked out ahead of time.

*Consider Four himself.*

Are there some general things that will make kindergarten easier that can be worked on this year? Certainly, Four should be able to separate easily from his mother and father. Most Fours have had years of experience doing so. If not, this is the year for it. Four needs practice taking turns, sharing, and following directions promptly and agreeably. Preschool and daycare teachers can let you know how Four is doing in these respects.

Four also needs permitted ways of being assertive. Many Fours need encouragement to stand up for themselves rather than always calling on the teacher, caving in, or whining. Parents who can model and respond to forthright assertive expression do Four a big favor. Preschools may not specifically address this issue, except to encourage children not to hit and to call the teacher when there are disputes.

No matter how much parents may work against sexual stereotyping, it appears to be one of the aims of Four to seek out and act on such stereotypes.

Fours need to learn to be assertive without being aggressive. Roughhousing is one thing; bullying is quite another. A father who can model civilized behavior, and a mother who does not allow herself to be bullied by anyone, make a big impression.

Are there specific things Four needs to learn this year to make next year easier? The best way to find out is to talk with Four's preschool teacher or daycare mother, his putative kindergarten teacher, and the parents of somebody in Four's anticipated kindergarten. The alphabet? Printing his own name? Counting to how many? Throwing a ball? Catching one? Kicking a ball? What about swings, slides, monkey bars?

Can Four not only use the toilet, but wipe and flush and wash her hands?

Can Four make it through preschool without his security object? Are security objects allowed in the anticipated kindergarten?

Does Four take after a parent who had a particularly wonderful or a particularly dreadful kindergarten or first-grade experience? And if so, is the prospective school enough like the parent's school that the experiences are likely to be comparable? If so, Four's experience may well replicate that of the parent.

## What if?

### STARTING DAYCARE OR PRESCHOOL

Starting Daycare is discussed in the previous chapter in *Window of Opportunity*.

### EXTENDED SEPARATION FROM THE PARENTS

### TRAVELING WITH THE CHILD

### CONCEIVING A SECOND CHILD NOW

All the above are covered for this year in the What If? section of the previous chapter.

## Arrival of a New Sibling

For many families, having a new baby when the first is Four or older is easier than if they'd done so earlier. Four is independent and can separate easily. He has a life of his own with plots and subplots to keep his eye on and his mind occupied. He can talk about feelings instead of acting them out, make logical connections, and has a certain degree of self-control.

Parents have had a break from the intensity of the first three years, and they get to have one child finish college before the next begins. (The advantage here isn't just financial. The sibling left behind gets to go through high school with less of a burden; the older one might have been a hard or notorious act to follow.)

But there are a few special aspects to take into account when Four becomes an older sibling.

- Fours have more intense and detailed fears about childbirth. They absorb overheard conversations about labor, hemorrhaging, and danger in general. A Four whose younger sister has Down syndrome worried all through her mother's third pregnancy and was able to discuss amniocentesis in a sophisticated manner.
- Fours also have fears about hospitals in general. They may be convinced (largely through TV) that hospitals are for emergencies and dying.
- No matter how clearly or how often the facts of life have been presented to Four, he is still likely to have misconceptions.

Descriptions of normal vaginal birth or of a cesarean section, if one is anticipated, are both likely to be met with a suspicious eye.

- Four is still very self-focused, but has a better grasp of the future than Three. (Many Threes have trouble even using the future tense.) Four may start worrying way ahead of time about where he will stay while his mother is giving birth, where the baby will sleep, and what will happen to him in the new family constellation.
- Four's propensity to ask "Why?" all the time can get parents into difficulty. "Why are you having another baby? Why are you having that baby right now? Why are you so tired? Why is Mommy throwing up? Why can't Mommy play with me? Why do we have to shop for the baby? Why can't I go to the hospital with you?" It can make parents a bit defensive, and it's easy to find yourself backed into a guilt-ridden corner.
- Four suffers much more than Three when she falls short in a role. Three has only a vague notion of the duties of an Older Sibling. Four does too, but she doesn't realize it. Any comment or criticism can be devastating: "Terry, PLEASE stop that loud singing; you'll wake the baby." (This to a young person whose voice had always been a matter of pride to her whole family.) When devastated in her feelings, Four tends to rage. This makes matters worse.

Ah, but all these are compensated for by two wonderful things about Four.

Four is beginning to be aware of standing with peers. The arrival of a new baby is a real coup. "I have to hang up now," Jason tells a friend. "I have to go burp the baby."

And Four can memorize an oft-told story and tell it to himself later on. One of the most helpful things parents can do way ahead of time is to make a series of "books" for Four and "read" them with her often.

One is what will happen to Mommy when she goes to the hospital. It has pictures of the hospital, a room and bed, giving birth, holding the baby. Copies of Four's own birth pictures are fine.

Another is what will happen to Four during the same time period: pictures of Grandma and Grandpa, who are coming down from Chicago to stay, or of the neighbors Four will be taken to. Then pictures of some activities Four will take part in while his parents are otherwise occupied: going to preschool, going to the park.

There is the exciting picture book about going to the hospital to see the new baby, with tactful reminders about hand washing, gentle touching, and not bouncing on Mommy's belly.

Finally, there is the coming home book, with pictures about where the new baby will sleep, and of the usual activities of life with a newborn. These pictures can dramatize the rather vegetative life of a neonate so that Four can rehearse the fact that Baby isn't going to smile and play, much less join Four in the sandbox.

There is a back-up picture book also, about "What if I get one of my colds and can't see Mommy and the new baby right away?" This features talking on the phone to Mommy, seeing pictures of the new baby, watching forbidden TV programs on the couch, and playing lots of board games while drinking lemonade. It includes Mommy and the baby coming home and waving from the doorway, hearing the baby cry a lot, and ends with a glad reconciliation.

Making such books at home works better than buying them, not only because they are personalized but because Four can help create them. Each step can be presented as an interesting problem for Four to solve: "What kind of bed should Mommy have in the hospital? Some beds go up and down. What would a good bed be able to do?" And books are better than videos because you can dwell over them and turn back pages and point at things. Sources for pictures can be snapshots, magazines, and hospital brochures.

Four may also be able to come up with her own solutions to her worries. A Four wanting reassurance is told, "I love you and I love the new baby. I love you just as much as I did before the new baby came." As if this was exactly what Four had been thinking, she chimes in, "And you love me with the pink love and you love Baby Jenna with the blue love." "Yes, that's right," the alert mother agrees without missing a beat. "I love you with the pink love and Baby Jenna with the blue love."

## DEALING WITH A YOUNGER SIBLING

1. Is Four getting enough satisfying one-to-one time with each parent?

2. Does Four have a satisfying, challenging, and rewarding life with her peers, on a regular and daily basis?

3. Is there strife between the parents?

Fours are often dramatically more able to cope with a younger sibling who is aggressive, nonverbal, and attention-getting than they were at Three. However, Four still cannot "baby-sit"—entertain and protect—anyone, even herself, safely. It's a mistake that can be tragic to leave Four alone, even for a minute, with an infant. "But she *told* me she wanted a peanut," Four howls, appalled, as the paramedics arrive.

If Younger Sibling is old enough (around Two) for both to play together, it's the right age to nip Sibling Battles and toxic rivalry in the bud. See the essay in Section III on *Sibling Battles*.

## DIVORCE AND CUSTODY ISSUES

For a discussion of joint versus sole custody, and frequency of visitation, see this section in the previous chapter on Threes.

Fours are very vulnerable to family discord and divorce. They are in the tough position of understanding a great deal of what is said (or shouted) in their hearing, and also of feeling as if whatever is happening is their own fault.

Such stress on Four makes Four very difficult to live with: ornery, anxious, clinging, bitter, frightened, stubborn. This makes any marital stress that much more intense, and Four reacts with an escalation of desperate behavior, and a vicious cycle can spin out of control.

It is worth every effort to keep Four away from violence, even the silent bitter violence that goes unspoken. Unfortunately, getting Four out of the house on a prolonged visit to Grandma, for instance, may backfire, as Four is likely to worry that things will fall apart further in his absence. He may regard himself as the force that is holding his parents together, as well as the engineer of their unhappiness.

Almost universally, Fours will make up a story to satisfy themselves about why a divorce takes place, and the story will feature Four as the main actor. This is another reason why the most popular videos, in which children participate in a parent's loss, should be avoided now (see section on Separation Issues in the previous chapter).

Equally universally, Fours announce that parents are going to reunite, or get married again. They may tell family and friends this. They may ask *when* is, not *why isn't*, the other parent going to come back. It is not unusual for a Four in this situation to need formal therapy. Because Four is old enough to construct a scenario, to tell a story, to dramatize his pain, such therapy can be very effective.

Some things to understand about Fours facing divorce include:

• Fours have a lot of trouble with time concepts, and the anxiety of going back and forth on a frequent basis may be overwhelming. "I don't understand," Chris says over and over and over. "Will

I sleep here this night and tomorrow night? Or will I sleep at home with you?"

- Some preschools aren't very tactful. Children without mothers are asked to make Mother's Day cards and those without fathers to produce Father's Day clay dishes.
- Like Threes, Fours may find it very traumatic to go directly from one parent to the other. Pickups and drop-offs are usually easier on Four if they occur on neutral territory.
- Four is just mature enough to make it tempting to use him or her as a go-between or tale-bearer. Unfortunately for Four, the same maturity, the ability to tell a story, to remember and reproduce conversation, and to judge adult behavior makes it terrifying for him to have to take on this role. It only makes Four feel more powerful and responsible. Such feelings may be so unbearable that they overwhelm Four's attempts at developing conscience-directed behavior.
- Four is very likely to try to manipulate parents, telling Mommy that Daddy bought him this or Daddy that Mommy always lets him do that. The thing to keep in mind is that what Four really wants is not more gifts or more privileges, but the comforting knowledge that he or she is not in charge and not able to manipulate his powerful parents. Say to yourself, "This is a test. This is only a test. In the event of a real desire, Four will come up with a better reason than 'Mommy lets me' or 'Daddy gives me.'"

- Loving, nonmanipulative relatives and friends that form an extended family can be most helpful for Four. They should be counseled about all the special features of Four and how to handle "made-up" reconciliations, attempts at manipulation, and the occasional Four who has been asked to spy, and who discusses parental secrets with anyone and everyone.
- Both parents need to address the issues of good touch/bad touch and strangers in a civilized fashion. Accusations of molestation and attempts at child abduction by a parent are common. If the divorce is not amicable, and if both parents can agree on a therapist for Four, such a person should be involved as early in the proceedings as possible. This person should be the arbitrator in the event of accusations and counteraccusations. It is nearly bootless to introduce such a therapist after the hostilities have reached this point, but involvement early on can prevent much pain and devastation to all concerned.
- Be careful about storybooks about divorce. Fours may take things so literally that they think what happens in a given book is exactly what will happen to them. Avoid like the plague any book with a "happy ending" that reunites the parents.
- When a parent has a new friend of the opposite sex, it's kindest to Four to keep that person out of the picture until it is certain that the new friend will be a permanent attachment. Fours who see a mommy or daddy with many different

friends, expressing physical and verbal affection, can become confused, enraged, resentful, and depressed.

## SURGERY AND HOSPITALIZATION

Since Four is still solidifying body image, elective procedures should be postponed for another year or two if possible. Many times this isn't possible. In particular, it isn't uncommon for Fours to require tests or procedures that involve the genital region: bladder and kidney X rays that involve catheterization, treatment of hernias, and so on.

It often helps to prepare Four, explaining in simple terms what has to be done and why, and showing the procedure on a doll. The more Four can be persuaded that a procedure is normal, the more comfortable Four will be.

It's particularly helpful to tell Four explicitly that:

- A scar is stronger than regular skin, not weaker, and won't come open afterwards.
- His body makes new blood all the time, and when the doctor takes some out, his body doesn't even miss it. You might even say that this blood is special blood that his body makes extra, just for doctors to use for testing.
- The surgery doesn't change anything fundamental about him or her, especially not anything about being a boy or a girl.
- Many Fours have to go to the hospital to have surgery; Four is not alone. Find a picture book at the library to reinforce

this. Books are useful here because the trauma of going to the hospital and having surgery is very much the same experience no matter where you live or what your family situation is. That's not true of other situations, such as parental divorce.

## MOVING

Because Four is more able to think ahead and worry about the future, Four benefits from as much preparation for a move as possible. Pictures of the new home and neighborhood are helpful; pictures of the new school, and if possible of teacher and classmates, are brilliant. A preschool teacher may make a project with her class of welcoming the new child, sending the class picture and drawings from the children and writing down questions and comments that they have.

Leaving old friends is just as hard as adjusting to new ones. Four will like to give special friends a batch of preaddressed, prestamped envelopes. Parents of left-behind friends can send photos and transcribed letters to Four in the new place.

Books about moving should be approached cautiously with Four, who may think (as with books about divorce) that they accurately predict the new home, neighborhood, school.

Having Four elsewhere and supervised on moving day is absolutely crucial. Not only does Four feel mature and able to be in charge of everything, Four is likely to be in a state of high emotional arousal. Fours on moving day get lost, stow away in the moving van, fly out of windows and off roofs, cut and burn themselves, and throw up.

As always, when you settle down after the move, turn your hot water heater down to 120° and find out the level of fluoridation of the local tap water.

## DEATH OF A FRIEND, RELATIVE, OR PET

Please see the discussion in the previous chapter. The only difference between Three and Four is that Four is more able to tell the difference between real and pretend, and may be more aware of the permanence of death, in that the person who has died will not come back in body.

# The Five-Year-Old Well-Child Visit

This is the Big One, the prekindergarten checkup, and everyone goes into it with both anticipation and apprehension. Five (or Soon-to-Be-Five) often has heard all about the "five-year-old shots." Parents know that part of the visit is a discussion of kindergarten readiness. The pediatrician is aware of how important it is to steer parents in the right direction at this time, whether to hold back for another year, or to encourage kindergarten entry.

So there are some big issues here. While each pediatrician's style is different, I would suggest a few parental strategies to make the visit a happy one.

- Don't bill it to Five as a prekindergarten visit. It may be tempting to do so, making it a memorable milestone. But this can backfire. Five is very vulnerable to performance anxiety, and such a build-up can result in Five, like Three, retiring under the examining table. Even when the reaction isn't this severe, it takes some of the play out of the visit.

- If you are uncertain about whether Five is kindergarten ready, or are in fact certain that Five is not, don't discuss this in Five's hearing. Slip a note to the pediatrician, asking for a phone call or separate visit to discuss the question. The same strategy is a good idea if you have any concerns about Five that could affect self-esteem: clarity of speech, over or underweight, short or tall, toileting issues.

- If your Five is obsessing over the "kindergarten shots," consider making a separate visit to get them over with beforehand or asking to bring Five in for a separate visit after the examination, so that you can promise "no sticks" for the visit. (Be sure to include the blood test and TB test that are usually given at this visit in your request.) If Five has not heard about the shots, but asks on the way to the office whether there will be "any pokes," answer, "I'm not sure. We'll see what the doctor says." Forewarning Five often backfires, producing tears and preoccupation rather than preparedness.

- Ask your pediatrician about preparing Five for the vision and hearing tests.

Many offices use a chart that asks the child to tell in which direction a letter *E* is pointing. Most Fives are able to do this, but a little practicing at home doesn't hurt. It's a good idea to prepare Five for these examinations, but don't call them "tests"; Five understands about tests, that there are right and wrong answers. Call them hearing and seeing games. If your Five is notoriously tired and cranky by mid-afternoon, book the appointment in the morning if you can.

Most pediatricians look for the following, rather than specific skills, as an indication that Five is kindergarten-ready.

- An attention span of at least twenty minutes.
- Ability, and willingness, to follow directions of a liked adult.
- Ability to get adult attention, and keep it, in socially appropriate ways rather than by aggression, whining, or balky refusal.
- Ability to interact with other Fives in pleasurable ways, sometimes as leader, sometimes as follower; ability to take turns, to stand up for oneself assertively, not to bully. Fives can share, but usually only with someone they like. Sharing is not yet seen as justice or fairness (unless the person on the receiving end is Five himself, and feels left out).

If Five has these abilities well in hand, specific skills or the absence of them usually won't make kindergarten easier or harder.

However, entering kindergartners are able to show the following accomplishments:

- *Intellectual:* Most know numbers up to ten, at least; colors, and the alphabet. Most can tell back a story just read, with the beginning, middle, and end in place. Most can recount an experience coherently.
- *Fine motor:* Most hold a pencil or crayon in the adult, "tripod" manner. Most can draw a circle, cross, and square without having to imitate someone else. Most can draw a person that has a body with legs coming out of it rather than just a head with arms and legs emerging.
- *Gross motor:* Most can hop; many can skip. When shown how, most can walk heel-toe, as if on a balance beam. Most can pedal a three-wheeler, pump a swing, and turn a somersault.
- *Self-care:* All who are physically able are dry in the daytime except for occasional accidents. Most can pick out their clothes and dress themselves, wash hands, brush teeth.
- *Manners:* Most refrain from masturbating, though not from occasionally touching their genitals, in public. Most can use "excuse me," "please" and "thank you" appropriately. Very few can refrain from commenting on differences, such as disabilities, in others, or refrain from staring.

Every Five cries with, and often before, the immunizations and blood tests. Even the Five who tumbles from heights and scrapes knees and undergoes bee stings with nary a peep decompensates when pain is inflicted by someone he knows, trusts, and may even love, no matter how cogent the reasons for the pokes. And a good thing, too.

## Looking Ahead

Four is starting to form a conscience, and Five takes a fairly well formed one off to kindergarten. It's a very special conscience. For instance, Five can share, but generally does so because sharing is rewarded, not because it is "right." Five believes sharing is fine when Five wants to share, and usually shares only with friends, especially those of the same sex. Sharing is a social action, not a moral rule.

Four is starting to judge her own behavior, and is just beginning to be able to see her own actions through the eyes of another person. But Five goes a step further. Five has an idea of himself as a Public Figure, someone with an image to uphold. Five understands what it is to be a sissy, a bully, a nut, a wuss.

Four tends to be a rocky age. Five, for all the new challenges, tends to be easier to get along with, more likely to listen rather than chatter, less likely to panic and lash out after a mistake. Five is starting to perceive more clearly the difference between pretend and real. Five can whisper an embarrassing observation rather than shouting and pointing.

Best of all, Five laughs at your jokes, and you'll be able to understand the point of Five's jokes too, and laugh with him.

# Part Two

___

# Illness and Injury

# Introduction:
# When Your Child Is Ill

For most of us, it feels like an outrageous surprise when a baby or child shows signs of sickness. "He never had a fever before!" "She never throws up!"

The very unexpectedness of the event, and the fact that it always seems to occur at the most inopportune moment, tend to scatter one's wits.

Fortunately, most childhood illnesses aren't serious. But they are fraught with emotion. Parents worry that the illness bodes real danger. They are repelled by mess, annoyed at interrupted lives, guilty that something they did or didn't do caused the illness, and apprehensive that care will be painful, time-consuming, inconvenient, and expensive.

For children, an illness can be painful or uncomfortable and often frightening. If the child is of an age to be trying hard for control over his or her body, vomiting and diarrhea can be humiliating. If the child is old enough to be developing a conscience, being ill may feel like a punishment for being naughty or for having angry feelings.

Having been on both sides of the telephone, exam table, and hospital bed, so to speak, I have some suggestions that I think help parents to cope when their child is ill.

1. *Pretend to be calm even if you aren't.*
   Remember that most frightening behaviors are symptoms of innocent illnesses. Your calm behavior will be a huge source of comfort to your child during the event and a huge source of pride to you afterwards. And if the symptom is truly serious, staying calm will increase the chances of getting help promptly and following crucial instructions.

2. *Anticipate in your mind the most scary or disgusting symptoms likely to occur at your child's age. Rehearse how you would react.*
   This goes double for real emergencies, in which you may need to remember how to clear a blocked airway, control bleeding, or even perform CPR. But it

is also true of the minor crises of child-hood, which can seem pretty awful at the time.

Here are the symptoms I have found most likely to cause parents and care-takers to decompensate. And the descriptions I get over the phone tell you why.

*Croupy breathing that "sounds like Darth Vader."*

*Vomiting "out her nose like Linda Blair in* The Exorcist.*"*

*Diarrhea "all down his leg onto the couch and all over me."*

*Night terrors: "Screaming and looking right through me like Linda Blair in* The Exorcist.*"*

*Febrile (or fever) convulsions: "I thought he was going to die. He looked horrible, like Linda Blair in* The Exorcist.*"*

*Breath-holding spells: "She screamed and then just went limp and turned blue and jerked, like Linda Blair in* The Exorcist.*"*

*Pinworms: "They keep coming in and going out and coming in and going out!"*

*Subluxed or nursemaid's elbow: "I must have broken her arm when I pulled her away from the cat!"*

I feel certain that in some distant sequel, perhaps *Exorcist VII,* Linda will manifest diarrhea, a subluxed elbow, and pinworms.

At any rate, this is a short list, only eight items. It doesn't cover all the really disconcerting symptoms of child-hood, but by far the most common ones that take parents by surprise. All are covered in this section, and are dis-cussed as well in the relevant chapters by age. For instance, febrile seizures and croup can start at six months or older, so are first mentioned in the well-baby visit section of the chapter *Six Months to Nine Months.* Vomiting and diarrhea can occur at any age.

In reading, envision your own child showing such a symptom, and your own reaction. Practice it in your mind. If you scream and dodge away from the vomi-tus, your preschooler may absolutely panic; if you are obviously horrified at the diarrhea, your toddler may confuse your reaction with anger at him for a potty failure. *And if you react as if your child were possessed she will reward you by really getting into the part.*

3. ***Assess the symptom or problem the way the pediatrician does.***

- Is this finding or condition or activity normal or not? Children are different from adults. For instance, is it normal for vomitus to come out the nose as well as the mouth? Should breathing be perfectly regular? How high can a normal temperature go?

- Is the symptom itself endangering the child *right now,* and, if so, what do I do about it? Trouble breathing, the threat of dehydration from diarrhea, or a nosebleed that won't stop are some examples.

- Is the symptom a possible sign of an underlying serious problem? If so, how urgently do I need to get help?

High fever, severe headache, and abdominal pain are all symptoms that can be innocent, but might spell serious trouble.

- If the symptom persists, even though no underlying serious problem exists, will it endanger the child, and if so, how can I stop the symptom? Persistent diarrhea can produce dehydration; croupy breathing can exhaust a child.
- Can I identify the illness, so I have some idea of its usual course and characteristics? How much should I worry about the future of this problem? How do I get help most efficiently?

All the chapters in this section are set up to answer these questions.

4. ***Have a rough idea of what the normal and expected illnesses and injuries are for a child this age.***

Preschoolers are supposed to catch many illnesses and sustain a fair number of bumps and bruises. It's good to know ahead of time about such rashy diseases as hand-foot-and-mouth and fifth diseases. It's reassuring to know that all bitten tongues and lacerated scalps bleed profusely, that bumps on the head swell up like goose eggs, and that the most common reason for a toddler to stop using an arm is nursemaid's elbow.

5. ***Try to have the right medical equipment on hand.*** Of course, this will vary according to the age of the child. Part I describes the basic medical chest and how to update it. It also offers updated injury prevention for each age.

## Calling the Pediatrician

If you are expecting the doctor to phone back, turn off the answering machine. Don't let the two-year-old answer the phone. If you are going out and have asked your pediatrician to leave a message on your machine, check the recording you have left. Pediatricians make upwards of thirty calls a day. We greatly appreciate a recording that gives some indication that we've reached the right number, and are additionally thankful when that recording doesn't go on, and on, and on.

Have a pencil and paper ready. One of my partners says that the Lord inscribed the Ten Commandments on stone tablets with lightning because He was so annoyed with Moses for coming up the mountain without a pencil and paper.

Try to be somewhat organized and concise. This is especially important when calling at night. My heart sinks when a parent begins, "He woke up at seven o'clock and I gave him a little oatmeal and some apple juice and he sat down to watch cartoons in the family room on the sofa. Then the phone rang, and while I was on the phone he came and said his ear felt funny but I didn't believe him because I thought he was just trying to get my attention, so then . . ."

What you really want to have down pat are:

- What are the main symptoms right now?
- What objective descriptions can you give? Temperature, skin color, activity level and responsiveness, color of vomitus or stool, distribution of the rash, etc.

- When did the child first show signs of illness, and how has his or her condition changed since?
- What have you done already: change of diet, medications, visited an emergency room or another doctor?
- What are your hidden worries? (For instance, if you are secretly worried that your three-year-old's headache stems from the fall he sustained at six months from his stroller, tell the doctor. You can always preface it by saying, "I know this is neurotic, but . . ." But get it out in the open.)
- If you are getting advice from someone else that differs from what the doctor says, ask. No sense in alienating a grandparent or endangering the child.
- What is the child doing right now? Is he screaming and writhing? Pale and whimpering? Insisting on a tuna melt and root beer? Gone to the zoo with friends?

Try not to alienate the pediatrician when you call about nonemergencies after hours. With very few exceptions, your pediatrician will be awakened when you call at night. Very few are paid to stay awake for emergencies and phone calls. Most calls after office hours are not true emergencies. However, the pediatrician is programmed to respond, fully adrenalized, whenever the phone rings because it might just be one of those true life-and-death situations.

So when you call after office hours about a probable nonemergency condition that didn't just start, explain why you didn't call earlier. Or if you call about something that is obviously not anywhere near an emergency, but is worrying and exasperating you, apologize and explain how upset you are and why you are calling in the middle of the night, or on Christmas morning, or whatever. (Even if the doctor doesn't observe Christmas. After all, it's a legal holiday.)

This isn't just for politeness; it's to focus the attention of a just-awakened, adrenalized physician.

Imagine: it is two A.M., you have been sleeping exhaustedly but worriedly because you know there is a woman in early labor with premature twins. The phone rings, you answer, and the person at the other end says, "Yes. This is Mr. Smith. Jason's been vomiting for three days now and he just did it again." Long pause.

Better scenario: "This is Mr. Smith. I am terribly sorry to call at such an ungodly hour, but my wife is having hysterics she's so worried about Jason. He's two, and he's been vomiting off and on for a couple days, not anything really bad, but now he's terribly pale and says his stomach hurts even *after* he throws up. We're worried he could have appendicitis or something. We would have called in the morning, but this pain and pallor are new just the last hour or so."

Or even, "This is Mr. Smith. I am terribly sorry to call at such an ungodly hour, especially since Jason probably doesn't have anything really serious, and especially since we should have called earlier. But he's been vomiting off and on for a couple days, and things always seem more serious at night, and my wife just remembered there's this disease called Reye syndrome, and I think she's going to go tearing off with Jason to the E.R. if we don't get some advice."

Another hint: Have the parent who has been caring for the child the most do the talking, and try to have the conversation in a room away from the screaming infant, sibling, grandparent, parrot, whatever, so the pediatrician can hear what you are saying.

If your child has a special problem or condition, state it right off the bat and if the doctor doesn't seem to grasp what you are saying, persist. If Jordan takes a medication to prevent seizures, or Monica has had her spleen removed, or Petey is on low-flow oxygen and a monitor, SAY SO, even if the doctor you are arousing has known Jordan or Monica or Petey since they were born. Especially if this is the doctor's partner you're talking to.

I am assuming, of course, that parents never call at night frivolously, just because they know that they'll be able to talk with the pediatrician right away. I am equally sure that parents would never call at night just because they are, right now, at this particular wee hour of the morning, fed up with a behavior (a stuffy nose, a tendency to wake up at two A.M., an itchy rash) that has gone on for days.

Finally, when you receive your charge for an office visit or other service, bear in mind that you are paying the pediatrician and partners to be on call at night and on holidays, willing and able to respond to such summonses. Don't take me wrong. Some of my warmest memories are of conversations in the wee hours about problems that were not really emergencies. Part of the warmth stems from a parent who recognized, and let me know that he recognized, that such a call was a bit of an imposition on me, and a necessity for him, and a real benefit for the child.

# Preface to Part II:
# Illness and Injury

The pages that follow deal with the most common symptoms pediatricians see in children from about two months to five years of age.

They are designed to help parents judge three things:

1. When and how urgently to call the pediatrician
2. What to tell the pediatrician
3. How to take care of the child until, or instead of, seeing the pediatrician

Bear in mind: these are the most common symptoms we encounter. Your child may well have something that isn't included: persistent hiccups; say, or pains in the calves while walking but not while running; or a thumb that clicks.

Of course, in this situation, or if the picture isn't clear from consulting the book, ask your pediatrician.

**A special warning: CPR instructions are included in Frightening Behaviors,** **but will not be useful unless you have already taken a CPR course and are looking at the pages merely to refresh your memory. Do not rely on them for advice but only as reminders.**

Chapter 13 covers frightening behaviors that demand immediate decision making and action including:

Acting, looking or smelling sick; choking; trouble breathing; convulsions, seizures, and fits; anaphylaxis; breath-holding spells; dehydration; fever; night crying and night terrors (includes pinworms).

Chapter 14 covers first aid for usually minor crises, including: head bonks; neck injuries; eye injuries (and foreign bodies in the eye); nose injuries, nosebleeds, and foreign bodies in the nose; mouth and tooth injuries; arm and hand injuries; leg and foot injuries; cuts and bleeding; heat burns and abrasions; human and animal bites; poisoning and ingestions; and insect stings.

Chapter 15 covers medical symptoms presenting in specific parts of the body from the head down, including: head; neck; eyes and eyelids; ears; nose; mouth; tongue; throat and voice; airway and lungs; abdomen/intestinal tract (includes vomiting, diarrhea, and constipation); genitals; urinary tract; skin; hips, legs, and feet; and "Smelling Sick."

Chapter 16 discusses various illnesses, some that are common but have unfamiliar names and some that are uncommon but have familiar names. The first category covered includes thrush; oral herpes; hand-foot-and-mouth; strep throat and scarlet fever; viral sore throats; infectious mononucleosis; herpangina; fifth disease; erythema multiforme; erythema subitum (roseola); bronchiolitis; rotavirus; bacterial diarrhea; giardia lambia. The second category covers measles (rubeola); mumps; German measles (rubella); and chicken pox.

# { THIRTEEN }

# Frightening Behaviors

## Not Acting Right

*If your child is under four months of age, please consult the specific age-related chapters.*

Every moment you are with your infant, toddler, or preschooler, you have a global impression of whether he's "OK" or not. When he's not OK, you also have a gut feeling that tells you whether to really worry or not to really worry. I'm not talking about the presence or absence of specific symptoms, but rather whether your child is acting "like himself" despite, say, a cough or fever or diarrhea.

This is an important impression. You should trust it. If you have a sinking feeling about how your child seems, but you can't quite put your finger on it, take a moment to put a name to your worry. Is he lethargic? Irritable? Looking really dry?

There are two kinds of acting sick: Acting Normally Sick, as with a cold or virus that gives fever and aches; and Scary Sick, which may mean an underlying serious illness.

Pediatricians call this kind of appearance "toxic" or "septic." The younger the baby, the harder it is to tell the difference between normally sick and scary sick. That means, the younger the baby, especially under a year of age, the sooner you need to get advice for any signs that make you worry.

### ACTING "NORMALLY" SICK

A child who is acting "normally" sick does, of course, act sick.

- Her appetite diminishes. She may refuse foods altogether for a day or two, though continuing to take fluids well.
- When her temperature goes up, she becomes flushed, and her breathing and heart rate speed up. During the fever times, she may grunt a little as she breathes, but when the fever goes away, the grunting stops and her breathing is smooth.
- She needs an extra nap.

- Nothing really pleases her. Her temper is short, and she is A Real Pill. She is fussy and her tone of voice is querulous and annoyed. Her toys amuse her for a few minutes and then she becomes annoyed or bored with them.

Not fun, but nothing here that sounds as if her problem is a serious illness.

## ACTING SCARY SICK

A child who is described as looking "toxic" looks as if all she can do is concentrate on, or be at the mercy of, the illness. Her energy sources are all used up.

A toxic child may muster a weak smile, almost as if she's trying to humor the adults, but she is otherwise not social. She won't giggle or laugh, won't reach out for a toy, or try to get to it; she just lies there, her eyes watchful and anxious, or completely uninterested. With pain, she may whimper, or cry a shrill cry; she doesn't bellow, using all her lung power and voice.

So what you want to do is to get help at once if you think your child is Acting Scary Sick. You also want to intervene if your child who is Acting Normally Sick seems to be getting worse. But sometimes it's hard to tell when an urgent call is appropriate.

If your child over the age of a year is acting only moderately sick and has a fever, it's helpful to give her a dose of acetaminophen or ibuprofen, wait an hour, and see if her activity and mood improves. (Unless, of course, she seems too ill to try this test.) You should still call the pediatrician, but with the added information that she's feeling better after the fever went down.

But don't assume your child is OK just because she has *no* fever.

## WHEN YOU GET HELP

Don't describe the child as "irritable" or "lethargic" or "fussy" and leave it at that. Be specific. The terms irritable and lethargic have lost currency in medicine because they mean such different things to different people. (I once had to pull over out of traffic on I-5 to answer a page about a "lethargic" child. When I asked "What is he doing right now?" the answer was, "He's gone to the zoo with his father.") Instead, say that you think your child has signs of a worrying illness, and then be specific about what you see.

Name your concerns. For instance:

- *Activity:* How much can he play? Intermittently? Only with a lot of coaxing, and then for only a minute or two? Not at all?
- *Crying and complaining:* Loud forceful crying when in pain, then plays again? Cries, and then lies still without playing until starts to cry again? Whimpering all the time?
- *Appearance:* Is he pale—and if so, are his lips also pale? Any blueness around his mouth? Do his eyes look sunken in, or his lips dry? "He doesn't look like my baby," say mothers of very dehydrated infants.

For specific worries, such as trouble breathing or skin rashes, see the subsequent chapters.

# CPR: Cardiopulmonary Resuscitation

*Any adult caring for children should take, and update yearly, a CPR course that includes infant and child resuscitation. To find a course in your area, contact the American Heart Association: 877–AHA–4CPR*
*www.cpr-ecc.americanheart.org*

*If you have not been able to do so before, get help after performing CPR for one minute, leaving the child briefly. If you can carry the child and there is no concern for neck or spine injury, carry him with you to the phone. Resume CPR* *after you phone.* DO NOT STOP *CPR until the Child is breathing on his own or help arrives.*

## CHOKING, OBSTRUCTED AIRWAY

- If the child is red and can make a noise with his voice, let him try to get the choked-on item up himself. Don't pat his back or turn him upside down.
- **If the child is pale or blue and/or clearly has a completely blocked airway:**

*Under one year of age*
- Place the baby on belly head downward, over your arm or lap. Give five blows

with the heel of your hand between the shoulder blades.

- If this doesn't work, put him on a firm surface on his back. Put your two fingers, lined up, down the lower half of the breastbone. Press the breastbone in, towards the spine, sharply, rapidly five times.
- If he still can't breathe, cup his chin in your hand with your thumb over his tongue. Pull the mouth open, looking for an object. *If you can see it,* SWEEP it out with your pinkie finger, going from one side to the other. Don't reach in trying to pull it out.
- Repeat steps one and two until he starts to breathe *or* until he becomes unresponsive. If the latter, then start CPR.
- If the object is dislodged, but he doesn't breathe on his own, start CPR while waiting for help.

*Over one year of age*

- Instead of a slap on the back, do the Heimlich maneuver. Embrace him from behind. Make a fist with one hand and put it, thumb up, just under his breastbone. Be sure you're not touching the breastbone or ribs. Cover your fist with your other hand. Thrust sharply *back* and *up*. Make sure you are in the midline, not one side or the other.
- If this doesn't work, open the jaw and look inside, just as in the instructions for *Under One Year*, above. If you see something, sweep with your index finger. Don't poke the object farther back by trying to pull it out.

- If the object doesn't come out and the child becomes unconscious, put him on his back on the floor. Start CPR (see below).

## CPR: NOT BREATHING AND NO SIGNS OF CIRCULATION, AND/OR NO PULSE

If the child is not breathing and he is not responsive:

- Put him on his back on a firm surface. If you think there could be a neck injury, do so carefully, moving head and body as a stiff unit.
- Open the airway: pressing on his forehead with one hand and lifting his chin with the other. If you have any reason to suspect a neck injury, do not tilt his head back.
- Give two "rescue" breaths. If you can fit your mouth over his nose and mouth, do so. If you can't, put your mouth over his and pinch his nostrils. In either case, make a tight seal. Blow in hard enough to make the chest rise; in a small baby, cheek puffs will do so. Big breaths are required for preschoolers.
- If the chest still does not rise, go through the "choking" steps again. Open the airway. Blow in a bit harder.
- Start chest compressions ("heart message") as follows:

  **Under one year of age:**
  Place two or three fingers down the breastbone, in a line. The top one should be one finger-width below the

nipple line. Press in ½ to 1 inch. Press rapidly, about 100 times a minute (about twice a second.)

**Over one year of age:**

Place the heel of one hand over the lower third of the breastbone. Press in about 1 to 1½ inches. Press at about 80 times a minute.

- Breathe and give chest compressions like this:

Every five compressions, give one breath, using the technique you used for the "rescue breath." If choking is a possibility, look in his mouth at each rescue breath to see if the choked-on object is visible. If so, sweep it out.

Keep watching to make sure the chest rises when you give the breath. If not, do the "Open the airway" step. Keep watching for this.

*If the child starts to breathe and regains consciousness, he still needs an immediate evaluation.*

## Trouble Breathing

### BASICS

When you encounter a child with trouble breathing, your first worry is not what's causing it—not even if you know the child choked on something—but how much trouble the child is in.

This means you need to look before you listen. Looking will tell you more about how severe the problem is than listening will.

*If you need emergency care, take this book with you. If you are told that epiglottitis is suspected, read at once the discussion of this in Body Parts: Airway and Lungs.*

### ASSESSMENT

*Severe distress*

*Behavior:* A child in severe distress concentrates on breathing to the exclusion of all else. She may fight any attempt to change her position: she may need to sit up, leaning forward, or sit with her nose tilted up as if "sniffing" the air.

Or she may be frantic and agitated, "hungry" for air.

She can barely make eye contact, or respond to questions, and won't show an interest in anything except where her next breath is coming from.

*Signs of effort to breathe:* Severe trouble breathing means extra muscles have to work to get air in and out. The stomach muscles, the muscles between the ribs, and those above the collarbone suck in and out.

*Color:* Her tongue and lips, and the area around them, are likely to be blue; her nails may be blue also. And the rest of her is likely to be pale or mottled, like marble.

*A child who has ANY of these signs needs immediate attention, no matter what the breathing sounds like. This is the time to call 911 or to take the child quickly to the emergency room, whichever is faster. If you call 911, tell them that the child is going into respiratory arrest. Keep the child in her position* of comfort. *Don't make her lie down. Try not to agitate her further by getting out of control yourself.*

### Moderate distress
*Behavior:* A child in moderate distress can be restless or exhausted, but makes eye contact and can take a brief interest in something. He may even smile and reach out for a toy, but don't be deceived; if he doesn't play but drops it or just holds it, he is having serious difficulty. He may be able to utter two or three words at a time, but can't get out a whole sentence comfortably.

*Effort to breathe:* You may see his stomach muscles and the lower muscles between the ribs sucking in and out when he breathes, but you shouldn't be seeing the muscles above the collarbone sucking in and out.

*Color:* He is likely to look pale. A child who has normal color, but other signs of moderate distress, is still in trouble. A child who has blue lips and tongue, or who is mottled, is in SEVERE, not moderate distress.

**A child in moderate distress can deteriorate rapidly to severe distress.** Call your pediatrician at once for instructions, or get the child promptly to a nearby emergency room. Don't take the child to the pediatric office unless your pediatrician tells you to. (Perhaps your doctor has been called to the hospital, or the office has closed for a holiday or meeting.) Don't waste time. Keep the child as calm as you can, and let him assume the position most comfortable to him.

### Mild distress
*Behavior:* A child in mild distress can play, move around the house, and talk with you, but doesn't have anything like her usual energy. She is unlikely to get into trouble or test limits. She may be clingy and anxious.

*Effort to breathe:* A child in mild distress may show that she's working harder than usual to breathe by the stomach muscles sucking in and out. But the muscles between the ribs shouldn't be sucking in and out, nor should the muscles above the collarbone.

*Color:* Her lips and tongue and nailbeds should be pink, and she shouldn't look mottled, pale, or blue.

If your child is in only mild distress, you can take the time to assess the situation further **unless** you suspect she has choked on a foreign object or has gotten into medicine or poison. If you suspect this, she needs help at once, even though the distress is mild.

For assessing the child in mild distress, see the discussion of *Lungs and Airway* in the chapter *Body Parts, Bodily Functions, and What Ails Them.*

# Convulsions, Seizures, and Fits

## BASICS
**Parents who see their child having a first-time seizure almost always think the child is going to die. But seizures are not uncommon in early childhood, and nearly all of them, no matter how frightening to watch, do no damage to the child and do**

not mean that the child has a dire or handicapping disease.

If your child is having a seizure right now, he is blue around the lips, there is foamy saliva in his mouth, and his body moves in jerks. He may seem not to breathe or to breathe in funny gasps. His eyes are open, which may give you the feeling that he is conscious, but he is not; and his eyes are probably rolled way back or to one side. He may well have peed or pooped in his pants, even though he hasn't done so for months or years.

You are undoubtedly scared to death. Try to stay calm and read what to do.

## Treat First, Assess Later
Here is what to do:

- **Turn him on his side,** so that if he throws up he won't choke.
- **Look at your watch** and try to estimate how long the seizure has been going on. When the seizure stops, see how long it has lasted.
- **If it is your child's first seizure, or if you think the seizure is related to a head injury, or if you think the child has gotten into a medication or poison, call or have someone call 911.**
- **If you must leave the child to call, do so if the seizure lasts longer than three minutes. If you do not suspect head or neck trauma, pick up the child and go for help.**

Most of the time, the seizure will be over or finishing by the time the para-

medics arrive. The child still needs to be evaluated by your pediatrician or in the emergency room.

If the child is still seizing by the time medical people see him, they will probably want to give him oxygen and perhaps start an intravenous line for medication to stop the seizure.

- **If you didn't have to call the paramedics,** then, after the seizure is over, call your pediatrician at once. Even if your child has had seizures before, even if he is on anticonvulsant medicine, call the pediatrician.

## DON'TS

- **Don't try to put something in his mouth** to keep him from biting his tongue, or let anyone else do so. You could break his teeth or gag him to the point of vomiting, and besides, he won't bite his tongue.
- **Don't plunge him into a cold bath** if he feels feverish, even if he's stopped the seizure. It could shock him into another one and won't bring down the fever.
- **Don't try to give him medications or fluids or food by mouth,** even after the seizure is over. He could have another one, though that's unlikely, and vomit and choke.

After the seizure is over, the child will relax limply, and start to breathe heavily. His color will return. He will fall into a deep sleep and be very hard to awaken.

If you are the one taking him to the pediatrician or the emergency room, make sure he is in a car restraint. Try to have someone else drive; you will be shaky and will need to keep an eye on the child.

## ASSESSMENT

Seizures are chaotic electric impulses in the brain. They can be caused by all kinds of things: scar tissue, poisonings (like severe lead poisoning), medications, and chemical imbalances (like too much sodium or too little sugar). Sometimes they come just out of the blue, without explanation. Rarely, flickering lights (like Nintendo, or driving past a line of telephone poles against the setting sun) can bring on a seizure.

Most seizures in childhood look like the one described above. But there are other kinds: brief loss of consciousness without jerking, odd repetitive motions without loss of consciousness, staring spells, or brief loss of consciousness with a fall or stagger. A very specific and rare kind of seizure called "infantile spasms" occurs only in young babies. It is a repeated jerking of the head, arms, and body without loss of consciousness, and almost looks as if the child is bowing or "salaaming." Such a spell needs prompt pediatric evaluation.

Most seizures in children from six months to five years of age are fever convulsions or are caused by breath-holding spells.

### Fever Convulsions

About four children in a hundred will have at least one fever convulsion. Most of the time, a fever convulsion occurs when the temperature rises very rapidly; often it's the very first sign that the child is sick, which certainly is a dirty trick on parents and caregivers. Fortunately, most fever convulsions last well under ten minutes, though it's clearly a very long ten minutes to the worried adults.

If your child has a fever convulsion, or febrile seizure, your pediatrician needs to make sure that there isn't any sinister cause for the fever, such as meningitis. Sometimes lab tests are needed; sometimes not.

If the diagnosis is "simple febrile seizure" due to a virus infection or ear infection, most pediatricians don't put the child on seizure-preventing medications. Half of the children who had one febrile seizure will have another—but then again, half won't. And since the seizures are innocent and don't cause damage, it seems a pity to put a child on medication that may cause side effects.

Epilepsy is the term for seizures that occur with some frequency. It's rare that a child with a febrile seizure turns out to have epilepsy. And there's absolutely no reason to think that febrile seizures will cause epilepsy.

If your child has a seizure without fever, she needs medical attention and diagnosis at least as urgently as the child who has a seizure with fever. In most cases, such a seizure without fever means that the child has a "focus" in the brain that can trigger seizures, and may need medication to suppress that electrical "focus." But rarely it can mean that the child has a serious underlying problem that needs rapid diagnosis and treatment.

If you are reading this in anticipation of being with a child (your own or any child) who is having a seizure, try out the scenario in your mind, envisioning what you would do in such a situation. What if you were out in public? Or in a car?

The best thing you can do for anyone having a seizure is to protect him from hurting himself, summon help if necessary, and be there calm and collected when he wakes up. Children and adults who have seizures don't remember them at all, and can be very frightened and embarrassed to find that they have been out of control in such a way. The demeanor of the person they are with can make a big difference.

# Anaphylaxis

## BASICS

Anaphylaxis is an explosive, very scary allergic reaction. It is rare, thankfully, but life-threatening. Typically, it comes on suddenly after a child is exposed to a food or medication, or is stung by an insect, to which the child has a severe allergy. The most common triggers in childhood are foods, especially peanuts and tree nuts.

Here are the signs of anaphylaxis:

- Puffy lips, eyes, face (there may be hives, but hives by themselves are not indicative of anaphylaxis).
- Drooling, because the child can't swallow saliva through her swollen throat.
- Trouble breathing, from swelling of the upper airway (a croup-like noise) or of the lower airway (wheezing).

- Poor color—mottled, pale, gray: a sign of shock (low blood pressure).
- Rapid heartbeat: another sign of shock.
- Terrified, apprehensive, or unresponsive child.

## ASSESSMENT

If a child comes down with any of the scary, sudden, and mysterious symptoms listed above you should suspect anaphylaxis. But you need to look at the whole picture. A child who is *just* drooling, as if she's just too preoccupied to swallow, or *just* has hives (a common, innocent expression of many childhood viruses), does not have anaphylaxis and does not need emergency attention. A child with one puffy eye from a reaction to a mosquito bite does not have anaphylaxis. It is the scary signs that you need to look for: sudden facial/lip swelling, drooling because she can't swallow her saliva, breathing trouble, signs of shock.

Be alert to the possibility of anaphylaxis if a child encounters a substance to which you know she is severely allergic, or is stung by an insect. In this case, just drooling or just hives may indeed be the first sign of anaphylaxis.

## TREATMENT

The life-saving treatment for anaphylaxis is an injection of epinephrine. Anyone who has had one episode of anaphylaxis should carry along, everywhere, a prescribed kit that contains such an injection. In our Under-Five age group, pediatricians usually prescribe a kit called Epipen Junior. It is very easy to give, even through clothing; you do not need to clean the skin first. You do not need to see or manipulate a syringe and needle.

### Child with history of previous anaphylaxis

- If you suspect anaphylaxis is starting, whether or not you have witnessed the child encounter the offending food, medication, or insect, give the injection at once. Then call 911, or head for medical care at once, because the shot may wear off in fifteen minutes or so, and the symptoms may return.
- If you do not have the epinephrine kit with you, call 911, or drive to the nearest emergency room with your horn honking and with emergency flashers on.
- If you have an outdated kit, go ahead and give it anyway; it may have enough potency left to help until you can reach medical care or medical help arrives.

### Child with no history of previous anaphylaxis

This is the more typical situation. Call 911 at once, or if it's close, head for the nearest emergency room. Don't worry about any Managed Care repercussions; get to help at once—don't call for authorization first. If Cherub seems fine when you get to the emergency room, you can always sit out in the waiting room or parking lot, and phone your pediatrician for instructions or authorization.

### PREVENTION

For stinging insect prevention, see the section in *First Aid*.

For food or medication allergy, you will need the most scrupulous measures to keep Cherub safe. That means careful instruction to every single caregiver, from grandparents to daycare or preschool adults, to parents of friends: they must not allow Cherub even to touch the allergic substance. Whoever cares for Cherub must have, and know when and how to use, the Epipen Junior kit. This means that parents need hands-on instruction with a "trainer" kit, and need to repeat the instruction with caregivers.

If you are traveling, take along three Epipen Junior kits; if you need to use the injection, it may wear off before you can get to a safe place, and you may need to repeat it again and again. Do not pack the Epipen in airplane luggage that you check: it may freeze in the baggage compartment, and the spring that makes it operate may be damaged.

In my view, a child who has had a true episode of anaphylaxis should have one visit at least with a pediatric allergist, to reinforce instructions and to help to identify foods that should be avoided. For additional information, a good website is *www.foodallergies.com*.

# Breath-holding Spells

Breath-holding spells can occur any time from six months to six years, but most happen between twelve and eighteen months of age. About 5 percent of all children will have one. In a breath-holding spell, a minor event (like a scolding or a fall) triggers an involuntary spell of not breathing. The child may cry out or not, may turn blue or pale, and may or may not have a seizure. Like simple fever convulsions, these are innocent events. Breath-holding spells come in two versions.

1. *"I'm going to hold my breath until I turn blue."*

   Blue, or cyanotic, breath-holding spells in fact are mostly associated with anger and frustration, but the child doesn't then purposely hold his breath. It's involuntary. Generally, he gives a loud yell, then takes a big breath aaaaannnndd—*holds it.* Sure enough, after about half a minute of this, he turns blue. After this little performance, he either resumes screaming or will go limp or rigid, and will pass out. Then he may have a real seizure.

2. *"Something bad happened and I have decided to collapse."*

   Pale breath-holding spells generally occur after fear or a minor injury. One child I know has them routinely when she is jostled in line, trips on a step, or hears anyone scold her. In a pale spell, the child often doesn't cry, or just makes a tiny yelp. Then she goes limp or rigid and passes out, finishing up, sometimes, with a seizure. Again, the whole process is involuntary.

**There are three crucial things to do if your child has such spells.**

1. Satisfy yourself, with EEG or EKG or whatever you and your pediatrician decide you require, that there is really, truly, nothing wrong with your child's brain and heart. Many times, such studies need not be ordered. But if you have to have them done for your own peace of mind, talk with your pediatrician.

2. Become a calm pro at witnessing and handling passing out and seizures. You will do your child an enormous favor and get a reputation for heroism.

3. Don't try to prevent the spells by trying to make sure your child is never angered, frustrated, frightened, or suffers a minor injury. First, that is a sure-fire way to raise a barbarian and become a martyr. Second, you couldn't do it if you tried, and it is only time and repeated experiences that seem to help a child grow out of these spells. If you cater to him, they may persist far beyond what they should.

# Dehydration

## BASICS

"My child doesn't look like himself anymore." That's a statement from a parent that rings alarm bells. It means that a child is *really sick*. On the rare occasions I've heard a parent say this, in each case the child has been severely dehydrated, and required urgent intravenous rehydration.

Fortunately, most children who are at risk for dehydration never get that sick. Dehydration means a loss of bodily fluids, usually from diarrhea or a combination of vomiting and diarrhea.

Mild dehydration makes a child cranky and thirsty; severe dehydration can result in fatal shock. With dehydration, many children build up acids in their bodies; this *acidosis* can make them nauseated and unable to take more fluids, producing a vicious cycle.

Besides losing fluids, children with vomiting and diarrhea can lose salts or *electrolytes*. This loss can produce problems of its own, especially severe behavior changes.

The treatment for acidosis is sugar, the treatment for electrolyte loss is salts, and the treatment for fluid loss is water.

**So any fluid given to prevent dehydration or to treat the mild forms must contain sugar, salts, and water.** But the type of such a fluid is also important. If the ingredients aren't balanced correctly, the body chemistry can be thrown off and make the child much sicker.

**Before giving any fluids, however, assess the state of hydration of the child.**

## Assessment

**The younger the child, the more quickly and severely dehydration occurs. A baby under six months is at very high risk; under two years, at high risk.**

### Severe dehydration

**If your child has any of these signs, you need to get very urgent attention. If there will be more than a ten-minute delay getting to an emergency room—or to the office if your pediatrician has specifically told you to go there—call 911.**

- Extremely dry mouth (tacky or sticky when you touch the roof with your finger)
- Sunken eyes
- Mottled skin and cool hands and feet
- Warm, dry skin, cherry red lips, rapid pulse
- Intense irritability, or lethargy, or disorientation
- Diminished urine output. **It may be very difficult to tell urine from watery stool, especially in diaper-clad children. Therefore, the signs above are more important when assessing severe dehydration.**

### Mild dehydration

Many children with ordinary simple diarrhea become mildly dehydrated, and most recover just fine with the right fluids. See the sections on vomiting and diarrhea in *Body Parts, Bodily Functions, and What Ails Them.* Signs of mild dehydration are:

- Dry lips but a moist mouth. A drooling baby will have LESS drool, not NO drool. If you put your (clean) finger in the mouth of a child of any age, it shouldn't have any trouble getting wet—the roof of the mouth shouldn't be sticky or tacky (like drying paint).
- Clingy, whiny, or fussy behavior. However, the child should still be able to make sustained eye contact, smile, and

be interested in a toy. Babbling children will babble, and talking children will talk; crawling babies crawl, and walking children walk—but not enthusiastically.

- Decreased frequency of urination, but normal amounts. The urine looks dark yellow and smells strong.
- Good skin color, even if skin seems dry.
- I wouldn't concentrate on presence of tears; even very dehydrated babies may have tears. Or whether the soft spot is sunken; it will always look sunken compared to the rest of the head because there is no skull there.

*Signs that dehydration is worsening*

If your child seems to be mildly dehydrated, or not dehydrated at all, it is safe to go ahead and treat the diarrhea and vomiting, BUT KEEP MONITORING HIM. If it is bedtime, check on him often during the night. The younger the child, the more rapidly things can change. Signs of worsening include:

- Inability to keep down fluids
- Extremely profuse and frequent stools despite dietary changes
- Increasing lethargy and irritability, with decreased ability to play and take an interest
- Tacky or sticky roof of mouth when you touch with your finger
- Decreased urination

**If these are occurring, the next thing that may happen is severe dehydration, so you need to get help promptly.**

For treatment of vomiting and diarrhea, See *Bodiy Parts, Bodily Functions, and What Ails Them* in Part II.

# Fever

Fever in babies under Four Months of age is addressed in the age-specific chapters.

## BASICS

High body temperature needs to be lowered fast, urgently, only if it is due to rare, abnormal causes. If you don't treat a fever, the temperature will not continue to rise above 106°. (Unless the cause of the fever is heatstroke rather than infection: see below if this is the case.)

Even a high fever, from 104° to 106°, is not dangerous in and of itself. The main worry with a high fever is whether the underlying illness is a serious one, such as meningitis or pneumonia. A very rapidly rising fever can cause a fever convulsion, but there's no evidence that aggressively treating a rapidly rising fever prevents such a convulsion.

So even a high fever in a child who is otherwise not acting "scary sick" should not be a cause of panic—though it needs to be investigated.

It is not necessary or desirable to lower the temperature right away: in fact, putting the child in a cold bath is likely to make matters worse. It is not necessary for the body temperature to go all the way back down to normal, either.

> Treating the fever will not shorten the underlying illness, nor prevent complications. Instead, the goal of treating a fever is primarily to make the child more comfortable, so that she'll be able to take fluids and rest. Also, so that we can judge her behavior.

The medical concern with fever is that it is a clue. What is causing it? An ordinary virus? A fairly common bacterial infection, such as strep throat or urinary tract infection? A dangerous infection, such as meningitis, sepsis (infection in the bloodstream), or tuberculosis?

A child age two or younger may not show other symptoms of a serious illness (such as a stiff neck, in meningitis) that would be obvious in an older one. So a fever over 104° in Two or younger always needs prompt attention.

## ASSESSMENT

### Heat stroke

Since the only reason to act very, very quickly in the face of fever is when it is due to heat stroke, we'll address this first. Heat stroke is much more easy to prevent than it is to treat. For instance, don't dress a baby warmly, then put him in a car seat beside a sun-filled window for a nice long ride in the heated car.

HEAT STROKE IS AN EMERGENCY. If the fever is due to heat exposure, the fever itself is very dangerous because the body's thermostat is no longer in control. In that situation, the body temperature **can indeed** rise to over 106°.

### SIGNS OF HEAT STROKE

Suppose that you are in a very warm environment, or your child seems to you to be extremely overdressed for the temperature, or is on a medication that you have been warned can predispose him to heat intolerance. Now, suppose Cherub starts to get a fever and act sick. How can you tell if it is due to heat or to a contagious illness? You can't know for sure. However, you need to assume that it is due to overheating, because if that is the case, you need to act at once. Giving fever-reducing medication won't hurt, but it also won't treat the fever of heat stroke. This is one case where sponging will work, because heat stroke is due to external heat, not to an internal "turned up thermostat."

The symptoms and signs of heat exhaustion and heat stroke overlap with the symptoms of a sudden-onset infection. They include headache, nausea, and vomiting, hot dry skin, looking pale, and a rectal temperature over 101°.

If your overheated child exhibits these symptoms, do not assume that this is just a virus. Instead, follow the measures described below, and then call your pediatrician. Or, if your child seems to be taking a turn for the worse, call 911 or head for the emergency room.

### TREATMENT
Call your pediatrician, head for an emergency room, or call 911 immediately after you've taken these measures:

- Undress the child.
- Move the child to a cooler place.
- Sponge the child with cool water, one limb at a time. Don't use alcohol. Do not plunge him into a cold bath. Rapid cooling can make things worse.
- Fan him so the water evaporates to cool the skin.
- Take his rectal temperature. If it is over 102°, or if the child is acting sick, you need help right away.

*Ordinary fever not due to heat stroke*
The reason to treat ordinary fever—fever from infectious illnesses—is to make the child comfortable, to assess his behavior when he is reasonably comfortable, and to prevent fever making other symptoms worse.

### FEVERS IN CHILDREN WITH COMPLICATING MEDICAL PROBLEMS
Children with certain underlying medical problems have additional reasons to have their fevers treated. You will want to discuss the topic of fever with your pediatrician or with other specialists:

1. If your child tends to have fever convulsions, In this case, it's probably helpful to try to keep the temperature lower than 104°. However, many fever convulsions occur not because of how high the temperature is, but because of how rapidly the temperature goes up. So the seizure is often the first sign of the illness. See *Fever Convulsions* in *Frightening Behaviors*.
2. If your child has an underlying condition which predisposes her to serious infection. Examples include, but are not limited to: a shunt in the brain, for hydrocephalus; congenital heart disease; kidney abnormalities; lung cysts; immune deficiencies such as AIDS.
3. If your child has an underlying condition in which fever can cause other complications, as is the case with insulin-dependent diabetes or skin abnormalities that prevent sweating, for instance.

### FEVERS IN CHILDREN WITHOUT COMPLICATING MEDICAL PROBLEMS
Rectal temperatures are the most accurate. Do not rely on armpit, oral, ear, skin patch, or pacifier-thermometer temperatures for a definitive reading.

- 100.4° to 102°: Low fever
- 102° to 104°: Moderate fever
- 104° to 106°: High fever

It is always appropriate to treat a child with fever just to make her more comfortable, to make it easier for her to rest, play, and take fluids, and to improve her mood and behavior.

Fevers come and go for the duration of an illness. Many fevers due to viruses go away in the morning and return at night. This means that a child should have a normal temperature for a full twenty-four hours before you assume that she is well and no longer contagious.

The most effective treatment for fever is medication that turns down the body's thermostat in the brain. These medications are acetaminophen and ibuprofen, which are available under a myriad of brand names. Read the label!

These medications may come as single-ingredient drops, suspended in liquids, or in chewable tablets, or they may be combined with other medications that are designed to treat congestion or coughs. In either case, the dosages by weight and age are on the label. For medicolegal reasons, the dosages for babies under age Two Years are not given on the label; for the same reason, I am not including them here.

Two warnings about fever medications:

- Do not use real aspirin (*salicylic acid*), even though it is labeled Baby Aspirin or Children's Aspirin. Aspirin is only appropriate if your pediatrician has specifically told you to give it for a special reason. Aspirin has been implicated as a contributing cause to the rare brain and liver disease, Reye syndrome (see Glossary).

- Make sure you give an accurate dose. Infant drops are much *stronger* than children's liquids. If you give a teaspoon of the drops, mistaking them for the liquid form, you can give a dangerous overdose.

## COOLING DOWN

Bathing or sponging a child with fever due to infection doesn't make much sense, and may make the child very uncomfortable or even precipitate a seizure in a child prone to convulsions.

When you have fever, what has happened is that the thermostat in your brain has been turned up—from a normal of, say, 99.6° rectal to, say, 103°. Now, say that this happened in your house: you turn the thermostat up. Now you want to cool the house down. Would you run around spraying water on the outside so that it would evaporate and cool the house down? No, because the thermostat would still be telling the furnace to produce more heat.

So it's much more rational to turn down the thermostat. Which is what acetaminophen and ibuprofen (and the forbidden aspirin) accomplish.

Most children with fever hate cool baths and sponging; it may even make them worse or seem to precipitate a fever convulsion. And putting alcohol in bath water or rubbing a child with alcohol is dangerous: the alcohol can be absorbed through both skin and lungs, and send the child into a coma.

## POSITIONS FOR TAKING A CHILD'S RECTAL TEMPERATURE

Use an electronic, digital reading thermometer if at all possible, and dispose of any mercury glass thermometers by treating them as toxic waste. When disposed of improperly, they can contribute to mercury contamination of ocean and freshwater fish.

- Have her lie on her back. Babies on their tummies can really clamp their buttocks together. You can try it if you want, though.
- Give a baby a toy to occupy her hands. For an older infant, put a (large) sticker on each of her hands; she'll try to take them off. For a toddler or preschooler, try a puzzle-toy, a computer game, or tell a story.
- Hold her feet or legs up, if she's on her back; press a hand on her lower back if she's on her tummy.
- Insert the business end of the digital thermometer. Some give readings in both Centigrade and Fahrenheit, so make sure you press the button you want.

## CALLING OR SEEING THE PEDIATRICIAN URGENTLY

*See age-appropriate chapters for children six months and younger.*

Call *urgently* if:

- The fever is above 104°.
- There are worrisome signs besides the fever.

Call during office hours if:

- The fever lasts longer than three days, even though the child seems fine otherwise.
- There are any signs other than an ordinary cold.

## HYGIENE

Assume that your child with fever has a contagious illness. Wash your hands well before touching your own face, especially eyes and nose. Keep the child away from others until the temperature has been normal for at least twenty-four hours.

# Night Crying and Night Terrors

## BASICS

When your child wakes up in the middle of the night screaming, it is a crisis. There are three kinds of night-crying crises:

- Crying for a physical reason
- Crying from a bad dream or separation anxiety
- Night terrors

## ASSESSMENT

The child who is crying from a bad dream or separation anxiety can be calmed if parents stay calm themselves. Children who are crying for a physical reason also feel better with a calm parent. Children with night terrors are unaware of the environment, including their parents.

Even if the screams are heart-rending, try to appear calm.

First, see if the child seems to recognize you and hold out arms for comforting. If so, this is not a night terror. If not, it is; go directly to the discussion below on Night Terrors.

Second, if the child does recognize you, or if you are unsure about it, first try to calm and reassure, holding and talking softly and singing. Consciously try to make your body relaxed; communicate that relaxation. Note the time so you can tell how long the crying lasts.

Third, while you are holding the child, try to estimate fever by touching your lips to his or her forehead. See if the child is working hard to breathe: a good hint is how long each cry is. If the child can take a big breath and let out a prolonged howl, or utter a full sentence, trouble breathing is not likely. A child who is flushed or pale and seems half-scared and half-angry is probably going to vomit. Be prepared

Fourth, examine the child as a whole. If the child doesn't calm down with reassurance and cuddling, then you need to look at him or her all over. Gently look at how the child is holding his or her body, whether one part is being held in a protective manner. Make sure the head and neck move freely. See if the belly looks bulging and feels hard or hurts when touched.

Fifth, undress the child entirely. Look for a rash, an insect sting or bat, rat, mouse, or spider bite, or a tourniquet of hair or thread wrapped around a finger, toe, or penis. Examine the genitals and anal area. Is the foreskin OK, or is it pulled back and "tourniqueting" the penis? Is the scrotum enlarged, as with a hernia? Is the vagina red? Do you see little white worms, pinworms, around the vagina or anus?

## TREATMENT

### Non-night terrors

If you discover a medical reason for the awakening, consult the appropriate section in Part II for guidance.

If you can't figure out what, if anything, is wrong, time the crying while trying to calm the child. (This is best done by rocking or a warm bath in a parent's lap.) Sustained crying for over an hour warrants a call to the pediatrician, even if it is three A.M. Take the temperature first.

### Night Terrors

It's more of a terror for the parents than for the child, usually. Real night terrors tend to remind parents, inevitably, of *The Exorcist*. Their child screams, and it's a scream they haven't heard before. They run in, and the child stares through them, or, worse, shrieks with increased terror as they approach. The child may rage and yell incoherent threats and entreaties.

If they try to calm him down, things get worse. But if they back off, holding their breath and wringing their hands, the child calms down, falls back to sleep, and the next morning is his usually cheerful self. If he's old enough to talk, it's clear he remembers nothing of the escapade. (Night terrors can start at least as young as nine months of age.)

Night terrors happen when a child is brought to full awakeness directly from the deepest stage of sleep. He doesn't pass through lighter and lighter stages: it's like a scuba diver surfacing rapidly. No wonder the child shows a kind of emotional form of "the bends."

The two questions about night terrors are:

**1. What can you do to comfort your child?**

When your child is having a night terror, he is not in the same world you are. He's in the deep-sleep world. If you try to wake him up, it will be a long and harrowing endeavor. When he does awaken, he won't be grateful. He'll grumble and whine and may have trouble falling back asleep. He may be confused and frightened, especially if you're obviously distraught. He'll remember in the morning.

Instead, do one of those difficult tricks of parenting: back off and observe. Within fifteen or twenty minutes he'll calm down, curl up, and fall asleep again. And he won't remember a thing in the morning.

**2. What causes night terrors and how can you prevent them?**

Something awakens the child abruptly while he's in that very deepest stage of sleep, and most often we never find out what it is. Perhaps his own internal alarm clock got set wrong and went off prematurely; this can happen if he's sleep-deprived in general. Watch out for skipped naps and late-night videos. Or perhaps a car honked, or an explosion rang out, or someone bumped him or the bed.

But sudden pain can also semi-rouse a child in deep sleep and cause a night terror. Most likely suspects are earaches, the gas pain of constipation, and pinworms, little white worms that live in the intestine.

*Pinworms*

Pinworms can cause itchy rectums, it's true, but they are razor-sharp little critters and can also cause pain. Female pinworms exit from the rectum at night to lay their eggs on the anal and vaginal tissues. Night terrors from pinworms may be more frequent in little girls than little boys, because pinworms really hurt if they move onto the membranes surrounding the vaginal opening.

Looking for pinworms at night in a child who is experiencing a night terror is tricky.

The method is to shine a flashlight on rectal and vaginal areas and look for worms. Even if you manage to tackle the child and look, the sight of the worms is likely to further disturb your mental attitude. So gird your loins ahead of time, if your pediatrician advises this maneuver.

Another method is called the Scotch tape test, in which you or your pediatrician try to catch the eggs on sticky transparent tape, placed over the anus. The tape then is placed on a microscope slide. This can be done in the daytime.

Consult with your pediatrician. If you are advised to perform either procedure, make sure to explain to your child beforehand and again afterward what you were doing, or her re-told story may produce a visit from the child abuse team. However, DON'T tell a young child you are looking for *worms*. This will terrorize most preschoolers, or fascinate them, and you will never hear the end of it.

Pinworms are fairly repellent, but as parasites go, they possess some redeeming features. They don't travel throughout the body, wreaking havoc with organs; they stay put in the intestine. They are easy to treat with medication. Most authorities don't recommend the soul-destroying house-cleaning measures you need with lice or scabies.

Children catch pinworms by swallowing the microscopic eggs, either by touching a child with them or an object with eggs on it. They can reinfect themselves by scratching their itchy bottoms, and then putting their fingers in mouth or nose. (Pinworms don't cause nose-picking; they result from it.)

If night terrors occur frequently enough that you become exhausted because Cherub keeps waking up screaming, talk with your pediatrician. Together you may be able to spot sources of stress or fatigue, or a reinfestation of pinworms. Rarely, pediatricians prescribe sleeping medicine until the spate of night terrors diminishes.

# First Aid: Assessing and Handling Usually Minor Injuries

## Head Bonks

### BASICS

Bonks on the head start no later than nine months of age and usually earlier. Rolling off the bed is popular for the baby set, as is bouncing up suddenly while on a lap and smashing the adult's chin. Older infants learning to pull to stand fall backwards and stun themselves. Then there's the Coffee Table Collision, the Counter Crash, the Shower Slip, all the way up to the four-year-old specialty: the Car Door Doozie (exiting the car as the door is just being slammed).

My favorite recent minor head injury was the two-year-old reported to have fallen off the pool table. How, I asked the mother of five, had that happened? "Oh," she replied airily, "he caught his heel in the pocket."

When is it an ordinary bonk on the head, and when is it a worrying head injury?

Deciding which is which depends on the age of the child, how the injury occurred, and how the child acted right after the injury and thereafter.

A bonk is a worrisome injury if there is any sign that there could be increased pressure on the brain, from bleeding inside the brain, or from fluid building up inside the brain, or from brain swelling. It is amazing how rarely this happens, considering how often and with what élan children fall on their heads.

Falls and direct blows to the head have different implications. A fall usually causes, if anything, a concussion. This is when the brain bumps against the skull from within; it kind of bounces forward, then back inside the skull, causing a brain-bruise. Fractures usually don't bear much relationship to the degree of injury inside the head if caused by falls.

A direct blow to the head is a different story. The injury isn't a bruise from a bouncing brain, but rather a direct hit. A fracture from a direct blow is more likely to reflect the severity of the underlying injury. A fracture from a direct blow that is depressed, or presses into the brain, may have to be urgently treated.

## ASSESSMENT

Five factors must be considered in assessing whether the child should see a doctor following a fall.

*The age of the child.* The younger the child, the harder it is to assess symptoms. Any baby under six months who has had a fall or sustained a blow to the head is of concern, and the pediatrician needs to be notified.

*The nature of the fall.* When an older infant, toddler, or preschooler falls from his or her own height, serious consequences are extremely rare. Most pediatricians want to see, immediately, a child who has fallen from four feet or more. Even if the child is acting fine, and even if the surface he fell on was soft.

*The nature of the blow.* Any significant direct blow to the head, not from a fall, that causes swelling or bleeding needs to be seen rapidly.

*The nature of the head swelling.* A "goose egg" swelling isn't unusual after a bonk on the head. However, if there's a goose egg-sized bump in a baby under a year of age, or if the lump is located just above the ear—the location of a rather fragile artery to the brain, inside the skull—your pediatrician probably will want to examine, and possibly x-ray or do a CT scan, on the child.

*The behavior of the child.* Behavior is worrisome:

- Most importantly, if there was any loss of consciousness after the injury. Some children will be so startled and insulted after a minor injury, they will cry and then have a "breath-holding spell." If this is the child's first breath-holding spell, or if you have suspicion that the fall itself caused the loss of consciousness, the child should be seen at once.

- If the child falls asleep after an injury. Now, many children with minor head trauma do want to take a nap, probably just from stress. But this symptom always demands a phone call, and often an examination.

- If the child's behavior doesn't return perfectly to normal after she's finished crying. If she doesn't seem to know or be comforted by you, won't smile, can't be entertained; or if she's "not acting right" in other ways. If the child is examined and found to be "fine" but goes home and acts irritable, sleepy, or strange, check back with the pediatrician immediately.

- If the child vomits. Again, many children with innocent minor injuries will vomit, but this sign must always be reported. If a child is examined and found to be "fine" but goes home and vomits some more, check back with the pediatrician immediately.

- If the child who is old enough to recount experiences can't remember what happened, or has a gap in memory before and during the time when the injury took place.

- If there are any signs of serious persistent, worsening headache (see section on

*Head* in *Body Parts* chapter) at any time in the two months after a significant head injury.

## TREATMENT

If your toddler or preschool child just fell, hitting his head, but never lost consciousness and is screaming lustily, he probably has an "ordinary" head injury. If so, he either has a huge goose egg swelling up before your very eyes, or he is bleeding profusely and you can't see the cut because of all the blood.

Do this:

- Pretend to be calm. Admire how splendidly he fell, and tell him you know he must be scared but that everything is fine. Say that it probably hurts but will feel better soon.
- To halt the swelling of a goose egg, either get your ice pack or fill a small plastic cloth or bag with ice chips or frozen vegetables, secure the opening, wrap a thin towel around it, and gently apply it to the bump. Your child won't like it. Tell him that the bump wants a cold hat and it is his job to keep the cold hat on the bump.
- If he is bleeding, tell him that the blood he sees is "extra" blood, special blood just for cuts, and is supposed to come out. Take a clean cloth (preferably red, so he won't see the blood) and press it on the area that is bleeding. Hold it firmly for seven minutes without peeking. Tell him that the cut on the head wants to be pressed on, and that it is his

job to hold still. Tell him how wonderful his aim is and how only big boys (or girls, of course) get such excellent head bumps.

In such ordinary head injuries:

- There is no loss of consciousness, no passing out.
- Once the child calms down, he starts to play again.
- There is no blood or fluid from ears or nose.
- A child who's old enough to recount experiences can tell you, and remembers, the entire episode.

### Fracture

Most head injuries don't need ordinary X rays, so don't be surprised if your pediatrician doesn't order any. The times you worry about fractures are:

- If the fall was from such a height, or by such a young or fragile baby, that fractures might be extensive. (Even so, this doesn't mean brain injury necessarily.)
- If the fracture could be depressed, or smashed in, so that it presses on the brain. This can happen if the child falls on something protruding or if he is hit over the head rather than falling.
- If the fracture could have broken the barrier that seals in the spinal fluid that bathes the brain, so that there is risk of the fluid becoming infected. These are called basilar skull fractures. Signs are blood or fluid from nose or ears.

(Remember that streaming tears will come out the nose. It's fluid persisting after the tears stop that you look for.)

- If there's a laceration over the fracture that could let bacteria into the brain—especially from an animal bite.
- If there is a chance there could be a lawsuit about the injury.

More sophisticated X rays, CT or MRI scans, are needed to determine whether or not there is bleeding or swelling of the brain.

## Neck Injuries

Most cases of stiff neck in childhood aren't from injuries but are a side effect of upper respiratory infections and muscle spasm. They are covered in *Body Parts, Bodily Functions, and What Ails Them* of this section, under *Necks*.

True neck injuries come in three varieties: a direct blow from a fall; whiplash, from a sudden stop while facing forward in a vehicle, or from shaking; and strangling injuries, either accidental or, tragically, on purpose.

If your child has Down syndrome, her neck may be specially at risk for dislocation. Discuss this with your pediatrician.

### FALLING AND HURTING THE NECK
The overriding concern here is damage to the spinal cord. This is why the American Academy of Pediatrics has made a statement urging that trampolines be regarded not as toys, but as dangerous weapons.

If you think your child has sustained a neck injury in a fall, do not move her until you have made sure her head and neck are stabilized—that there is no chance the neck can be twisted, flexed, or moved in any way. If she needs CPR, do not hesitate to apply it, but don't manipulate her neck.

If the child is up and moving before you can restrain him, still notify the pediatrician. Depending on the nature of the fall, X rays may be required. Sometimes vertebrae can be knocked out of line with not much in the way of immediate symptoms.

### WHIPLASH AND SHAKING
Whiplash and shaking can be either accidental, as when a vehicle (even a forward-facing jogger) stops suddenly or a child is playfully shaken by an older playmate; or inflicted. A crying baby, and a stressed adult, perhaps lacking control because of drink or drugs, can be the recipe for "shaken baby syndrome." The concern with these injuries is not just damage to the bones or muscles of the spine, but rupture of the blood vessels that line the brain. Severely shaken babies, infants, and young children may not show any signs of stiff neck or wry neck, but rather signs associated with severe head injury (see *Head Bonks*, above). If your baby or infant has been shaken and shows any behavioral changes at all, get help at once.

Older preschoolers may, however, complain of a stiff neck after a whiplash-type injury, most often after the car in which they are a forward-facing passenger stops suddenly. The concern with young children is accidental dislocation of the neck bones ("subluxed cervical vertebrae"), which could cause spinal cord damage. Fortunately, this is very uncommon.

However, if you think your child has whiplash, make a "cervical collar" from thickly twisted towels (or if you are very organized, you may have a commercial one on hand). Make sure it is applied without moving her neck, and is applied properly. Then call your pediatrician for instructions.

### STRANGLING-TYPE INJURIES

These can happen if a child gets tangled in something (a button catching on the net of a mesh playpen; a necklace catching on a hook; a curtain pull). They can also occur if a child is riding a tricycle and is flung neck-first onto the handlebars, of if a child runs full-tilt into a rope or cord stretched in the path. They can also be inflicted with hands or cords or ropes.

If the strangling is severe, the only recourse is CPR and dialing 911.

Any child with visible bruising on the neck, or who is hoarse, after a strangling-type injury needs pediatric attention promptly. If strangling by a person is suspected but not witnessed, do not contaminate the child's testimony by asking leading questions such as, "Did he put a rope around your neck?" or even "How did he hurt you?"

Also, don't show extreme emotion, though you may wish to, or the child may become more fearful about telling the truth.

## Eye Injuries and Foreign Bodies

### BASICS

Injuries, foreign bodies, and chemicals attacking the eyes need prompt, full examination. Very young children can't tell you exactly what happened; older ones may not want to. You want to be sure that no foreign bodies are trapped under the upper eyelid or have scratched the cornea; that there's no bleeding into the front chamber of the eye (*"hyphema"*—see Glossary), and that the retina hasn't become wrinkled (detached) or the lens detached by a blow. And any injury that could have actually penetrated the eye needs prompt, full examination.[1]

### ASSESSMENT

Sometimes there isn't time, or need, to fully assess the situation before taking immediate action. If the child has sustained a direct blow to the eye, an injury or an animal bite that could penetrate (even if it is just close to the eye and not into the eye), or chemicals in the eye, give first aid treatment as described below and then get immediate help. If the eye injury was part of a general head injury that could be serious, you need immediate

---

1. I always think of the humorist James Thurber, who lost the sight in one, then eventually both, eyes after a penetrating injury from a crust of bread hurled by his brother.

help. If the child is in severe pain and can't open the eye, you need immediate help.

## TREATMENT

### Chemical spills

Chemicals, and this includes some cosmetics, can produce serious burns. It is never a mistake to immediately flush out the eye with plain tap water. If you strongly suspect the chemical is caustic, the eye should be flushed out for at least fifteen minutes. Then call the pediatrician or Poison Control. You must hold the lids open. If the chemical is documented to be innocent, and if the child has no complaints of pain, redness of the eye, or sign of vision problems, a call to the pediatrician may be all that is needed. In any other case, an immediate assessment is warranted. Do not be deceived by a very red eye but no pain: a badly burned eye may be numb.

### Blows to the eye

Never try to force the eyelids open. Such pressure can make matters much worse. Place a clean cloth over the eye and tape it in place without putting any pressure on the eye itself. This helps the child to keep from trying to focus on close objects—an action that can further stress an injured eye. Get help at once.

### Foreign bodies

Any sharp foreign body, or anything that entered with force, could have penetrated the eye, and should be dealt with as if it were a blow to the eye. Ordinary foreign bodies, like sand, can often be washed out. If you can see a foreign body in the eye, you can try to remove it with a clean cotton swab or tissue curled to a point, but be alert to symptoms that may persist afterwards. The cornea could be scratched. Or there could be more than one foreign body. If you are isolated with the child and desperate, you can try to look under the flipped up (everted) upper lid for foreign bodies stuck there.

To evert an eyelid:

1. Have good lighting and an assistant standing by.
2. Have the child lie down.
3. Put something down by his feet for him to stare at. While he keeps looking down, place a cotton swab on top of the affected eyelid. This allows you to grasp his eyelashes and pull the lid back up over the cotton swab.
4. This will expose the underside of the lid. Now you can discard the swab as you hold the eyelid in its everted position.
5. If you see the foreign body, your assistant can swab it off the underside of the lid with another cotton swab. When you let go, smooth the eyelid back into place.
6. If the eye remains red, painful, or teary, or if a bright light causes discomfort, the child must be seen by a physician.

# Nose Injuries, Nosebleeds, Foreign Bodies in Noses

## FOREIGN BODIES

### Basics

Don't, the mother in the fable says, Don't, dear children, put beans up your nose while I go to market.

Of course they do. They had to, didn't they? But why do children who are not so invited put things up their noses? I always look up noses even during routine exams, even if the child comes in with an irrelevant complaint. If she got into one kind of trouble, the reasoning goes, she might have put something up her nose, as well.

You may discover such a foreign body because the child confesses, or because you see him insert the object, or because you glimpse something protruding from a nostril. Or you may smell a terrible smell that washing doesn't get rid of. You may see foul bloody discharge, usually just from one nostril, though sometimes a young person will stick something up both nostrils.

### Assessment

If you can see the foreign body, and it looks soft and fluffy, you may be able to grab it.

If you smell a terrible smell, there is a foreign body in either the child's nose or her vagina. Either way, the terrible smell means that it has been there for a while and is probably stuck because the tissues have become inflamed. Usually a terrible smell means that the pediatrician needs to find and remove the object.

### Treatment

If you are far from medical care and your cooperative older toddler or preschooler informs you that she has just put an object up her nose, there is one ploy you can try.

Look up the nose to see the foreign body. If you can see it and it is a soft, cushy, or stringy object, or has a protruding part, try to grasp it with tweezers.

If you can see it but it is up too far or is too smooth to grasp, try to blow it out yourself. Do not ask a young child to blow it out because she is most likely to prepare for this by taking in a deep breath through her nose, which can suck the object way back, or even cause her to choke on it.

To blow it out yourself:

- Read these instructions all the way through first. Then tell the child what you are going to do. Be a bit playful, but not hilarious; you don't want her to snort and choke.
- Have the child lie on her back.
- Plug the nostril that has no object in it by pressing it into the septum.
- Have the child open her mouth big and wide.
- Take a big breath yourself, press your mouth tightly over hers, and blow.
- With luck the object will pop out, or at least get within your reach.
- When it pops out, it could pop into her open mouth, and she could choke on it! Prevent this by putting your hand over

her open mouth as soon as you remove your mouth from hers.

- Then look up both nostrils carefully with a flashlight to make sure there's not something else stuck up there.

## PREVENTION

Don't tell your child not to put something up his nose. Why give him the idea in the first place?

Sometimes objects are put up the nose to quell a constant itch. A child with a constant runny nose or a child who repeatedly performs the "allergic salute," rubbing up the nostrils with the palm of the hand, or the child who picks and picks, needs pediatric help.

## NOSEBLEEDS

### Basics

Some of these are from falls or blows. Many others are due to dry or irritated mucous membranes and nose picking. They can be prevented by humidifying the air (cool mist humidifier), making sure the child drinks plenty of water, cutting his nails short, and applying a little petroleum jelly inside the nostrils morning and night.

Uncommonly, a bloody nose occurs because the child has stuck something up the nose. Usually the first sign of this is a terrible smell, however.

Rarely, a bloody nose is one of several signs that the blood can't clot properly. This is often a temporary condition or a sign of a mild inherited blood clotting problem. Only exceedingly rarely is a recurrent bloody nose a sign of something more serious, such as high blood pressure, serious blood disorder, or tumor in the nose.

### Assessment

Concern that a bloody nose isn't just ordinary occurs when:

- The nosebleed doesn't stop despite following the treatment directions below exactly.
- The child has an underlying chronic illness or known high blood pressure, or is on medication that can cause either increased blood pressure or trouble clotting the blood.
- There is a family history of bleeding disorders or trouble clotting the blood.
- Nosebleeds recur despite preventive methods.
- The child shows other indications that blood clotting or other blood functions aren't working well:

  *Bruises are unusual: very dark, on parts of the body that don't usually get bruised, or without history of injury.*

  *A rash looks like red dots that don't blanch or small purple splotches or bruises.*

  *The child is on a medication that could cause trouble with formation of blood cells. If you don't know, call the pediatrician.*

  *The child is pale, listless, or acting sick.*

### Treatment

1. Get some (preferably red) tissues and a clock with a minute hand. Turn on the

TV or get a picture book to look at to distract the child.

2. Have the child sit up leaning forward so that blood doesn't trickle down his throat. If he swallows much of it, he will throw up.

3. If the child is cooperative, ask him to blow, once, to get rid of clots.

4. Wipe away blood and clots.

5. Pinch the nostrils together tightly for five minutes. Don't keep peeking to see if the bleeding has stopped.

6. After five minutes, cautiously let go. If the bleeding has stopped, apply a little petroleum jelly or antibiotic ointment, and distract the child so that he doesn't pick. If the bleeding hasn't stopped, call your pediatrician, making sure that the staff knows you have followed the above directions to the letter and the nose still bleeds.

7. Don't have the child lie back, or put ice under his nose or on his neck. This won't work and may make things worse.

## NOSE INJURIES

Well they stick right out there, so of course things are going to happen to them. The three main concerns about nose injuries are:

- Has the septum, the cartilage that separates the nostrils, been bruised? If it has and goes untreated, the cartilage can degenerate and the nose can "collapse" into an extreme pug. (Perhaps this is what happened to Socrates.)

- Has the nose been broken?
- Could there be damage to the back of the nose, where a fragile barrier separates the nose from the spinal fluid?

*Assessment*

The pediatrician ought to be called about an injured nose if there is an associated *worrisome* head bonk, or if:

- Profuse bleeding occurs.
- The nostrils look asymmetrical.
- Bloody or clear fluid leaks from the nose.

The main concern about bleeding is a hematoma, or bruised swelling, of the septum or partition down the middle of the nose. If such a bruise occurs, and isn't relieved right away, it can put a lot of pressure on the septum, which can then crumble under it, causing major cosmetic problems and an ensuing intimate relationship with a plastic surgeon or ear, nose, and throat specialist.

Many pediatricians feel that such a septal hematoma can be pretty much ruled out if there wasn't much bleeding and if the child will allow you to press up on the end of his nose, making it look like that of a pug dog. This implies that there isn't severe pain, and also allows you to get a bit of a look inside. If both nostrils look pretty clear, the septum is almost without doubt not in any danger.

But what about a broken nose? Shouldn't the nose be x-rayed right away?

No. Two reasons. First, the nose of a child this age is made of cartilage, which does not show up on X rays. Second, because even if it is broken it doesn't have to be diagnosed right away; there's time to remold it. The only urgent questions are those discussed above. If the nose is still asymmetrical after the immediate swelling goes down, then your pediatrician may want to see, or see and refer, that nose.

*Rarely, after a head injury, a break can occur in the delicate barrier at the back of the nose, so that spinal fluid leaks out. This can happen after a blow to the head, even if the child otherwise seems fine, without any sign of concussion. If your child was fine before the head injury and suddenly has a runny nose that drips clear or blood-tinged fluid afterwards, call your pediatrician. (Make sure that the child has stopped crying, though; otherwise the runny nose is probably just tears.)*

# Mouth and Tooth Injuries

## MOUTH INJURIES

### Basics

Mouth injuries always bleed like crazy. Try to remain calm; toddlers and preschoolers confronted with lakes of blood and parental hysteria usually go right over the edge. No matter how much blood there is, if your child seems otherwise all right, tell him or her that you admire how much blood there is, that there is supposed to be a lot of blood, and that you are proud of him or her for bleeding so well. Your child may not entirely believe you, but may be so fascinated by your calm and positive reaction that the panic subsides.

Also, many mouth injuries, no matter how much they bleed, do not need suturing. Don't promise this to the child, but don't forecast the worst either.

### Assessment

- Assess the rest of him first. Did he pass out? Is he breathing all right? (If he's howling, he's breathing all right.) Is he pale or blue? Tend to this first.
- If he seems all right otherwise, then keep calm. He'll be scared and angry and shocked enough as it is. Besides, as he realizes he's not badly hurt he may take a sneaky pleasure in keeping you upset.
- Have him sit up, lean forward, and bleed comfortably onto or into something. Assure him the bleeding will stop soon, and even if it doesn't he has plenty more blood. It is really rare that a healthy child loses a dangerous amount of blood from a mouth injury: it always looks like more than it really is.
- Look to see if all his teeth seem to be in his mouth. Don't touch them yet. If one is missing, look for it but don't make a big thing about it: he may have swallowed it, and the new affront of a lost tooth will upset him all over again. If you find it, say you are going to save it for the Tooth Fairy.
- Give the child a drink of very cold water but not a piece of ice; he could choke on it.
- Be prepared for the child to vomit. Swallowed blood can cause that.

- When the bleeding and the child have calmed down, wash gently and try to assess the damage.

Very young children tend to injure their mouths in four common ways:

- They bite through their tongues.
- They bite through their lower lips.
- They cut that little flap of tissue that connects the upper lip to the gum, called the frenulum. I believe Nature put the frenulum there in order for children to lacerate it. Very few grow up with an intact one.
- They dislodge, loosen, or break a tooth (see *Teeth*, below).

*Treatment*

It's usually a good idea to call and touch base with your pediatrician unless the cut is very small and stops bleeding promptly. If you can't do so, here is a guide:

- A bitten tongue doesn't usually have to be sutured unless the bite is very deep or jagged, or unless it makes a flap at the border.
- A cut frenulum doesn't have to be sewn. But if the gum is deeply lacerated, exposing bone, the child must see a dentist or oral surgeon promptly.
- A bit or cut lip usually doesn't have to be sutured, but it does if the cut creates a big gap or flap or if it crosses the "vermilion border" between lip and skin. Especially in the latter case, it must be carefully sutured, or the child's smile will be off in a surprisingly obvious manner. Many pediatricians refer such cuts to a plastic surgeon.

## TEETH

*Basics*

Injuries to the baby teeth are very common. Even though the baby teeth may seem disposable, they really are important. First, each reserves gum space for the permanent tooth. Second, an infection of the baby tooth can ascend and damage the nerve of the developing permanent tooth. Infection is more likely if the nerve of the baby tooth is damaged.

*Assessment*

If you can't find a missing tooth, check the child for trouble breathing. Baby teeth are so small that the child may not choke but may inhale the tooth right down into one of the small bronchial tubes. Over the passage of a few hours this will cause wheezing and coughing. This is an urgent problem, because the longer the tooth stays stuck the more swelling there is of the lining of the tube, and the harder it is to get the tooth out. If you suspect that this has happened, remember not to give the child anything by mouth as general anesthesia will be needed to get the tooth out of the lung. Get in touch with your pediatrician or go to the emergency room right away.

After any kind of blow or fall that involves the mouth, check the baby teeth for chips, looseness, or displacement.

If a tooth is knocked completely out, make sure Cherub hasn't inhaled it; if she's coughing or wheezing, and you can't find the tooth, she needs to be seen and to have a chest X ray. If she's swallowed it, or you simply can't find the tooth but Cherub is fine without any breathing problems, forget it. Pediatric dentists do not reimplant "baby teeth." If Cherub asks about the Tooth Fairy (unlikely below age Five), make up any plausible "rule" about missing teeth that you want to.

If a tooth is chipped, assess the damage. If the chip is large, or the edge very sharp, this could interfere with eating. The larger the chip, the more likely that the nerve was injured. Either way, the dentist should be involved.

Make a mental note that a couple of weeks after any mouth injury you need to check the teeth for discoloration that might indicate damage to the nerve of the tooth. This needs a dental assessment.

*Treatment*
When a tooth is loosened or nudged up into the gum, the child needs to see the dentist. Call your dentist or your pediatrician for guidance.

# Arm and Hand Injuries

## BASICS
You might actually see the injury occur. Or you might discover swelling and bruising, or notice that Cherub is simply not using the hand or arm normally.

A fracture of the clavicle can make a child not use the arm, so your exam should include the "collar bone," not just the arm and hand.

The most common arm injuries are nursemaid's elbow, in which the elbow joint is dislocated when the child's arm is yanked; and fractures of the wrist or elbow from falls. The most common hand injury is catching the fingers in a slammed door.

## ASSESSMENT

*Witnessed accident*
When a toddler or preschooler falls and then complains of pain or doesn't use the arm and hand perfectly normally, a fracture is not just possible but likely.

A fractured clavicle (collarbone) is likely to have a bruise or lump that is very tender. A fractured arm or hand may show obvious deformity (bending) of the involved bones, or bruising and swelling, but the absence of such finding does not rule out a fracture.

If you suspect a fracture of the clavicle, don't panic. The clavicle is the most frequently fractured bone in the body, and it almost always heals by itself without lasting problems. A visit to the pediatrician and most likely an X ray are in order.

If the elbow is injured, make sure to check the hand on that side for circulation—color and warmth—and the ability to move fingers. If the hand is cold, blue, or the child won't move it, it's more urgent to get prompt care. Once in a great while, the blood vessels or nerves can become trapped by an elbow fracture. This may require fixing under anesthesia.

If you suspect a fracture of the hand, elbow or arm, find a way to keep the arm and hand resting in the position of most comfort. Sometimes the child will allow you to make a splint, using an ace bandage to gently splint the arm to something bulky, like a roll of paper towels. A cold pack to the injured area can help keep swelling to a minimum.

A fracture that only occurs in children is a "growth plate" fracture, which injures the site at the end of the bone that makes it grow. A fracture of the growth plate may not show up on an X ray. If the X ray is normal but the symptoms persist for longer than a couple of days, a growth plate fracture is a strong possibility. If the pediatrician or orthopedic surgeon suspects a growth plate fracture, a cast is needed—just as if there were a fracture obvious on the X ray.

Fractures need casts to stabilize the bones while they heal. This is important to make the arm look good, and often to protect the growth plate so that the bones don't stop growing, or grow unevenly.

Fingers crushed in a door often survive without breakage. It is amazing how infrequently fractures occur. However, if there is a cut in the skin, blood under the nail, or the base of the nail is injured, the child needs to be seen. That's because if there is a fracture, the cut may make it open to the air and likely to get infected.

If any of the fingers look bent, or if she isn't using all fingers normally after half an hour, the child must be seen. Never put ice on fingers; they can become frostbitten. A cold pack, though, is fine.

A yank on the hand followed by not using that arm usually means that the child has suffered a "nursemaid's elbow," see below. This is true even though the child complains of pain at the wrist. The treatment of a nursemaid's elbow is a maneuver putting it back in place.

### Swelling, without any known cause

A swollen wrist, hand, or arm may mean trauma, or it may mean an allergic reaction to an insect sting, or even infection. In any event, such swelling that persists needs pediatric evaluation.

### Bruising

Bruises that are clearly caused by minor trauma usually can be soothed by a kiss. A *swollen* bruise in a small child makes a fracture more likely. Have it evaluated.

### Not using an arm

Look for swelling or bruising. Ask the child to point where it hurts, and see if it hurts when you tap on the bone at the site. If so, the child needs a visit to the pediatrician: if a direct blow hurts enough not to use the arm, there is a high possibility of a fracture.

If you see no swelling or bruising, and you didn't witness a fall, the most likely cause of not using an arm is "nursemaid's elbow," in which the arm is yanked so that a little piece of tissue gets trapped in the joint. That makes it very painful when the child twists the arm from palm-down to palm-up. It is not serious but must be treated.

*Nursemaid's Elbow*

The child with nursemaid's elbow often is perfectly happy unless you try to move his arm, which he holds carefully with the palm down and the elbow slightly bent. He will point to the wrist as the site of the pain.

You may be able to elicit a good history of arm yank by an adult or older child; or he may have yanked it himself, grabbing to save himself from falling, or climbing, or whatever. Or he may simply come out of his room crying, after playing "quietly."

The maneuver to put the elbow back in place is simple and effective and hurts briefly. Parents feel devastated with guilt for yanking, or furious at the older sibling or sitter who yanked, but unless someone was truly abusive, chalk it up to accident and don't scold—yourself or anyone else.

### TREATMENT

Any injury that results in noticeable swelling of the hand or arm, obvious bending of a bone, or lack of normal functioning after an hour, requires a visit to the pediatrician or emergency room. X rays may not suffice to make a diagnosis: nursemaid's elbows don't show up on X rays. Neither do fractures of the parts of bones that aren't yet calcified. This includes the important growth centers. Children under Five may well need a cast even though the X ray shows no fracture.

Applying cold is always good first aid, but never apply ice directly to the skin. Ice packs help keep down swelling, but remember not to apply them to small, frostbite-prone extremities—not on:

Fingers and Toes
Penises and Noses
Or Ears,
My dears.

## Leg and Foot Injuries

### BASICS

A child who limps, or won't bear weight on a leg, or who won't move a leg, or who has a swollen painful joint or limb, has not necessarily suffered an injury. Unless you have actually witnessed a fall or other injury, don't assume that this is what has happened. Limps and swelling can be caused by infection and other medical conditions (See *Hips, Legs, Feet* in *Body Parts* chapter.)

If you strongly suspect or have seen that your child has injured a leg or foot, you're way ahead of the game. The most common injuries are bruises, possible fractures, possible sprains, and puncture wounds to the foot. Bee stings can cause mysterious pain and limps also (see *Stings,* this section). Puncture wounds to the foot are not all innocent. The first concern in a child who is not "up to date" on routine immunizations is not only tetanus. Bacterial infection (see below) can also occur.

### ASSESSMENT

As with all injuries, look at the whole child; don't be distracted by an obvious leg injury from, say, the fact that the child is having trouble breathing, or continues to be in a dangerous situation.

*Puncture wounds to the foot*
If a puncture wound is deep, it may carry infection into the foot. This is a major problem because the sole of the foot is a tight little place where infection can invade the bone. Evaluate whether the puncture wound:

- Is obviously deep and dirty.
- Occurred through a shoe or sock (this makes infection more likely).
- Is in the toe or the ball of the foot rather than the heel.

If any of these is the case, your pediatrician may want to irrigate the wound, and may wish to give antibiotics. Rarely a puncture of the foot requires hospital admission.

*Fractures and sprains*
An obviously swollen or distorted leg, knee, ankle, or foot could harbor either a fracture or a sprain. You may not be able to tell the difference, even with an X ray, at this age, as parts of the bones are not yet calcified.

## Treatment

*Puncture wounds*
Wash the foot in warm soapy water and let it soak. Don't use antiseptics such as hydrogen peroxide. Assess the damage and if you are concerned call the pediatrician. If the wound looks innocent but the child continues to limp, or you see redness or a streak of red developing, notify your pediatrician promptly.

*Suspected sprain or fracture*
The rule for this kind of injury is to try to reduce swelling, not so much to prevent further injury as to make the child more comfortable and to allow a better evaluation of how much damage has been done.

The way to reduce swelling is to elevate the leg so that the foot is higher than the knee, knee higher than hip, hip higher than heart.

Applying cold will help. An ice pack or a bag of frozen vegetables will help also, but not on little feet—on ankles and up. And never put ice on the skin.

A dose of acetaminophen or ibuproten is unlikely to mask a serious injury and may help a child recover more promptly from an innocent one.

# Cuts and Bleeding

## Basics
The tendency we all share is to look at the blood, not at the child. There are two problems with this: first, it's possible to miss a much more serious, nonbleeding injury. Second, the child becomes even more panicky when an adult focuses on the bleeding.

Any child with a penetrating wound needs to have tetanus immunization status assessed and, if necessary, updated right away.

For puncture wounds of the foot, see previous section on *Leg and Foot Injuries*.

## Assessment
So first address the child. Is she in any danger of being hurt further by where she is? Is

she alert? Breathing without problems? Is an arm or leg looking bent out of whack? Could she have injured her neck or back?

Next, given that all else is safe, address the bleeding. Nearly every bleeding injury of childhood will stop if you apply firm pressure to the site (if possible, with a fairly clean cloth) for five minutes without letting up the pressure. This is not easy when the child is frantic, so it's important to remain calm. Children seem to do best when the adult treats the bleeding as routine: expressing a degree of moderate sympathy, tinged perhaps with a look of resignation, as if you were expecting this to happen and have seen such injuries daily for the last twenty years. However, don't try to distract a child from staring at and screaming at the sight. It won't work, and the child may get more frantic because he thinks you're ignoring an obvious problem.

Once the bleeding has stopped, and you have reassured yourself that the child is safe and intact, you can decide whether or not the child needs to have the wound taken care of by a medical professional. This will be the case if:

- There could be foreign bodies, clean or dirty, embedded in the wound. Even little bits of dirt, if not carefully removed, can cause infection. Or they can become healed permanently into the wound, causing a dirty appearance. This is known as a "road tattoo."
- The wound is deep enough that the skin gapes apart.
- The wound is deep, into the fat or mus-cle, even though the skin does not gape apart.
- The wound is located in a cosmetically sensitive area, or on a part of the body, like the finger or sole of the foot, where underlying nerves or tendons could be cut, or where scarring could impair function.
- Bleeding can't be stopped, or stops and then starts again.
- A cut or puncture wound could have penetrated the eye or the skull—a deep animal bite, a dart, a pencil, etc.
- Legal action is contemplated because of the nature of the injury.

If the wound doesn't fit any of these categories, it can be cleaned with plain soap and water. Antiseptics such as iodine or hydrogen peroxide aren't a good idea. If the cut was from an animal bite, see that discussion later in *First Aid*.

Once it's washed, assess the wound again, and if you are still comfortable that it doesn't need medical attention, apply a dressing. A nonstick Band-Aid or Telfa pad is fine, but beware their use in children under about two-and-a-half. Such young children tend to ignore the symbolic significance of the bandage and instead eat or choke on it. Antiseptic ointments help prevent infection. Get one that does not contain neomycin, as this ingredient can produce an allergic response in some children.

# Sunburns, Heat Burns, and Abrasions

## BASICS

Either of these injuries can affect only the top layer of skin or can go down very deep. Both hurt a lot, unless so deep as to injure the nerves. Both are prone to infection and to subsequent scarring.

A child with any degree of injury should be checked to be sure that tetanus immunization is up to date.

## ASSESSMENT

Check the child's general condition first. Get the child away from the sun, and from any danger of further burns from other sources. Make sure no other, more serious injury exists.

A first-degree burn is a red flush on the skin. Sunburn is a common cause of first-degree burns. A first-degree abrasion is a mild scrape that doesn't actively bleed.

A second-degree burn has blisters. A second-degree abrasion bleeds actively.

A third-degree burn is so deep that tissue looks white or grey, and the area does not hurt. A third-degree abrasion is deep enough to see tissue below the skin.

## TREATMENT

The first step is to flush the area with cool or cold water, not ice. If there is clothing, soak it off or remove it as quickly as possible.

Never break the blisters of a second-degree burn; it is Nature's way of providing a sterile protection to the burned skin.

Never put butter, lard, or other home remedies on the burn; they seal in heat.

An extensive sunburn, even if it is all first-degree, can endanger a baby or young child; take the child's rectal temperature, assess her general condition, and call the pediatrician.

Second-degree burns and abrasions need to be medically assessed and treated to avoid their becoming infected and turning into third-degree ones, which scar. Some second- and third-degree burns (if they are extensive, or on face, soles, palms, or genitals) need hospital treatment.

# Human and Animal Bites

## BASICS

Your first concern is serious injury to tissues of the body. Animal bites are more likely to inflict bad injury. The most likely scenarios for this to happen are:

- Bites by a pet dog everybody thought was safe, but who was disturbed while eating or who was otherwise provoked.
- Bites by a cat. Cats' long pointy teeth can inflict serious injury, even if the puncture wound looks small. If the bite breaks the skin and is on the head or face, Cherub needs to be evaluated to be sure it didn't cause damage to underlying structures—especially brain or eye.
- Bites by a dog trained to bite or bred to be vicious.

- Bites by a wild animal, such as a bear or mountain lion.

Prevention is the only way real way to deal with these possibilities.

With both animal and human bites, your second concern is infection. Local infection occurs when bacteria get into deep structures and cause redness and swelling, and is common with both animal bites and human bites.

Transmission of systemic disease is, thankfully, rare. Tetanus is a possibility with both human and animal bites, but can be prevented by immunization. If a child has had at least four tetanus shots, she is fully immunized: this is usually accomplished by age Two. She needs a booster dose at Five. But between Two and Five, if the bite is an ordinary and minor one by a domestic pet, she doesn't need a booster.

Animal bites can also transmit rabies, a viral disease that is 100% fatal unless prevented by rabies shots. Cat bites can transmit Cat Bite Fever, a bacterial disease (see Glossary). Human bites theoretically could transmit HIV, the AIDS virus, and Hepatitis B and C, but the reports of this kind of transmission are extremely rare. Moreover, most children are immunized in babyhood against Hepatitis B.

## ASSESSMENT

### Animal bites and infection:

#### 1. Rabies

Rabies is a virus infection that is 100% fatal once symptoms set in. It must be prevented, either by preventing exposure, or, once exposure is a possibility, by a series of injections.

Rabies can be passed on by the saliva of any infected mammal. Wild animals are most likely to infect a person with rabies; raccoons, skunks, foxes, coyotes, and especially bats are particularly risk-laden. Of domestic pets, the most likely are cats, dogs, and ferrets.

To assess the danger of Rabies from a particular bite, several things must be taken into account. Rabies is more of a possibility:
- if the attacking animal was not provoked into biting, or if it was acting strange or sick; or
- if animals of the type causing the bite have a record of testing positive for rabies in your area of the country.

The most frequent animal source of rabies in people, these days, is bats.

You do not need to be bitten by a bat to contract rabies. It's been shown that the virus can be airborne in caves containing bats. Also, it may be that bat saliva can transmit the virus by getting onto mucous membranes, without a bite. We know this from cases of rabies in which no possibility of a bite occurred; just contact with a bat. For this reason, any bat exposure—one that flies around a child's bedroom, for instance—is taken seriously.

*Any* animal bite, even a nip, needs to be assessed as a possible cause of rabies. Most such bites are perfectly innocent. But you really need to run it by your pediatrician to be sure. In the case of domestic animals, the pet can be observed for ten days for signs of illness. It is usually safe to delay starting rabies shots until that period is over.

If there is a question that rabies might have been transmitted, rabies shots (a gamma globulin shot, plus a series of vaccinations) should be given as soon as possible. This is the only treatment available, and must be given before symptoms begin.

Rabies shots have the reputation of being dangerous and painful. In the past, this might have been true. Today, however, they are much less painful and much safer, though just as expensive as ever.

If you are traveling to an area where rabies is widespread in the dog population, and you won't be having immediate access to good medical care, you should consider obtaining the vaccine before leaving.

### 2. *Tetanus*

Tetanus is a potentially fatal disease. Tetanus bacteria do not cause a serious infection. Rather, they produce a toxin, or poison, which can lead to lockjaw (a severe spasm that prevents opening the mouth), and excruciating, uncontrollable muscle spasms. It is very difficult to treat, and deaths can occur with the best treatment.

Tetanus bacteria inhabit the intestines of animals and man. The spores live in the soil, and can contaminate any wound. An animal bite may be more likely to cause tetanus because animals like to root around in soil, which may be contaminated.

Once a child is immunized fully, however, a booster isn't needed for minor bites.

### 3. *Localized infections*

An animal bite can cause a local infection with redness and swelling. Several bacteria are possible culprits.

If the bite was from a domestic animal, identify the owner. If a wild animal, note the location and behavior of the animal. In both cases, call Animal Control even if you know (or are!) the owner of the animal. Most domestic animals can be examined and then observed closely in quarantine to set to rest concerns about rabies.

If there is any concern about rabies, and shots are suggested, by all means have the child undergo the shots: modern rabies shots are both safe and much less painful than the old ones, which had a terrible reputation.

*Human bites*
1. Localized infection from human bites is common. Look for redness and swelling.
2. Some serious viral diseases can be transmitted by bites.
3. In all the literature, there is only one case of possible transfer of AIDS by biting.

Human and animal mouths contain lots and lots of bacteria, and the bite can carry these germs deep into tissue. Certainly, however, you will want to take prompt first-aid action (see below) and consult your pediatrician right away if you have real reason for concern.

First, of course, assess the whole child. Is the bite an isolated injury? Don't be distracted by your outrage; make sure that the child is not otherwise injured or still in danger.

Second, assess the circumstances of the biting: who or what did it, under what circumstances. "Tame" bites are from friendly, known humans who didn't really mean to inflict harm, or did mean to but only as an irresistible impulse. "Wild" bites are from humans you don't know, or whom you do know but who acted in scary fashion. For an animal, this would be an unprovoked attack, or any attack from a wild animal. For humans, this would be an older child, teenaged, or adult biter (no exceptions). Any "wild" bite needs pediatric attention. At the least, it will need to be reported to the proper agency.

Third, assess the bite itself. Did it break the skin? Is it located near the eyes, nose or mouth; at a joint; or on hands, feet, genitals? Is the bite likely to have legal implications? If your answer is Yes to any of these, you need to see the pediatrician.

## TREATMENT

- Wash the bite promptly and with warm soapy water. Irrigate it thoroughly with warm water. Don't use antiseptics like iodine (Betadine), alcohol, or hydrogen peroxide.
- If the bite is only skin deep, and there aren't any complicating factors, you can then apply an antibacterial ointment,

preferably one without neomycin in it, and a Band-Aid.
- Any bite that has broken the skin needs to be seen by the pediatrician, to determine whether it should be sutured, and whether the child needs systemic oral or injected antibiotics.
- If legal action is a possibility, take a Polaroid picture of the bite and have your pediatrician or the emergency room doctor sign and date it.
- Continue to monitor the bite area for redness, swelling, and streaking.
- After assuring the safety of the child, get information about the biter.

If it was an older child, teenager, or adult who bit, report the incident to Child Protective Services and to your pediatrician.

It is theoretically possible for rabies to be passed on by human saliva, but this would occur only after the human was ill with rabies. Only people in contact with the sick patient's saliva, spinal fluid, or brain tissue—almost always medical personnel—are at risk and need to be immunized.

## Poisoning and Ingestions

**A single toll-free phone number: 800-222-1222 will connect anyone anywhere in the United States to the nearest poison control center.**

## BASICS

"Ingestion" is a medical term that usually means a child has eaten or drunk something that he wasn't supposed to, but which is not itself considered a poison, such as Grandma's heart medication. The medication is fine for Grandma, but not for Cherub.

The term Poisoning, on the other hand, usually means that the child ate or touched or inhaled something toxic, something that has no "normal" use in human beings.

Bear in mind that seemingly innocuous diet supplements and medications can be very toxic. Iron supplements are particularly dangerous, as the liquid iron is sweet, and iron pills are often sugar-coated. Aspirin and acetaminophen are another pair of potential dangers.

Be sure to save and take to the hospital with you any clues about what Cherub got into, and how much he might have taken: containers, clothing or objects the substance spilled on, whatever he might have vomited, and whatever he vomited onto or into.

## ASSESSMENT

No matter what kind of poisoning has occurred, your initial assessment consists of answering the following questions:

- Is there further danger? Get the stuff out of Cherub's hands, make him spit it out or get it out with your fingers. Move the container way out of reach.
- Is Cherub in immediate danger? Having trouble breathing? Falling asleep? Severe throat pain or drooling? If so, summon

the fastest way to get care—911 or a race to the emergency room. Take the container, and whatever evidence you can find, with you to the hospital.
- If Cherub is in no or only mild distress, call Poison Control. If you don't have the number taped to the phone, call information. If Cherub got into a household product, there may be instructions on the label, in case of ingestion: **Do not follow these directions! Call Poison Control first.**

## TREATMENT
**Treatment consists of:**

- Preventing further absorption of the substance.
- When appropriate, administering an antidote.

There are three ways to prevent further absorption of the substance: Vomiting, pumping out the stomach, and having the child take a substance that absorbs the poison in the stomach, so that it can't get into the bloodstream.

### Syrup of Ipecac
Some pediatricians and others advise parents to keep a bottle of Syrup of Ipecac in the house. This medication usually induces vomiting within about twenty minutes after it is given. When it is given appropriately, it may indeed keep a child from absorbing whatever it is he just ate or drank: he'll vomit it up before absorption can take place.

But Syrup of Ipecac can be dangerous. It can cause vomiting in a child who is rapidly going into a coma, resulting in severe aspiration. It can be given mistakenly to a child who drank something caustic, and the substance will burn his mouth and esophagus twice: once going down, and a second time coming back up. It can be given mistakenly before the child is given the real antidote, and cause the child to vomit the appropriate medication.

So most pediatricians and Poison Control centers recommend that you keep Syrup of Ipecac *only* if you live far from medical care, and *only* if you understand thoroughly when to use it and when not to do so.

**DO NOT USE Syrup of Ipecac if:**

- The child is unconscious or getting sleepy.
- The substance ingested is caustic and will burn as much coming up as it did going down. If you think this is the case, but are told by Poison Control to induce vomiting, ask to talk to a Supervisor.
- The substance the child ingested has as its main danger aspiration into the lungs, such as oil, paint thinners, turpentine, and so on. If you think this is the case, but are told by Poison Control to induce vomiting, ask to talk to a supervisor.

If you do give Syrup of Ipecac on the advice of Poison Control or your pediatrician, it will work fastest and most effectively if you follow it with a glass of water, soda, or juice, and then have the child move about. You can follow him around, bearing the bucket for him to vomit in. Save the vomitus.

**One last caution:**

Never try to "neutralize" a caustic substance, even though you learned in chemistry class that acids and bases neutralize each other. A byproduct of that neutralizing reaction is heat—that is, burns.

# Insect Stings: Bee, Wasp, Hornet, Yellowjacket, and Stinging Ants

## BASICS

Most often, bee or beelike stings and ant bites cause only a local reaction, with redness, pain, swelling and itching around the sting or bite. Usually the swelling is less than three inches in diameter, and lasts less than twenty-four hours.

Sometimes a reaction will be larger, last longer, and may be accompanied by hives that are limited to the area around the sting.

Rarely (though much more commonly in adults), a child with a sting from either the bee family or from stinging ants can go into a generalized reaction called *anaphylaxis*. This consists of swelling of the airway, wheezing, swelling of lips and face, and a drop in blood pressure (shock). Such a reaction can be fatal. In the United States, about 40 people a year die of insect stings (though this statistic does not reflect deaths from the Africanized or "killer" bees).

A child who has had one severe episode after a sting, or who has suffered from true anaphylaxis, is more likely to have another severe reaction. So is a child with a biological parent who has an allergy to bee stings. If a child has such a history, it's necessary to take careful precautions against bee stings, and to always carry an allergy kit containing an injection of epinephrine, such as Epipen Junior or Anakit. If you are likely to be some distance from medical care, you need to carry one kit for every fifteen minutes it would take you to get help, should a sting occur. A child with such a history also needs to wear a MedicAlert bracelet.

To prevent beestings, especially in an allergy-prone child:

- Avoid perfumes or perfumed lotions;
- Avoid bright and pastel colors;
- Wear long pants and shoes when in high grass or woods. Insect repellants are useless against beelike stings.

The Africanized bees tend to build nests in dark hidden places, such as discarded tires, holes of trees, under the eaves of buildings. Your public health department can give you advice on how to prevent, check for, and get rid of them safely. Don't go poking around there yourself without careful precautions, even if you're just tidying up.

## ASSESSMENT

Sometimes you are there when the sting occurs; often, though, it's unwitnessed. Young children may not know they were stung. Perhaps they stepped on a bee and didn't see it. Children old enough to talk will make up a story to explain the sting to themselves: They may say they stepped on

something sharp. But when you examine the area, you'll see a white raised area surrounded by a red circle with a vague, not definite, border. If the insect was a bumblebee or honeybee, you may see the stinger embedded in the white area.

In the rare event of anaphylaxis, the child becomes very sick very fast.

## TREATMENT

1. ***Get away from any further danger of being stung.*** Honeybees and bumblebees lose their stinger after they sting, and they die. You don't have to worry about being stung again unless the hive is aroused.

   Other stinging insects like wasps, hornets, and yellow jackets don't lose their stingers, and don't die, and can come round and sting you again and again. Your best bet is to get you and Cherub away from the area rapidly, away from food that might be attracting them.

   Africanized bees don't sting as individuals: they come out in clouds. They can fly long distances. Run and try to find a vehicle or building in which to secure yourselves. Don't go into water; they will wait for you to come up.

2. ***Get the stinger out, right away, if you can see it in the sting.*** The key is to get it out in the first 10 to 20 seconds, and it doesn't matter what you take it out with. It used to be thought that pinching or squeezing the stinger released more venom, and that the best instrument was a blunt-edged knife or a

credit card. This is not the case. The only thing that helps is speed.

3. ***If you are with a child who has been stung, and you have the epinephrine kit with you***, give the shot at the *very first sign* that indicates more than a simple local reaction. Here are some signs: the child doesn't feel right and starts whining and looking scared; looks pale, has a hive or two, starts drooling, has swollen lips; starts to take sighing breaths or wheezes or can't seem to get the breath in—anything that is out of the ordinary. You will not hurt him if you give it unnecessarily; but you can risk his life if you wait. Even if he then seems fine, immediately head to or call for medical help; go directly to the emergency room. The shot of epinephrine will wear off in about fifteen minutes, and the anaphylaxis may well return.

4. ***If you do not have the epinephrine kit with you***, but a stung child shows any signs other than a local reaction, pick up the child and hasten to, or call for, emergency help.

5. ***For simple local reactions:*** Put something cold on the sting; elevate the stung foot or hand; give a dose of the antihistamine *diphenhydramine* (Benadryl); and one of acetaminophen (Tylenol) or ibuprofen (Motrin or Advil).

## PREVENTION

If a child, or anyone, has had a severe local reaction or a generalized reaction to a sting, urge that he or she have allergy testing and desensitization, and always carry a "bee sting kit" and know how to use it. Teach your children not to provoke bees; especially, consult your pediatrician for guidance about whether "killer bees" may be approaching your area. Avoid loose colorful clothing and wear shoes during bee season.

# Body Parts, Bodily Functions, and What Ails Them

## Headaches

Headaches come in several patterns:

1. A single, acute headache, coming on over minutes or hours;

2. Acute significant headaches that recur, with the child feeling well in between;

3. Annoying, mild, recurrent headaches with the child feeling well in between; *or* a mild, annoying headache that is present practically all the time, without getting any worse. The child's activities and behavior are not affected. She just complains a lot;

4. A headache that is present all (or practically all) the time that *is* getting worse, as you can tell by Cherub's constant complaints and decreased play. *This kind of headache may reflect pressure on the brain, from internal bleeding, brain swelling or edema, a mass such as a tumor, or other cause. Get Cherub to medical care right away.*

### ACUTE HEADACHE, RIGHT NOW, OUT OF THE BLUE

Most acute headaches are innocent and due to ordinary common conditions. Fever all by itself causes headaches. Many upper respiratory viruses, such as influenza, feature acute headaches, and strep throat is famous for doing so. In fact, many children complain of a headache rather than of a sore throat.

When faced with a young child with an acute headache, the first thing to do is to assess whether she has any signs of acting "Scary Sick." (See *Frightening Behaviors.*) In this unlikely event, she needs to be seen right away.

If there are no such signs, give Cherub an age-appropriate dose of ibuprofen. Don't forget to give her something to coat her stomach, such as a banana or ice cream. Also, take her temperature *before* you give her the medication, since it will mask any fever. If the headache improves over the two hours after the ibuprofen is given, it is safe to observe and monitor her for other symptoms

## SIGNS OF POTENTIALLY SERIOUS HEADACHE: GET HELP NOW!

- Acting "Scary Sick" (see section in *Frightening Behaviors*).
- Headache significant enough to cause crying or behavior changes (not just a sore place from a bruise) following a head injury.
- Increasing pain over the two hours after ibuprofen or acetaminophen.
- Behaving very unlike himself: Disoriented, talking without making sense.
- Recurrent vomiting along with persistent headache.
- Any neurological signs: seizure (convulsion), one pupil larger than the other.
- A stiff neck that prevents the child from bringing chin to chest. This can be a sign of meningitis, because such a movement puts a stretch on the meninges— the lining of the spinal cord. If the meninges are inflamed, the stretch hurts.

A stiff neck when turning the head to one side or the other is *not* a sign of meningitis. See *torticollis* in the Glossary.

that give a clue to its cause. An ordinary headache that lasts longer than twenty-four hours, coming and going despite doses of ibuprofen, deserves a phone call to the pediatrician, even if there are no other signs or symptoms.

For signs that a headache is not ordinary, and may have a serious cause needing prompt treatment, see the box above.

## ACUTE, SIGNIFICANT HEADACHES THAT RECUR WITH CHERUB PERFECTLY FINE BETWEEN EPISODES

In childhood, the most frequent cause of recurring acute headaches is migraine, a headache that appears to originate in spasms of the blood vessels around the brain, although the exact mechanism is unclear.

Migraine headaches can be triggered by practically anything: fatigue, chocolate, sunlight, stress. Besides headaches, there is often a tummy ache, nausea, or vomiting. The headache is increased by noise, light, and exercise; it's helped by rest or sleep. There is usually a family history of migraine. Cherub is completely well between attacks.

There is no specific examination or X ray or lab test to diagnose migraine; it is diagnosed by excluding other causes of recurring headache. Acute headaches that recur need an evaluation by the pediatrician, and sometimes by a pediatric neurologist. It may be necessary for Cherub to undergo an MRI, or other imaging test, to rule out repeated small bleeds into the brain, or high blood pressure, or other extremely rare causes.

## PERSISTING OR RECURRING MILD HEADACHES

Age Four is a fuzzy but real dividing line for this kind of headache. In children Four and older, such headaches are common complaints. Most often, these are migraine headaches, or headaches secondary to Life Itself ("attentional" or "functional" or "tension" headaches). You can safely observe for awhile, gathering information that will help identify the cause of such headaches.

However, in children under Four, especially those Two and Under, chronic or recurring headaches are uncommon and much more worrisome. A child this young with persistent or recurring headaches over a week's time needs to be evaluated, even if there are no other serious signs (see *Signs of Potentially Serious Headache*, above).

### Children four and over

Children Four and Over complain of headaches frequently. It's a term that they are comfortable using. This means that you are likely to hear Four and Older complain of headaches that are obviously very innocent and mild.

That's because at around Four, children start to understand the meaning of the term. "Headache" is a rather sophisticated concept. It's hard for children under Four to think of their head as a body part, like an ear or a tooth. As adults know, when you have a headache, it's more as if *you*, yourself are in pain. "My Soul hurts," George, diagnosed with migraines at Four and a Half, told me.

By Four, most children have heard enough adults complain of headaches to start to grasp the concept. That means that Four is able to use the term Headache when she is feeling generally sick and her head is uncomfortable. She is also able to use the term (as adults sometimes do) when she is feeling out of sorts and can't come up with a different explanation for whining and complaining. She can also use the complaint of a headache to get attention, just as adults do: It doesn't take Four and Older long to figure out that "headache" is entirely subjective. Nobody can prove whether you really have it or not.

If Cherub shows no serious signs (see *Signs of Potentially Serious Headache*) and acts and looks fine, keep a tactful diary of the headaches. Don't keep asking her if she has a headache, or let her overhear you worrying about it, or see you writing things down. You may well start to see a pattern. Perhaps they occur when she is overly tired, has watched too much TV, or is constipated.

### Children three and under

The year from Two to Three is an in-between age; I'd err on the side of caution.

Such very young children usually only localize pain to the head if it REALLY hurts. They usually can't describe what hurts verbally. At most, they can say Yes or No if you ask them. So you have to go by what you observe.

A very young child who can talk may point to his head and/or say it hurts, or sim-

- How often do they occur? Weekends as well as weekdays?
- What time of day?
- How bad are they? Does she have to stop playing, and lie down, or does she go on with the game?
- How long does the headache last?
- Can you think of a trigger, such as fatigue, stress, or specific food?
- What helps, and what makes it worse?

If even such mild headaches show increasing frequency or intensity, or continue to recur over more than a three week period, or other symptoms appear, your pediatrician needs to see her. Find out first, if you can, if there is anyone on either side of the family with a diagnosis of migraine headaches. Take your diary with you to the appointment.

ply may hold his head with his hands and look as if he's in pain. It may be obvious that certain activities cause pain: being bounced, jumping, or other sudden changes of position may cause him to cry and, if a year of age or older, to hold his head. Another sign of headache is a dramatic increase in daytime sleep, accompanied by a change in behavior (irritable, lethargic, seeming in pain) while awake.

He needs to be seen URGENTLY if he has any of the serious signs below.

# Neck

## THE QUESTION OF MENINGITIS

Parents' biggest worry about stiff necks is meningitis. However, meningitis in young children OFTEN doesn't give a stiff neck, and in young babies ALMOST NEVER gives a stiff neck.

When meningitis is what is causing the stiff neck, the child is usually sick in other impressive ways. He usually has fever, vomiting, is "not acting right" in a serious manner, and often has a rash. The stiff neck of meningitis occurs because the spinal fluid is inflamed, and anything that stretches the spinal cord really hurts. If the child resists or shows pain when you have her curl up with nose to knee, chin to chest—call your pediatrician.

## SIGNS THAT A HEADACHE IN A YOUNG CHILD MAY BE SERIOUS

- Any history, or suspicion, of having been shaken.
- Holding onto the head for periods of time. Reluctant to turn the head to look at something. Moving about carefully, and looking anxious or in pain. Not eager to jump or be bounced.
- Signs that the skull is enlarging too rapidly: an enlarging soft spot, a forehead that bulges out with distended veins.
- If the fontanel is still open: a soft spot that is bulging, making an obvious protrusion, while Cherub is sitting quietly. (When a baby sits, the soft spot ought to make a very subtle dent or a smooth line with the skull, not a bulge. When a baby cries, the soft spot will bulge up momentarily and then soften again.)

### STIFF OR WRY NECK IN YOUNG BABIES

Babies under six months of age who look at you askew either have head tilt or wry neck, and it is sometimes hard to tell which is which. The causes range from birth position and tight muscles to spine abnormalities to vision problems: a lazy eye can make a baby compensate by tilting her head.

Either way, the baby needs to be checked by the pediatrician. If there is a wry neck (torticollis), your pediatrician will also want to check the hips carefully, as there is an association between torticollis and hips that haven't gotten firmly in the sockets.

### WRY NECK IN TODDLERS AND PRESCHOOLERS

A painful neck, with the head turned to right or left, is pretty common in young children.

Sometimes it's associated with a sore throat or earache (which needs pediatric evaluation). Sometimes it's from overactivity or sleeping on it wrong.

If there's no sign of other problems, and no history of injury, acetaminophen (or ibuprofen, if your pediatrician advises it) plus ice massage usually relieves the problem. For ice massage, use a plastic sandwich bag filled with frozen vegetables or a high-tech cold pack. If that hasn't helped in a couple of hours, you need more help.

## Eyes and Eyelids

*Babies Four Months Old and younger have special eye issues. See the chapter on* Eyes *in* Birth to Two Weeks.

## SIGNS THAT A CHRONIC OR RECURRENT HEADACHE MAY BE SERIOUS

- Vomiting as an *unexpected* event in a child who has not been nauseated. "All of a sudden, out of a clear blue sky." Recurrent vomiting early in the morning, without other signs of intestinal illness.
- Signs that the brain isn't governing the body well. An eye that starts turning in or out. A smile that is weaker on one side, making it asymmetric. One pupil seeming much bigger than the other. Weakness of an arm or leg. Walking with legs apart, as if he's having trouble keeping his balance.
- A recurring headache that is always located in the same spot of the head.
- A recurring headache that is located in the back of the head, and that is increased by coughing or vomiting.
- Headaches that awaken the child at night, or that are present when first waking up and worsen with lying down; headache that gets worse with activities that involve holding the breath and bearing down, which increases pressure inside the head: coughing, sneezing, straining to poop.
- Headache in a child who had a significant head injury within the last two months, especially if he passed out when injured.
- Headaches that may mean the pituitary gland isn't functioning well: great thirst and lots of urine (loaded diapers, new wetting accidents, having to drink water or get up to pee—or newly wetting the bed—at night.) Not growing well.

## BASICS

The white of the eye is called the *conjunctiva*. The colored part of the eye (the part that you mean when you say He's got brown eyes) is the *iris*. The black circle in the center is the *pupil*. The *cornea* is a transparent membrane that covers the iris and pupil; it's what hurts so badly if it gets scratched.

*The white of the eye is not white*
If the white of the eye is bluish, this is most often a normal accompaniment of very blue eyes. However, it can also be a sign of iron deficiency. It can also be a sign of Fragile Bones, an inherited genetic disorder.

Very dark-eyed children, especially those of Oriental, Mediterranean, Hispanic, or African descent, often have brown or black spots or smears in the whites of the eyes, and these are almost always normal as well.

Yellowness of the whites of the eyes (jaundice) that appears for the first time after five days of age always is abnormal, though not always serious; it always needs prompt evaluation.

A child with yellow skin, but no yellowness of the whites of the eyes, does not have jaundice but rather carotenemia (see Glossary).

If the white of the eye is pink—please, please, don't call and say Cherub has pink eye. I wish nobody had invented the term "pink eye." The people who use it seem to know exactly what they mean by it; but I never do. The term *conjunctivitis* isn't any better; it too just means the white of the eye is pink or red. Such redness can mean infection (virus or bacteria), injury, foreign body, allergy, or sun exposure, and a few dozen very rare entities as well.

### Goop in the eye
Tearing or matter or "sleep" or any other kind of discharge from the eyes can mean a blocked tear duct, or infection, either viral or bacterial.

The same goes for puffy eyelids. Are they swollen from infection, or irritated and puffy from allergy, or waterlogged from edema— fluid accumulating in the tissues? "Goop" and puffy eyelids need pediatric assessment.

### The color of one iris is different from the color in the other
This is called *heterochromia*, and always needs an evaluation for rare underlying problems. But most often, it's normal.

### Pupil size
Pupils become large when the environment is dark, or when a child is excited or apprehensive. They become pinpoint in bright light. (And what is dark or light may seem quite different to each child.)

Moreover, nobody is symmetrical, and often one pupil will seem *slightly* larger (or smaller!) compared to the other. This is almost always normal.

But if one pupil does become much larger than the other and stays that way, this needs a prompt evaluation even though the child seems perfectly well.

### Pupil color
A pupil that looks white, say in a flash photograph, or in any lighting where the other pupil reflects red, needs prompt attention as well. It means that something is interfering with light getting to the retina: a cataract, a tumor, something.

### Eye turns in or out
When an eye turns in or out, not quite in sync with the other eye, it is usually a condition called *strabismus*, in which the eye that turns is weaker than the one that is doing the focusing. In strabismus, the "lazy" or "wandering" eye is able to move in all directions.

However, an eye that is *unable* to move in all directions, that is stuck in one position, may signal a true emergency such as a brain tumor. Such a "stuck" eye needs an immediate evaluation.

### Miscellaneous
Eyelashes with white specks on them may be infested with lice. This can happen in the cleanest of families. And needs to be treated with prescription ointment.

Strange eye movements, like jerking eyes or "dancing" eyes, always need checking, whether they come on suddenly or gradually. One exception: Normal jerking eyes (*nystagmus*) occurs for a few moments after a child has been spinning round and round, say on the wooden carousel in the playground, usually right before she throws up on your shoe.

## ASSESSMENT

Usually, the point of assessment is to figure out whether you need an office visit, and if you do, how urgently. Sometimes an eye infection can be treated with prescription drops that your pediatrician can call into the pharmacy; have the pharmacy phone number handy, just in case. Make sure the pharmacy is open, and that it takes your insurance.

## TREATMENT

How do you get medicine into the eyes? I suspect that this was the true, unanswerable riddle of the sphinx. After all, when you live around all that sand . . .

Most parents find drops easier than creams or ointments, if you are offered a choice.

## SOME EYE PROBLEMS THAT NEED *IMMEDIATE* ATTENTION

- Sores on the eyelid. (The concern is a herpes infection, which can damage the cornea of the eye.)
- One pupil looks markedly larger than the other. This can be a neurological problem or a problem restricted to the eye.
- An eyelid that is swollen, rather than just puffy—red, painful to the touch, so swollen that the child can't fully open the eye. This can mean an infection called *periorbital cellulitis* that can be serious if it goes untreated.
- A red eye sensitive to bright light—Cherub winces and tears and won't open the eye. This means something has irritated the cornea: a scratch or abrasion, a foreign body, an infection, a chemical—something.
- Eye that is bulging out. This can mean a mass behind the eye.
- A white pupil, when the other one is red, in a photo.

*Babies*

*First try:* Give the baby a pacifier; eyelids relax when the mouth is occupied. A very placid young baby may keep the eyes open and allow you to put the drops or cream in. You may even be able to gently pull down the lower lid, exposing the lining of it, and give the medicine directly into the little pocket thus exposed.

*Second try:* Wait until the baby is asleep, then pull the lower lid down. You may have to do one eye, then wait until the baby falls asleep again to do the other.

*Third try:* If the baby has eyes clenched shut, squeeze the lids together even more tightly but without pressing on the eyes themselves: you'll find that when you do they pop out so that they make a little pocket for you to drop the medicine into. Then close the eyelids so that the drops are delivered to the eye surface. You may have to swaddle the baby or get someone to hold the rest of him still.

*Toddlers*

The techniques are the same, but you'll probably have to hold down a toddler.

*Two-person hold:* Put the toddler on her back on the floor. One adult hugs her still from the neck down restraining her arms. The other kneels with the toddler's head between the adult's knees, and applies the medication.

*One-person hold:* The toddler lies on her back, arms outstretched like a T. The adult sits at her head with one adult thigh over each toddler arm.

*Preschoolers*

Preschoolers are likely to be more receptive to any medicine if they can experiment first. It may waste a little medicine, but try putting a drop on the child's (freshly washed) finger, and let her smell it, taste it (warn her that it is Yucky!) and touch it to her eye. Tell her ahead of time what you are going to do, and practice with a drop of WARM water first. Let her put a drop in her doll's eyes. It is worth taking your time with this, because holding down a preschooler is quite a feat.

# Ears

Very young ears are prone to:

- Infection of the middle ear (*otitis media*).
- Infection of the ear canal (*otitis externa*; Swimmer's Ear).
- Injury and foreign bodies. (My favorite ear foreign body was a slip of paper in a two-year-old's, which I pulled out as though from a fortune cookie and read aloud: "Inspected by #10873.")
- Too much ear wax (*cerumen*), which is pushed up against the eardrum.

## MIDDLE-EAR INFECTIONS AND FLUID IN THE MIDDLE EAR

*Basics*

Middle-ear infections are so common your child is almost bound to have at least one. This means there will be swelling of the lin-ing of the middle ear, pus in the space behind the eardrum, and pressure on the drum itself. Symptoms vary, in large part by how rapidly the pus accumulates and whether the drum has a chance to stretch slowly or rapidly.

Fluid, rather than pus, in the middle ear can occur after or without infection. If it stays in the ear for a long time, it can cause damage to the middle ear. Even if it doesn't, subtle or moderate hearing loss from fluid in the ear can impair a child's language development, behavior, and ability to learn.

The faster an ear infection comes on, the more it hurts, because the eardrum has to stretch rapidly to accommodate the pus and fluid and blood. Once it has stretched to the maximum, many young children say that the ear no longer hurts. An older child or adult will hate the pressure and muffling, but little ones don't seem to care. So when the child complains of acute pain and then seems better, there still may be a middle ear infection.

Rarely, a middle ear infection can spread to cause *meningitis* (infection in the spinal fluid and lining of the spinal cord) or *mastoiditis* (infection of the spaces in the bone behind the ear). A child who is not just in pain but is acting much sicker (see *Acting Sick* in *Frightening Behaviors*) than would be expected with an ordinary ear infection must be seen promptly or urgently. Other signs include:

- A neck that is stiff or painful when the child tries to touch chin to chest or nose to knee.
- A severe headache.
- Redness, swelling, and pain over the bone

behind the ear. (This is different from the common finding of a lymph node in this area. A lymph node feels like a little pea or ball bearing that rolls under the fingers. It might be a little tender to touch.) It may be mastoiditis—a bone infection.

Once in a while, the pus builds up so rapidly and with such pressure that the eardrum ruptures and the pus drains out. This instantly relieves the pain in the middle ear, a benefit that usually outweighs the discomfort of having stuff leak out of the ear. Thankfully, most of the time such a rupture of the eardrum spares the delicate structures that conduct sound, and heals without a scar. But you don't want to tempt fate by having this happen again and again. You also need to be sure that the underlying infection is treated, that the pus itself does not infect the ear canal, and that the perforation does indeed heal. So even though your child is perfectly happy with her draining ear, the ear needs to be examined, along with the attached child herself; medication prescribed; and careful follow-up scheduled.

*Assessment*

### BABIES

A baby with a fever and cold who seems to be in pain may well have a middle-ear infection. But so may a baby who cries when sucking or lying flat (because of pressure changes in the ear), or who has been on an airplane flight during which she cried briefly but now seems to be babbling less and to be less alert to sounds. A perfectly healthy baby who plays with her own ears is not likely to have an ear infection; but be alert for crankiness, waking at night, decreased appetite.

### TODDLERS

A toddler with a fever and cold who grabs her ear may well have an ear infection. A toddler who is grumpy and fractious may have an infection or just fluid. A toddler whose language development is stalled may have fluid with hearing impairment.

### PRESCHOOLERS

A preschooler with fever and cold who says his ear hurts is probably correct, but many young children don't localize or describe pain very well. (I once spent half an hour with a three-year-old whose "toothache" turned out to be a sharp pain from a splinter in her left foot.) Complaints of headache or sore throat may really mean ear pain. Fluid in the middle ear can make a preschooler act disobedient, sulky, whiny, and babyish.

*Treatment*

Many ear infections first cause pain at night because of pressure changes when the child lies flat. So what can you do for your little one who is in agony at 2 A.M.?

Acute middle-ear infections are often complications of viral illness. Antibiotics are indicated for some ear infections, but they won't get rid of the runny nose, sore throat, or fever right away because those are caused by the virus. Moreover, antibiotics take time to work, and starting one in the middle of the night is not likely to produce immediate relief from pain.

Most often, home remedies (a very warm steamy washcloth to the ear, breathing steam, an over-the-counter decongestant) may seem to help.

True, fast-acting remedies can also help, however, to relieve pain. Numbing drops in the ear (by prescription) or even warmed cooking oil certainly can help,[2] as can oral analgesics like acetaminophen and ibuprofen. (Aspirin shouldn't be used in young children unless you talk with your pediatrician first.)

All too often, however, the remedy that *seems* to help gets false credit. Ear infections hurt the most when the eardrum is being stretched acutely as pus builds up fast. Once the drum is thoroughly stretched, many young (and some older) children seem to tolerate the muffling and pressure without any objection. This can occur within a couple of hours of the onset of pain, and can make it look as if medicine is working.

Certainly, if the child has worrisome other signs, such as acting sick or looking sick or a fever that is worrisome for age, an immediate call to the pediatrician is in order.

Any child who has such nighttime ear pain needs to be seen the next day, however, even though there isn't any more pain or complaint of any kind. It is amazing what awful-looking eardrums you see on children who haven't a complaint in the world.

When an antibiotic is given, be sure to give the entire course, every dose. A child with a draining ear is usually prescribed antibiotic eardrops.

A follow-up visit assures that infection is vanquished and that any rupture in the drum is healed. If sterile, uninfected fluid is still present in the middle ear, your pediatrician will want to discuss with you further follow-up and possibly treatment.

Such a visit also may reveal underlying causes for fierce, recurrent, or persistent otitis media (see the Essay in Part III called *Trouble in the Middle Ear*).

*Prevention*

The Essay in Part III goes into this in agonizing detail. Suffice to say that the mythical child with never an episode of otitis media has the following profile: parents who never had otitis media themselves, and passed on their genes; a family history free of allergies; no secondhand smoke; no exposure to viral illnesses such as one might get in daycare; doesn't take a bottle lying on her back, nor dunk her nose in the bath water; has minuscule adenoids; and has been breast-fed, with no formula or solid foods, for the first four months of life. She doesn't fly in planes or travel to mountains. I myself have met this infant wonder, but have promised her parents not to reveal her identity.

Ear infections are not contagious in the same way viruses are contagious. An infection doesn't hop out of one child's ear and into another's. But most ear infections start as viral colds, and viral colds are definitely contagious. Hence the increased incidence of otitis media in children who attend daycare. Especially large or drop-in daycares.

---

2. Such eardrops can cause pain if the eardrum perforates under pressure and the drops leak through into the middle ear.

## External Otitis, or Swimmer's Ear

### Basics

The ear canal, where the wax lives, is a cul de sac that ends at the eardrum. When the ear canal becomes swollen and infected, it can hurt terribly. This usually happens when the ear canal becomes wet from swimming or bathing, and bacteria invade the lining of the canal. The canal swells and pus forms. Since this all happens in the cul de sac bounded by the eardrum, the middle ear is not affected. However, since swimming can predispose to middle-ear infection as well as to external ear infection, it's not rare to have both conditions simultaneously.

### Assessment

When the canal leading from the world outside to the eardrum becomes swollen or infected, the main symptom is that it hurts to wiggle or touch or move the ear or to press on the area called the tragus.

Rarely, a child with external otitis may have an infection that is overwhelming, often caused by a staph infection. If the child is feverish and acting sick and in great pain, he should be seen promptly or urgently, and may need a shot or hospitalization.

### Treatment

If the infection is mild or moderate and localized, drops that contain an antibiotic, with or without cortisone, usually work.

Sometimes the pediatrician or ear doctor will want to flush the ear or insert a little wick so that the drops go all the way into the ear canal.

If there is also otitis media, oral antibiotics may be needed.

### Prevention

There are commercial drops advertised to use after swimming to prevent swimmer's ear. They contain ingredients that dry the canal and leave it with a coating of acid, both of which discourage bacteria and fungus from growing.

You can make similar drops. The recipe is: 1 part white vinegar to 1 part rubbing alcohol. You should store it carefully so no one drinks it, though I can't imagine anyone would imbibe much.

Use these drops after a day in the water. They are for prevention only. Do *not* use them if there is ear pain, pus or fluid from the ear, or a perforation or a tube in the eardrum.

## Foreign Bodies in the Ear; Trauma to the Ear Canal

### Basics

The most dangerous foreign bodies in children's ears are inserted by adults trying to remove wax. They include cotton swabs on sticks, bobby pins, and long fingernails; these can scratch the ear canal, push wax back against the drum so that it can't come out naturally and so that hearing is blocked, and can even perforate the ear drum.

*Lesson*: you aren't cleaning the child's ears

when you use these instruments. You are merely pushing the wax dangerously back in so that it doesn't show. Tempting, but unwise. Your grandmother was right when she told you, never put anything in your (or your child's) ear that is smaller than your elbow.

Other foreign bodies are inserted by children themselves, either because the ear hurts or itches, or because they are "playing doctor," or simply because it seemed like a good idea at the time. I personally have removed, or tried to remove: pearls (one real one) and beads, popcorn kernels, paper, cotton, a tiny wheel, one of Barbie's pink slippers, and an M&M.

Another source of foreign bodies, alas, is bugs. Earwigs, ants, beetles, all like to crawl into little ears.

### Assessment

Often the child shows no symptoms except signs of a guilty conscience, if he's old enough. The more guileless ones will confess, or announce, or simply mention it casually, "Oh, I put a raisin in my ear yesterday." Sometimes you'll pick up signs of not hearing well, without any sign of a cold or ear pain. Sometimes, if the light catches it right, you'll see the object glinting.

If there is a bug in the ear, though, the child will let you know, either by screaming or writhing and holding the ear; if he's old enough, he'll tell you something is crawling around in there.

### Treatment

If the foreign body isn't alive and crawling, don't try to attack it yourself unless it is a soft fluffy object you can easily grasp and remove with tweezers. After you get it out, you need someone with an otoscope to look in both ears and make sure there isn't another foreign body in there. (Behind the pearl is where I found the little wheel.)

If it is a bug crawling around in there, and your child doesn't have a hole in the eardrum (from infection or a tube in the ear), drop in some oil—vegetable oil, olive oil—enough to kill the insect. If you have no oil, alcohol will do, but will be very cold and the child won't like it. If you have no rubbing alcohol, a bit of drinkable whiskey or vodka will do. Then call or see the pediatrician.

## WAX (CERUMEN)

Wax is there to protect the ear canal from crawling bugs and parents wielding Q-Tips. Many parents seem embarrassed about ear wax in children's ears. Don't be. It is supposed to be there.

### Assessment

You may suspect wax of causing pain or diminished hearing, but you won't know for sure without the pediatrician's assessing the situation with the otoscope. Wax is a problem when:

- It blocks the view of the pediatrician, hiding otitis media.
- It presses up against the eardrum and

causes pain, pressure, and diminished hearing.

- There is so much of it that hearing is diminished. Since children don't know that their hearing is "off," this can go undiscovered for a long time.

Wax comes in two types: crumbly and gooey. Both types can become hard and impacted if pushed back into the ear canal in an attempt to present a tidy appearance.

### Treatment

Wax can be removed by your pediatrician using a little instrument. (Yes, I know I said not to put anything smaller than your elbow in the ear canal, but this is a special case. It is like the dentist telling you not to use metal instruments on your teeth.) Or it can be flushed out with water.

You can get the wax out yourself, and keep major buildups from occurring, by using a nonprescription wax-softening drop at bedtime. Use it nightly until the wax is flowing out naturally, and then once a week or so to keep it from building up. Usually, you don't need to flush it out. Once the wax is soft, the normal jaw movements when chewing help it to emerge to the outside, where it can be wiped away with either a washcloth or, of course, your elbow.

## Nose

*Nose injuries, foreign bodies in noses, and nosebleeds are covered in First Aid.*

Noses are invitations for trouble. They stick right out there, inviting injury, picking, and the insertion of foreign bodies. They get stuffy or runny. They turn green. (It still startles me to be relayed the message that a parent has called because the child has a green nose. Too much science fiction at an impressionable age.)

## Stuffy Nose: Plugged But No Discharge

### Basics

Most stuffy noses are related to viral infections, allergies, and dry air, but sometimes what seems to be a plugged nose is really enlarged adenoids that block the back of the nose.

### Assessment

You probably want to consult the child's pediatrician if the nose is very stuffy a lot of the time, or if the child complains about it or it impairs eating. Most pediatricians don't recommend over-the-counter nose drops, as they can cause "rebound" swelling. Studies also indicate that oral decongestants don't have much effect in children under age six.

If there are signs that the stuffy nose is really an obstructed upper airway, a visit to the pediatrician is warranted. Allergy, adenoids, and rarely a tumor, can block the passages. Signs are:

- Mouth-breathing consistently.
- Loud snoring, especially with restless sleep; especially if the child makes more

than one snore-noise during one intake of breath.

If the airway stays obstructed for a long time, the child can develop problems with jaw and chin development and even tooth decay because the teeth are exposed to air instead of to saliva.

A stuffy nose also can be due to a deviated septum. If the child's nostrils are of different shape and size from each other, check with your pediatrician.

### Treatment

Usually stuffy noses are best helped by judiciously humidifying the air[3] and making sure the child is drinking plenty of water. Salt-water nose drops or sprays may help. Cutting the fingernails short and applying petroleum jelly to the inside of the nose will help prevent picking and consequent nosebleeds.

## RUNNY NOSE; GREEN NOSE; A NOSE WITH LIQUID OF SOME SORT DRAINING OUT OF IT

### Basics

I believe it was the same Melinda Willett who coined the term Whipperdill (see Glossary) who also proposed designing little tiny tampons for little tiny noses. Most older toddlers and preschoolers have ten runny noses a year. If you have more than one such child in the house, that does make a lot of, er, nose run. Some runny noses are from colds, some from allergy, some from sinusitis. Occasionally, a foreign body will produce a runny nose.

### Assessment

#### COLDS

Most social children get about ten colds a year, with runny nose, low fever, some cough, scratchy throat. By a week, a cold should show improvement, and should be nearly gone by ten days. As the cold goes away, *the discharge may become green.* If the child seems to be getting better, the color of the discharge doesn't matter.

#### ALLERGIES

Signs of allergy include "allergic shiners," dark circles under the eyes; sneezing and watery eyes; clear nasal discharge; and an itchy nose. A child with an itchy nose tends to perform the "allergic salute"—he rubs upward against the nose with the palm of his hand. After many months of this, the nose can get a little crease from being crinkled so often.

#### SINUSITIS

A thick runny nose that isn't getting better after ten days, or a very thick yellow or green discharge in a child with fever can be a sign of sinus infection. Sinuses are hollow spaces (what other kind of space is there?) in the bones that surround eyes. When your

---

3. Too much moisture encourages dust mites, to which many children are allergic.

pediatrician suspects sinusitis, he or she may want to obtain X rays of the sinuses. Sinus X rays give very little radiation, but it is important to be sure that proper positioning and shielding is done to protect the eyes from irradiation. Sometimes a CT scan is advised. This could be because sinusitis is strongly suspected but X rays are normal, or because sinusitis is so persistent that an anatomic cause, like a deviated or crooked nasal septum, is suspected.

*Treatment*

### COLDS

They generally have to run their course. Antihistamines and decongestants may help symptomatically, or they may not, or they may produce more side effects than the original cold. Talk it over with your pediatrician. At any rate, not one study has ever shown that treating a cold with such medications prevents the development of otitis media.

### ALLERGIES

There are lots of ways to help allergic runny noses (*allergic rhinitis*), from prescription nasal sprays to antihistamines to allergy testing. Why suffer? Why run the risk of a crinkled nose? See your pediatrician.

### SINUSITIS

Most episodes of sinusitis in young children get better without antibiotics. However, if the symptoms haven't improved at all after fourteen days, your pediatrician needs to assess the problem.

You want to treat a sinus infection completely, or it's likely to come back again. Some infections take more than four weeks of antibiotics to eradicate. The sinuses are crafty little spaces, with poor blood supply and drainage, and it takes patience to get them clear.

# Mouth, Tongue, Throat, and Voice

*Mouth and teeth injuries are covered in* First Aid.

*Basics*
You can make life easier for yourself, your preschooler, and the pediatrician's office staff if you teach your child how to open his mouth and say *AH*. The key points are:

- Have him look up and open his mouth really wide.
- Have him say AA, not AH. This is hard to explain due to regional accent variations, but what you are trying for is the AA sound in "Annie." Look and see the difference for yourself: the AA sound opens up the back of the throat, while the AH sound raises the back of the tongue and blocks the view.
- Some children offer a better view with tongues plastered to the floor of the mouth, rather than out.

Children's throats have more parts visible than adults' generally do. When you look in your child's throat, you will probably see seven items of interest:

- Tongue.
- Tongue frenulum, the little band that holds the tongue down to the floor of the mouth.
- Upper lip frenulum, the little band that holds the upper lip to the gum.
- Uvula, the thing that dangles down from the center in the back of the throat. (Sometimes it is split, or bifid.)
- Two tonsils, little balls, one on each side of the uvula.
- Epiglottis, or lid to the airway, that just peeks up over the tongue. In adults, it's located farther down in back, and you don't see it.

When a structure in the body is infected, inflamed, or swollen, the term for such a problem is one that combines the name of the structure with the suffix "itis." So you can have *tonsillitis, uvulitis,* or *epiglottitis. Glossitis* is an infection, inflammation, or swelling of the tongue. *Gingivitis* is inflammation of the gums. *Stomatitis* is ditto for the mouth in general. *Gingivostomatitis* is inflammation of both the gums and the mouth.

Note, though, that these are merely descriptions of what you, or I, or anyone sees. The mere name does not tell you what is causing the "itis"—the swelling, infection, inflammation.

*Assessment*
It's not a bad idea to take a look in your child's mouth this way a few times when she is perfectly healthy, both to get her used to opening up and to see what her version of normal is. For instance:

- Many young children have fallen and ripped the frenulum of the upper lip sometime in the past, so that it looks broken-then-healed. This shouldn't cause any problem.
- Many young children's upper lip frenulum extends down to cause a big separation of the two front teeth. This should be assessed, around age Two or Three, by the pediatric dentist, to make sure the baby teeth aren't being crowded.
- Some young children have a very short tongue frenulum, which can even cause the tongue to dimple in at the tip. This is often called tongue-tie. Most of the time it is normal, and will stretch as the child grows. If not, it is fixed under general anesthesia by clipping the frenulum. This is done if the frenulum is so short the child has, or is thought likely to have in the future, problems speaking clearly.
- Many young children have splotchy "geographic tongues." This, too, is normal; it is just a patterning of the surface of the tongue.
- Those bumps you see at the back of the tongue are normal taste buds.
- Some children have a uvula that looks like an upside-down valentine or even an upside-down Y. This is normal, but may increase the chance of middle-ear infections. If adenoids are removed in a child with a bifid uvula, the voice quality may change, becoming flat and less pleasant.
- Some children have little teeny tonsils and others have ones the size of large marbles. Large tonsils are normal unless

they cause problems breathing or swallowing.

When your child is sick, and either complains of a sore throat or acts as if the throat is sore, assess the following:

- Does the child look or act "scary" sick?
- Is the child having trouble breathing?

If either is the case, assess this condition first, before perusing the mouth and throat (see *Frightening Behaviors*). Once assured that the child is not having significant problems, it is safe to investigate further.

- What do you see inside the mouth and throat?
- What is the child's voice like?

(Rarely, a child who is usually cooperative, and who now looks or says his throat is sore, can't open the mouth wide. Most often this is because he is afraid you are going to stick something down there, but sometimes it is because there is an abscess or other problem limiting his ability to open the mouth. In such a situation, the child should be seen. Don't try to pry the mouth open yourself.)

## MOUTH

A white coating in the mouth of a baby under a year is most likely thrush. (See *Common Illnesses with [More or Less] Unfamiliar Names* in Chapter 15 of Part II.)

Ulcers or sores in the mouth or on the tongue could be either herpes or hand-foot-and-mouth. (See *Common Illnesses with [More or Less] Unfamiliar Names*.)

## TEETH

*Basics*
More than half of all children will have at least one cavity by age Three.

*Assessment*
Brown or white spots on the baby teeth can be due to cavities, injuries, or, rarely, other causes such as an inherited defect of enamel or medications taken during pregnancy. No matter what the age of the child, such findings warrant a visit to a pediatric dentist.

*Treatment*
The dentist may fill a tooth or perform a root canal and crown, or even pull, a very decayed tooth.

*Prevention*
Babies whose teeth are constantly exposed to sugar, in many sticky snacks, in milk (even mother's milk), or juice (even juice that says "no sugar added") are prone to devastating decay, known as "nursing-bottle mouth." Sometimes all the baby teeth have to be pulled and replaced with crowns.

A baby who takes a bottle to bed with her, or a baby who carries a bottle around all day, or who nurses frequently after age one year, is a candidate for nursing-bottle mouth. A toddler who is always snacking or sipping is also at risk. Raisins are especially effective at producing cavities, as are sticky candies like Gummy Bears. It's not the quantity of sugar consumed; it's the amount of time in contact with the teeth.

Ask your pediatrician or dentist about fluoride. Every now and then there is a flurry of bad press reports about fluoride: that it is poisonous or carcinogenic. In great concentration, fluoride is toxic; but in the right amount, it confers excellent protection against cavities without causing side effects. Your pediatrician will know how much fluoride is in the local tap water, but bear in mind that if your child doesn't drink the local tap water, she's not getting fluoride from that source.

Try to think of brushing teeth as quality time. Start cleaning the teeth before they even come through, and keep the habit so regular the child thinks it is part of the natural order of things. Toddlers and preschoolers may like to name their molars, and then parents can take an interest in the hygiene of (or pretend to be the voice of) George the Tooth without sounding like a nag.

If your child is a mouth-breather, ask your pediatrician and dentist if this is affecting her jaw development or causing cavities. Teeth exposed to air rather than saliva are more prone to decay.

## Throat

Tonsils that look bigger than they usually do in your particular child are probably infected, with either a virus or bacteria. If they are beefy red, with gooey creamy pus on them, you may be looking at strep throat (see *Streptococcus* in Glossary). This is especially likely if the uvula is red and swollen.

Sores on the tonsils and back of the mouth that look like little red pits may herald herpangina. (See *Common Illnesses with [More or Less] Unfamiliar Names* in Part II.)

Pus on the tonsils that looks like little white dots may be caused by strep or, more likely, by viruses, even by infectious mononucleosis (mono) (see *Common Illnesses with [More or Less] Unfamiliar Names*).

Sore throats are prone to myths and legends. Here is the straight poop:

- You can have strep throat even though you have no tonsils.
- Most strep throats in childhood present with headache or tummy ache, not with sore throat.
- You can have strep throat with no pus on the tonsils.
- Most of the time, a sore throat and/or pus on the tonsils is due to viral infection.

## Voice Quality

**Hot potato voice:** If the child sounds as if he's trying to talk around a piece of hot food in the back of the throat, this is a sign of tonsillitis, often strep.

**Hoarse voice:** Most causes of hoarseness, if the child isn't having trouble breathing or swallowing, are innocent. Too much yelling and screaming at play can swell the vocal cords. Mild viral infections can cause hoarseness (laryngitis).

The time to worry about a hoarse voice is:

- If the hoarseness is accompanied by noisy breathing, it may be the beginning of significant croup, which often gets worse at night. This deserves a call to the pediatrician.
- If the hoarseness hasn't improved significantly after two weeks, there may be benign growths on the vocal cords, called vocal cord nodules. Many children with this problem require speech therapy to learn to talk in a way that helps the nodules shrink.
- A baby under the age of three months with hoarseness may have a congenital problem with the vocal cords that needs special attention, and needs a visit to the pediatrician.

### Treatment

Some mouth and throat complaints, like thrush and strep, require specific prescription medications. Once in a while, tonsils need to be removed surgically, but this is never done lightly or just because they look big or even because the child has more than his share of ordinary sore throats. Your pediatrician will advise you if the tonsils seem to be causing significant obstruction or, rarely, the specific virus syndrome called "PFAPA" (see Glossary). The main concern about strep is the complication called *rheumatic fever* (see Glossary) and this is exceedingly rare under the age of Five, no matter how many times the young child has strep.

For uncomplicated hoarse voice, the vocal cords need to be rested. Tell the baby or child not to yell and scream. (Ho, ho ho! Pediatricians get an unholy pleasure from dealing out such instructions.) You may have to change their playmates or play style temporarily also, if your child's best friends are screamers. Breathing humidified cool air is helpful. Warm humidified air from a hot shower may do some good, also.

Be wary of medications designed to numb or soothe throat pain.

- The prescription jelly that contains lidocaine, a topical anesthetic, can be dangerous if overused. Lidocaine can be absorbed through sores in the mouth and enter the bloodstream. Too high a blood concentration can lead to serious, even fatal, heart rhythm disturbances.
- Children under Five, especially those under Three, are at high risk of choking on and aspirating throat lozenges.
- Some throat lozenges, gargles, and sprays contain ingredients, such as phenol, to which some children are highly sensitive. The ingredient numbs the throat but also irritates it, so that when the numbness wears off the throat is even more sore than before—so you give another dose.
- A combination of one part honey to one part lemon juice is soothing and safe for children over one year of age. (You don't want to give honey to younger babies on account of the worry about infant botulism.) You can give a teaspoon every ten or fifteen minutes to coat the painful area.

# Airway and Lungs

*New babies breathe differently, and have different problems with the airway, than do older babies and children. If your baby is four months old or younger, see the chapter for his or her age.*

Normally young children breathe without effort and without making much noise. When they sleep, they breathe irregularly, with pauses and then a sigh. They don't normally snore.

When children have trouble breathing, they show effort and make noises. You can tell most about how much trouble a child is having by watching her breathe. See *Trouble Breathing* in *Frightening Behaviors*. You can tell most about where and what the trouble is by listening.

Of course, if a child is having severe or even moderate trouble breathing, you don't want to take the time to try to figure out what's going on. You get help, fast. But if looking at the child reassures you, you can take a little time to assess where the problem is originating.

We begin with a view of the airway and lungs. The nose warms and humidifies air and even filters out some germ particles.

Then the air has to bypass the little lid called the epiglottis, whistle through the vocal cords, and hurtle through the wind pipe (trachea) into the large tubes called bronchi. There is one main bronchus for each lung, and each divides into smaller and smaller tubes. The very tiniest tubes, like grape stems, are called bronchioles. The grapes are the air sacs, or alveoli: the true lung tissue itself. The alveoli are where oxygen goes into the blood and carbon dioxide comes out. Then the air has to hurtle back up out the same pathways.

Breathing problems that originate in the chest[4] generally come in two varieties. First, there's the kind that blocks passage of air through all those tubes. Second, there's the kind that fills up the air sacs with pus or fluid or blood.

## BLOCKAGE OF THE UPPER AIRWAY

Trouble in the upper airway, from the nose to the place where the trachea splits into the two bronchial tubes, usually gives a child more trouble breathing *in* than breathing *out*. This is commonly caused by enlarged adenoids, croup, and rarely by a swollen epiglottis or a foreign body.

### Enlarged Adenoids

#### *Basics*
Children with big lumps behind the nose tend to snore, mouth-breathe, and talk as if they have a wad of peanut butter stuck to the roof of the mouth.

#### *Assessment*
Sometimes an X ray will show objectively how big the adenoids are; occasionally an

---

4. There are breathing problems, like hyperventilation, that originate in the soul; and others, like the sighing of diabetes, that originate in the acid balance of the blood.

ear/nose/throat doctor will look with a fiberoptic tube. Sometimes your pediatrician will ask you to bring in a videotape of the child sleeping to hear and see how bad the snoring is, and to see if the child may be having "sleep apnea," with the airway intermittently being completely blocked. Rarely a special "sleep study" or pneumocardiogram (see Glossary) is suggested.

*Treatment*
Sometimes time and growing solve the problem. Sometimes antibiotics and cortisone shrink the adenoids. Sometimes treating an underlying allergy helps. Sometimes the adenoids need to be surgically removed.

## Croup

*Basics*
Croup is a syndrome, that is, a collection of symptoms and signs: a hoarse voice, a cough that sounds like a foghorn or a goose honking or seal barking, and stridor—a raspy or honking sound when the child inhales. Most of the time, croup starts in the middle of the night.

*Assessment*
Croup comes in three varieties:

- Viral croup starts with a cold caused by any one of several viruses. There is a low-grade fever as well as all the other signs of croup.
- Spasmotic croup tends to recur with change of seasons or weather. There is no fever.

- *Laryngotracheobronchitis* is a bacterial form of croup that is quite rare and much more likely to be serious than the other forms. It features fever, low or high, and a severe struggle to breathe, with the child thrashing about for air.

It's rare to see croup under the age of six months, and unusual to see it after four years.

*Treatment*
Treatment of viral and spasmodic croup is directed at soothing the air passages with cool humid air. Sometimes just a cool mist humidifier works. When croup gets worse at night, going outside into cool night air or going into a steam-filled bathroom both help. (Remember steam rises. A toddler on the floor won't get much.) Sometimes cortisone is given by mouth or injection to reduce the swelling of croup, but this takes several hours to work. Breathing treatments given by a machine that produces a jet of medicated steam may be given in the office or hospital.

Treatment of bacterial croup, laryngotracheobronchitis, requires hospitalization with antibiotics and oxygen and assistance with the airway.

*Prevention*
Preventing viral croup means avoiding the virus in crowds, daycare, and so on; and careful, frequent hand washing.

A child with recurrent spasmodic croup may have fewer attacks when you humidify the bedroom with cool mist, especially during a dry spell or when the central heat is on.

## Epiglottitis "EH-pee-glot EYE-tis"

*Basics*

This is the most feared of disease-caused airway obstructions, because it acts just like a foreign body that could totally block the airway at any moment. The epiglottis, that little lid protecting the airway, can get infected and swell to the size of a cherry. That's bad enough, but it also gets sticky. So if it gets stuck, it stays stuck; even a Heimlich maneuver won't free it.

*Assessment*

A child with epiglottitis looks very frightened. She makes a little high-pitched noise on inhaling. She may fight to sit up, her nose up in the "sniffing" position, because that keeps the airway open better. She doesn't cough. She may well have fever. She drools and has no voice.

*Treatment*

A child in this situation needs urgent care. Calling 911 is appropriate, or if you are very close, a trip, with another adult along, to the nearest emergency room. Do not go to a freestanding urgent care facility or to a private office, even that of your own pediatrician, unless you have determined that the office is prepared to deal with this severe emergency. If the child has epiglottitis, she needs a tube placed into her airway to protect it from the blockage produced by the swollen epiglottis. This is a difficult procedure, requiring special skills, available oxygen and other equipment, and support personnel.

Don't try, or let anyone else try, to make the child lie down until she has had her breathing relieved by having a tube placed in her airway. Don't let your fear make her more anxious. Insist on instant evaluation by the most pediatrically experienced physician, and make sure that doctor either knows how to establish an emergency airway or immediately calls in one who does.

If an X ray is ordered before a tube is placed, do not allow your child to be sent to X ray unless accompanied by a physician skilled in managing her airway and by a well-equipped emergency cart. Make sure they won't make the child lie down or agitate her. Don't let any procedure, like temperature, weight, blood pressure, blood drawing, or IV, delay the evaluation of the airway. Ask to see that there is child-sized equipment. This may all make you feel like a nag or a fool, but do it anyway.

If you are accused of nagging or being out of line, demand that the staff explain to you why they do **not** think the child has epiglottitis, or, if they think she does, why they are delaying airway management.

Once a tube is placed for the child to breathe, usually all goes well. Hospitalization, usually in an ICU or Pediatric ICU or Special Care, is mandatory. Antibiotics are indicated.

*Prevention*

Fortunately, most such infections are caused by H. flu (hemophilus influenza) bacteria—

the bacteria protected against by the HIB vaccine. Thankfully, epiglottitis is much more rare than it used to be. Be sure your child gets the full series of immunizations, starting at age two months!

## BLOCKAGE OF THE LOWER AIRWAY

### Asthma, Reactive Airway Disorder (RAD), Wheezing

*Basics*

Asthma (or Reactive Airway Disorder) is a condition in which various stimuli cause a child to wheeze. Wheezing is the whistling noise that shows that breathing air out is causing effort. In fact, the very first sign of asthma or RAD can be an extra effort, or extra long time needed, to exhale.

What causes this effort is a combination of things. First the little muscles encircling the bronchial tubes clamp down. Perhaps this is Nature's effort to tell an allergic person to get out of this particular environment, fast. Second, the lining of the bronchial tubes swells up and mucous becomes thick and sticky.

Asthma can be triggered by a virus, or by allergy to anything from foods to bees to dust. It can be worsened, if not set off in the first place, by both emotional stress and tobacco smoke. Exercise, especially the kind that comes in spurts, or an encounter with cold or hot or dry air can also bring on an attack. So

can irritation caused by regurgitated food.[5] If there is a family history of asthma, a child is more likely to be susceptible.

*Assessment*

If your child should start to wheeze, the first problem is to assess how much difficulty he is having breathing: see the section called *Trouble Breathing* in *Frightening Behaviors*. The second problem is to determine whether the cause of wheezing is asthma or something else. Your pediatrician will need to take a careful personal and family history, examine the child, and may have to perform several kinds of tests, including a chest X ray, to be sure what is going on. Most pediatricians routinely test a child with frequent or recurrent wheezing for *cystic fibrosis* (see Glossary).

*Treatment*

The usual treatment for an acute attack of asthma is a medication that unclamps the muscles surrounding the bronchial tubes such as albuteral. Such medications are called bronchodilators. In an acute attack, these can be given in a mist produced by a machine called a nebulizer. The child breathes the mist through a mask or mouthpiece.

These medications also work well, but take longer to do so, when given by mouth. Once in a while, adrenalin must be given by injections. All bronchodilators are related to adrenalin, and may make a child shaky or nauseated; the heart will beat very rapidly.

---

5. See gastroesophageal reflux, under the Reflux entry in the Glossary.

But bronchodilators only take care of one phase of asthma, the first clamping down of the bronchial tubes. It's important that the second phase, inflammation, be considered also. Medications that treat or prevent swelling of the lining of the bronchial tubes are often necessary. These include cromolyn sodium, or Intal, and steroids (cortisone is another common term).

Indeed, the special caution in treating asthma is to make sure that the inflammation part of the disease is treated as well as the first, clamping down part. Rarely, the medications that relax the bronchial muscles are so powerful that they can make a child feel comfortable even though the bronchial tubes are really getting very swollen. Then suddenly the bronchodilators don't work anymore, and the child can be in real, terrible trouble.

### Prevention
Asthma itself is not contagious, but if it is triggered by an underlying viral infection, that may be. Allergy testing may help a child avoid things that trigger the attacks. If a child has moderately frequent or severe attacks, daily preventive medicine, and a special machine to detect early worsening of attacks, are often prescribed. Secondhand smoke should be absolutely eliminated. Children with asthma should receive a yearly flu (influenza) vaccine.

## Bronchitis

### Basics
Bronchitis means pus or fluid in the big bronchial tubes. Some researchers think it isn't really a disease in children; it's more of a "spill-over" from infection higher up in the sinuses or throat. Some feel that bronchitis really is a defined separate illness, caused by viruses. Some feel that it's really a subtle form of asthma.

### Assessment
A child with bronchitis has a deep cough often, from his toes, but doesn't usually have trouble breathing.

### Treatment
If the pediatrician thinks that bronchitis is a secondary manifestation of, say, a sinus infection, usually antibiotics are prescribed. If not, cool, moist, clean air, time, and a cough loosener like guanefisin or a suppressant like dextramethorphan are usually suggested. If asthma is suspected, a bronchodilator may be tried.

## Bronchiolitis (RSV: Respiratory Syncytial Virus)

### Basics
Bronchiolitis is different from bronchitis.

Bronchiolitis is an infection of the little tiny tubes ("grape stems") that attach to the air sacs. In its characteristic form, this is a disease of infants and toddlers under age Three. It is not a disease to take lightly. Babies with bronchiolitis often don't have fever and may not cough. They do struggle to breathe, with their tummies going in and out. They wheeze.

They wheeze because of trouble pushing the air out. When these little tubes become infected, they make a kind of valve: it's not so hard to get the air into the air sacs, but it's awfully hard to push it out. If the little air sacs become completely blocked, the air inside gets absorbed out of the sac and the sac collapses. And it's terribly hard to open up again once it is stuck shut.

Bronchiolitis can be a mild disease, but it also can be a real danger. The degree of severity depends partly on the virus causing the disease—the usual culprits being RSV and influenza (see Glossary)—and partly on the age (and strength, immune system, etc.) of the baby. In general, babies under six months of age, or who have any underlying medical problem, are at highest risk.

### Assessment

As the illness goes on, the baby needs to be watched for signs of trouble breathing: increased work by the muscles that help the air in and out of the lungs. Flaring nostrils, a tummy that pushes up and down with each breath, and sucking in of the muscles between the ribs and above the collarbone are signs that the baby needs immediate help. A baby who is pale, blue, or mottled is a baby with an emergency.

Exhaustion is another problem. A baby who is too tired to feed well and can't sleep comfortably is getting into serious trouble. A baby old enough to smile and play should do so, and if she can't, is becoming seriously fatigued.

### Treatment

Home treatment, with your pediatrician keeping close track of the baby's progress, usually consists of cool mist from a humidifier, elevating the head of the crib, and frequent feedings. Close monitoring is essential.

Some babies, especially those younger than six months and most especially those under two months, may need hospitalization. Most of the time, extra oxygen and intravenous fluids provide the help needed, but on occasion a special inhaled medication called ribavirin is recommended, though its use is controversial.

### Prevention

Most of the time, the viruses that cause bronchiolitis are passed on by children who have no sign of illness whatsoever. Moreover, these viruses continue to be spread by sick children long after they are well. The virus lives for a long time on inanimate objects, like toys and doorknobs. Finally, many people become reinfected with the same virus again and again, at any age; an adult with a "cold" may really have, and be passing on, RSV.

By the age of Three, nearly all children have had RSV, even if they have not attended day care, ever. So it's best to resign oneself to the inevitable.

## Obstruction of the Air Sacs

### Pneumonia

#### Basics
Pneumonia occurs when fluid or pus fills the air sacs of the lungs. It can be caused by viruses, by bacteria, by the funny little bug called mycoplasma, which is sort of halfway between (see Glossary), and by aspirating liquids, powders, or foreign bodies.

#### Assessment
Babies and children with pneumonia breathe rapidly and shallowly and they sometimes grunt with each breath. The grunting helps to keep the little air sacs open instead of collapsing. Often a baby will show signs of working to breathe, with "retractions" of muscles of the abdomen and between the ribs, and flaring of the nostrils. Usually, there is fever and a persistent cough.

Sometimes children with pneumonia will wheeze because the bronchi are also involved.

Sometimes, a very young baby with pneumonia will have fever without obvious trouble breathing. And sometimes a toddler or preschooler will complain only of a tummy ache, not trouble breathing or chest pain, even though you can tell by looking that she is short of breath.

Pneumonia can be so mild that you barely notice it ("walking pneumonia") or very acute and serious. It can come on slowly or rapidly.

#### Treatment
Treatment of pneumonia is partly supportive, making sure the child is getting enough

oxygen and breathing out enough carbon dioxide, getting plenty of fluids, and not becoming exhausted. And treatment is partly directed at the underlying cause if it is treatable, as in the case of giving antibiotics for bacterial and mycoplasma pneumonia. Even though there is no specific remedy for most viral or aspiration pneumonia, children who receive good support during the illness often do beautifully and get over the disease "on their own."

*Prevention*

Since pneumonia can follow a common childhood illness like chicken pox or measles, immunizations are important. The vaccine called Prevnar is very effective against pneumococcus, the most common cause of bacterial pneumonia. And since a very severe pneumonia can follow the aspiration of certain chemicals, a baby-proofed house is crucial. One of the worst kinds of pneumonia follows the aspiration of oil, so children receiving mineral oil for constipation should not be forced to take it if they really refuse, nor given it when coughing, vomiting, or lying down.

There is no proof that treating an upper respiratory infection or cold with antibiotics will prevent the development of pneumonia. And there isn't any proof that going without a hat or boots, or getting chilled, soaked, or rejected in love brings on pneumonia either, but I don't recommend any of them.

## Cough

**Newborn babies normally don't cough. A persistent cough in an infant under two months of age always warrants a visit to the pediatrician. So does a cough in a child who is known to have choked, or is suspected of having choked, on something.**

*Basics*

An English proverb states that anything can be hidden except Love and a Cough. In my experience, children don't even try to camouflage either one.

Children cough differently from adults. Children never, or almost never, "expectorate." This is a polite term for getting rid of coughed up phlegm into a hankie (or just spitting it out). Children swallow what they cough up. This isn't a disaster; after all, they do get the stuff out of the lungs, and the stomach is a pretty rugged organ. Sometimes, however, all that swallowing can produce a belly ache or vomiting.

Most of the time when parents worry about coughs, they think of pneumonia or asthma. In most of these conditions, however, there is significant trouble breathing. The trouble breathing is the real problem with the disease, and the cough is Nature's way of trying to help.

Bronchitis, another parental worry, is more of a nuisance than a danger: see above.

*Assessment*

Your first goal is to see whether or not the cough needs medical attention, and if so, how urgently. If there is any suggestion that a cough is from an inhaled foreign body, the child needs to be seen at once.

Any baby under the age of two months with a cough needs to be evaluated by the pediatrician. The younger the baby, the more rapidly a cough can exhaust a child, and the more likely it is that the cough is due to an infection (or other problem) that has to be treated. A baby under six months of age with a cough needs medical attention, even if there isn't any trouble breathing or fever. (If the baby is at the older end of the age range, and the cough is infrequent, a phone call may do.)

Any cough that accompanies visible signs of trouble breathing requires medical attention. How promptly depends on the severity of the breathing problem: see *Trouble Breathing* in *Frightening Behaviors*. Chest pain and exhaustion are also signs that attention is needed.

Coughs that last longer than three weeks need attention. Their causes can range from common diseases like sinusitis and allergies to rare ones like cystic fibrosis or tuberculosis. Not infrequently, such a cough is *psychogenic*, caused by a tic, or stress, or just habit: but even these need, and benefit from, medical diagnosis and attention.

Your second goal is to see if the cough that doesn't require urgent attention right now fits into a pattern of a particular illness that might need attention later on. The two that are most easily suspected are croup and whooping cough.

## CROUP

"Croooop! Croooop!" The cough of croup sounds like a seal barking or a goose honking, with a metallic/musical tone. The child, usually between six months and six years, is hoarse, and often makes a noise breathing called *stridor*, which sounds like a rasp or a honk, except it's made breathing *in,* not breathing *out.* Croup always gets worse at night, and may start at night. For the complete scoop on croup, see *Croup* in *Airway and Lungs*, above.

## WHOOPING COUGH (USUALLY CAUSED BY PERTUSSIS)

Yes, we still see it—sometimes in a baby who hasn't been fully immunized against pertussis, or whose parents have refused to allow the immunization; rarely, in a child in whom the full series of immunizations "didn't take." And there are other causes of whooping cough, too: a virus called *adenovirus*, and choking on a foreign body, and even tuberculosis.

Babies under six months of age usually don't whoop, but have a spasmodic, exhausting cough, which makes them turn red or purple. Most older children with whooping cough have a nonspecific, irritating, prolonged cough, and get seen because the cough has lasted longer than three weeks.

But now and then a child turns up with the classic whoop.

In this case, the child probably started with a "cold" that lasted for about two weeks, and then developed an increasingly bad cough. At first, it's just very persistent, especially at night. But then, it changes. She coughs diabolically, frighteningly, turning red or purple, looking scared; and then to catch her breath, sucks in air with a loud "WHOOP!" Afterwards, she may vomit profusely and/or fall into a deep exhausted sleep.

Many children do fine with whooping cough, but any child with this set of symptoms needs to be evaluated. The younger the child, the more urgent the need for assessment. Whooping cough can result in brain injury from the pressures produced by the coughing; it can result in pneumonia, and it can indeed be fatal. Moreover, treating the child with whooping cough, and any other family members carrying the pertussis bacteria, can halt the spread of the disease.

Don't try to "loosen the phlegm" or jiggle or upset a child with the signs of whooping cough. Keep her as calm as possible, and get help.

*Treatment*

Any of the coughs requiring medical attention will need treatment addressed at relieving the cough itself and/or getting rid of the underlying problem.

Most ordinary coughs, the kind that you treat at home, are caused by viral infections, so antibiotics won't usually help. Cough medicines purchased without prescription for children this age generally contain any of the following ingredients:

- Guanefisin, which may help to liquefy mucous and help the child get rid of it.
- Antihistamines, which dry secretions,

suppress allergic components of coughing, and sedate the child and help him sleep. Even though antihistamines counteract allergy, they aren't a good idea to use in an asthmatic child because of their drying effect.
- Dextramethorphan, often abbreviated as DM and added to the brand name of a product, such as Robitussin DM or Dimetapp DM or Triaminic DM. DM suppresses the cough reflex.

Labels often direct parents to call for advice before using these medications in a very young child. This is a good idea, but try to do so during office hours, if possible!

Young children with coughs are strong candidates for choking on and aspirating things, including cough drops. Don't use cough drops.

A session in steam, in the bathroom (remember steam rises; don't assume a child sitting or standing on the floor is getting any) or with a cool mist humidifier may help an ordinary cough. The sore throat remedy of equal parts honey and lemon juice, given at a teaspoonful every ten minutes as needed, may help in children over the age of a year, who can safely take honey.

*Prevention*

Since many coughs are caused by viruses that are shed most vigorously right before the child shows any symptoms, it is difficult to prevent spreading these common illnesses.

Moreover, children rarely cover their mouths when they cough, and those around them get sprayed. If you sustain a direct hit, you may indeed catch the virus or bacterial source of the cough. (Remember, just because you "catch" the bug doesn't mean you'll automatically get sick with it.) But most of the time, you won't catch it unless you get the child's mucous on your fingers and then touch your own eyes or nose. Rather than get into the habit of washing your hands *after* you touch the child, get into the habit of washing your hands *before* you touch your face. If you think you never touch your face, put some colored chalk on your fingers and look in the mirror half an hour later.

Should the child go to daycare? If he has had a fever in the last twenty-four hours, definitely not. If the cough has just started, and you don't know what will develop, he should stay home. If there are any indications for visiting the pediatrician, the child should stay home until seen.

And be sure your child is fully immunized.

## Abdomen/Intestinal Tract

### STOMACHACHES

#### Basics

Perhaps someday Oprah will have a show, "Children who never have tummyaches." There will be perhaps three children, found after an intensive national search. I myself will be sure to watch: I want to know who those three children are. I've never met any myself.

Most abdominal pain in childhood has one of four causes:

- Constipation
- Mild viral infection
- Food intolerance
- Distress of the soul

But, of course, there are other causes of abdominal pain: appendicitis, bowel obstruction, infection in the intestine or elsewhere (strep throat, sinusitis), pneumonia, and even highly unusual diseases like juvenile onset diabetes.

You'd think it would be easy to tell when abdominal pain is serious and when it isn't, but sometimes it's tricky. In a young baby, you may not be able to tell what part of the body hurts. In a toddler, you may not be able to tell what part of the abdomen hurts. In a preschooler, you may not be able to tell how badly it hurts; sometimes the loudest wails go with the mildest symptoms.

Moreover, some of the innocent causes can themselves need medical attention. A toddler who is constipated to the gills from withholding stool isn't going to be "fixed" by a simple suppository and change of diet; he needs an assessment and ongoing care. A preschooler whose tummy aches terribly because his parents are getting divorced or—God forbid—he's being sexually molested needs professional care.

#### Assessment

Your first goal is to decide if the child needs immediate, urgent attention.

Signs of an urgent problem, such as a

bowel obstruction, overwhelming infection, or metabolic disease needing immediate treatment include:

- **Behavior:** Unable to play, disoriented, or less responsive. An infant or young child who intermittently falls asleep after a crisis of pain is very worrisome.
- **Activity:** In a child who is old enough to walk, refusing to do so, bending over to splint the pain, limping, or showing pain when jumping.
- **Appearance:** Pale or mottled skin, during or between attacks of pain.
- **Other symptoms** that, when analyzed, show the need for urgent help (see *Frightening Behaviors*):
  *trouble breathing*
  *acting sick*
  *smelling sick*
  *complex vomiting (see below)*
  *complex diarrhea (see below)*
- **Worrisome history:** A fall, even if not directly on the belly; question of having ingested a medication or poison or foreign body.

Where in the abdomen the pain seems to be may or may not be significant. Pain focused right at the belly button is much less likely to be caused by a serious problem than pain way off to one side, or much higher or lower, or radiating to or from the back.

**Signs of problems that will need medical attention soon (same day):**

- Simple vomiting or diarrhea that doesn't get better with treatment.

- Pain that gets more intense or constant as time goes on.
- Persistent loss of appetite.
- Symptoms such as persistent cough, trouble or pain when urinating, fever, sore throat.

**"Innocent" causes of pain that still need treatment:**
*Suspect simple constipation:*

- If the child doesn't have any of the worrying symptoms or signs listed above.
- If the stools are almost always harder than firm peanut butter when the child is well.
- If he goes for several days without a stool and then it is huge or little marbles, or if there is just staining.
- If he gets hysterical at the sight of the potty.
- If you see him "withholding": grunting or straining as if trying to go, but nothing comes out.

*Suspect viral illness or food intolerance:*

- If the symptoms just started in the last twenty-four hours.
- If your child has none of the serious or worrisome symptoms above.
- If there's no sign of constipation.
- If he seems happy otherwise, without unusual separation anxiety, school phobia, or other signs of distress.
- If there haven't been any crises or emotional trauma: family discord, a move, a change in daycare or preschool (including a change in friends, teacher, or classes).

Treatment here is to give the child the diet for diarrhea, avoiding both dairy products and juices. Make sure that you give him a source of sugar (no diet drinks), and don't give just water without carbohydrates. Gatorade or sodas are fine for mildly ill children two and older.

If the child gets worse on treatment, something else is going on, and he needs pediatric attention.

*Suspect "distress of the soul" if the child has none of the other symptoms listed above, but:*

- Is acting sad, obnoxious, withdrawn, sullen, clingy, or aggressive.
- Has developed nervous habits, like biting his nails.
- Is having trouble sleeping.
- There is discord or trouble at home. Even if you think you have kept something secret—marital fighting, illness or death of a relative, financial worries. Children pick up lots of clues.
- There has been a traumatic event, even one on TV, that has scared your child. When little Jessica fell down the well and was rescued on national news, my schedule was packed with preschoolers complaining of tummyaches.
- If there is a change of any kind occurring or impending at daycare or preschool— even a minor one. Change of teachers, classrooms, or a friend moving away or ceasing to be a friend are all biggies.
- If there is any suspicion the child could have been sexually molested.

*Treatment*

Extra time one-to-one with each parent, a talk with the daycare provider or preschool teacher, making sure that schedules are regular and the child is getting enough sleep will help with simple disorders.

If there is a family secret that you haven't told your child, the child may be very aware that something is wrong. You may want to discuss with your pediatrician how to handle this situation.

If you have any suspicion your child could have been molested (acting odd after being with a babysitter, relative, etc; masturbating more or without pleasure, as if compelled; clinging behavior; genital irritation) of course see the pediatrician promptly.

## VOMITING

**If you think your child could have eaten something poisonous, call Poison Control or your pediatrician immediately. If your child is deteriorating rapidly, call 911.**

If your baby is less than six months old, read the vomiting sections in Part I.

Do not give prescription medicine to stop vomiting unless your pediatrician has ordered it specifically. Never give Compazine to a young child unless you are SURE your doctor wants you to for a special reason: it can be very dangerous.

*Basics*

Why are parents so upset when a baby or young child vomits? I have lots of theories. It's a rejection of the nurturing food we've offered. It's an unappealing act, to say the least—one that should be reserved for adults. It's unpremeditated. And children are not interested in finding a suitable receptacle.

All of those things. But let me set one thing straight: Yes, I saw Linda Blair in *The Exorcist*, too, but young children often vomit through the nose as well as through the mouth. It doesn't mean they're possessed.

Most vomiting is due to innocent causes: a too-full stomach, food sensitivity, mild viruses. But once in a while, vomiting can indicate a serious and severe problem.

*Assessment*

There are four main serious causes of vomiting that need immediate attention. All are *very* uncommon.

- An obstruction or perforation of the intestine. Signs of this include: vomitus that is green, brown, or bloody; pain that is not relieved by vomiting; stool that looks like red jelly; recurrent projectile vomiting; and spells between vomiting in which the baby is pale and sleeping exhaustedly, or acting very unlike herself. An older child will walk doubled over, or limp. See Stomachaches, above.
- An underlying infection in the blood stream (*sepsis*) or spinal fluid (*meningitis*). Signs include a child acting very sick (see *Acting Sick* and *Fever* in *Frightening Behaviors* and *Ominous Rash* or *Spots Like Bruises* in *Looking Sick*) and in older children, a stiff neck (see *Neck* in *Body Parts*, above.)
- Increased pressure within the brain, either from bleeding into the brain,

swelling from infection, or a growth or tumor. Signs include: history of a head injury including having been shaken within the last two months, especially if the child was unconscious; headache, if it is severe, persistent, or is present first thing in the morning or awakens the child at night; recurrent sudden vomiting without nausea; signs of neurological problems, like weakness, an unsteady gait, loss of symmetry of the face, or loss of coordination; acting odd—irritable, subdued, lethargic, disoriented; seizures or convulsions.

- A toxic or metabolic problem. Signs include: irritability or sleepiness; odd behavior—not recognizing people, calling things the wrong name or seeing things that aren't there; severe headache. (A history of recent flu-like illness or chicken pox raises the suspicion of Reye syndrome (see Glossary).

*Treatment*

**Treating simple vomiting:**

When children vomit, they build up acids in their blood that make them even more uncomfortable and nauseated. This vicious cycle can almost always be fixed by the following method, since the antidote is sugar:

1. Let the stomach rest. Wait one hour after the last vomiting episode, the one that empties the stomach. Don't give anything by mouth for this hour—not water, crackers, ice cube, lollypop— nothing!
2. Then give very small amounts of a

sugar-containing fluid such as Pedialyte, Gatorade, or the colas (not diet cola). Juices may further upset the stomach because of fiber or acid content and aren't recommended as a first choice. Give a tablespoon of the soda (measure it!) every ten minutes for two hours. By this time, the child with simple nausea and vomiting often feels much better, and the diet can be very gradually expanded to include crackers, soup, applesauce, and other bland foods.

3. If the child goes to sleep (comfortable, normal sleep) during any phase of this treatment, don't wake her. But when she does wake up, start where you left off. That is, if you had given an hour's worth of a-tablespoon-every-ten-minutes, give her another hour of the same before you advance the diet.

Hint: If you do need to visit the pediatric office or emergency room, take along a small garbage bag with twist ties for the inevitable vomiting in the car. Normally, plastic is dangerous for a child to put over the face, but you will, no doubt, be supervising, and the child is unlikely to be in a mood to play with the bag.

## DIARRHEA
(Loose, watery, or frequent stools "All down his leg.")

**If your baby is six months old or younger and has diarrhea, read the illness section of the chapter covering his age. The younger the baby, the easier it is for the child to become dehydrated. Monitor any child with diarrhea closely for dehydration.**

Never use Lomotil or paregoric to treat diarrhea unless you are positive that your doctor specifically ordered it for this case of diarrhea in this particular child. Side effects, even without overdose, can be serious. The same goes for Pepto-Bismol. Pepto-Bismol contains an aspirin derivative, and has been associated with Reye syndrome (see Glossary).

If your child is acting sick with diarrhea, check the sections on *Dehydration* and *Acting Sick* in *Frightening Behaviors* before reading this section.

*Basics*
Pediatricians get to hear a lot about diarrhea, mostly about where it went (it all starts off down the leg, and then generally onto something white, uncleanable, expensive, and belonging to someone other than the parent).

Most acute-onset diarrhea in young children in this country starts out from relatively innocent causes: viruses, food intolerances, and forms of bacterial infection that often give mild symptoms and that run their course without antibiotics or other medication. Treatment usually consists of giving the child the right oral fluids and diet. When this is done, usually the child can be kept hydrated and comfortable while the condition runs (!) its course. But rarely diarrhea can endanger a child.

## COMPLEX SERIOUS DIARRHEA

Once in a while, diarrhea requires an urgent visit to the pediatrician or emergency room. This happens when:

- The child is becoming dehydrated, or losing salts that cause an imbalance in body chemistry (see *Dehydration*, in *Frightening Behaviors*).
- The underlying cause could be serious. Such serious conditions include an overwhelming infection, a toxin, or an obstruction in the intestine. A child with such an underlying cause for diarrhea usually acts or looks sick, has abdominal pain, fever, or is vomiting (see appropriate chapters).
- The diarrhea stool shows worrisome signs that indicate the need for very prompt medical care. Such stools as the following may indicate intermittent bowel obstruction, or bleeding from a very irritated bowel, or a rampant form of diarrhea that rapidly results in dehydration:
  *Stools that look like reddish (currant) jelly.*
  *Red blood in the stools.*
  *Stools that have no texture or color at all, but are just cloudy water.*
  *Profuse, frequent diarrhea that occurs after a trip to an area where cholera is a problem.*

## SIMPLE DIARRHEA

A child who has none of these ominous signs or symptoms but just loose or watery poops has simple diarrhea. Usually, simple diarrhea responds to the diet treatment given below. But sometimes you still need advice from the pediatrician.

If simple diarrhea hasn't abated in forty-eight to seventy-two hours, call the pediatrician, and try to have the following information ready: the child's current state of behavior, history of fever or other symptoms besides diarrhea, state of hydration, and the number and characteristics of stools.

Three common problems in communication tend to occur with such phone calls:

- It is hard to tell sometimes which is urine and which is stool in a wet diaper. If your pediatrician asks you about "wet diapers," this is really a question about peeing. If you aren't sure if the wetness in the diaper is urine or stool, say so.
- When your pediatrician asks how often the child is having a loose stool, don't say "every time I change his diaper." The next question will be, "And how often is that?" And you will say, "Every time he has a loose stool." Instead, count the number of loose stools in a twenty-four-hour or twelve-hour period. Or how many stools per hour!
- Use words that distinguish between urine and stool. Sometimes a parent will use euphemisms such as "went to the bathroom" or "went to the restroom." A busy pediatrician may assume you mean one thing when you mean quite another.

*Treatment*

The treatment of simple diarrhea in children consists entirely of modifying the diet and making sure the child stays well hydrated. Diarrhea medicines usually do no good and may be very dangerous in young children. The following is the "diarrhea diet" used by most pediatricians.

Many people think that a diarrhea diet consists of "clear fluids" and give apple juice or Jell-O water. This is not a good idea. Fruit sugars and table sugar usually make diarrhea much worse, no matter what the cause, because they aren't absorbed from the intestine—they actually pull water out of the body and into the stool. Worse, such fluids don't contain the balance of salts, such as sodium, chloride, and potassium that children lose in the diarrhea stool.

Recommended diet for simple, mild diarrhea without dehydration:

- If the baby is nursing, it's usually ideal just to continue nursing but to nurse more often than usual.
- For babies who aren't nursing, special fluids such as Pedialyte or Ricelyte are best. These are sold over the counter at grocery stores and pharmacies. They are specially balanced with the right mixes of easily absorbed sugars and salts, so as to rest the intestine and keep the child hydrated. Don't mix them with water or juices!
- Gatorade and other "sports" fluids are not well balanced, and are not appropriate hydrating fluids.
- Formulas that contain no milk sugar are often helpful. Sometimes the diarrhea stool "strips" the intestine, temporarily, of the ability to absorb milk sugar. When

this happens, the milk sugar in the stool pulls out more water from the intestine and makes even looser stools. Once the child is over the diarrhea, most pediatricians suggest returning to the milk or milk-based formula.

- After a couple of feedings of whichever fluid you have chosen, if there is no vomiting, you can start the "BRAT" expanded diet. The so-called brat diet does not refer to the child's behavior, but to the traditional foods for diarrhea: Bananas, Rice, Applesauce, and Toast. We now recommend, as well, pasta, potatoes, lean meats, and starchy vegetables like yellow squash and sweet potatoes. Stick to the above beverages and foods until the child has had normal stools for twenty-four hours; then gradually return to a normal diet.
- Water by itself doesn't have salt or sugar, and too much can cause dangerous chemical imbalance. It should never be used alone for rehydration. It's OK if the child is eating plenty of solid foods.
- Monitor the child frequently for signs of dehydration or complex diarrhea.

*Finally:* Wash your hands, wash your hands, wash your hands, not just after changing diapers but before you touch your face or prepare food. Viruses can live on inanimate objects, get on your fingers, and thence into your mouth or food.

## Trouble Pooping

*Basics*

When a child can't, or won't, or hates to poop, the whole world becomes bleak. "It's crowning," one obstetrician mother shrieked at me over the phone, "but she can't deliver it!" Or the child is hysterical and hiding in the closet. Or he will only poop while sitting in a warm bath, listening to a Raffi tape and clutching G.I. Joe.

Rarely, there's a medical cause for such a problem, but most often its cause can be traced to a disorder of diet, toilet habits, and developmental attitudes about self versus nonself, power, control, and authority. You'd think that diagnosis would come as a big relief, and it does. But even "nonmedical" constipation can become a big deal.

*Assessment*

If the child is otherwise healthy, growing, developing, playing, and learning normally, the chances are very great that this problem is the one called "simple stool retention." If you have any concerns that the child is not doing well otherwise, skip this and skim the rare medical causes section at the end.

The problem with simple constipation is that it can become a vicious cycle.

Basically, what happens is this:

1. ***For some reason, the baby, infant, toddler, or child decides that pooping is an unpleasant, undesirable activity. The reasons can be legion:***
   - The stool is big or hard and is difficult or actually hurts to pass. Why?

Our nutritional lifestyle. Some formulas are more constipating than others. Most children don't like to drink water, and the stool becomes dry. Nearly every food beloved of children, from bananas to chocolate, from low-fiber cereals (like the Chocolate Frosted Sugar Bombs enjoyed by Calvin of "Calvin and Hobbes") to pizza, is constipating.
   - The anus or rectum is irritated by diaper rash or pinworms, or maybe a lone hard stool scratched the lining (causing an "anal fissure"), and that makes it hurt to poop.
   - It's difficult to poop because of the way toileting is set up. If a child tries to poop but he's afraid of falling into the toilet, or if he has no place for his feet to rest and they just dangle, he won't be able to push hard enough for the stool to come out. If he's rushed or eager to get off the pot, he won't stay after the first splash.
   - Pooping is a big deal, developmentally speaking. It's about exploring where your body ends and the outside world begins, about control over where you get rid of (or give up or give away) something, and about the difference between clean and dirty, nice and not-nice, good and bad. It's about authority and power and independence. If things aren't going well in other areas where these issues are important, the bright and perceptive child will move the battlefield to an area where he can always win. You

can lead a child to the potty, but you cannot make him poop.

2. **Once the child has held back even one poop, the stage is set.** That stool blocks the stool coming from behind. As the mass of poop builds up, the muscle of the intestine stretches. It loses its tone and strength. Even the nerves become less alert, and send weaker signals to the brain that it is time to poop.

3. **Since the muscle is weakened, it's even harder to push out the poop and more tempting to try to hold back.** That would be a fine vicious cycle all by itself, but usually it's speeded up when parents and caretakers, aware of the problem, introduce even more factors as they try to make the child poop. They chase him from room to room with the potty chair. Offer rewards for delivering. Make him sit there until he goes, or until it's time to dash off to daycare or preschool. And they ask advice from everybody, within the child's hearing. So that even if there weren't any developmental issues in the first place, there are now.

*Treatment*
- At the first sign of stool withholding, galvanize. A very young baby may need a suppository, a change of formula, or a stool softening dose of juice: ask your pediatrician. Older infants, six months and above, generally don't mind the insertion of a pediatric suppository or a special bulb filled with glycerine, such as Babylax. Such an older baby, if otherwise absolutely healthy, could also be given an

ounce or so of prune juice or white grape juice to produce a soft stool.
- Examine the physical aspects of pooping. Pooping shouldn't be rushed. It shouldn't be stressed, with parents and older siblings harassing the child. It shouldn't be scary; there should be either a potty on the floor or a secure small seat on the toilet so the child doesn't worry about falling in and being flushed down. The area should be clean and homey. Make sure that there is a place for the child to rest her feet, a stool or bench or box that braces them so that her knees are up closer to her chest, so that she can push out the poop.
- Examine the psychological aspects of pooping. If the child is old enough for control, power, shame, etc., to be issues, examine his whole life-style. Is he being stressed too much in his daycare situation? Is he getting enough one-to-one time with each parent? Does he have the feeling that he controls a good part of his life, or that he's powerless? See the age-related chapters.
- Introduce nonconstipating foods early and often, but don't fall into the juice-addiction trap. At a year or so, start using high-fiber foods like bran. Give water early as a beverage, and model drinking water. Go very easy on the constipating foods, such as starches, cheese, bananas.

*Prevention*
- Encourage loosening fruits, veggies, and grains. Easy on the constipating ones.

- Make toilet training developmentally appropriate, low-stress, and fun. Let it be the child's achievement, not yours.[6]
- Make sure the toileting area is child-friendly: see *Treatment*, above.
- Be alert for the signs of stool retention: a baby or child of any age who gives all the signs of trying to poop but nothing comes out. He isn't trying to poop at all, he's trying **not** to!

**If the problem of constipation, stool withholding, or infrequent hard stools persists despite treatment,** your pediatrician really can come to the rescue, and I urge you to make that appointment early on, before the whole family becomes enmeshed in the child's poop problems. Untreated, it is very likely to get worse.

It is not rare for a child caught up in this problem for months or years to have it culminate in encopresis, or pooping in places other than the toilet. This can mean anything from stained underpants to full-blown stooling in them—or elsewhere. Often children with this problem seem well adjusted and successful, but once the problem is unmasked and they can talk about it, they reveal they have been depressed and shamed and preoccupied. Severely. Long-term.

Your pediatrician will give you instructions for emptying the distended, weakened muscle of the large intestine, and for toning it. If psychotherapy or counseling is needed, your pediatrician will tell you, and will help you find it.

*Rare medical problems*

Very rarely, the problem isn't that of stool withholding, but reflects an underlying medical condition. The key here is that nearly always the child displays other signs that something is wrong.

### INFANT BOTULISM

Sudden onset of constipation in a young baby under a year of age can mean a disorder called infant botulism, which is considered an emergency. In this disease, babies somehow (maybe through honey or corn syrup) ingest the spores of Botulinis bacteria. These wouldn't hurt older children, but in babies affect the nervous system. Along with constipation, babies with this suffer inability to suck and nurse, and are floppy. They don't usually have any fever.

### TIGHT ANUS

Once in a while, a baby will have trouble pooping because the anus is too tight. The baby grows and develops well in all other respects. Usually, he shows great effort and discomfort when he goes to stool, however. Sometimes the anus looks funny, kind of underdeveloped and dimplelike; sometimes it takes a rectal examination to make the diagnosis. It may take "loosening" in the pediatric office or help from a pediatric surgeon to correct a tight anus (*anal stenosis*).

---

6. See the well-child visit of the appropriate chapter.

### HIRSCHPRUNG'S DISEASE
(*Aganglionic megacolon*)

In this rare condition, the baby is born with an absence of nerve cells in part of the intestine, so he can't push out the stool. Most babies with this have trouble pooping right from birth: they don't stool in the first twenty-four hours of life, they have to be helped, and they continue to have problems thereafter. Often, they don't grow and thrive well, and have skinny bodies with protuberant abdomens. During the rectal exam, there is *no* stool—not hard stool.

### LOW THYROID
### (HYPOTHYROID)

The baby can be born with this, or it can happen later. Almost always, the child fails to grow well. Children with low thyroid often have developmental delay, thick dry skin, heavy thick tongues, and can't keep themselves warm when it's cold.

### HIGH PARATHYROID
### (HYPERPARATHYROID)

A child with this disorder may have been getting too much vitamin D, perhaps from vitamin supplements or from drinking lots of milk. Sometimes there is more vitamin D in milk than the label indicates, and sometimes children love milk and overdose. (Rarely, the parathyroid glands just overproduce all by themselves.) At any rate, a child with this problem is irritable, hasn't much energy, doesn't grow well, and pees a lot—and is constipated.

### LEAD POISONING

This gives not only constipation, but slow development and sometimes seizures. This is one case where you really don't want to wait until your child has any symptoms at all. If you suspect your child could be exposed to lead in any way, have your pediatrician order blood tests or, if the exposure has been long-term, bone X rays. (Lead might be in chipped paint, exhaust fumes, water pipes, ceramic cookware, or contaminated clothing.)

### GIARDIA

A stool parasite; see *Common Illnesses with (More or Less) Unfamiliar Names* in Part II.

### GSE OR CELIAC DISEASE

This inherited inability to digest gluten, which is found in wheat and other cereals, can cause severe constipation, failure to thrive, and the symptoms of depression. It's most common in the Irish and Jewish populations, and only shows up after the child starts eating the foods that contain gluten. (GSE stands for *Gluten Sensitive Enteropathy*.)

## Genitals in General

*If your baby is under two months old, see the chapter pertaining to his or her age.*

### Basics

You can't tell the players without a handbook, so here is a brief tour of the private parts, or Down There, as we say in pediatrics. Actually, the medical name is perineum. Your pediatrician will probably be jolted if you use it, and may not even recall what it means. In pediatrics, perineum is considered something belonging only to mothers giving birth; it's where, in the baby's emergence from the womb, the nose and mouth get suctioned.

### FEMALE

When you look at the baby, the major opening you see is not the vagina. It is the opening bordered by the lips or labia.

Look inside that obvious opening and you'll see there are two openings inside it. The more noticeable one, at the bottom, has a little ruffle around it: that's the hymen and the opening is the vagina. Just above the vagina, below the clitoris, is the tiny, pinhead-sized opening for the urine to come out, called the urethra.

In newborn girls, maternal hormones make the whole area puffy and swollen and everything stands out, no doubt to help new parents locate anatomy even though they can't find their reading glasses. Later on, the tissues shrink and the inner structures become more hidden.

### MALE

The penis is composed of a shaft topped by the glans, which is velvety textured and often lavender colored. The glans is covered by the foreskin, which is stuck very tightly to the glans at birth. If the baby isn't circumcised, the foreskin eventually unsticks and can be retracted like a sweater cuff, but it takes time: sometimes years. If the baby is cir-

cumcised, the foreskin is surgically removed, leaving the glans exposed. Pee comes out of the urethra right at the tip of the penis.

The sac that hangs down is the scrotum, and contains, at most, two testicles. The testicles are attached to the biological equivalent of springs, and tend to zip up into the body, disappearing from sight and feel at the slightest whim. Your pediatrician will make sure both can be persuaded down into the scrotum.

### ACTIVITIES

All children masturbate (play with themselves), starting as early as eight or nine months. This activity peaks usually between Two and Four years of age; or maybe it's just doing it in public that peaks then. See the *Well-Child Visit* sections in Part I for chapters on each age.

All boys have erections, starting before birth. But they may not notice them until later, and it's not uncommon for toddlers or preschoolers to say an erection "hurts." Have the child checked, of course, anyhow.

### HYGIENE

An uncircumcised penis should generally be left alone until the foreskin retracts. Don't yank on it and don't let any well-meaning health professional forcefully and painfully retract it. Once it does retract, teach the little boy to gently pull it back, wash and rinse the glans, and then pull the foreskin up to cover the glans again. If he doesn't, and has an erection with the glans pulled back, it can cut off the circulation. OUCH. A preventable emergency.

Soap bubbles in the bath water and powder or sand in the crotch irritate both girls and boys (circumcised or not). Keep bubble baths for special fun, not routine; don't make a habit of letting your child sit in shampoo water; and try to keep the bar of soap from floating in the water. Rinse with clear water after bubbles and beach.

Girls need to be cleaned from front to back so as not to bring poop germs into the area of the vagina and urethra. Little girls' arms aren't long enough to do this in one swipe: teach them to wipe first in front, drop the toilet tissue; then in back.

*Assessment*

SOMETHING LOOKS ODD BUT THE CHILD ACTS FINE

### GIRLS

There's not a lot of variety here. Some clitorises are more prominent than others; some hymens rufflier; sometimes there is a little tag of mucous membrane. Anything that strikes you as markedly unusual needs to be brought to the attention of the pediatrician.

Newborn girls have a discharge from the vagina that is usually thick and mucousy, but can be creamy white. Blood from the vagina in a newborn up to two weeks of age is normal, due to the abrupt stopping of maternal hormones.

After a few weeks, you shouldn't see any but the tiniest bit of clear discharge. Blood or discharge isn't normal, and can mean infection or a foreign body or other problem, and needs to be checked. Even if the child seems oblivious to it.

Little girls can get hernias. A lump in the groin area that comes and goes could be a hernia or could be a lymph node. One good way to assist your pediatrician is to take a photograph of the lump when it makes its appearance.

Labial adhesions can make the vagina look as if it has disappeared and the whole area has closed up. In fact, what has happened is that the inner lips of the vagina have glued themselves together. This is not at all uncommon. Most of the time, they will unglue themselves after a few weeks, and no intervention is needed. However, your pediatrician needs to take a look to see what kind of irritation could have caused them to glue together in the first place. If the urine is trapped behind the adhesions, or the child is uncomfortable, or has had a history of urinary tract infections, your pediatrician may want to treat the adhesions. If the adhesions have just started to form and are very delicate, your pediatrician may separate the labia very gently; otherwise, you may have to apply hormone cream for a couple of weeks.

### BOYS

- Once in a while, the opening for the urethra isn't at the tip but lower down the underside of the shaft. This variation is called *hypospadias*. If the opening is far enough down on the lower part of the glans, or on the shaft, so that the child might have trouble peeing standing up, it may need to be fixed. If so, skin from the foreskin will need to be used—so don't let a baby with this particular anatomy be circumcised. And, of course, ask your pediatrician.

- Sometimes after circumcision a little foreskin is left. Sometimes the skin of the shaft is a bit loose, and it comes up and laps over the glans. Usually the boy will grow into it, but if it looks very shaggy ask your pediatrician to check it.

- Sometimes the penis is quite inconspicuous. But it is rarely "too small." A newborn's penis should measure at least 2.5 centimeters, stretched out, from the pubic bone where the shaft begins to the tip of the glans. But much of the shaft may be buried, and has to be felt rather than seen. This happens a lot, so the chances are that a small-looking penis is really just dandy. Your pediatrician will check carefully; ask, if you are concerned.

- Sometimes you can't find both testicles in the scrotum. Since the testicles start life up in the abdomen and descend during fetal development, one or both might not have completed the journey by the time the baby is born. If both were present at birth but you can't find them now, it's most likely that they've bounced up into the body on their "springs." Rarely, though, that's not the case: instead, the testicle descended through the wrong channel, and as the child grew, got trapped up inside the body. *Whether at birth or later, unfindable testicles need a pediatric assessment.* In an older child, whose testicles have previously been in the scrotum but now cannot be brought down, the pediatrician

may ask the parent to look for the testicles while the child is relaxed in a warm bath. I came across a note in a child's chart: "Parents found testicles in Jacuzzi."

- Public hair is definitely not normal below the age of nine, and your pediatrician needs to figure out why it is there.

- If one side of the scrotum looks larger than the other, or larger than it used to, it could be because there is extra fluid in the sac (hydrocoele) or that there is a weakness in the abdominal wall, allowing a loop of intestine to escape into the scrotum (hernia). Either way, the child may not show any discomfort. Take a picture so that you can show the pediatrician if the bulge goes away before your appointment.

SOMETHING LOOKS ODD AND
THE CHILD SMELLS BAD, ACTS
SICK, OR IS IN PAIN

### GIRLS

Vaginal irritation, redness, and discharge is the big category here. If it is mild and clearly due to bubble baths, try some baths with no soap but just enough baking soda to make the water cloudy. Rinse the area well, pat dry.

If she's really uncomfortable and it looks inflamed, see the pediatrician. A very common cause is strep. We usually think of strep as a germ that lives in the throat, but it can infect every part of the body, including vaginas and rectums. (Miss Manners may not approve, but my own observation is that little girls tend to pick their noses and then, well, masturbate. I suppose, as John Updike says in a different context, it's an endeavor in which the delight has to be experienced to be forgiven.)

Yeast infections inside the vagina (as opposed to on the skin) are very unlikely to occur in young children. Only when the hormones of puberty make the vagina friendly to yeast does this occur. A child with a yeasty discharge as opposed to a yeast rash needs a visit to the pediatrician.

A foreign body in the vagina usually produces bloody discharge and generally makes the child smell just terrible. Get her to the pediatrician at once. I promise you, you'll be seen without delay unless there's a true emergency or the office staff has anosmia.[7]

### BOYS

A red, swollen penis, or pus from the urethra usually mean a bacterial infection (balanitis—see Glossary) that needs attention promptly.

Irritation at the tip of the penis is usually caused by the opening of the urethra rubbing against the diaper. If there aren't any other symptoms, try an over-the-counter antibacterial ointment—apply frequently. After he's over it, keep the end of the penis coated with petroleum jelly.

Any swelling of the scrotum that hurts needs pediatric attention at once. Most often it's a hernia, which means that a little piece of intestinal contents has gotten trapped in the scrotum: being trapped means that it can't get back up into the abdomen again.

Rarely, painful swelling means a testicle has twisted and must be untwisted urgently before it cuts off its own circulation. Often the scrotum looks bluish red in this case. This is very, very rare in children Five and under.

## Pee Problems

*Basics*

Normal urine is sterile, without germs. So when your newborn pees in your, or his, face, don't panic. It's a bit startling, that's all.

Peeing should be effortless and painless for both sexes, all the time, right from birth. (Peeing on command, in an older toddler or preschooler, can bring on performance anxiety. That's not the same thing as effort or pain.)

Normal urine doesn't contain red blood cells, bacteria, protein, or sugar.

When a child pees, the bladder squeezes to force all the urine out the urethra. None of it should be squeezed back up the ureters into the kidneys. And very little should be left in the bladder after the child has finished peeing.

In the first weeks of life, urine generally looks like water or is just slightly yellow, because babies are taking in so much fluid. After they start eating solid foods and are sleeping through the night, the first morning urine should look somewhat yellow, because it isn't as diluted.

7. Lack of sense of smell.

Many healthy children don't stay dry at night until age Five (and sometimes later). Once they are dry, they generally stay dry except for occasional lapses. If a child who once was dry starts wetting, on more than an occasional basis, or if a child over Five has never been dry, consult the pediatrician.

The medical concerns about problems peeing mostly have to do with protecting the kidneys. You don't want any infection in the kidneys. Infection can destroy kidney tissue, especially if it occurs over and over. Bladder infections can turn into kidney infections.

Similarly, you don't want a situation where the urine goes back up into the kidneys when the bladder squeezes during the act of urination. This is called reflux. If urine keeps going back up, the pressure can damage the kidneys. This can be more devastating if the urine is infected.

Also, you don't want to have urine obstructed anywhere on its journey out of the body. An enlarged kidney felt during a well exam, or signs of effort when a child urinates—especially a boy—can be signs of obstruction.

Finally, you want the kidneys to do what they are supposed to do: keep protein, sugar, and blood cells in the blood, and filter out excesses of everything else they are supposed to filter out—water, body wastes, products of metabolic processes.

*Assessment*

### PAINFUL URINATION
Babies and young toddlers cry when they pee if it hurts; you have to catch them in the act. Older toddlers and preschoolers will complain and hold their genitals. This may not necessarily indicate a urinary tract infection. Little girls may have sore vaginal areas, and the urine burns. Little boys may have an irritation of the meatus, or opening for the urine. All episodes of painful urination suggest bladder infection until proven otherwise, and must be evaluated by a pediatrician.

### FREQUENT URINATION
The most frequent, and innocent, causes of frequent urination are dietary: drinking caffeine-containing sodas and lots of sugary juice. Sometimes a child will drink a great deal of water because his nose is stuffy and his mouth gets dry.

Constipation can also cause frequent urination, as the filled-up rectum presses against the bladder, keeping it from emptying completely.

Urinary tract infections and anxiety also can cause frequent urination.

**Children with juvenile onset diabetes urinate frequently, as the sugar in the urine draws with it large amounts of water. Notify your pediatrician at once!**

Finally, children who can't concentrate the urine, whose urine is always as diluted as water, urinate frequently. This problem can originate in the kidneys, or in the pituitary gland in the brain. Have your pediatrician check a first-morning urine to see how well it's concentrated.

### INFREQUENT URINATION
If it hurts to go, some children will hold back for hours and hours. If they are consti-

pated from withholding stool, urinating may seem frightening because it is another act of letting go. If the child isn't taking in fluids, or is losing more than he's taking in, infrequent urine can mean dehydration.

Rarely, infrequent urination may mean that the kidneys aren't functioning properly. The most common—but still rare—cause of this is the *nephrotic syndrome* (see Glossary). A child with this disorder accumulates fluids in the tissues, has puffy eyes and feet and a distended belly, and acts sick.

### ABNORMAL URINE

Pink urine in a brand-new newborn usually indicates uric acid crystals, normal ingredients of newborn urine that disappear as the baby's milk intake increases.

Urine that is red may indicate blood, and has to be assumed to do so even though there may be another, and far more innocent, cause (such as eating beets or drinking Easter-egg dye).

Dark urine that looks like cola or tea may indicate a liver problem.

Cloudy urine may indicate infection, but often merely reflects crystals that form when the urine sits around in the bladder or a specimen cup, for a while.

Urine that smells bad often indicates a urinary tract infection.

To assess any of these problems, the first step is to look at the whole child. Is she acting sick or looking sick or smelling sick? Does she have fever, abdominal or back pain? Sighing breathing and lethargy and listlessness? Any of these require a prompt call or visit to the pediatrician.

If not, what has she been eating and drinking recently? Could she be constipated without your knowing it? Has there been emotional trauma, even something scary witnessed on TV?

Next, see if there is any irritation of the outlet of the urine. A red, irritated vagina or a sore meatus are usually obvious problems. Even if you think one of these problems is the cause of the abnormal peeing, the child still needs to be evaluated by the pediatrician.

Finally, nearly always a urine specimen will be needed. If infection is suspected, the urine sample has to be as clean as possible. If the situation is urgent, your pediatrician may need to obtain it by inserting a thin, flexible catheter into the child's bladder.

To catch the cleanest-possible urine specimen from a potty-trained child:

*Girls*

1. Clean from front to back.
2. Have her sit backwards on the adult toilet. This gives the urine a clean shot into the cup.

*Boys and Girls*

Catch the urine midstream as it flows.

If the child has started antibiotics by the time the urine culture is ready, you are then committed to assuming that the urinary tract was infected, because you have no way of proving that it wasn't. This may commit the child to having follow-up studies unnecessarily.

In a typical urinalysis, you can get an idea

of whether or not there is a urinary tract infection—and if so, whether the kidneys or only the bladder is involved. Juvenile onset diabetes, kidney damage, liver problems, and the condition called the nephrotic syndrome (see Glossary) can also be ruled out or strongly suggested through the urinalysis. By ruling out these things, other diagnoses are assumed: the red in the urine was not blood but beets; the blue urine was Easter-egg dye.

The urine is cultured by putting drops of it on various kinds of media that bacteria like to grow on, then watching over time what grows out. Such a culture for bacterial infection takes at least twenty-four hours, sometimes longer. When bacteria grow, they are then tested to see which antibiotic is likely to get rid of them most effectively.

## Treatment

Well, of course, treatment depends on the underlying problem, which indeed may bear no relationship with the urinary tract, such as constipation or anxiety or juvenile onset diabetes. (See "Diabetes" in Glossary.)

If the child does have a urinary tract infection, she requires fastidious treatment. The worry about urinary tract infection centers on protecting the kidneys. It's crucial to give the right antibiotic for the fully prescribed course. Most pediatricians ask that the urine be checked both on and off the antibiotic to make sure the infection is eradicated.

Then there arises the consideration of a problem called reflux, mentioned above. Reflux occurs when, as the bladder squeezes, some urine is pushed back up the ureters to the kidneys. This can be dangerous for two reasons: if the urine pushes all the way up to the kidneys, the pressure can damage the kidneys. If the urine is infected, even mildly so that the child has no real symptoms, the kidneys can be damaged by the bacteria.

If your child has had a urinary tract infection, your pediatrician may want to order a special X ray looking for reflux. This is because in some studies 25 percent of children under Five who have had a urinary tract infection have been found to have reflux, though mostly the reflux is mild, with the urine going only part way up towards the kidneys.

The test is called a VCUG or an RUG (*voiding cysto-urethrogram*, or *retrograde urethrogram*). If your pediatrician suggests it, I urge you to have it done, even though it's not a delightful experience.

In this test, a little catheter is inserted into the bladder, which is then filled with dye. As the child pees out the dye, X rays are taken to see if it goes back up the ureters to the kidneys.

There's no symptom of reflux, and you can't tell it's there by looking at the urine. The only way to find out is by doing the VCUG.

At this time, many pediatricians and pediatric urologists feel reflux should be followed and treated aggressively, but this is becoming more controversial. Discuss it with your own doctors.

## Prevention

I always feel like an ogre when talking about preventing urinary tract infection because so many of the dictates take fun out of life. Certainly, you don't want to center your whole parenting existence around discouraging urinary tract infection. Well, here they are:

- Keep the vagina and end of the penis unirritated. This means avoiding bubble baths and soap crayons (and other sources of bubbles such as shampoo water and letting the soap float in the water).
- In little girls, avoid tight clothing and nonbreathing materials in underpants.
- Encourage children who have learned to use the potty to go at the first urge so that the urine doesn't sit around in the bladder.
- Prevent and treat constipation.
- Teach little girls to wipe front first and then back to avoid carrying bacteria from rectum to urethra.
- Encourage water drinking.

Ha. Easy for me to say, right?

## ADDENDUM: A WORD ON DAMP PANTS

Little boys with damp pants usually are just in a hurry. They wait too long to go, or pull on their pants while they are still peeing.

Little girls with damp pants usually have the same tendency. But there is another cause. Sometimes a few drops, say a couple of teaspoons, of urine exits from the urethra and then goes back up into the vagina. This happens because as a little girl sits on an adult toilet the vagina is lower than the urethra. When she stands up, the urine from the vagina exits into the pants. Both diagnosis and treatment are to have her pee sitting backwards, facing the toilet tank. This position moves the vagina higher than the urethra.

Of course, if the pants are always wet, and gentle investigation and correction of habits doesn't help, this phenomenon needs evaluation by your pediatrician.

# Skin

*Newborn babies have an entirely different array of skin problems than do older ones. Please turn to the chapters on Birth to Two Weeks and Two Weeks to Two Months.*

*Basics*

I mentioned earlier that someday Oprah might have a show on children who never have tummyaches, and that it would most likely feature three children discovered after an exhaustive nationwide search. Would these be the same three children who never have had a skin rash?

Let me be brutally honest: about half of all skin rashes in children are totally mysterious. We never find out what caused them and they go away as mysteriously as they appeared. However, that doesn't mean you can ignore all welts, pimples, blisters, bumps, streaks, spots, or whatever. The underlying cause could be allergy to something touched or ingested; irritation from something caustic; infection—viral, bacterial, or fungal; infestation with a variety of what the Scots call "leggedy beasties" such as scabies or lice. Rarely, a skin rash can be a symptom of an underlying disorder elsewhere in the body.

When your child has a skin problem, you first need to decide if there's an urgent problem. If not, you may simply make an appointment with the pediatrician. If you want to discuss the skin problem on the phone, you need to decide how to describe it. This is somewhat of a mission impossible. One person's blister is another person's pimple. What one person describes as a "little whitehead" can be seen by another as a big red blotch. Many pediatricians call any rash in the diaper area a diaper rash; many parents think a diaper rash is one caused by diapers.

*Assessment*

### URGENT SKIN PROBLEMS

Usually a child with a rash or spots or other skin problem doesn't have an urgent problem. But rarely the rash signals a very serious condition that needs immediate attention.

1. If the child is acting sick, look at the guidelines under *Acting Sick* in *Frightening Behaviors* before paying more attention to the rash. If the guidelines don't indicate an urgent problem, then you have time to look at the rash.
2. If the rash looks like little red dots or bumps, or red-purple bruises or brown spots that do not blanch out when you press on them, you need prompt contact (even in the middle of the night), even if the child is acting fine. To find

out if they blanch, press and rub with your finger so that it covers the bump and a bit of surrounding skin. Some of the skin color will fade away dramatically. Does the spot disappear from view also and then reappear an instant later? Or doesn't it fade at all?

Spots that don't fade may be signs of serious infection or problems with blood cell formation. They are often tiny and red, and are called petechiae. Or they can look like bruises, purple or brown. These are called purpura. If you see these, you need to call your doctor at once, even if your child seems fine.

3. Many viruses, and a few bacteria, cause rashes that are supposed to peel. However, a rash that peels might mean either a serious underlying infection, such as toxic shock, or a subtler but also serious illness such as Kawasaki's disease. Rashes that peel need a same-day call or visit, even if the child feels just fine; urgent care if the child is acting sick.

### DESCRIBING THE RASH

This means getting the child completely undressed, shoes and socks and diaper also.

Where does he have the redness or the spots, or bumps, or whatever? Mostly on his trunk, or confined pretty much to arms and legs? If on the trunk, are the spots or blotches or whatever located especially around the buttocks, or the front bikini area? Is his face spared? Or just the area around his mouth? Or are just his cheeks involved?

What about his scalp? His palms and soles? Is it mostly the creases that are involved? Which creases: knees and elbows? Genitals and buttocks?

Can you see individual spots, or just a generalized redness, or roughness, or both? Are the spots separated by normal skin? How many spots are there? Just one, say on the cheek or eyelid? Or so many tiny ones you can't count them, so that the child looks like a portrait by Seurat or Warhol?

How big are they? The size of a pin prick or of the head of a pin, or a pencil eraser, or a dime, or what? Are they all the same size, or different sizes? Does it look as if the spots are changing, the longer they are present? Do they start out looking, say, like pimples, then turn into blisters, then crust? Or do all the spots you see belong to the same stage? Do they disappear in one area and reappear in another?

Are the spots raised or flat or a mixture? If you can't tell, close your eyes and rub them gently. If some are raised:

> *Solid bumps are pimples, and if they are shiny red and inflamed, they are boils or abscesses.*
> *Bumps filled with clear fluid are blisters.*
> *Bumps filled with pus are pustules, or you can just say blisters filled with pus.*
> *Oddly shaped, pink raised areas, often with a very fine white line outlining them, are welts or hives.*

Are the bumps in a certain pattern? Like chains of three or more (often flea bites) or just between the fingers or just on palms and soles? Is there crusting or oozing, or is the skin red and peeling? Sometimes you don't see anything else, just crust or ooze or peeling. Sometimes you see spots or bumps also.

Are there streaks? *A streak from an obviously infected area means that infection is spreading along a lymph channel, and this infection needs prompt attention.* If it is just a lone streak, that starts out of nowhere, and the child acts fine, it could be magic marker. Try to wash it off with rubbing alcohol. If it doesn't wash off, and the child is *entirely* well, it's still probably some indelible pigment. If it persists or the child becomes ill, see the pediatrician.

If there are streaks and splashes of pigment change, try to see if your child could have been playing with lime juice. If she has, and then gets exposed to sun, the places where she's gotten the juice make a kind of chemical burn. It is temporary, but takes a long time to go away.

## Treatment

Of course, treatment depends on the condition. Since skin rashes without other symptoms are very rarely urgent problems, you may wish to guess just what condition is causing them before calling or seeing the pediatrician.

## Bumps

**Flea bites** are pimples, but sometimes can look very red and inflamed because an allergic reaction sets in. Fleas usually only bite one person in a family or group. The bites tend to be in lines of at least three: that is breakfast, lunch, and dinner for the flea.

**Chicken pox** (Varicella) starts out with pimples that turn into blisters, each on a red patch, often described as a dew drop on a rose petal. (See *Common Illnesses with Familiar Names.*)

**Hand-foot-and-mouth** (*Coxsackie*) gives hard purplish pimples on palms and soles. It may give a rash of same-sized inflamed pimples, that crust over, on the groin, buttocks and ankles. (See *Common Illnesses with Unfamiliar Names.*)

**Hives** (*Urticaria*) are welts. Each welt is a red splotch with a very thin white line "drawn" around outlining it. Some are tiny, some huge, all on the same child. They come and go, appearing on different parts of the body. (See *Common Illnesses with Familiar Names.*)

**Erythema Multiforme:** This rash looks like hives, but there's a difference. Some of the spots look like bulls eyes, with a central area that looks brown or clear or blistered. Also, the joints get puffy. (See *Common Illnesses with Unfamiliar Names*) and *Mycoplasma* (see Glossary.)

**Cellulitis** is a red, shiny, swollen area that can be as small as a dime or bigger than a Susan B. Anthony silver dollar: it's a bacterial infection of the deep skin tissues and, sometimes, below the skin. Common sites are cheeks, eyelids, buttocks, but cellulitis can occur anywhere. It always needs prompt medical attention.

**Lyme disease:** The rash of lyme disease starts out with a red spot or bump (where the tick bit) which gets bigger and bigger, sometimes huge. The outside of the big splotch is bright red in a rim, and the middle of the splotch looks clear. It lasts for several weeks. Usually the child has a fever, is achy, and has "swollen glands" or lymph nodes. See your pediatrician at once if you think your child has this. (See Glossary.)

**Scabies:** This delightful (!) disorder gives terribly itchy bumps, classically between fingers, on hands, on wrists, or around the waist. But they can occur anywhere, even giving blisters on palms and soles in infants. The bumps usually itch more at night. This teeny tiny mite gets under the skin and burrows. Your pediatrician needs to diagnose and treat. Most recommend that everyone at home and at daycare be treated as well. What fun!

## FLAT SPOTS OR RED AREAS OR SPOTS TOO TINY TO COUNT

**Measles** (*rubeola*) starts with spots and some bumps around the hairline, then spreads ever downward. As it goes, it gets blotchy—the spots sort of merge and blend. Usually the rash starts after three or four days of a runny nose. The child is often quite sick, with red eyes and cough. (See *Uncommon Illnesses with Familiar Names.*)

**Three-day or German measles or Rubella** is an innocent disease for almost everyone except fetuses. The rash is pink, with flat spots and a few bumps, and starts on the face, then spreads down and to arms and legs. There are usually big lymph nodes at the nape of the neck. (See *Uncommon Illnesses with Familiar Names.*)

**Roseola**, (*exanthem subitum, herpesvirus 6*), which is so common in children 6 months to 2 years, doesn't produce a rash until after 3 to 5 days of very high fever (sometimes with a runny nose and other symptoms). (See *Common Illnesses with Unfamiliar Names.*)

**Fifth disease** (*examanthem infectiosum; parvovirus B 19*) is also called Slaps, because it starts off with bright red cheeks, as if they'd been slapped. Then the child may get a distinctive flat rash especially on arms and thighs: it looks lacy, like a red doily. (See *Common Illnesses with Unfamiliar Names.*)

**Scarlet fever** is a strep throat with a rash. Often the rash is inconspicuous and looks like a sunburn just in the bikini area. (See *Common Illnesses with Familiar Names.*)

**Eczema** (*atopic dermatitis*) in children over infancy usually looks like dry patches on the creases of elbows and knees, on wrists and ankles. It often goes along with rough dry skin.

Children can get into a vicious cycle of scratching which engenders more eczema which then increases the scratching. Infrequent baths and water exposure, plus pediatric advice, is indicated.

**Ringworm:** (*tinea corporis*) is not a worm. It's a little maligned fungus that doesn't deserve all the nasty things said about it. It's scaly little red patches that enlarge, leaving a clear place in the center, so that you get rings that look, I suppose, like worms.

## SCALP AND HAIR

**Seborrhea** happens when excess oil clings to skin cells and gives crusty patches in the scalp that don't ooze. They can be softened with baby oil or an over-the-counter cream containing cortisone, and shampooed away.

**Tinea capitis** is a fungus that attacks the scalp, giving spots of hair loss. Usually you see little black dots where the hairs are broken off, even in blonds and redheads. Sometimes these really get out of hand and cause unpleasant, boggy, growths called kerion. It is pretty common under age five, and then becomes less frequent. It needs heavy-duty prescription medicine by mouth.

**Lice** (by now, you ought to be scratching in sympathy), of course, are little bugs that lay eggs, or nits, on the hair shaft. First you see your child scratching her scalp viciously. Look more closely, and you see little white eggs glued to the hair shaft so that you can't remove them. If you are unlucky you will also see a louse or two, greyish brown with six legs, wandering about. The treatment is the over-the-counter 1 percent permethrin (Nix). Read the directions carefully. The eggs can be loosened after Nix by rinsing with a solution of white vinegar with a little water so that it doesn't sting. But you still have to try to remove every nit. And be a detective: lice are passed by head-to-head contact. Sleep-overs? Sharing hats or bike helmets? Combs and hair ribbons?

## Hips, Legs, and Feet

*Injuries to legs and feet are covered in* First Aid.

### Basics

Right after birth, a baby's legs are molded from being scrunched up in the womb. It's normal for breech babies to keep their feet up around their ears, and for all babies to have bowing of the lower legs and a slight curve of the foot, so that when viewed from below they look like a pair of parentheses: ( ).

The question is, are the legs developing and straightening normally as the child grows?

### Assessment

#### HIPS

At birth, and at every well visit until the child starts to walk without a limp or waddle, it's most important to make sure that the hip sockets are developing normally. They form while the baby is in the womb, by the ball of the leg bone pressing into the pelvis. The goal is to have a nice deep socket in just the right place. Even after birth, a baby can develop sockets that aren't quite perfect. So pediatricians check the hips carefully at each well visit until the baby is walking. The earlier a problem is found, the easier it is to treat, and the less likely, in most cases, to need surgery.

If the hip sockets aren't forming right, this is called developmental hip dysplasia. Some conditions are so much associated with this condition that they ought to be red flags for it. These include: a fetus in the breech position; an infant with head tilt or wry neck or with very inturned feet.

If you think that the cheeks of your baby's buttocks aren't symmetrical, or the leg creases don't line up with each other, or one leg is longer than the other, or you can't get the baby's legs to spread way out in the frog position, ask your pediatrician to check the hips. Or, if when your child starts to walk, he limps or has an exaggerated waddle.

#### LOWER LEGS

The lower legs are almost always "bowed" and stay that way until after the child has been walking for quite a while. It's called tibial torsion, because the tibia (weight bearing bone) is torsed (spiraled). Ask your pediatrician if the curving seems extreme, or if one leg is much more bowed than the other, or if it seems to be getting worse, not better, after the baby turns Two.

There's no evidence the baby who loves to be pulled to stand has any increased chance of hip or leg deformity.

#### FEET

Feet that really curve in sometimes need exercises, shoes, or even casting to help correct them, and always should call for a careful hip examination.

Toenails are microscopic at birth and seem not to grow at all until about age Two. One less thing to worry about.

### Treatment

A child with developmental hip dysplasia is usually treated by using a harness made of Velcro straps to keep the hips in the right position for the sockets to develop properly.[8] This is called a Pavlik harness, but patients in my practice call it a Rhino kicker. It isn't bulky or painful or uncosmetic, and babies get used to it rapidly. If your baby needs one, be grateful the condition was found early on. Try not to think of it as a big deal.

Shoes aren't necessary or recommended for babies and infants with normal anatomy, until they start walking outside and need shoes for protection and warmth. And when you do get shoes for children with normal feet, they needn't be expensive. Flexible shoes that don't rub the heel and are roomy for toes are all that's needed.

Children used to get "rickets," which produced very bowed legs, because they didn't get enough vitamin D. Today, most babies get plenty: from sunshine, from supplemented formulas and milk, and from vitamin drops. If your baby is premature, or if you breastfeed in a climate where your baby doesn't get at least fifteen minutes of sunlight a week if he's light-skinned, or if he's dark-skinned and not out in the sun every day, you may need to give vitamin D drops. Babies with rickets have soft skulls and bumps on their ribs, as well as bowed legs.

### LIMPING

Limping is a symptom pediatricians regard with respect. Most often, the reason for the limp is innocent: a splinter, tight shoe, blister, very minor sprain.

There are other causes that can be worrisome, and if they are suspected, pediatric care is needed right away. Of course, a suspected fracture always needs evaluation and treatment, but that's usually straightforward. You suspect by history or by appearance, get an X ray, and there you are.

But there are limps that are sneaky and hard to diagnose sometimes. **The most common urgent cause in young children is bacterial infection of the hip joint ("septic hip").** Sometimes this is absolutely obvious: the child has fever, acts sick, screams with pain when you touch the hip or move the leg. But sometimes the child just limps, maybe has a little fever, and likes to lie on his back with the knee on that side pointed out. If you even suspect this, you need medical help right away. The hip is a tight joint, and if it gets infected the bone can be destroyed!

Often, a child with these symptoms turns out NOT to have a septic hip, but a condition called toxic synovitis, which is quite innocent. This happens when a viral infection inflames the hip joint. The main concern with toxic synovitis is making sure it's not a septic hip. So your child may have to undergo lab work, X rays, even special procedures—but it's crucial to know one way or the other.

Rarely, a limp may turn out to be a symptom of a condition that is neither traumatic nor infectious, like arthritis or a blood disorder. But that is, indeed, rare.

---

8. Sometimes the harness needs to be used only for a few weeks, sometimes for months and months. It does not inhibit a baby from doing anything—and I mean anything.

We don't know what they are, but we sure hear about them a lot. Growing pains, by definition, are pains in the legs that don't have any known underlying cause. (Clever of us pediatricians, eh?)

They're usually in the calves, shins, or thighs, not in knees, ankles, or hips. Sometimes they're in one leg, sometimes in the other, sometimes both. They come on when the child is at rest, often at night. There's no limp, redness, swelling, heat, or fever. They feel better when rubbed. They tend to be related to jumping activity. Of course, call or see the pediatrician to be sure. And remember that the American Academy of Pediatrics has made a strong statement against trampolines (of any size)—not because of growing pains, but because of spinal cord injuries.

## Smelling Sick

All parents know what their young child smells like, even if they can't describe it. Sometimes deviations are easily explained: Jason ate all the garlic croutons in your Caesar salad; Jessica just vomited. When our daughter Sara, at fifteen months, drank the Herbal Essence Shampoo she smelled like flowers for several days.

But there are some smells that indicate problems. The most common are:

- Smelling sickly-sweet, like rotting apples or fingernail polish; this is usually from acetone on the breath of a child who is acting sick. Most often it is from the *ketosis* that builds up with vomiting and diarrhea. See these sections in *Body Parts and What Ails Them*.

Rarely, it can be a sign of juvenile diabetes: the child in this case will be very listless, have sighing breathing, complain often of a stomachache, and will be thirsty and urinating more than usual. She may vomit repeatedly. If this is the scenario, you need help right away.

- Bad breath. The most likely cause of bad breath is just like the cause in adults: bacteria on the tongue, way in back, producing evil odors. But here are some more serious causes to consider: viral mouth infections like herpes, strep throat, sinusitis, dental problems. If the child has a very chronic cough, lung problems may give bad breath. Sometimes what seems like bad breath is really more pervasive, however: see "smelling awful," below.

- Smelling awful and a bath doesn't get rid of it. The awful smell sometimes seems to be from her breath, but it's really hard to tell. This is usually from a foreign body that's gotten stuck and is starting to fester.

If the child is a boy, chances are great he's put a foreign body up his nose. If the child is a girl, the foreign body is either in her nose or her vagina. Sometimes there will be bloody discharge from nose or vagina; but sometimes there won't be—the foreign body is blocking it. By the time such a foreign body smells, it has inflamed the surrounding tissues and getting it out is likely to be a dicey task, so get medical attention.

The good side: you'll be seen *promptly*. People in the waiting room will *urge* you to go ahead of them. The staff will fetch the doctor *right away*.

- Smelly feet. Even the youngest child can have smelly feet from wearing footwear that seals in moisture and bacteria. Cure: go barefoot, wear sandals or ventilated footwear, cotton socks; or live with it. Rarely, it can mean something else: see below.
- Genital odors. Even normal, clean babies of both sexes will have genital odors, mostly not terribly unpleasant. Unpleasant smells can mean yeast or strep or other infection, or a foreign body. Avoiding bubble bath and soapy water, giving a bath with baking soda added— just enough to make the water cloudy—

may help. If not, something more than hygiene may be going on, and a trip to the pediatrician is in order.
- Smelling like something very specific, but not garlic.

Occasionally, I will hear from a worried parent who has heard of a devastating disease in which the baby smells like "sweaty feet" or "maple syrup" or even "cat urine." Yes, there are such diseases; but they are very rare, and the child acts sick, usually with recurrent severe vomiting and neurological problems. If you are worried, by all means ask your pediatrician.

It's also true that ingesting a toxic chemical can give an odd smell, and this can help trace down what it was the child got into.

**Moral:** If your child smells peculiar and is acting sick, notify your pediatrician promptly.

# {SIXTEEN}

# Illnesses, Both Common and Uncommon

## Common Illnesses with (More or Less) Unfamiliar Names

Time and again, parents are startled to discover that their child has an illness with a name they never heard of. They are even more startled when informed that "everybody in town has it" and that "most people had it in childhood and are immune to it as adults." How can this be? Who remembers having had, say, herpangina or rotavirus? I cringe when I see that look of disbelief on a parent's face.

So here are the most common of the common illnesses with unfamiliar names.

### MOUTH PAIN

There are three common mouth infections that outnumber all the others. Thrush, a yeast infection especially of babies under six months of age and of those on antibiotics; oral herpes (*gingivostomatitis*); and hand-foot-and-mouth disease (*coxsackievirus*).

### Thrush

*Basics*

Thrush is a yeast infection that gives a white, cottage-cheesey coating to tongue, gums, and insides of cheeks. (The yeast is called candida or monilia.) It is most common in young babies, especially if they share rubber nipples or toys or if they are on antibiotics (which get rid of bacteria that protect against yeast). A baby with thrush may cry and not eat well, or may not seem to notice.

Yeast is all around us, and it loves babies' mouths and diaper areas. Usually, thrush is innocent; an annoyance at most.

It does not have anything to do with the controversial condition "systemic candidiasis" in adults. Babies only get systemic candidiasis if they are especially fragile; especially at risk are premature babies with what is called an "indwelling catheter" or tube into a major blood vessel for a long time. People caring for such babies must

518

avoid fungal infections that fingernails can harbor, especially artificial nails.

## Assessment

Sometimes thrush just looks like a coating of milk on the tongue that persists long after a feeding. Sometimes it looks like white streaks and clumps on the inside of the lips and cheeks and even the roof of the mouth. Sometimes a nursing mother realizes her baby has thrush only after her own nipples become painful and red and swollen.

If thrush becomes severe, it WILL hurt and the baby won't eat. If a baby has severe resistant recurrent thrush, it can be a symptom of a serious problem with the immune system. If a child over six months of age who is not on antibiotics has *significant*, resistant, recurrent thrush, it can be a sign of an immune problem or of diabetes. If you are following all the suggestions for preventing and treating thrush, and it doesn't go away, ask your pediatrician specifically about these concerns.

## Treatment

When you treat thrush with the medication your pediatrician prescribes, apply it all over the inside of the baby's mouth. Use a cotton swab on a stick, or wrap your finger in a gauze pad and scrub. Don't put the dropper in the baby's mouth and then back in the medicine bottle; the dropper will contaminate the rest of the medicine.

## Prevention

- Boil rubber nipples once a day. (Don't forget to set a timer. Nothing smells worse than boiled-dry nipples, and they'll ruin the pot.)
- If your baby's on antibiotics, pay attention to mouth hygiene, rinsing her mouth with water after feedings.
- Don't refeed a bottle more than an hour after the baby has fed from it.

## Oral Herpes (Gingivostomatitis)

### Basics

This is a virus that everybody (pretty much) gets sooner or later. Many children will catch the virus, make immunity to it, and never show symptoms at all. Others will have a few sores in their mouth. A few will have a severe infection and a mouth coated with little ulcers.

This is not the herpes that is sexually transmitted. Oral herpes is a cousin of genital herpes. It's called herpesvirus type 1. The genital form is herpesvirus type 2.

*Gingivo* means gums, *stoma* means mouth, and *itis* means infection of. So this nasty illness consists of painful and plentiful ulcers on and under the tongue, on the lips, and inside the mouth. They start out as dirty-looking blisters, then rupture and look red on a white base. Sometimes ulcers spread to the throat.

It's highly contagious, and can be passed on by an infected child with no symptoms whatsoever. There's no vaccine in the offing.

Pain, fever, drooling, and refusal to eat may start before you can see the ulcers forming, and the child is likely to be very uncomfortable for between four and nine days.

- **Hydration.** It is rare that a child with oral herpes becomes dehydrated in the absence of diarrhea and vomiting, but this is one situation where high fever and drooling deplete a child's fluids and she may refuse to take enough to make up for it.

- **Ketosis.** When a child with high fever doesn't take sugar, she can build up ketones (from metabolizing fat) that make her feel even more miserable, so she takes even less sugar.

- **Rare serious spread of herpes.** Very rarely, herpes can cause infection of the brain or spread so that the virus attacks the whole body. If anyone with, or exposed to, oral herpes becomes very ill, seek attention promptly, as the early use of acyclovir can be of great benefit.

*Treatment*

Most pediatricians do not treat this common childhood illness with acyclovir (Zovirax), the medication for serious herpes infections, unless the child is at special risk for the virus spreading to the brain or to the body generally. Acyclovir hasn't been well studied yet in very young children, so risks must be balanced.

Treatment thus usually consists of sympathy, acetaminophen (or maybe ibuprofen or even codeine if your pediatrician thinks these are warranted) and whatever fluids the child will take. (But not diet drinks. Give the sugar-loaded sodas.) Sweet fluids at room temperature seem to be most welcomed, but some children like them cold.

*Prevention*

The only good thing to be said about oral herpes is that many, if not most, people who "catch" it don't get sick at all, but they do acquire immunity to it. Moreover, the virus does not pass from one person to another except by intimate contact. You can't catch it from inanimate objects, like toys and doorknobs.

Some authorities feel that healthy young children actually benefit from being exposed to this virus early in childhood, but not in the first six months, because there isn't likely to be a vaccine for this illness for a long, long time, and because it's more severe in older people.

Any person whose immune system is immature or impaired, however, can catch a devastating case of herpes. So they need to be protected. Certainly, a very young infant, pregnant women, and anyone with AIDS or on chemotherapy should be protected from herpes.

Does this mean that a child with herpes should be kept home from daycare? Experts are split on this. It seems to me that since many children who are contagious show no symptoms, the safest thing is to keep the child or adult with immunity problems home from daycare rather than the mildly ill child with herpes.

At any rate, people at high risk can be protected by avoiding all exposure to other people's saliva and nasal secretions. This means excellent hand washing after touching any child who might be contagious and before touching one's own mucous membranes (eyes, nose, mouth, and genitals).

The incubation period, from exposure to

getting sick, is three to five days. If someone in a high-risk category knows she has been exposed, she should notify her own physician immediately.

### Hand-Foot-and-Mouth

*Basics*
No, not hoof and mouth. That's cows. This is hand-foot-and-mouth, and it is caused by a virus named coxsackie. Only people carry this virus; you can't catch it from pets, much less from cattle.

*Assessment*
In HFM, the child not only gets blisters in the mouth, but raised painful purplish blisters on the palms and soles as well, and often a rash of bumps in the diaper area. Most children have fever. Often the sores in the mouth hurt so much that it's painful to swallow even saliva, and drooling is common.

As with oral herpes, the main concern is being sure the child is well hydrated and comfortable. As with oral herpes, serious complications are rare, but if any child looks or acts very sick, get prompt attention.

*Treatment*
There is no specific treatment for HFM. Acyclovir doesn't work for it. It's a totally different virus. So the treatment for HFM is just the same as that for uncomplicated oral herpes: fluids, pain medicine, sympathy, and time.

*Prevention*
Most experts agree that children with HFM need *not* be removed from daycare unless they are so sick that they are better off at home. This is because the virus hangs around for days before the child is sick until weeks after he gets well, so nothing would be gained. Also, most infections occur without symptoms, so that the other children who are apparently well in the daycare are also probably passing around the virus.

Adults taking care of children do well to wash their hands fastidiously and avoid kissing even well children on the face or hands—not just sick children, but children in general.

## SORE THROAT

### Strep Throat and Scarlet Fever

*Basics*
Strep throat is an infection caused by the bacteria group A hemolytic strep. No other germ qualifies you as having strep throat. You can have strep throat even though you have no tonsils.

Strep throat with a characteristic rash is called scarlet fever. It's usually not any more serious than plain strep throat.

The worry about strep throat is that in children over the age of Five it is rarely associated with the disease called rheumatic fever. RF results when the strep starts the body making antibodies against its own tissue. It can cause joint and more permanently heart disease.

*Assessment*
The sad truth is that any sore throat could be strep. You can't tell by looking. You can sus-

pect strep most strongly if the tonsils (if present) are beefy red, the uvula is red and swollen, there are big lymph nodes in the neck, there is a rash that looks like scarlet fever, and the child talks as if he's got a hot potato in the back of the throat.

But even then you could be wrong. It could be a virus (like mono) or another bacteria causing all these signs and symptoms.

Therefore, most pediatricians either perform a rapid test for strep (which is based on an analysis of its biological "fingerprints") or a throat culture or sometimes both. Some pediatricians treat, awaiting results. Sometimes a throat is so "strep-looking" that it gets treated even though the tests are negative—there are always a few false negatives with every test devised by humankind.

### Treatment

If you didn't treat a strep throat with an antibiotic, the child would probably eventually get well without complications. But very few pediatricians and parents choose this option. One reason is that rarely the strep can overwhelm the child, causing pneumonia or sepsis. Second, antibiotics shorten the course of strep throat, helping the child get back to normal activities. Third, children under Five may not be prone to rheumatic fever, but they may pass on strep to an older child who is a candidate.

It's particularly important that the antibiotic be one that strep is sensitive to: penicillin is the classic choice, but many others do the job. It's equally important that the medicine be taken for the full prescribed course. On the last day of medication, discard the old toothbrush and get a new one.

To prevent rheumatic fever, the antibiotic need not be given immediately after symptoms appear; indeed, it will have effect if given even a week or more after the child becomes ill.

### Prevention

Strep is most contagious during the early days of symptoms, and stays contagious until the child has been on antibiotics for twenty-four hours at least. You catch it through hand contact, mostly, or by being in the direct line of fire of a strep-infected sneeze or cough.

## Viral Sore Throats

Most viral sore throats are "nonspecific," that is, they don't give a specific picture that lets you know what virus is causing the discomfort. Two kinds of viruses that give sore throats common in young children often can be specifically diagnosed, and deserve to be more famous.

## Infectious Mononucleosis (Mono, the Epstein-Barr Virus)

### Basics

Good old mono, which used to be called the "kissing disease" because it is spread among teenagers and young adults by close contact, is really a disease many children contract in early childhood.

When very young children contract the mono virus (EBV), they generally don't get very sick, or don't get sick at all. They just build up immunity. Teenagers with mono often suffer through weeks of fatigue and sore throat and jaundice, but young children with normal immunity tend to just have a mild sore throat and "swollen glands" (lymph nodes).

So in a way it's good to have "mono" as a child. However, most young children who have mono have such nonspecific signs and symptoms that they are never tested for the illness. This is not really a problem, since there isn't a great deal you'd do about it if you knew the child had mono.

### Assessment

Most young children with mono have a sore throat and fever and fatigue. Often their eyelids are puffy but not red. If they are tested for strep throat, or the throat is cultured for bacterial infection, results are normal. This is because viruses don't show up on these tests.

If the illness persists for more than a few days, or if the child has a rash, enlarged spleen, or other specific signs that suggest mono, most pediatricians order a blood test. Sometimes the blood test, called the monospot, which is quite accurate in adults, is falsely negative in young children, and the diagnosis has to be made by further blood testing.

Once in a while a young child, more often an older one, will have strange vision phenomena with mono. Things look bigger or smaller or closer or farther away than they should. This is called the "Alice in Wonderland syndrome" and goes away by itself.

It's very unusual for a young child to be severely ill with mono, but of course if the child refuses fluids and becomes dehydrated, or the swollen tonsils give the child trouble breathing or swallowing, or if the child just looks and/or acts very sick, prompt attention is warranted.

### Treatment

Mostly, the treatment is just rest and comforting. Rarely, medication may be needed for very swollen tonsils or other complications. Most physicians ask parents to take care to protect the enlarged spleen that always accompanies mono. The spleen is a blood-filled organ, and you don't want it injured. That means avoiding activities like stomach-kicking and straining to lift heavy objects, like a younger sibling.

### Prevention

Oddly enough, it makes no sense to isolate the child (or anyone with a normal immune system) with mono because the virus is shed before the child gets sick and for weeks and weeks after he's well. Moreover, most children who are contagious for mono don't show any signs of sickness. Of course, a child who needs "nursing care" is better off at home. Pregnant women who think they haven't had mono should be careful about hand washing before touching their own faces—but that's true whether or not they are exposed to a child with mono.

## Herpangina: Coxsackie Virus Again

### Basics

Painful ulcers on the throat and tonsils but not in the mouth is often called herpangina, which means pain from herpes. But the virus that usually causes it isn't herpes, but rather a member of the enterovirus family. Often this is one of the coxsackie virus, our old friend that also causes hand-foot-and-mouth.

### Assessment

When you look way in the back of the throat, you'll see ulcers, little sores that look indented into the tissue. These are limited to the throat; you won't see them on the tongue or mouth. Usually the child, if old enough, complains about a very sore throat.

### Treatment

Once again: sweet fluids, sympathy, and pain control are about all that can be done.

### Prevention

At the risk of being repetitious, it does little good to isolate the sick child, since most of the time the virus is passed on by children who have no symptoms. Careful hand washing, careful hand washing, careful hand washing.

## BODY RASHES

The big names here are hand-foot-and-mouth (see above under Mouth Pain), fifth disease (also called slaps, erythema infectiosum, and parvovirus B 19), erythema multiforme, and roseola (also called herpesvirus 6). Also, see Kawasaki disease in the Glossary.

## Fifth Disease (Slaps; Erythema Infectiosum; Parvovirus B 19)

### Basics

This is a usually mild disease that gets its name as the fifth named childhood disease with rash. It is most common in school-age children, but young children and adults can catch it also.

### Assessment

Fifth disease consists mostly of a very distinctive and rather pretty rash. It starts with very red cheeks (hence the nickname "slaps"). Then a lacy rash appears that often is most prominent on arms and thighs. The rash can come and go for weeks. Sometimes a child will have a runny nose or other mild complaints, perhaps a low-grade fever. But mostly the whole illness consists of the rash.

A few children, and many adults, will have joint pain and even swollen joints with Fifth disease. If you find yourself with a swollen joint problem four days to twenty days after your child has been diagnosed with Fifth disease, you may save yourself and your physician a complex work-up if you remember this fact.

Rarely, the disease can cause serious problems, primarily in people with underlying blood diseases that cause destruction of the red blood cells, such as sickle cell anemia. It can also cause serious problems in people with immune deficiencies.

Exceedingly rarely, a miscarriage can occur if a woman in the first trimester has Fifth disease. This happens in fewer than 10 percent of such cases. In the last trimester, the fetus can be severely affected.

### Treatment

The only time treatment is necessary is when someone with an underlying problem such as Sickle Cell or AIDS becomes very ill and requires intravenous gamma globulin.

## Prevention

Fifth disease is most contagious right before the rash. Once the rash appears, the child no longer spreads the virus. Pregnant women who know they have been exposed to Fifth disease should discuss this matter with their obstetrician.

## Erythema Multiforme

Erythema multiforme is a skin outbreak that looks a little like hives. However, besides blotches, there are spots that look like bullseyes or targets or donuts or bagels. Often the hands and feet and joints swell. E. M. can be caused by infections (viral and bacterial) and allergies to medications, especially the sulfas. Pediatric assessment is always needed.

## Roseola (Exanthema Subitum, or "Sudden Rash"; or Herpesvirus 6, or HHV6)

## Basics

A child with a classic case of roseola is between six months and two years of age, with a high fever and runny nose. Despite the high fever, which may go over 105°, the child is remarkably cheerful. After three to five days, the fever goes and stays down; a rash comes out, and during the twenty-four to forty-eight hours the rash is present, the child is incredibly grumpy. Nearly everybody catches roseola before the age of Two, certainly by Three. However, most children who catch, and make immunity to, the virus don't have any symptoms at all.

## Assessment

The main problem with assessing roseola is that there is nothing special to distinguish it from other causes of high fever in infants. Naturally, one has to suspect, and rule out, serious causes: sepsis, meningitis, urinary tract infection. Of course, just as all these test results come back from the lab, the child drops the fever, breaks out with the rash, and the diagnosis is obvious.

It's not rare for a child who is prone to fever convulsions to have one during roseola. Since there are no other physical findings to make the diagnosis, nor any specific blood tests, such a child almost always winds up having a lumbar puncture to rule out meningitis.

## Treatment

There is no specific treatment for roseola, and even controlling the temperature isn't necessary if the child is happy and playful.

## Prevention

Since the diagnosis is only made after the fever disappears, a child with roseola needs to stay home from daycare until the diagnosis is assured. The chance is too great that the child with high fever could have a dangerous illness, such as sepsis, and either get much worse at daycare or pass on the dangerous illness there. Obviously, since all his daycare buddies who appear well are probably shedding the virus themselves, this does not arrest the spread of the virus.

## Trouble Breathing; Wheezing

### Bronchiolitis; Respiratory Syncytial Virus (RSV)

Bronchiolitis makes infants wheeze and struggle to breathe and toddlers and pre-schoolers wheeze and cough. (Actually, a person of any age can catch bronchiolitis.) RSV is the name of the virus that causes most of the cases. The disease is very frequent in the wintertime. It's described more fully in the section on *Airway* and *Lungs* in the previous chapter.

### Mycoplasma

Mycoplasma is the name of a bacteria, not of a disease. This little bug has features of both viruses and of bacteria, and it can cause a number of illnesses, including tonsillitis, ear infections, erythema multiforme (see above), walking pneumonia, and a form of croup. I put it in here so you wouldn't feel surprised that it's not in this chapter. Mycoplasma responds only to the antibiotics erythromycin (and its cousins) and tetracycline. You can't use tetracycline in young children; it stains the teeth.

## Diarrhea

There are several brand-name diarrheas that a young child is likely to encounter. Foremost among them is rotavirus. Other popular contenders include the bacteria shigella, salmonella, campylobacter, yersinia, E. coli, and the parasite giardia lambia.

## Rotavirus

### Basics

This is the most common cause of winter-time diarrhea in young children in the United States. It usually lasts from three to eight days. Some children are quite or severely ill, with fever, vomiting, and profuse watery stools. Other children, and most adults who become reinfected with the virus, have no symptoms at all and simply shed the virus, causing the infection to spread.

### Assessment

A young child with diarrhea in the winter-time often is simply assumed to have rotavirus. The stool frequently has a classic look—fluorescent green and watery—and a classic smell, which I forebear to try to describe. Usually there is no blood in the stool.

The major worry with rotavirus is dehydration. The younger the child, the greater the worry. See the discussion of *Dehydration* in *Frightening Behaviors*.

If there is blood in the stool, or the clinical picture is complicated, diagnosis can be made in the lab by testing the stool.

### Treatment

Keeping the baby or child well hydrated is the biggest challenge of rotavirus. Staying patient through its prolonged course is another.

### Prevention

Since most of the virus is passed along by children without symptoms, and since the virus can contaminate objects such as toys and be passed on when children handle or mouth them, keeping a child home from daycare is mostly for the benefit of the sick child herself, who needs care and monitoring. Moreover, a child whose stool can't be contained in a diaper or toilet is a much more potent source of the virus than a child who can be kept clean. Most authorities recommend keeping a child home from daycare or preschool until the diarrhea has abated. All authorities recommend hand washing, careful diapering, and separate areas for food handling and toileting.

## Bacterial Diarrhea: Salmonella, Shigella, Campylobacter, Yersinia, E. Coli

### Basics
All these causes of diarrhea can cause other symptoms such as fever and vomiting. In any, there may be blood and mucous in the stool. Each can be passed on when fecal material of an infected person, or sometimes an animal, contaminates food or water; people transmit them through hand contact. All can be serious.

### Assessment
There are just three things to assess in a child with diarrhea. First, is the child dehydrated? See *Dehydration* in *Frightening Behaviors*. Second, is the diarrhea caused by something that tends to cause serious problems? Rarely, some bacterial infections can spread to the bloodstream and give a serious infection; sepsis; or produce a toxin that damages the lin-

ing of the intestine. Even more rarely, some bacterial infections such as E. coli can cause a reaction that damages the kidneys; it's called hemolytic uremic syndrome, or H.U.S.

A child with frank blood in the stool needs to be assessed by the pediatrician for these problems.

Third, is the diarrhea due to a persistent underlying problem? If it isn't better after five days of dietary treatment, even simple diarrhea needs to be assessed.

### Treatment
You would think that antibiotic treatment would always be indicated for these illnesses. However, for minor salmonella infections, treatment may prolong the carrying of and passing on of the bacteria. On the other hand, treatment shortens both the illness and the spread of campylobacter, shigella, yersinia. Antibiotic use in E. coli is dangerous, increasing the chance of H.U.S.

### Prevention
Hand washing by all adults who care for children and who prepare food could reduce considerably these bacterial strains of diarrhea. Besides hand washing, all foods that could have been contaminated at any stage (e.g., from the laying of the egg or slaughter of the cow or pig to the cooking in the microwave) should be well cooked. Meat should not be served rare, and should reach a high internal temperature. Eggs should be hardboiled, not soft. No beef tartare, ever. Not even if you knew the cow personally. Unreliable water should be boiled before it goes into the mouth,

whether as drinking water, in an ice cube, or on a toothbrush.

### Giardia Lambia

Giardia is a common cause of diarrhea year-round.

*Basics*

Giardia is a parasite, but a microscopic one, not a worm you can see. Most people with giardia never know they have it, and so can't take special precautions not to pass it on. Those who have symptoms have diarrhea and cramps and gas, and sometimes constipation alternating with diarrhea. The symptoms can go on and on. It's easy to get reinfected right after treatment, too. You can catch diarrhea from ingesting anything contaminated with the parasite. Both people and animals can catch and give each other giardia. Contaminated water, the hands of other people, and foods are common sources of giardia.

*Assessment*

Often the stool isn't really watery, but bulky and foul-smelling, and it floats on the surface of the water in the toilet. Children with giardia aren't so likely to become dehydrated as they are with other kinds of diarrhea, but may complain of tummyaches, pass a lot of gas, and not gain weight well.

Diagnosis is usually made by examining stool samples under the microscope and looking for the little critter. But this is often very tricky, as many stools can be passed that don't have the parasite in them even though the child is infected. Sometimes a high-tech test will reveal giardia where previous microscopic examinations were negative.

*Treatment*

Antiparasitic medicines are the treatment for giardia, and there are several. You will want to discuss them, their side effects, efficacy, how good they taste, and how much they cost with your pediatrician.

*Prevention*

Hand washing, fastidious care during diapering and food preparation, and boiling water from unreliable sources are the best ways of not catching or spreading giardia. Once a child in daycare is known to have giardia, he or she should stay home until treatment is completed and diarrhea has stopped. Most experts do not encourage routine testing of the child's buddies who have no symptoms, as such testing is expensive and unreliable. But all experts agree on the importance of hand washing and careful diaper, toilet, and food hygiene.

## Uncommon Illnesses with Familiar Names, Including Chicken Pox

These should be uncommon because children should be immunized against them. Outbreaks occur in unimmunized communities, however. Moreover, the names are so familiar that it is easy to suspect that a child with symptoms that resemble these illnesses has one of them.

## MEASLES (RUBEOLA)
Children are immunized against measles when they receive their MMR vaccine, usually at twelve to fifteen months of age (earlier if there is an epidemic in the community), with a booster either at five years or just before puberty (entrance to junior high or middle school), or both.

Measles can be a mild disease; it can also be very serious, even fatal. A child with a classic case of measles has a terrifically runny nose, so runny that it is dignified with the name *coryza*; red eyes (or bloodshot eyes, anyhow, redness of the white of the eyes), a rash, a cough, and fever.

The rash of measles starts on around the fourth day of the fever, cough, and red eyes. The rash is flat, not bumpy, and looks like a very red blotchiness that spreads down from the face to the trunk and arms and legs. After two or three days, the rash starts to fade and peels in little shreds of skin.

It is sometimes hard to tell measles from other illnesses, such as *Kawasaki syndrome* or *Stevens-Johnson syndrome* (see Glossary) or even *scarlet fever* (see *Common Illnesses with [More or Less] Unfamiliar Names*).

If your child has been exposed to measles, call your pediatrician right away about a preventive dose of gamma globulin and/or immunization and/or vitamin A supplements.

## MUMPS
Mumps is a viral illness also prevented by the MMR vaccine (see above). It usually consists of pain and swelling of the glands that produce saliva. Usually this means a swelling in the angle of the jaw; sometimes the glands under the chin swell, giving the child a very swollen face, bad temper, and poor appetite. This makes the act of producing saliva very painful. (For a comical view of mumps, you and your children aged six and up will enjoy the storybook *Lentil* by Robert McCloskey.)

But mumps isn't always mild and funny. Ten percent of children with mumps have a form of viral meningitis with headache, neck stiffness, and vomiting. When boys or men get mumps, they can have pain in the testes and sometimes become sterile.

*Moral:* immunize. If you or your child become exposed to mumps without having been immunized, call your pediatrician. Prompt immunization may give protection or at least a milder case, because the incubation period (see Glossary) is long enough so that you may build up immunity before you show symptoms.

Children with enlarged lymph nodes, or those with bacterial infections or other viral infections of the saliva-producing glands, may be falsely suspected of having mumps.

## RUBELLA (GERMAN MEASLES; THREE-DAY MEASLES)
This is the R of the MMR, the third viral illness prevented by this excellent vaccine.

This is a mild disease (usually) with a rash

and mild fever and swollen lymph nodes. The rash starts on the face and spreads down; it is composed of little blanching pink dots and can be flamboyant or mild. Teenagers and adults may get swollen painful joints temporarily.

The people in danger from rubella are mostly unborn babies. If a pregnant woman catches rubella, she may suffer a miscarriage or the baby may become infected. The earlier in the pregnancy, the more likely and the more serious the infection of the baby, but problems can occur any time during the pregnancy. Such infection can leave the baby with mental retardation, deafness, blindness, poor growth, heart deformities—indeed, with problems affecting any organ of the body.

*Moral:* it's particularly important that every female child be immunized, once in infancy and again when starting kindergarten. However, since no immunization is perfect, it behooves *every* child, male or female, to be immunized to prevent the spread of rubella.

Children with scarlet fever or with non-specific viruses may be falsely suspected of having rubella.

## Chicken Pox (Varicella; Herpes Zoster)

Chicken pox is a usually mild viral illness; many of us have fond memories of sharing the itchy disease with siblings or friends.

### HOW IT SPREADS

Chicken pox, like most viruses, is mostly spread right before the child with the virus gets sick. It's a terribly contagious virus because the virus particles actually get into the air we breathe. It also can be passed by touching the person with varicella. It can't be passed by touching objects, however, because the virus doesn't survive very long on objects.

The person with chicken pox stays contagious until all spots have scabbed, usually about five days after the start of the rash; however, the contagiousness rapidly decreases after the first couple of days. Since it's the not-yet-sick child who is most contagious, it's hard to control chicken pox by isolating sick children.

### COURSE OF THE ILLNESS

Chicken pox usually starts with a fever and rash, though some children don't develop much of a fever. The rash looks like bug bites (especially like flea bites) on the scalp, neck, and trunk. After a day or so each "bug bite" turns into a tiny clear blister on a red patch: a "dew drop on a rose petal." Very pretty image, but they itch and hurt. After the blister stage, the spots crust over into scabs. The scabs gradually heal and usually don't scar, unless they've gotten really dug at with little fingernails or infected.

Most children complain most intensely about spots in the vagina and anus.

### TREATMENT

There is a specific medication, acyclovir (Zovirax) for chicken pox. If it is given starting in the first twenty-four hours of the rash, it can decrease the number of spots and the days the child is sick. Most pediatricians reserve it for children with underlying medical problems, or whose family members are at special risk for chicken pox, or for special circumstances (a child model who fears scarring, or a child whose family has major travel plans that can't easily be changed, or a child in a cast, for instance). This is because the effects of acyclovir aren't very dramatic, and because we fear that the varicella virus could develop resistance to it if it is used too often—and we really need acyclovir for the serious cases of varicella, those that occur rarely in normal children and frequently in those with immune problems. There is also some concern about whether children given Zovirax will make as good immunity to chicken pox as those who don't take it, and about whether "shingles" might occur later on more frequently in treated children (see below).

Then there are lots of remedies that can make a child with chicken pox feel better.

- Fever control with acetaminophen or ibuprofen.
- Keeping the child dressed very lightly.
- Antihistamines by mouth for itching, such as Benadryl (diphenhydramine). Benadryl in lotions or spray (like Caladryl lotion) shouldn't be used, as it can be absorbed into the blood through the little sores. This can give a child an overdose, which can be serious.
- Calamine lotion, nonmentholated shaving cream, and the touch of an ice cube may control itching or at least distract the child for a bit.
- Oatmeal baths such as Aveeno baths can be soothing.

### COMPLICATIONS

Rarely are there major complications of chicken pox. However, there are rare cases of pneumonia or encephalitis (see Glossary) that can complicate chicken pox. Infection with a bacterial illness, such as strep or staph, can occur and, especially if untreated, this can become serious. Notify your pediatrician right away if your child:

- Is acting or looking very sick (see *Acting Sick*).
- Is having trouble breathing (see *Trouble Breathing*).
- Complains of a severe headache.
- Complains of a very sore throat.
- Is disoriented or incoherent, talking about things that don't make sense, not recognizing familiar people or things.
- Has a fever of 104° or over.
- Has a rash that is looking a lot worse than the dew drops on the rose petals. Great big blisters, pus, big areas of redness are of serious concern. The appearance of spots that look like bruises, or weird-looking spots that don't blanche and lose their color when you press on them, is ominous. Cold hands and cold feet are also a serious sign needing quick attention. These are all signs of an infection with a bacteria on top of the chicken pox virus.

*Prevention*
The chicken pox vaccine, ideally given at age one year.

### Reye Syndrome

Reye syndrome is a mysterious disease that affects the brain and the liver. If diagnosed and treated aggressively in the early stages, the outlook is much better than if you wait. Waiting can be fatal.

There is (somewhat controversial) evidence that aspirin can trigger Reye syndrome in children with one of the underlying viruses, like chicken pox and influenza A and others that are associated with the syndrome.

The signs of Reye syndrome are vomiting, sleepiness, headache, and disoriented bizarre behavior. The child may be combative, then lethargic. You may notice the child breathing rapidly and deeply—hyperventilating.

Infants with Reye syndrome may not show any of this, but may have diarrhea and seizures.

If you suspect Reye syndrome, get help right away, and say that this is what you suspect.

#### PEOPLE AT SPECIAL RISK

Chicken pox can be exceedingly serious, even fatal, for people who can't mount a good immune response to it. This is why we try to isolate contagious children.

Anyone at high risk who is exposed to chicken pox should notify his physician as soon as possible. There is a gamma globulin shot that can protect him, and acyclovir should he become sick. People at risk include:

- Anyone with an impaired immune system. This includes people on chemotherapy for cancer, on long-term steroids for asthma or other conditions, or with AIDS.
- Fetuses and newborn babies of mothers who contract chicken pox.

### Shingles (Herpes Zoster)

Once you have chicken pox, the varicella virus—also known as herpes zoster—stays in your nerves for the rest of your life. When it flares up, it causes a syndrome called shingles, I suppose because the skin break-out looks like a wind-blown roof. Sort of. This is a painful outbreak of blisters in one area of the body—an area supplied by the particular nerve that got triggered to release its herpes zoster virus.

We don't know what triggers an attack, but exposure to a child with chicken pox appears to be one of the factors. Whether this is the virus, or the stress of being with such a child, who knows.

People with shingles can spread the chicken pox virus to nonimmune people through the liquid in the blisters. So they should either stay isolated or keep the blisters covered and not touch them.

# Part Three

---

# Pediatric Concerns and Controversies

# {SEVENTEEN}

# Growing in All Directions

*". . . where all the women are strong, and all the men are good looking, and all the children are above average."*

—*Garrison Keillor*

If you ask me, it's mostly parents who have the growing pains.

*"I've heard that you can predict adult height by multiplying the height at the age of two. Does this really mean that Jessica is going to be six feet three, and Jason four eleven?"*

*"I'm five feet two, and Henry is five seven; now we've got this baby who's in the ninetieth percentile for height and everybody keeps teasing me about maybe it was the mailman?"*

*"She's as thin as a rail and she only gained five pounds over the whole last year!"*

*"Everybody thinks she's three so I finally put a label on her shirt that says, 'The reason I don't talk yet is I'm only fourteen months old."*

*"Since he turned a year, he hardly eats a thing. Is he too thin?"*

*"I don't know why she's so fat. She gets so much exercise. I take her for a ride on my bike for two hours every day!"*

Here you are at your child's well visit. You have so many questions, the entire time could be devoted to issues of height and weight. What usually happens instead is that you "put the child on the growth chart," sort of like pinning the tail on the donkey.

Many pediatricians don't like to emphasize growth charts. The charts certainly aren't very user-friendly. And it's not good for parents to get all caught up in percentiles, and to feel that growth is a competition. Many pediatricians fear that if they tell a parent the child is too chubby, the parent will put the child on an ill-advised diet, or be highly critical of the child and sabotage his or her self-esteem.

I think parents can rise above these temptations.

So I think the charts are valuable, provided you use the right ones and know what you're looking for. The reason I like to use growth charts is that we can get such a false impression from merely eyeballing the children around us. For instance, our daughter Sara, taking after her parents, is on the short side. When we lived in New England, that's just what she looked like: on the short side. When we moved to California, State of

Giants, she didn't look just short; she looked teeny weeny. (So did her parents.)

More worrisome, when we look at children today, is the fact that many of them (30 percent, in fact) are overweight. Of these, about ⅓ are actually obese. This means that we get used to seeing children who are overweight. When this happens, the normal-weight child looks slender, and the slender child emaciated.

Many parents don't have a mental image of what a child the age of their own *ought* to look like. But growth charts can give you a pretty good idea of whether you ought to consider your child of normal weight, though maybe slightly chubby or slender or worrisomely over- or underweight.

Because this is such a common concern today, I've added a separate essay on the problem: *Chubby or Not, Here We Come.*

So in this essay, I'm only going to address questions about height. I've also put the growth charts, and instructions for use, in the "Chubby" essay, because that's where parents (and medical students, medical assistants, and many of us pediatricians) need the most help in figuring them out.

Here are some questions that parents ask about height:

**Is my child taller or shorter than most children his or her age and sex?** The charts can answer this very well, bearing in mind that they represent children from all racial and genetic backgrounds, from tall Scandinavians and Masai to short Asians and Welsh.

**Is my child growing consistently along one curve, or jumping to a higher or lower curve?** Again, the charts are very help-ful. When a child does jump to a higher or lower curve, the charts won't tell you why. But your pediatrician ought to be able to.

**Can this curve be used to predict my child's adult height, or are there other factors which make that prediction low in accuracy?** There are, indeed, other factors.

So, what does it mean when your pediatrician tells you that Cherub is in, say, the 50th percentile for height? It means that if you took a hundred children Cherub's age and sex, and lined them up by height, with the tallest at one end and the shortest at the other, then Cherub would be right in the middle, with fifty on either side of her.

If Cherub is in the 90th percentile, Cherub is taller than ninety of them, and shorter than ten. If at the 10th percentile, taller than ten of them and shorter than 90.

So you'd think this would be straightforward, but it's not.

Even if your child is in the fiftieth percentile for height by age, Cherub's friends all may tower over Cherub, or Cherub over them. This could be because the percentile curves are derived from normal healthy children of all racial, ethnic, and nutritional backgrounds, and your child's friends are of a different mix from his or her ancestry. Maybe if you plotted *the friends,* they'd all be in the 90th percentile, or in the 5th percentile. Or it could be that you are assuming that everybody in your child's kindergarten is Five, but in fact, when you ask, you discover that most of them are nearly Six. Or still Four.

Moreover, the age of the child does not necessarily reflect the maturity of his or her stage of growth. The key to final height is in large part the age at which the child experiences puberty. Most of us can remember the girl who towered over all the kids in fourth grade, started periods in fifth grade, and by high school was the shortest in the class. Or the short, wiry clown of the class who grew nine inches in the summer after tenth grade and wound up in the back row, head and shoulders over everybody else, in all the graduation pictures.

## PREDICTING ADULT HEIGHT

Doubling the child's height at the age of Two, or following his or her height curve up through all the future years to adulthood, may predict fairly accurately the adult height, but there are many pitfalls. IF the child has grown consistently along one percentile curve, and IF the predicted height seems reasonable in considering the parents' height, and IF the same-sex parent had a similar growth curve with puberty at about the age of the national average, and IF no illness or nutritional or hormonal problem intervenes, THEN it's a pretty good rule of thumb. But, as discussed above, don't count on it.

If you or your pediatrician are very concerned about your child's being at one or the other extremes for height, an X ray of the hands and wrists can show *bone age*. This is an assessment of how mature the bones are. After all, you stop growing when your bones stop lengthening, and immature bones grow for a longer time than mature bones. If your child is two years old but his bone age is sixteen months, plotting his height for a two-year-old is deceptive. His bones think he is eight months younger, and plotting him as a sixteen-month-old will show his real percentile.

## EXTREMES OF HEIGHT

### Very short

This is such a complicated question that it shouldn't be discussed here but rather with your pediatrician. There are many innocent, normal reasons for a child to be very short in stature. For one thing, somebody's got to be in the shortest five percent, or the concept wouldn't exist. Very short parents will most likely have very short children. (And I, as one of the Short, say more power to us: we take up less room in elevators, consume less bath water, and don't block the view.) But there are some conditions in which very slow growth can signal a problem. So take it up with your pediatrician.

### Very tall

Parents rarely express concern if such a child is male, but worry for girls. Medically, extremely tall children inspire concern in several rare situations:

- If the child is showing signs of early puberty, such as height that crosses percentile curves, breast buds, body odor, pubic hair.
- If there is evidence that the excessive height could be associated with an inborn problem, such as a chromosomal abnormality or a problem with bone and connective tissue.

- If there is a sign that the pituitary gland is producing too much growth hormone.

But these are rare conditions.

Parents of girls are often relieved to discover that a girl sailing along at the 95th percentile for height has an estimated adult height not of six and a half feet but of five feet nine.

## CHANGING PERCENTILE LINES

It is often normal for a child to "cross the percentiles." When short parents have a very tall newborn, it's expected for the baby to slow down height growth, sometimes gradually descending from the 90th percentile for height down to the 25th or lower, over the first year or two of life. Conversely, when a small baby is born to very big parents, the baby may ascend the height curves in the first year or two, in the opposite direction.

Crossing the percentile lines is of concern when:

- The height percentile has stayed constant for over a year and then plateaus—say from the 75th to the 50th. This slowing of vertical growth can be a sign of hormonal or other problems.
- The height curve remains the same but the weight-for-height goes up or down the percentiles. This may mean only that the child is attaining his or her "genetic potential," but the pediatrician needs to look at the situation carefully.

## VARYING MEASUREMENTS THAT DON'T MAKE SENSE

Can a child shrink in the interval between one measurement and the next?

Only in the movies. There's a good excuse for error at every age. Newborns, having been curled up for months, don't want to straighten out and many kind measurers don't want to force them to. From two to six months, anybody lying naked on her back pees on the exam table, obliterating the markings and generating chaos. From six to nine months, anybody lying on her back has to fight to turn over, generating more chaos. From nine months on, they have to be tackled and wrestled down. When they're old enough to stand, they spook at the whole idea. They wear hair ornaments and headgear and refuse to take them off. They stand on tiptoe, slump, rock back and forth on their heels, hold their breath, stare at the ceiling. It's a miracle we get as close to accuracy as we do.

## FINALLY

Sometimes it's very hard to cope with the difference between a child's growth pattern and a parent's hopes and dreams for the child. It's rare that the discrepancy is extreme: a mother brokenhearted because her daughter is destined to be too tall for ballet or too short for basketball or modeling; a father devastated because his son is too small for football, too tall for jockeying, too stocky for track. (Another reminder that growth issues tend to be very gender-related.)

It's not uncommon for issues of growth and nutrition and self-esteem to get all tangled up with parents' feelings about their children and their competence as parents. Sometimes a disturbed growth pattern is really nature's way of revealing to us that the parent-child relationship is askew. An honest soul-searching, perhaps with spouse or pediatrician or friend or even therapist, can prevent more extreme problems later on.

## {EIGHTEEN}

# Chubby or Not, Here We Come!

## The Big Questions

1. How can you tell if Cherub is over-weight?
2. How can you be sure it's not a hormone problem?
3. What's our goal?
4. Why is it important to intervene early, rather than waiting?
5. But aren't you worried about making Cherub overly preoccupied with her weight? Of triggering an eating disorder like anorexia or bulimia?
6. What do we need to do to help Cherub?
7. What if there's a family history of over-weight?
8. How can I change Cherub's diet and TV habits without making the whole family change?
9. But what about the Skinny Minnies?

1. ***How can you tell if Cherub is over-weight?***

When I tell parents that Cherub has

gained more weight than is good for her, they often look at me with disbe-lief. I don't blame them. *"You must be joking!"* they say. *"Cherub doesn't look chubby, she looks adorable!"*

I agree. You often can't tell by look-ing whether a child is gaining too much weight.

As a nation, we have gotten so used to seeing chubby children—on the media, in real life—that Chubby has come to seem Normal. A chubby child looks fine; a normal-weight child looks really skinny. A *much* too chubby child looks—well, just chubby.

We have forgotten that children, starting at age Two, should look more slender each year until about age Six. At age Six, normal-weight children look skinny, with elbows and knees sticking out. It's not just parents and grandpar-ents who are fooled: pediatricians and medical students alike can be flum-moxed. When a child is very *very* chubby, it may be obvious. But for most children, you need to use the growth charts.

2. ***How can you be sure it's not a hormone problem? Her thyroid, maybe?***

Fewer than five percent of all cases of childhood overweight are caused by hormone problems. Weight gain is never the *only* symptom of a hormone problem. For one thing, nearly all hormonal or other medical causes of overweight slow down growth in height. Low thyroid gives rough skin, constipation, an enlarged tongue, and decreased energy and brightness. Chromosomal problems that cause excess weight each have their own set of physical findings as a clue. Nearly always, chubbiness due to a medical problem goes along with a slowing of height growth.

But do check it out with your pediatrician. If there is any suspicion of a problem, your pediatrician will want to perform blood tests and perhaps an X ray of Cherub's hands to judge the degree of bone maturity.

3. ***So what's our goal? You want every kid to be a Skinny Minny? Our society worships thinness. We need to accept the fact that some kids, and adults, are just bigger than others. It's the teasing that's the problem, not the chubbiness.***

I'm all for diversity of appearance, style, abilities, and beliefs. I'm very much against teasing, whether from within the family or from outsiders.

But the lifestyle that has caused this

epidemic of overweight in our children has nothing to do with innate, genetic diversity of body size. It has to do with greed: Greed on the part of the fast-food businesses, juice and soda manufacturers, and purveyors of high calorie, low nutrition snacks. Greed on the part of the media, feeding children and adults addicting, mindless fare, punctuated by ads for fast-food. Greed on the part of real estate developers, who design neighborhoods and shopping areas to be Car-Friendly, with no thought to healthy and safe places to walk (especially to school and back) and exercise.

Parents need to dig in their heels and fight this coalition of Greed. This is not easy. No wonder it's tempting to avoid the problem altogether. But the truth is, chubby toddlers and preschoolers have a great chance of slimming down if their lifestyle is redesigned for them. The older they grow, the less likely this is to happen.

4. ***Why is it important to intervene early, rather than waiting? She'll grow out of it.***

You certainly can't count on it. Not today. We live in a culture that is designed to make children gain extra weight. If we'd sat down as a committee, with Childhood Obesity as our prized objective, this is the culture we would have come up with.

I especially worry about little girls. If a little girl stays extra chubby into first and second grade, she increases her chances of starting puberty much earlier than her peers. Bad enough to start periods in third or fourth grade. What's worse: The hormones of puberty for girls are designed to add even more extra weight! It's a myth that girls "slim down during their growth spurt."

The best strategy is to prevent extra chubbiness in the first place. That means knowing the normal, expected pattern of growth and weight gain for your child, and providing the lifestyle (exercise, activities, meals, beverages, and snack approach) that will keep Cherub in good shape.

The next best strategy is to keep an eye out for any change in the direction of chubbiness. As soon as you spot that happening, figure out what created the change and fix it.

5. *But aren't you worried about making Cherub overly preoccupied with her weight? Of triggering an eating disorder like anorexia or bulimia?*

The best way to prevent an eating disorder later on is to establish healthy eating habits, a healthy body image, and healthy enjoyment of activity and exercise early on. In fact, it's possible to argue that chubby preschoolers already *have* an eating disorder:

- They eat not because of hunger, but out of boredom or habit.
- They continue eating after they are full.
- They have learned to prefer foods that are full of fat or sugar, and low on nutrition.

Of course, we should never, ever tell a child she is fat, or allow teasing from anyone, including siblings.

But children know, themselves, when they are fat as early as age Four. Their self-esteem does not usually drop earlier than puberty, fortunately. There's plenty of time to make changes, a bit at a time, to form the habits Cherub will need to slim down and stay healthy.

6. *What do we need to do to help Cherub?*

We have to be detectives, and figure out what it is in her lifestyle that's creating the problem.

Becoming extra-chubby is very easy. All it takes is a daily, subtle accumulation of a few more calories than one needs. Calories that aren't used up in normal weight gain, activity, or metabolism turn into fat. Over a year, fifty extra daily calories translates into five extra pounds of fat.

That doesn't seem like much, but for a toddler and preschooler, an extra five pounds a year doubles the normal amount of weight she's supposed to gain. At the end of that year, she's gained all the weight (and maybe more) that she was predicted to gain for the *next* year.

This can happen in three ways:

- Perhaps Cherub has started watching too much television. Television watching has been shown to be one of the strongest influences on overweight in childhood.
- Perhaps Cherub is nibbling chips or crackers all day, or drinking juice

instead of water, or in love with milk and drinking much more than 16 ounces a day. Perhaps she gets huge helpings of rich foods.
- Perhaps Cherub is no longer getting as much physical exercise as she used to—running around, playing outside.

## AGE AND WEIGHT GAIN

*What's the most likely age for a young child to gain extra weight? And why?*

Every age has its own predisposition toward overweight built into it.

*Two months to age one*
- Caretakers respond to all or most crying by offering a bottle or snack.
- A baby over Four Months Old "insists" on being nursed or fed through the night.
- Caretakers aren't alert to signals that Cherub has had enough, and continue to give her spoonfuls of food.
- Cherub isn't encouraged to start feeding herself at Nine Months.

*One to two*
- TV becomes a regular pastime.
- Cherub drinks more than 16 ounces of whole milk, daily.
- Cherub isn't encouraged or allowed to feed herself entirely by Fifteen Months at the latest.
- Caretakers don't realize that the normal eating pattern is a good breakfast, fair lunch, practically no dinner. They urge or coax Cherub to eat more.

- Cherub gets "handout" snacks: baggies of chips, raisins, crackers throughout the day.
- Cherub drinks juice or soda instead of water for thirst.
- Cherub "insists" on second helpings of addicting foods such as macaroni and cheese.
- Cherub regularly eats fast food—burgers, fries, shakes.
- Caretakers try to reward or bribe good behavior by offering treats.

*Two to three*
All the features of One to Two, plus:

- Cherub starts to prefer TV to other activities, as she becomes more verbal and interested in stories. She may watch more than two hours each day.
- Cherub continues to drink whole or 2% milk; at Two, nonfat or 1% is the better choice.
- Caretakers encourage two naps a day after Cherub is ready to give up the morning nap.
- Cherub develops excellent whining skills, demanding juice and snacks.

*Three to five*
All the features of One to Three, plus:

- TV becomes even more appealing, because Cherub is able to understand the stories.
- TV becomes even more of a factor in choosing foods, because Cherub can also understand the commercials for snacks and sodas.

- Caretakers may not have a safe place outdoors for preschoolers to play.
- Caretakers encourage a nap after Cherub has outgrown the need for one.
- If Cherub has already gained extra weight, she may prefer inactivity to activity. She may also have "taught" her stomach to expect large meals and a constantly full feeling.
- Cherub needs more stimulation as she becomes more sophisticated; when bored, she may look to food for something to do.
- Girls and boys play separately. The boys play actively; the girls don't.

7. **What if there's a family history of overweight? Doesn't that mean that Cherub is automatically destined to be on the chubby side?**

   *"But everybody in our family is big. We're big people. Doesn't that mean that Cherub is, well, meant to be overweight? Designed that way? Listen, there's nothing wrong with being big."*

   It's true that genes play a role.

   If one parent is obese, the chance for the child being obese is about 40%. If both parents, the chance is 80%.

   But you'll notice that the chance is not 100%! Genetics play a role, but it's the combination of genes and environment that determines what happens.

   When there is a family history of overweight, it's important to be clear that overweight is neither inevitable nor desirable for Cherub. Sometimes adults who are overweight become so used to being overweight that they do not regard it as a problem. They may regard the social difficulties of overweight as being a problem of discrimination on the part of other people and institutions. They may feel that their own experience of life has not suffered because of their extra weight.

   However, it isn't fair for adults to make this judgment for children. Every child deserves a lifestyle that makes it likely that he or she will grow up free of the medical and other problems of overweight. (See *Problems of Overweight.*)

   *But hasn't Cherub just inherited a sluggish metabolism? And if that's the case, how can we do something about it?*

   Actually, metabolism in people (children and adults) who are overweight is a bit *faster* than the metabolism of those who are not. So it's not a simple problem of a sluggish metabolism.

   The inborn factors that make a person inclined to be overweight are complex, and still are being investigated. It might be that there is a problem with not feeling satiated, feeling as if one has never had enough to eat.

   However, there is good evidence that children can be trained to establish or reestablish accurate cues for when they are hungry and when they are full.

   That is, there is real hope that if you intervene early with Cherub, and keep a sharp eye out for future changes in her weight curve, you can help her attain an appropriate weight not just now, but all the way into adulthood.

8. ***You make it sound too easy. How can I change Cherub's diet and TV watching without making the whole family change?***

You probably can't. And I'm not saying it's easy; I'm just saying that it's important, and that it's part of being a responsible parent. I know our Greedy Culture is marshalled against you, and I know that many family members are held prisoner by that same culture. Set them free, too!

*"I know I shouldn't have junk in the house, but her Dad is skinny as a rail, and he loves that stuff."*

Ask Dad what he thinks should be done. He's a grown-up, and he loves Cherub, and he can find his own solution to the problem. One idea: he can just eat more of the healthy things in the house! Being skinny as a rail doesn't automatically mean you're heart-healthy.

*"I know I shouldn't have junk in the house, but her brother is skinny as a rail, and he loves that stuff. I'm afraid he'll resent her if we get rid of it. Already, he's teasing her about her weight. This'll just make it worse."*

My heart sinks when I hear this one. What, I ask myself, can this parent be thinking of? What kind of family spirit reigns in this household?

Let's imagine this. Cherub had cancer, and needed chemotherapy. Her oncologist says you can't have any pets in the house. Would you get rid of Cherub and keep Fido? Would you allow Brother to complain about getting rid of Fido? Or to tease Cherub about going bald? Being chubby is usually not as serious a condition as cancer, but it can become emotionally and physically devastating.

Mean teasing of a sibling is a serious violation of the Family Contract. (See the essay on *Oppositional Behavior*.) There needs to be a Zero Tolerance policy against all forms of teasing, including Rolling Eyeballs, Exasperated Sighs, Grimaces, and whispering nasty stuff to a visiting friend. Brother needs to be talked to in no uncertain terms and immediate, significant consequences carried out.

*"But I feel as if I'm depriving him. I don't want to have him have to sneak out of the house for chips and a Coke!"*

Brother is being deprived, right now, of learning to be a responsible family member in a harmonious family. He is being deprived of his right to decide what is important in life and what is not. He is being deprived of his right to develop scorn for the media and food giants trying to run his life.

If this does seem to be an unconquerable feeling on a parent's part, it is my strong suggestion that the parent seek counseling. I am perfectly serious. This attitude reveals a problem with parent/child boundaries that is likely to cause even more severe problems as the children attain adolescence.

9. But ***what about the skinny minnies?***

Most of the time in developed countries, a child who consistently has a weight for height in the "skinny" range is a healthy youngster with thin parents, an active metabolism, and an equally active lifestyle. By the skinny range, I mean a weight for height that is at or below the 25th percentile.

However, if the weight for height curve has consistently been higher than that, and then begins to fall, this is an indication to look for a reason. For instance, if a child was in the 25th percentile of weight for height up until age Three, and then at Three and a Half was only at the 10th, and then at the 5th percentiles, a work-up is indicated. This might be for problems absorbing food substances, for subtle kidney problems, for emotional or social difficulties, or for other entities depending on history and examination.

But what of the true Skinny Minnies, the ones whose weight for height percentiles are consistently, reliably, at the 10th percentile, or the 5th, or even below the chart? Oddly enough, there are not many studies on normal, healthy children who are very, very lean. However, if the child is active, healthy, happy, and developing normally, most pediatricians feel that this is normal for Cherub, and don't do much in the way of investigation.

## NAVIGATING THE GROWTH CHARTS

There are three kinds of USEFUL growth charts, but they are still a pain in the neck.

- The **Height for Age Chart** tells you how tall Cherub is, compared to other children the same sex and age. If he is in the 90th percentile, he's taller than 90% of them. If in the 50th percentile, he's average height. If in the 10th percentile, he's taller than 10%.
- The **Weight for Height Chart** tells you how much Cherub weighs compared to other children the same sex and **height**.

# GENERAL GUIDELINES FOR PREVENTING OVERWEIGHT

**TV Watching:**
- Don't put a TV set in Cherub's bedroom or playroom.
- Limit TV to an hour a day.
- Select a daycare/preschool that doesn't allow TV watching.
- Don't allow any snacking or eating while watching TV.

**Exercise:**
- At least one hour a day (adding it all up) of running, climbing, throwing, pedaling, and so on—out of breath and sweaty.
- Try to walk places rather than drive.
- Include Cherub in active chores.
- Make sure daycare/preschool encourages active play, involving children who, left to their own devices, prefer quiet activity.

**Snacks:**
- No "handouts": little baggies of chips, raisins, pretzels to gnash on during car rides, errands, and so on.
- A sit down, planned snack only when really needed, not as mid-morning of after-school *routine*. Good snack foods include fresh fruit and protein (such as lowfat dairy or lean meats). Avoid chips, crackers, fatty cheeses and lunch meats.

**Drinks:**

Age Two Years and Older: 16 ounces of 1% or Nonfat milk daily.

Juice: No beverage juices (apple, grape, cranberry, pear).

No sodas.

Water for thirst.

**Meals at Home:**

Serve appropriate portions: The size of the child's fist for bulky soft food; the size of Cherub's palm for meat, poultry, or bread.

Sit down together; talk about things other than food.

Prepare servings on plates in the kitchen—Don't put serving dishes on the table.

Children get second helpings only after adults have finished first helpings.

Strive for a healthy varied diet.

Rule: Nobody can complain about food. They can refuse it, but not complain about it.

Try to have at least one dish at each meal that you know Cherub likes. But if it is a rich dish, allow her to believe there's only enough for one portion (the size of her fist or palm) apiece. If she is still hungry, she can have the healthier foods to fill up.

Suppose Cherub is in the 90th percentile of weight for height. If you lined up 100 children as tall as Cherub, with the chubbiest at the top and the skinniest at the bottom, Cherub would be ninety up from the bottom and ten down from the top.

- The **BMI for Age Chart** tells you what Cherub's Body Mass Index is compared to other children of the same sex and age. It is like the Weight for Height Chart, and for children this young, the two can be considered interchangeable. The BMI for Age Chart does not include children under Age Two years.

  The BMI is a calculation of body fat which takes into account the role played by lean body mass (bones, muscles, organs) in the total weight. It roughly correlates with the percentage of fat in a person's body. If my BMI is 23, then 23% of Me is fat. More or less.

*There is one USELESS growth chart: the Weight for Age Chart. You may encounter this chart elsewhere. Be sure not to use it. I have no idea why it exists at all, except to confuse people.*

## PLOTTING NUMBERS ON THE CHART

You can just ask the pediatrician to do all this plotting for you. If so, see *Interpretation of Percentiles*, below.

But if you want to do it yourself, here's how:

- **Height for Age:** Find Cherub's Age on the scale along the bottom of the chart, and make a dot. Find Cherub's Height on the scale along the side of the chart. Make sure you do not confuse centimeters (CM) and inches (IN). Draw a perpendicular line from each point. Where

## START EARLY TO HELP CHUBBY CHERUBS

- If Cherub is 20% over ideal weight, it will take a year and a half of no weight gain whatsoever for her to "grow into" the extra weight. If 40% over ideal, it will take three years. That is the minimum amount of time needed. Since "no weight gain" is hard to achieve, you need to assume that it will take longer for her to "slim down."
- Up to Age Five, you have the most control over her environment, and the most say over her behavior. You decide what food comes into the house, what to offer her, when to offer it, and whether to give second helpings of rich foods. You decide how much TV she watches. Later on, you'll run into much more resistance.
- Once a child starts gaining weight, it's all too likely that the amount gained will increase every year, so that Cherub deviates more and more from her ideal weight. A modest problem can become a severe one.
- If a little girl is chubby by first or second grade, her chances of early puberty increase. It's no fun having breasts in third grade and menstruating in fourth grade. The hormones of puberty for girls ADD even more weight. It is a myth that a chubby girl will "slim down when she has her growth spurt."
- Children start teasing overweight children as early as Age Three, and by school age many chubby children are made to feel miserable.

they intersect, put a dot. The location of the dot will probably be *on* or *between* two percentile lines from 5% to 95%. Or it may be above the 95th percentile or below the 5th percentile.

- **Weight for Height:** Find Cherub's Stature (Height) on the scale along the bottom of the chart, and make a dot. Make sure you do not confuse centimeters (CM) and inches (IN). Find Cherub's Weight on the scale along the side of the chart. Make sure you do not confuse

pounds (LB) with kilograms (KG). Draw a perpendicular line from each point. Where they intersect, put a dot. The location of the dot will probably be *on* or *between* two percentile lines from 5% to 95%. Or it may be above the 95th percentile or below the 5th percentile.

- **BMI Chart for age:** First you need to calculate Cherub's BMI, or Body Mass Index. Let's say four year old Cherub is 40 inches tall, weighs 40 pounds, and is a girl. To calculate BMI:

1. First take Cherub's height in inches, and square it:

   40 × 40 = 1600

2. Now take her weight and divide it by the height squared:

   40 pounds divided by 1600 = .025

3. Now take that number and multiply by 705:

   .025 × 705 = 17.625

   That is her BMI: 17.625, rounded off to 17.6

   To plot her BMI, use the Girls' BMI chart:

1. Find her age on the scale along the bottom: Four years old.

2. Find her BMI on the scale along the side: 17.6—about halfway between 17 and 18.

3. Draw a perpendicular line from each point. Put a dot where they intersect.

   That dot will be close to one of the curving Percentile Lines labeled from 3rd to 97th percentile. In fact, it is fairly close to the 90th percentile curve. This means that Cherub's BMI is close to the 90th percentile for her age.

## Interpreting the Percentiles

If the Weight for Height or BMI percentile is at or below the 50th percentile, Cherub is not chubby.

If she used to be on the 50th or lower, but is now up to or close to the 75th on Weight for Age, or the 85th percentile on the BMI, pause for a moment. She's not overweight, but try to discover if something has changed in her daily life to add a little extra weight. Try to find it and fix it before you have a problem.

If she has *always* been on the 75th percentile on Weight for Age, or the 85th percentile of BMI, she is doing fine; but be alert. If she goes above these percentiles, find out why and fix it.

If she is below the 95th percentile, but above the "warning" percentile on either chart, Cherub needs help slowing down her rate of gain.

If she is at or above the 95th percentile on either chart, you will need help from your pediatrician. Many pediatricians check labs on a child this chubby, looking for a predisposition to type 2 diabetes and for elevated cholesterol and lipids. Cherub needs a lifestyle designed for her so that she comes as close as possible to gaining NO weight until she grows into some of the extra weight she already has.

If Cherub is way, *way* above the 95th percentile, she is at high risk for serious complications of overweight. She needs very special help—a full evaluation for the complications of extra weight, plus a restricted diet and supervised exercise to help her lose to a safe weight.

# CDC Growth Charts: United States

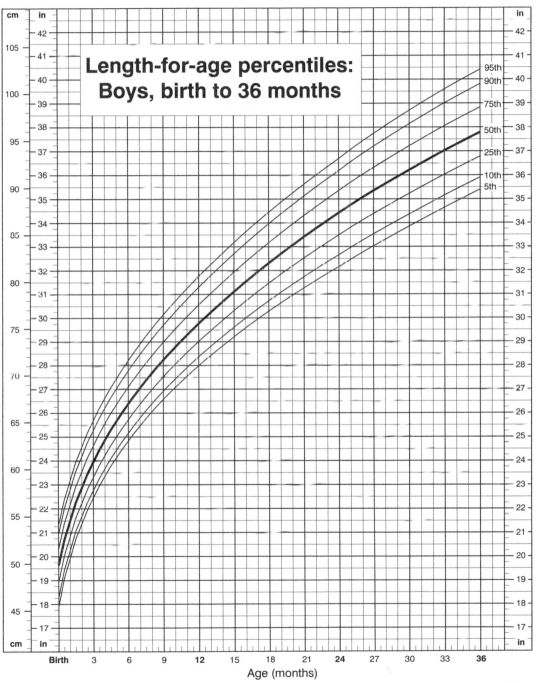

**Length-for-age percentiles: Boys, birth to 36 months**

Age (months)

SOURCE: Developed by the National Center for Health Statistics in collaboration with
the National Center for Chronic Disease Prevention and Health Promotion (2000).

# CDC Growth Charts: United States

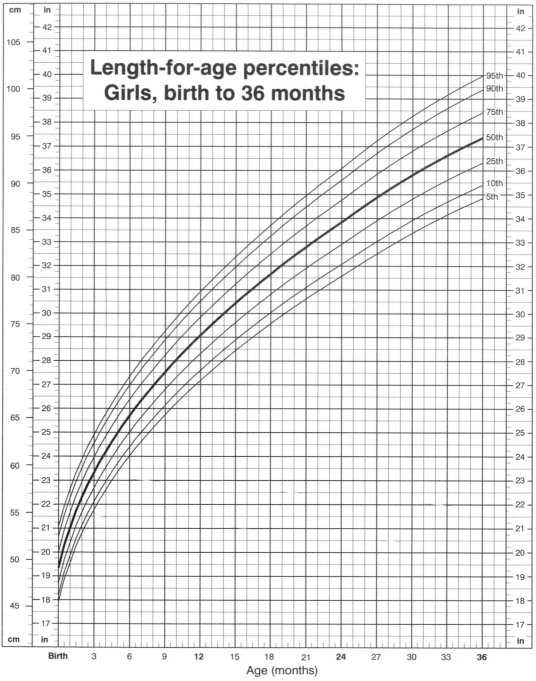

Length-for-age percentiles:
Girls, birth to 36 months

Age (months)

SOURCE: Developed by the National Center for Health Statistics in collaboration with
the National Center for Chronic Disease Prevention and Health Promotion (2000).

# CDC Growth Charts: United States

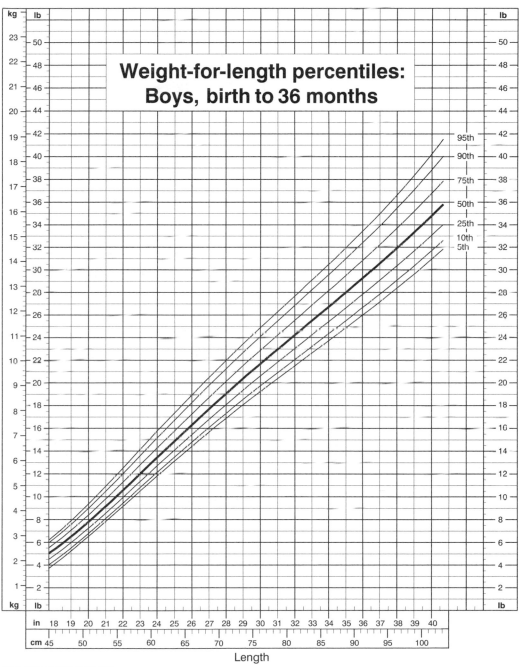

**Weight-for-length percentiles: Boys, birth to 36 months**

Length

Revised and corrected June 8, 2000.

SOURCE: Developed by the National Center for Health Statistics in collaboration with the National Center for Chronic Disease Prevention and Health Promotion (2000).

# CDC Growth Charts: United States

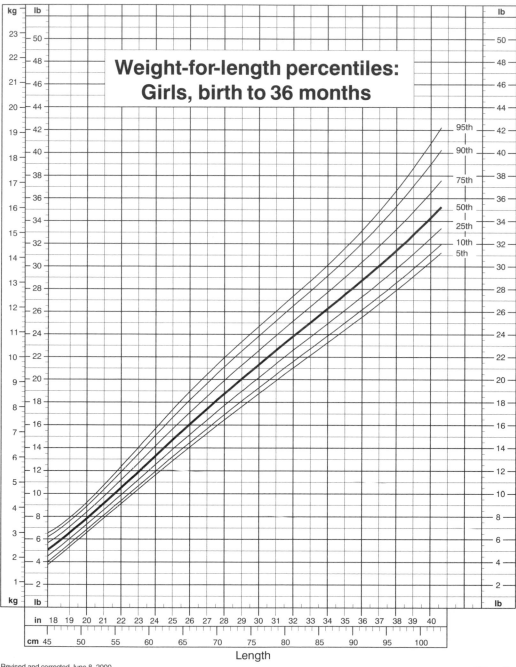

**Weight-for-length percentiles:
Girls, birth to 36 months**

Revised and corrected June 8, 2000.

SOURCE: Developed by the National Center for Health Statistics in collaboration with
the National Center for Chronic Disease Prevention and Health Promotion (2000).

# CDC Growth Charts: United States

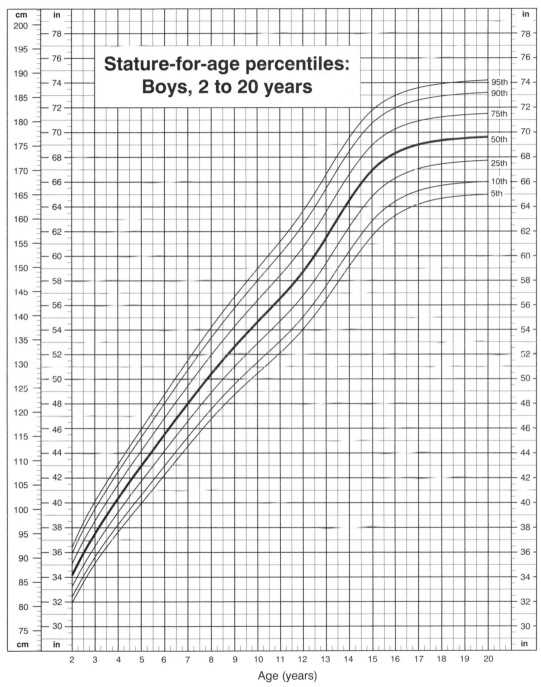

**Stature-for-age percentiles:
Boys, 2 to 20 years**

cm / in (left axis): 200/78, 195/76, 190/74, 185/72, 180/70, 175/68, 170/66, 165/64, 160/62, 155/60, 150/58, 145/56, 140/54, 135/52, 130/50, 125/48, 120/46, 115/44, 110/42, 105/40, 100/38, 95/36, 90/34, 85/32, 80/30, 75 / cm in

Percentile lines (right): 95th, 90th, 75th, 50th, 25th, 10th, 5th

Age (years): 2 3 4 5 6 7 8 9 10 11 12 13 14 15 16 17 18 19 20

SOURCE: Developed by the National Center for Health Statistics in collaboration with
the National Center for Chronic Disease Prevention and Health Promotion (2000).

# CDC Growth Charts: United States

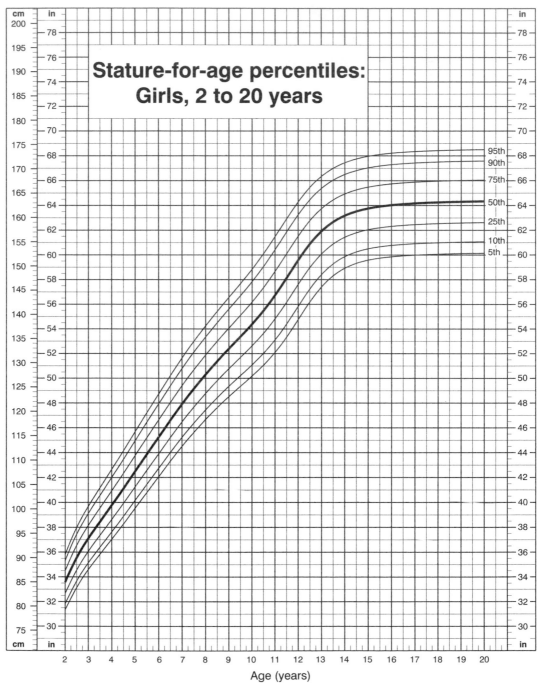

**Stature-for-age percentiles:
Girls, 2 to 20 years**

Age (years)

SOURCE: Developed by the National Center for Health Statistics in collaboration with
the National Center for Chronic Disease Prevention and Health Promotion (2000).

# CDC Growth Charts: United States

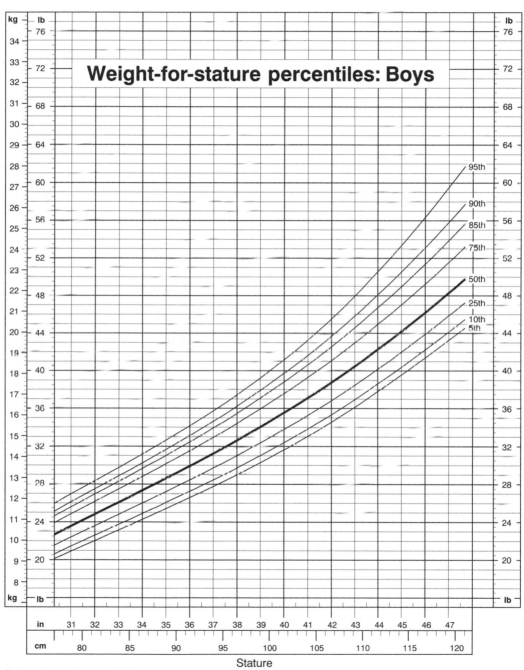

## Weight-for-stature percentiles: Boys

95th
90th
85th
75th
50th
25th
10th
5th

Stature

in: 31 32 33 34 35 36 37 38 39 40 41 42 43 44 45 46 47

cm: 80 85 90 95 100 105 110 115 120

Revised and corrected November 21, 2000.

SOURCE: Developed by the National Center for Health Statistics in collaboration with
the National Center for Chronic Disease Prevention and Health Promotion (2000).

CDC
CENTERS FOR DISEASE CONTROL
AND PREVENTION

# CDC Growth Charts: United States

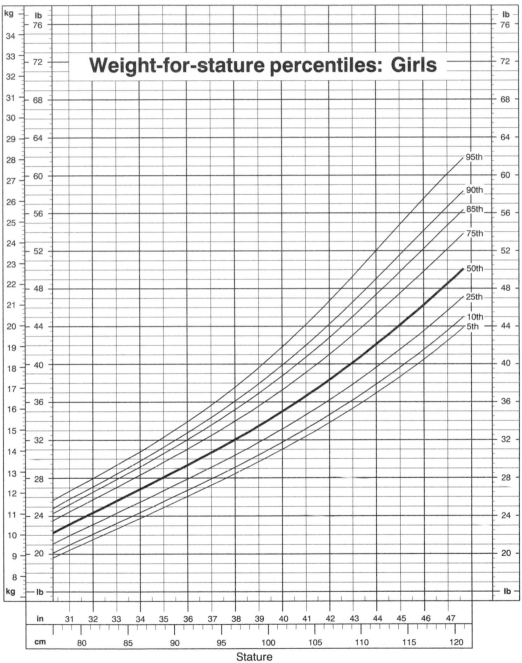

## Weight-for-stature percentiles: Girls

Stature

95th
90th
85th
75th
50th
25th
10th
5th

Revised and corrected November 21, 2000.

SOURCE: Developed by the National Center for Health Statistics in collaboration with
the National Center for Chronic Disease Prevention and Health Promotion (2000).

# CDC Growth Charts: United States

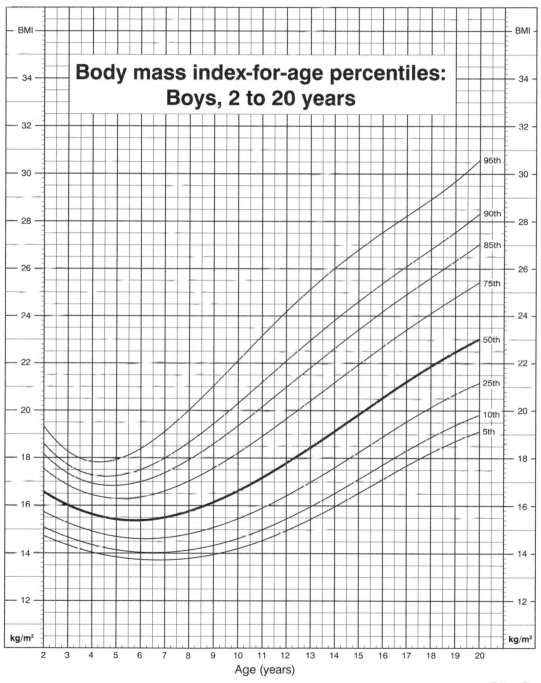

## Body mass index-for-age percentiles: Boys, 2 to 20 years

Age (years)

BMI — kg/m²

95th
90th
85th
75th
50th
25th
10th
5th

SOURCE: Developed by the National Center for Health Statistics in collaboration with
the National Center for Chronic Disease Prevention and Health Promotion (2000).

# CDC Growth Charts: United States

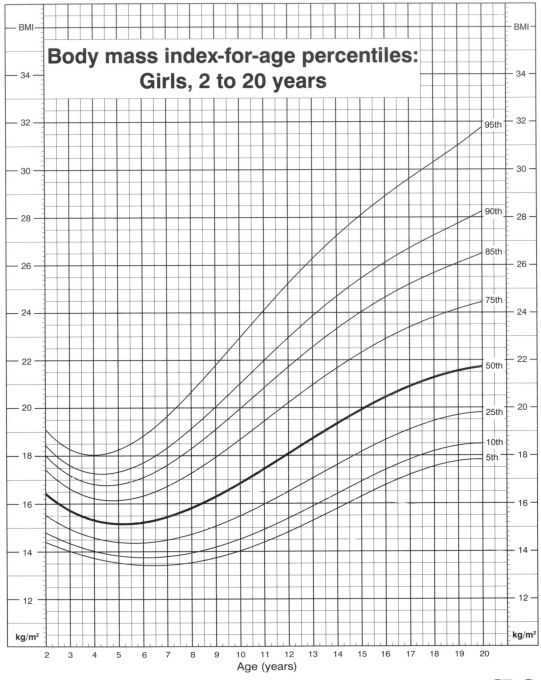

Body mass index-for-age percentiles:
Girls, 2 to 20 years

BMI

95th
90th
85th
75th
50th
25th
10th
5th

Age (years)

kg/m²

SOURCE: Developed by the National Center for Health Statistics in collaboration with
the National Center for Chronic Disease Prevention and Health Promotion (2000).

CENTERS FOR DISEASE CONTROL
AND PREVENTION

# NINETEEN

# Bacteria, Viruses, and Antibiotics

"Can't you just give an order for antibiotics over the phone?" "She took all her medicine; why does she still have the ear infection?" "Hasn't Jessie had too many antibiotics? I worry that all these medicines will damage her immune system." "Won't she get resistant to amoxicillin if we give it again?"

I hear these questions over and over again. But when I go to answer them, I am struck inarticulate. The parent and I are approaching infectious diseases from perspectives that are too different. The ideal solution would be to spend some time establishing a common ground. But somehow the office visit doesn't have the right intellectual ambience, what with Jessie muttering that she can't stand the white medicine and Jessie's brother throwing up in the corner.

So here are the things I'd like to say, if there were world enough and time during that sick visit.

1. ***Antibiotics do not "replace" the child's immune system and weaken it.*** They merely add ammunition to the child's own arsenal of defense. The reasons antibiotics have gotten this bad reputation are unclear, but I have some guesses. First, children get sick a lot; they're supposed to. Most of these illnesses are viruses, for which antibiotics don't work. If a child has had several illnesses that required antibiotics, and then catches a bunch of viral illnesses, parents may deduce that the antibiotics depleted the child's immune system. Not so. Second, if an antibiotic is given very early in the course of a bacterial illness, like strep throat, it is just barely possible (though studies conflict) that the antibiotic helps so effectively that the child's immune system doesn't have time to "memorize" that bacteria. In that case, a repeat infection may occur with the same bacteria. Third, there is some confusion about "resistant" bacteria. See below.

2. ***Antibiotics only work against bacterial illness, not against viral illness.***

*Bacteria* are very tiny one-celled animals. Since bacteria are self-contained little animals, they lead complicated lives. They have to construct a cell wall and machinery for producing toxins; they have to grow and reproduce. Bacteria thus engage in many different functions that can be interfered with, causing them to die or at least not to work well. Antibiotics are developed to interfere with these cellular functions, thus helping the child's immune system to work more effectively.

Bacteria are very clever. They can learn to outwit antibiotics. Some produce an enzyme that inactivates many antibiotics. Others evolve ways to hide, or even eliminate, the sites on their membranes that bind to antibiotics. Still others—well, they're thinking up clever new strategies every day of the week.

Not only that, but the bacteria that become drug resistant can pass that ability on to other bacteria—pass it on as a genetic ability, so that all future generations are drug-resistant, too.

That means that everybody needs to think through, and back off, using antibiotics when they're more dangerous than beneficial. Nobody should be prescribing, or asking for, an antibiotic for a viral illness. Nobody should be hastening to give or take an antibiotic if Nature alone is capable of healing a nondangerous bacterial infection: This means that a child with a normal immune system and an "ordinary" mid-dle ear infection might be able to go without an antibiotic.

However, we would not want a urinary tract infection, or a strep throat, or pneumonia to go untreated: the concern about complications and worsening is too great.

The silver lining of drug resistant bacteria is that the world of science is now looking harder at ways to support the immune system and at sophisticated substitutes for antibiotics—substitutes which may well be more effective and safer than anything we've had available in the past.

*Viruses* are just protein-coated bits of genetic material, tinier even than sperm—which also are protein-coated bits of genetic material. Viruses actually invade our body cells. They don't grow or produce toxins or cell walls. They take over our own genetic machinery in order to reproduce. By some definitions, they aren't even living things. They are so simple as to be impregnable: you can't interfere with viruses without also interfering with the human cells they attack.

Attacking bacteria is like attacking an invading army. You can bomb them, poison their water, and cut them off from supplies, metaphorically speaking. It's pretty straightforward.

Attacking viruses is more like confronting a terrorist movement that invades and corrupts the civilian population. All the armies and navies and air forces won't work here.

**Fortunately, most viral infections are mild. The body's own defenses get rid of the invaders quite well.** And there are a very few very clever medications that are effective against a few virus infections.

Most infections in childhood are viral. Sometimes you can tell by the clinical picture or specific lab tests which virus is causing the sickness: diarrhea from rotavirus, bronchiolitis from RSV, croup from parainfluenza, chicken pox from varicella, hand-foot-and-mouth from coxsackie, roseola from HHV-6. At other times, you have to be less specific: the cranky child with a low-grade fever and mild rash probably has enterovirus, but you aren't going to do blood work just to prove you're right. By the time you got the result, the child would be well and you wouldn't care any more.

3. *Each antibiotic has only a range of bacteria that it works on.*

Neither pediatricians nor parents like it when an antibiotic "doesn't work." Very depressing, especially when the antibiotic cost an arm and a leg, tasted like toad butter, and left a pink stain on the white couch.

When you do use an antibiotic, you need to be reasonably certain that the disease is caused by bacteria; you need to know which kind of bacteria cause this disease; and you need to know which antibiotics work on those bacteria.

Finding the bug isn't always easy. Some infections lend themselves to identification. If you can get a sample of the bacteria causing the sickness, you can perform a variety of tests to discover the name of the germ and the antibiotic ideally suited to get rid of it. Some tests are very rapid, and depend on "footprints" the bacteria leave. The classic test, a culture, takes longer. Here, a sample of the fluid or pus or whatever is planted on a plate containing the favorite food of the range of bacteria you regard as suspects. After the bacteria grow into a large crowd, samples of antibiotics are presented to them. Then you can see which antibiotic kills the most, or stops their growth best.

The problem, in many cases, is getting that sample. For instance, you can't get to the pus causing otitis media unless you insert a needle through the ear canal, piercing the ear drum. Now, this can be done; every pediatrician has performed the trick. But it's not something you want to do on every wiggly, friendly three-year-old with a sore ear. Nor can you gain access easily to the bacteria causing pneumonia or sinus infections.

So for many infections, pediatricians go by indirect assessments in prescribing antibiotics. How does the infection look? What bacteria usually cause an infection that looks like this?

4. *Giving the drug presents its own problems.*

When you don't know the exact antibiotic for a specific infection, you may have a selection of possible medications to choose from. And then you have to decide on a number of factors. The strongest antibiotic, the one that even the clever and wily bacteria aren't likely to outwit, may be bad tasting, expensive, and likely to produce side effects. The delicious one may not be effective.

So the pediatrician runs through a litany:

How sick is the child? How much does each appropriate medicine cost, and is the cost covered by the family's insurance? How good or bad does each medicine taste? How well is it absorbed with food; is it practical to ask the family to give four doses a day on an empty stomach? How frequently is it given, and how likely is it to be actually given that many times a day? What are the usual side effects, and how well would this particular child and family tolerate them?

Some antibiotics are very delicate and require a dark room or refrigeration. Some must be shaken thoroughly every single time you measure out a dose.

And some children are very clever, more clever even than bacteria. They spit it back. They hold it in their mouths and go to the bathroom five minutes later and regurgitate silently into the toilet. They hide the little blue tablets in Barbie's eelskin purse.

You also need to know whether there are any contraindications to a particular antibiotic. Is it in a family of medications to which the child might be aller-

gic? Does it have side effects, like permanent staining of the adult teeth or suppression of the bone marrow, that make it dangerous? Does it taste so bad you have never encountered one child who would swallow it? Is it much more expensive than medications that work equally well? Are there cautions that go with it, like avoiding sunshine or certain other medications, or drinking more water than usual?

Then you need to know how much of an antibiotic to use for a particular kind of infection. With any given bacteria, a mild skin infection usually responds to a lower dose than does a deep-seated infection like pneumonia.

And even if you do find the bug and give the right drug, the child's own anatomy and immune defenses have to engage in the battle as well. A child with allergies may have such swollen sinuses and nose and eustachian tubes that the infection can't drain. A child's huge tonsils or adenoids may be chronically infected.

5. **Once bacteria have figured out how to outwit an antibiotic, the medication loses a portion of its usefulness.**

An antibiotic may lose effectiveness in one particular child for one particular illness, as the clever bacteria chase out the sensitive ones. But the more serious implication is that a whole class of bacteria in a particular geographic area, or even nation or worldwide, will become resistant to an antibiotic that used to work very effectively. Pessimists forecast

that at some point the "age of antibiotics" will come to a close; there won't
be any left that the bacteria haven't gotten wise to.

6. *Every time you use an antibiotic on a*
   *child, you increase the chance of the*
   *child's becoming allergic to that*
   *antibiotic and to the family of antibi-*
   *otics to which it belongs.*

   The most common sign of an allergic
reaction to an antibiotic is a rash. However, some rashes, while caused by an
antibiotic, aren't really allergic and are
not an indication to stop that medication. Any rash from medication needs
to be evaluated from this point of view,
however.

   Moreover, there are nonallergic side
effects possible with any antibiotic.

   The most common nonallergic side
effect is diarrhea. This is most often
caused because the antibiotic kills the
normal intestinal bacteria that regulate
digestion, as well as the disease causing
bacteria. Usually modifying the diet
(stopping juices and dairy products)
and giving the medication with feedings
(if this is permitted) gets rid of the diarrhea; but in any event the pediatrician
should be notified.

   Rarely, an antibiotic may cause a side
effect outside both these categories,
affecting usually the liver, kidneys, or
blood-forming organs.

   Fortunately, stopping the antibiotic at
the first sign of a side effect, whether
allergic or not, usually brings the symptoms to a halt without danger to the

child. Very, very rarely does a child suffer severe damage.

   In rare cases, an antibiotic must be
used that is known to have a high rate
of serious side effects. Always, in this
case, the pediatrician informs the parents of the risks and benefits of such
antibiotic use before prescribing.

   Hence, the decision to use antibiotics
isn't always easy.

   Some bacterial infections MUST be
treated with antibiotics. Most of the
time, there is no disagreement here.
Meningitis, sepsis, pneumonia, urinary
tract infections, infections of bones and
joints—all these serious infections that
threaten life and limb and function
must be treated with antibiotics. These
are all infections that used to have close
to 100 percent mortality before there
were antibiotics.

   But what about the more common
respiratory and ear infections? Of course
people got over these on their own
before antibiotics existed. Otherwise the
human race would have died out long
ago. We treat these with antibiotics not
because they are immediately life-
threatening, but for three other reasons:

- To make the child more comfortable
  and the disease noncontagious faster,
  allowing her to get back to daycare or
  school and the parents back to work
  earlier.
- To decrease the possibility of a complication that could make this innocent infection immediately
  dangerous. For instance, ear infec-

tions have been known to seed the meninges and produce meningitis; tonsillitis can extend to become a peritonsillar abscess.

- To decrease the possibility of a delayed complication. For instance, we treat strep throat aggressively to reduce the chance, admittedly small but potentially devastating, that rheumatic fever could appear as a *sequela*.[1] Bacterial ear infections that go untreated have a higher chance of infecting the mastoid bone: this was a common and feared complication before antibiotics were developed.

*The Moral of the Story*

Antibiotics should be prescribed when a *bacterial* illness needs to be treated or prevented. Many bacterial illnesses (for example, many cases of sinusitis) resolve just fine without antibiotics. In some bacterial illnesses (E. coli, for example), antibiotics are actually dangerous. Antibiotics should not be prescribed for viral illnesses at all. When they are prescribed, antibiotics should be given exactly as directed.

---

1. Rheumatic fever, a late complication of strep that can cause heart damage, is almost unheard of in children under the age of Five. But very young children with strep can pass on the bacteria to older ones who are at risk, and so are treated anyway.

# Baby Shots and Grown-Up Worries

Explaining to a Four-year-old why he has to have shots before kindergarten is relatively easy: "This is medicine to keep you from getting some bad diseases. If you got one of these bad diseases, you would have to stay home from kindergarten."

If it turns out that staying home from kindergarten is just fine with Four, you can go on to say, "But these are very bad diseases. The president made a big rule that everybody who is Five must have these medicines. Even Batman (or Ariel) had these medicines when he (she) was Four."

Discussing immunizations with parents is another kettle of fish. There are so many immunizations now. The government is currently requiring parents to read a long, rather tedious booklet about every immunization at every inoculation. It's easy to become bored, frightened, confused, and irritated by the whole issue.

Yet the reasons for immunizing against disease remain constant.

Every **disease** for which immunization is urged has one or more of the following characteristics:

- **There is a high rate of serious complications or death.** You can't catch a mild disease of rabies: every case is fatal. Many cases of tetanus are also. The risk of a side effect from getting immunized is much lower than the risk of serious complications or death, should you catch the disease.
- **The maternal antibody to the disease either isn't passed to the baby at all, or wears off rapidly,** while the baby is still in a very high-risk age group. All the "baby shots" are in this category.
- **The risk of complications from the disease is low, but when complications do occur they are serious and a huge public health burden:**
  *Polio,* with paralysis and disability, respirators and misery; • *Measles* (rubeola) with ensuing loss of sight and hearing, and with seizures and mental

retardation; • *Mumps,* with encephalitis (infection of the brain) and *orchitis* (infection of the testicles, producing sterility); • *Hepatitis B* with fatal liver diseases in young adults, years after the initial, symptomless infection; • *Rubella*, devastating to unborn children, with loss of sight and hearing, and mental retardation; and • *Chicken pox* (varicella), dangerous to anyone without a competent mature immune system; rarely can maim or kill a perfectly normal child. • *Strep pneumococcus* can cause meningitis or sepsis (infection in the bloodstream), which can cause devastating complications or fatalities. • *H. flu,* cause of meningitis and epiglottitis.

**No immunization has a 100% rate of effectiveness for every individual. We depend on a high level of immunization in the community to keep us all safe.** A child can be immunized in good faith, but make less than excellent immunity. If that child is exposed to an unimmunized child who catches a disease, the child who has undergone the immunization is equally at risk. If this were your child, how would you feel about the parents of the unimmunized youngster?

Tragically, TV programs, books and articles, and word of mouth have given immunizations a bad reputation in some quarters. I suspect that some parents harbor a suspicion that immunizations are a hoax at best and a plot at worst. Ironically, this tendency increases as immunization becomes more

nearly universal. It's hard to get excited about diseases "nobody gets anymore." You don't see a show on "parents whose kids died of whooping cough." You're much more likely to see parents whose children suffered a devastating event after a DPT shot. And you are much more likely to be persuaded that the DPT caused the tragedy, even though the only evidence has to do with timing, not with any known cause-and-effect relationship.

Moreover, descriptions of the specific diseases in booklets handed to parents, even those issued by the government, aren't very vivid.

Then there are antiestablishment physicians. "The greatest threat of childhood diseases lies in the dangerous and ineffectual efforts made to prevent them through mass immunization." (Robert S. Mendelsohn, M.D., How to Raise a Healthy Child in Spite of your Doctor, Ballantine, 1984, p. 230.)

Dr. M. feels that some immunizations should be given only to a targeted population: for instance, rubella only to females, and mumps to males, just before puberty. He feels that others, like chicken pox, shouldn't be given at all.

But when does childhood end? Eight for some girls, sixteen for some boys. Should we go into the schools, starting with, say, fifth graders, looking for breast buds and pubic hair? Can you imagine the cruel comments, "Henry is fifteen, and he hasn't got his mumps shot yet!" And what if Julie turns up at eleven fully developed: Do we insist on a pregnancy test before we give her her rubella vaccine? I can feel another white hair popping out at the very thought.

As to chicken pox: yes, it's usually a mild childhood disease, except that it is highly dangerous to children with selected serious diseases. (And there are more such children now who live normal social lives and can easily be exposed.) These include everyone from youngsters on cortisone for asthma or kidney disease, on chemotherapy for leukemia or other malignancies, anybody with AIDS, newborns whose mothers catch the disease. Since chicken pox is most contagious just before it declares itself, and since most people don't go around announcing they are on cortisone or chemotherapy, exposure can be rampant and casual.

Chicken pox even in healthy children can be a dreadful experience. I remember a call: "Shannon spent the whole night sitting on the floor of our closet weeping, scratching her vagina with a paper towel. She wouldn't let us pick her up, so I moved the shoes and sat on the floor with her." Then there's Michael, sixteen, who missed taking his SATs and has permanent pox scars. And Jane, who spent three weeks at Children's Hospital with varicella encephalitis, but mercifully has sustained no permanent neurological damage.

Zovirax (acyclovir) can be life-saving in cases of chicken pox, and may be used ever more frequently. But it is only a matter of time, and exposure, before resistant strains of the virus emerge.

The DPT seems to be the most feared vaccine because of the bad press about the pertus-sis component. Parental fears are sometimes so intense that I want to say: *We aren't foisting this vaccine on children in order to increase our business and make money. We don't enjoy giving painful shots to healthy babies. If greed were the reason, we would make much more money not giving the shots and treating the ensuing diseases, all of which frequently require hospitalization and multiple, megabuck procedures.*

Studies have shown again and again that the DPT is not associated causally with crib death or brain damage even though these may be associated temporally: crib death does occur mostly in the first six months of life, when the first three DPT shots are given. And some cases may occur coincidentally soon after the DPT.

Media aside, however, I think the biggest hesitation parents have is one that is never really stated.

*If all the other kids are immunized, then my child really doesn't have much of a chance of catching these diseases. Look, smallpox has been eliminated; nobody gets vaccinated for smallpox anymore.[1] Now it's the same with these others, like polio and whooping cough. They're practically wiped out. So why should my particular child run even the tiniest risk of a side effect? Much less undergo the discomfort and expense?*

This is not merely unbecomingly selfish; it's a misconception. Even though you as a parent almost never encounter cases of diphtheria or polio, the bacteria and viruses that cause these and other diseases are still around. Why?

---

1. This was written prior to the bioterrorism concerns of September 11, 2001.

- Because a person can be immunized to some of these diseases, but still carry the bacteria or virus. Many cases of pertussis (whooping cough) occur in unimmunized babies who catch it from their immunized parents.
- Because germs like tetanus and diphtheria can be deposited in animal feces. You can catch the disease through a cut in your skin; even a tiny cut. (Rusty metal or not. Rust has nothing to do with it.)
- The world is a village. Polio is endemic in some parts of the world. You and your child can be at risk for many diseases even if you wash your hands consistently and never travel, and what a boring life that would be.
- Antibiotics and antiviral drugs aren't a true solution to all infectious diseases. The germs are too smart: they keep outwitting the medications. Better, if possible, to have your own home-grown immunized defenses than to rely on medication.

But there's another parental hesitation, more frequently voiced.

*Some of these shots are given just so that other people won't catch a disease. My little boy doesn't need a rubella shot: that shot is just to protect unborn babies. Why should he run the risk of a shot and why should I have to pay for it when he'll never get pregnant himself?*

That's easy: for every pregnant woman whose fetus is protected against rubella because little boys get rubella shots, there is a man protected against mumps of the testicles because little girls get mumps shots. Moreover, boys grow up to be daddies and girls to be wives. Finally, rubella and mumps both can cause serious and very inconveniencing disease in people of any age.

### The Moral of the Story

I'd be surprised if your pediatrician doesn't urge you to get your child immunized. On schedule. With a firm inner conviction that you are doing the right thing, a conviction that you will pass on to your child in your tone of voice, by the way you console her crying, in your manner towards the medical staff giving the immunizations.

And that, I promise, will make the shot hurt less.

{ T W E N T Y - O N E }

# Allergies, Asthma, and Eczema

## Allergy

*For allergic reactions to beestings and other stinging insects, see* First Aid *in Part II.*

*For a specific discussion of Asthma, see* Breathing Diseases *in the* Body Parts and Bodily *Functions part of Part II.*

Allergy makes whole families miserable much of the time.

When children have respiratory allergies, you're dealing with watery itchy eyes, stuffed or runny noses, skin rashes, wheezing, persistent cough, upset stomachs, and behavioral problems that may all stem from allergy. The more uncomfortable the child,

the less fun she is to be around. Playmates, caretakers, and parents naturally respond to her with less pleasure; and since emotions effect how bodies work (especially true with allergy) this may make matters worse.

It also can make pediatricians pretty miserable, for the following reasons:

1. ***It's tricky to figure out how much of a child's symptoms are due to allergy.*** Therefore, it's tricky to figure out how much testing to do. Even a child who tests allergic to, say, dust mites may have most of her symptoms due to recurrent colds. It's frustrating to get rid of all the dust

---

### HELPFUL SITES

The Food Allergy Network at 800–929–4040 or *www.foodallergy.org*

Asthma and Allergy Foundation of America: 800–7–ASTHMA (800–727–8462)

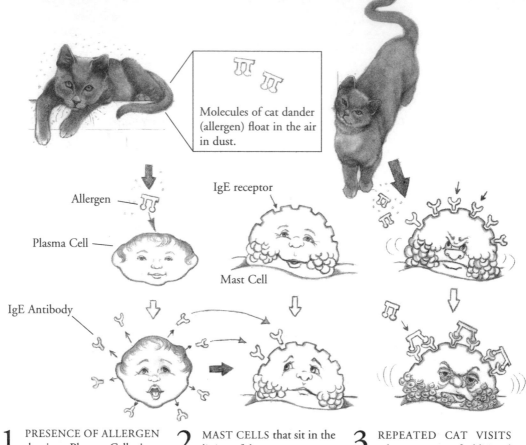

Molecules of cat dander (allergen) float in the air in dust.

Allergen

IgE receptor

Plasma Cell

Mast Cell

IgE Antibody

**1.** PRESENCE OF ALLERGEN deceives Plasma Cells into producing specific IgE antibodies against the specific allergen.

**2.** MAST CELLS that sit in the lining of the eyes, ears, nose, lungs, etc., acquire initial IgE response.

**3.** REPEATED CAT VISITS release an army of additional IgE antibodies, fully loading the Mast Cell. Pre-programmed IgE antibodies "glom" onto allergens.

**4.** BRIDGING TOGETHER, THE IgE ALLERGEN complexes cause Mast Cell membranes to break open, releasing chemicals onto the mucous membrane linings—chemicals causing allergy symptoms:

RUNNY NOSE, WHEEZING, ITCHING.

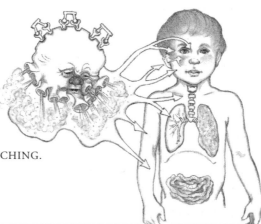

Mast Cells sit in the linings of EYES, EARS, NOSE, LUNGS, INTESTINE, SKIN.

Cat leaving too late!

mites and prescribe medications and not get rid of the symptoms. Also, blood and skin testing is often unreliable for "ordinary" food allergies. That's because many food allergies take a different route through the immune system, and don't leave a trail of increased allergic antibody (called IgE). The only really good test is to eliminate a suspected food from the diet, wait for symptoms to go away, and then challenge again with the food to see if it comes back. What a chore to impose upon a family! And then if it doesn't work . . .

2. **We have excellent therapies for certain kinds of allergy problems** such as itchy eyes and itchy runny noses; wheezing; and we have the life-saving rescue for anaphylaxis. But for others, treatment is often frustrating, as with eczema (also called *atopy*) or nonexistent (food allergies).

3. **We may have been giving the wrong advice about how to prevent allergies in a child with a strong family history.** We used to believe that if we could keep a child away from common allergens: furry pets; dust; peanuts and nuts; cow's milk, until her immune system matured at about age Three, we could help to prevent allergies. But recent studies seem to indicate the exact opposite: that exposing Cherub in the first year of life to cats, say, or to peanuts, helps to *reduce* the development of allergy. Clearly, our knowledge of the biochemistry of allergy is still emerging.

However, the stunning breakthroughs in some allergic conditions and hope for break-

throughs in others makes us smile again. Here is a rundown of some of these.

## DIAGNOSIS

It may be obvious that a child is allergic. Her eyes water, her nose runs every time Spring roles around. He wheezes around the cat, Grandpa, and when running.

But when the case is not so clear-cut, we need other hints.

- A *rhinoprobe* is not what you think it is but rather a smear from the inside of the nose. (Some people call it a *nasal cytogram*.) It's something like a Pap smear, but instead of looking for cancer cells it looks for allergic cells.
- A blood test can assess how much IgE is floating around in the blood, frequently a good indicator of how much "allergic potential" a child has.

These may help to confirm that Cherub is allergic to Something, but they won't tell you What. There are two basic ways of testing for specific allergies: skin tests and blood tests.

Both methods have two limitations. First, you can't tell whether a child is allergic to a particular allergen unless you test for it, and there are so many molds, pollens, danders, etc. that you may have several sessions of testing without hitting on the one thing that triggers the reaction. Second, it is possible to have all the symptoms of allergy but to have them mediated not through the IgE system described above, but by alternate pathways that cause the same symptoms.

Nevertheless, both skin tests and blood tests are very reliable when performed and interpreted by a skilled allergist or by a pediatrician with a specialty in allergy.

*Skin tests*

Skin tests involve injecting (either into or under the skin) tiny amounts of very dilute solutions of common Allergens like dust and dander. If the child is allergic to one of these allergens, she will get a hivelike reaction at the site it was injected. These tests are sensitive and accurate, but also somewhat painful. They must be done very carefully with pure and potent allergens. The staff must be skilled. Since, rarely, a severe allergic reaction can occur during or after testing, the office must be equipped to handle such emergencies. All these factors combine to make skin testing expensive.

*Blood tests*

Blood tests analyze how much allergic antibody (IgE) is present in the blood. Then the IgE in that blood sample is tested against a panel of possible triggers or allergens in a process called RAST (radioallergosorbent test). It's possible to do a predetermined panel of the Most Likely Suspects, such as dust, cats, dogs, June grass and so on, or to have a particular trigger tested, such as chinchilla.

These have to be performed by specially trained staff, take time, and are labor-intensive, so they too are expensive.

In both tests, you only get answers for the questions that you ask. If the child is allergic to feathers and you don't test for feathers, you won't get any specific information that will tell you that testing for feathers is in order. However, if the child is clearly allergic, and all the tests have come back negative, you may deduce that there is an allergen in the environment that you have not tested for.

Before your child has either kind of allergy testing, check with your insurance company. Some companies reimburse one kind but not the other, or will reimburse one or the other kind only under certain conditions. And sometimes they will only reimburse if the tests are done by an allergist, not a pediatrician, or vice versa.

So, you see, allergy is simple. It's insurance that's complicated.

TREATMENT

There have been two big breakthroughs in treatment of hay fever and asthma. The first is recognizing what a big role is played by inflammation (swelling of the lining and thickening of mucus). Inflammation can be reduced by high-dose steroids (cortisone), and can be prevented by several kinds of medication: low dose steroids, cromolyn sodium, which stabilizes the allergic cells so they don't let out their toxins, and medications that inhibit white blood cells from causing swelling and pus. The second is the development of products that allow even young children to benefit from these breakthroughs.

*Hay fever symptoms*

Children as young as Two can be given (usually prescription) antihistamines by mouth that have very few side effects, such as drowsiness or hyperness. These have such a long "lifespan" that most need to be given only once a day.

They can also be given through a nose spray that contains either a steroid (cortisone) or cromolyn sodium, to reduce and prevent swelling.

Nasal steroids need to be used judiciously. Some have been designed so that they are barely, if at all, absorbed into the blood stream. (And children are so designed that you're lucky if you can get the spray up that little nose in the first place.) Even so, it's best not to give extra steroids to a growing child, for fear of long-term side effects such as diminished growth or eye problems, such as cataracts. However, the studies on this have been very reassuring. Cromolyn sodium administered by nasal spray is very safe in both the long and short term.

To make the nose spray easier to take, let Cherub put a drop on her finger and sniff it and taste it. Have something good-tasting (lemonade or chocolate) for after the nose spray, since she'll interpret the smell as a bad taste.

For itchy watery eyes, prescription drops have been developed that reduce inflammation and also help to keep it from recurring. To give eye drops, put a drop on Cherub's finger and have her touch it to her own eye to prove they don't sting. Another suggestion: use one of the preschool picture books with holes in the pages—like Eric Carle's *The Hungry Caterpillar*—and have Cherub look at it while lying flat on his back. Then drop the drops through the holes into his eyes.

## Asthma or Reactive Airway Disorder

Reactive Airway disorder is the name given to wheezing attacks in young children who only wheeze when they catch a cold. Asthma is the name given when children wheeze on other occasions: from allergies, contact with dirty or dry air, exercise, cold air, and so on. They are terms that can be used interchangably. Reactive airway disorder doesn't sound serious, so many pediatricians prefer it when the problem is mild. (I should also say that wheezing can be caused by cystic fibrosis, foreign bodies stuck in the bronchi, and a bunch of other uncommon conditions.)

There are two things that happen with a wheezing attack: bronchospasm and inflammation.

- *Bronchospasm:* The little muscles that encircle the bronchial tubes (that carry air into the lung) clamp down, making the tubes narrow. This can happen suddenly and severely.
- *Inflammation:* At the same time, only more slowly, the linings of the tubes swell up and fill with thick mucous. This happens slowly and gradually, and left untreated it goes away *very* slowly and gradually.

*Medications for these two components of wheezing*

### ANTI-INFLAMMATORY MEDICATION

When a child has frequent wheezing episodes, you can assume that the tubes in her lungs are inflamed.

If a child is having a *severe* wheezing attack, you've got to reduce that inflammation fast, before the tubes are so swollen she can't breathe, or she becomes exhausted. The treatment for this is high doses of steroids, given by mouth, IV, or muscular injection. Soon, though, it may also be possible to treat severe attacks with a mist that Cherub can inhale.

If a child has frequent wheezing, even though it's not awfully severe, it's important to reduce the episodes. Partly, that's just to keep Cherub active and happy. Partly, it's to prevent real trouble. When lungs are constantly inflamed, she's at risk for a severe attack any time she has bronchospasm. Moreover, lungs that stay inflamed over years develop scarring that can cause problems as she gets older.

The treatment for chronic asthma is daily anti-inflammatory medication. That usually is either cromolyn sodium (Intal) or an inhaled steroid (lots of brand names, such as Asthmacort, Flovent, Pulmicort, Vanceril).

There is a third kind of medication that discourages white blood cells, thus reducing swelling and pus. As of now, these medications are not given to children under age Six, but this will probably change. They are given by mouth.

If your child is on an anti-inflammatory medication, you need to give it as directed every single day to help prevent attacks and to reduce scarring in the lungs.

### BRONCHODILATORS

This usually means albuterol, the most commonly used bronchodilator for children.

As I said above, when the little muscles that encircle the bronchial tubes go into spasm the tubes are suddenly narrowed and wheezing occurs. It comes on fast. Fortunately, many of the medications that relax the muscles—the bronchodilators—work fast, too. That is why they are called "rescue" medications.

The term "rescue medication" also emphasizes the fact that a bronchodilator is *not* intended to be used every day, all by itself. If that is happening, the bronchodilator may mask the fact that the real underlying problem is inflammation.

Albuterol, the most frequently used bronchodilator, comes as a liquid to be taken orally. It can also be given in a prescribed machine called a nebulizer, which turns a solution of albuterol into mist. The mist travels through tubing and out a mask over the child's mouth. A third way to give albuterol is with a hand-held inhaler. Inhalers are intended for adults: you have to shake it, coordinate your breathing with pushing a button, and hold your breath. A few Fours can learn to do this, but not many; and nobody younger than Four.

So when a child uses an inhaler, it is attached to a spacer a tube or other contraption that holds the albuterol solution in mist form. You put the inhaler on one end of the spacer and a mask on the other; squirt the inhaler into the spacer, put the mask over Cherub's face and keep it there as Cherub breathes in the mist. Sounds easy, but you actually need two adults, special skills in restraining, or a really smart octopus, the first few times at least.

**If your child needs to take albuterol, all by itself, more than three times a week, talk to your pediatrician about it. There may well be a need for daily anti-inflammatory medications.**

*The Moral of the Story*
Children with mild asthma or RAD do not need to use albuterol more than three times a week to live a normal life. If the albuterol is being used more often, Cherub needs to be evaluated for anti-inflammatory medications, given daily.

Today, nearly every child with reactive airway disorder or asthma ought to be close to living a normal life. That means Cherub ought to be able to participate normally in physical activities, sleep and grow well, and ought not to require emergency room visits. This may mean careful attention to Cherub's environment, daily medication, and occasional "rescue" medications.

## Eczema and Atopic Dermatitis

Whether you call it eczema or atopic dermatitis, this chronic rash is no fun at all: Cherub's smooth skin is covered with scaley bumps and open sores, and he tears at the itchy lesions with his nails, day and night.

It's not fun, but it certainly is common: with 5% to 7% of children affected and with 95% of them having their first attack under the age of five. The cause of eczema is still being elucidated. It is clear that the immune system is involved.

Eczema can be as mild as just dry skin, with a little redness in the creases of the knees and elbows; or it can be a full-blown, total body inflammation, with thickened red skin from head to toe.

The mainstays of treating eczema are moisturizing the skin, using topical steroids on the worst areas, and prescribing antihistamines to control itching. The more a child itches, the more he scratches; the more he scratches, the more he inflames his skin; the more inflamed the skin, the more he itches; and round and round we go.

Keeping the skin moisturized is most effective when the lotion or cream is applied right away after a bath—within three minutes of getting out of the tub—while the skin is still moist. Many parents have found the nonprescription nonmedicated moisturizing "barrier" cream called Triceram to be very effective, though not inexpensive. For information, go to *www.osmotics.com*. It may be available through your local pharmacy.

A child with very severe eczema may need high doses of steroids by mouth, powerful topical anti-inflammatory medications, and even chemicals used for chemotherapy. The prescription topical medication Tacrolinus, which works by affecting the immune system, is often remarkably successful.

When eczema is mild, it is usually not due to a food allergy. But when it is severe, requir-

> The moral of the story: If your child has mild eczema, it's unlikely it's due to foods. If the eczema is severe, discuss food allergy testing with your pediatrician or dermatologist.

ing daily use of topical steroids, food allergies may well be one of the causes. Such a child has Atopy, and is at risk for other signs of allergy, such as asthma and hay fever symptoms.

Fortunately, the RAST blood test described on page 574 can be very helpful in identifying which foods might be the culprits, and these are worth doing on any infant or young child with severe eczema. When you know which foods might be The Enemy, you can exclude them and see if symptoms do improve.

# Trouble in the Middle Ear

*"Of all sad words of tongue or pen, The saddest are these: 'Her ear, again!' "*
*(with apologies to John Greenleaf Whittier)*

## Basics

Sad words. Frequent words. Some days after office hours, if I close my eyes, all I see are little ears.

Indeed. In one large study, 71 percent of the children had at least one ear infection before they were three. In any given year, half of children under age five will have at least one ear infection. Many will have two or three.

Why so frequent a problem? "When I was a kid, I don't remember kids all the time having ear infections the way they do now," I hear parents say.

Perhaps it's because we are better able to diagnose ear infections now thanks to better and brighter instruments (if not doctors). Perhaps it's because so many children these days are in daycare or exposed to second-hand smoke, both of which increase the frequency of ear infections.

Or maybe it's heredity. Children whose parents had trouble with their ears as young-

sters have a greater risk of similar trouble. Before antibiotics, children with severe or persistent otitis were frighteningly liable to have serious complications. Perhaps those infection-prone children who, in the past, might have died of serious complications, have now survived to become parents, passing on the likelihood of ear problems to their own children.

Before antibiotics, acute ear infections used to strike fear into the hearts of parents. Infection could pass from the ear to cause meningitis (infection of the lining of the brain and spinal cord) or mastoiditis (an infection of the mastoid bone that surrounds the ear). Both were often fatal complications. You rarely see these as complications of treated ear infections, but once in a blue moon they still can appear.

Today, the woe caused by ear infections is more about lifestyle: missed sleep, missed daycare and school, missed parental work, the time and energy and expense of doctor visits and medications. Life is short: it isn't fun to have a child up at night in pain or crabby or to have to give up swimming lessons or to pay a penalty on airline tickets that can't be used. Giving a medicine three times a day for ten days can seem like forever, especially if you have to tackle Hughie and hold his nose for each dose, or if Brittany will take it only when you smuggle it into a Ben & Jerry's Cherry Garcia.

But the real worry, these days, is about hearing loss. Permanent damage to hearing due to ear infections is unusual, but can happen. It's more often the temporary, but sometimes prolonged, minor hearing loss that gives concern.

More than sixty studies looked at the effects of recurrent or chronic ear infections and middle ear fluid in young children. More than a few suggest, some strongly, that language and speech and even behavior can be significantly and permanently impaired when a child under about two years of age goes for a time with impaired hearing, even when the impairment is fairly subtle.

It's not just the toddlers who have trouble because of subtle hearing problems. Preschoolers and elementary school children usually *don't know* they can't hear. They think that their impaired hearing is quite normal, and that their problems at school or home or with friends stem from their own character flaws, or from bad luck, or are merely the human condition.

One of my colleagues treated a child with chronic fluid in the middle ear. The little girl came home from preschool next session just glowing: "Mommy," she said delightedly, "the children have started talking to me!"

Children who can't hear well, for whatever reason, act distractable, fidgety, crabby, disobedient, socially gauche. But they often don't get sympathy and a visit to the pediatrician: instead, they get scoldings and time out. Almost any undesirable behavior in a child, even bed wetting, can be a complication of diminished hearing.

(Of course, fluid in the middle ear isn't the only cause of temporarily diminished hearing. Sometimes there is incredible wax buildup, often because fastidious parents try to clean the child's ear with cotton swabs. Instead of removing the wax, this presses it up tight against the eardrums. *Permanent* hearing impairment can be caused by damage to the hearing nerve, often from infection in the fetus or as a complication of meningitis or problems of prematurity; or from congenital or acquired deformities of structures of the middle ear.)

## Know Thine Enemy

The eardrum is one wall of a tiny chamber called the middle ear. In this chamber lives that chain of little bones you remember from Trivial Pursuit: the malleus, incus, and stapes. This chain of bones is connected to the eardrum. When the eardrum vibrates to a sound, the vibrations pass through these little bones to the hearing nerve. (The hearing nerve transmits the vibrations in utterly mysterious fashion to the brain, which interprets them as specific sounds.)

For us to hear, then, the bones have to vibrate. That means that they have to be surrounded by air, not by fluid or pus. So, the middle ear is ventilated to the outside world through the eustachian tube, which connects to the back of the throat.

If the eustachian tube were open all the time, mucous and fluid could travel back, infecting the middle ear. So Nature arranges for it to open only during swallowing, pulsing air through. If your anatomy is normal, the tube opens in such a way that the food or fluid you swallow can't be pulsed through along with the air: a kind of valve action shuts it off. Very ingenious.

When a child gets an ear infection, something has gone wrong with this arrangement. Here are two of the more popular theories.

- Maybe the eustachian tube gets obstructed or collapses so that air can't get into the middle ear. Then the air that *is* in the middle ear gets absorbed into surrounding tissues, and a vacuum is created. The vacuum sucks in fluid from the surrounding tissues. If the fluid contains viruses and/or bacteria, infection gets started.
- Maybe something goes wrong with the ventilation mechanism, or it becomes overwhelmed, so that mucous or water or food or liquid, not air, gets sucked back through the eustachian tube into the middle ear.

# Assessment

## SUSPECTING AN EAR INFECTION

You'd think it would be easy: Pain, fever, runny nose. Not so.

- Very young babies with an ear infection just act sick. They may not have any signs of a cold.
- Older babies may give very indirect signs that their ears bother them. They may not sleep well, may cry when they try to suck or burp, may rub their ears against their bedding or a parental shoulder.
- Toddlers and preschoolers may ignore ear discomfort unless it is real pain. If the fluid in the ear builds up slowly, the eardrum has time to stretch and doesn't cause agony but only pressure and muffling. Many young children couldn't care less.
- Even if the pain comes on suddenly, as the eardrum gets used to being stretched, the pain dies down into pressure and muffling. A young child may complain only briefly, during the time the eardrum is being stretched rapidly.
- Fever, tugging at the ear, signs of a cold are useful indications of an ear infection, but none of these may be present.
- So sometimes the only sign in toddlers and preschoolers is a change in behavior: fussy, crabby, not babbling or talking as much. Sometimes, the change is so subtle you only notice it after the ear infec-

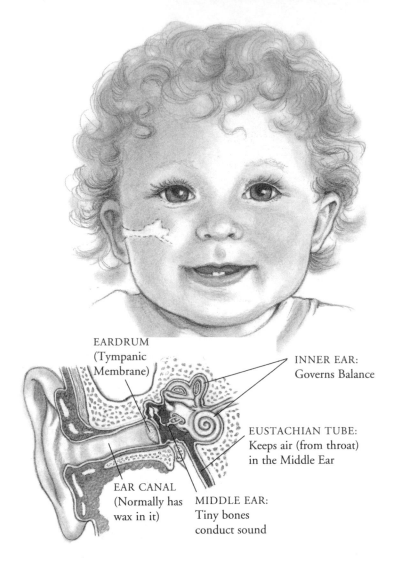

EARDRUM
(Tympanic
Membrane)

INNER EAR:
Governs Balance

EUSTACHIAN TUBE:
Keeps air (from throat)
in the Middle Ear

EAR CANAL
(Normally has
wax in it)

MIDDLE EAR:
Tiny bones
conduct sound

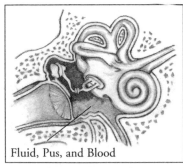

Fluid, Pus, and Blood

Acute Otitis Media (Moderate)

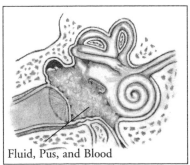

Fluid, Pus, and Blood

Acute Otitis Media (Severe)

tion has been diagnosed and treated, and the child suddenly becomes a ray of sunshine and a chatterbox.

## DIAGNOSIS

There are two common kinds of problems in the middle ear.

### Acute (suppurative) otitis media

The mucous membrane that lines the whole area of the middle ear and into the mastoid bone becomes infected, producing pus maybe mixed with blood. The eardrum stretches as the pus builds up. If it builds up rapidly it tends to cause fierce pain; if slowly, pressure and muffling.

### Serous otitis media

Clear serous fluid rather than pus fills the middle ear chamber, often keeping the eardrum from vibrating normally. Sometimes a child will complain of pressure or ears popping, and often hearing can be impaired temporarily. Sometimes the impairment is very subtle. Serous otitis usually disappears over a few weeks unless infection occurs.

### Looking with the otoscope

When you look into the ear with the otoscope, you see a straight-on view of the eardrum. If it is normal, it is pearly gray and so translucent you can see right through it to the first of the little bones, the malleus, where it's attached to the drum. You also see a nice clear glint of light where the drum reflects the light of the otoscope.

If the child is crying, the drum might be red just the way her face is, so redness by itself doesn't mean an ear infection. If the drum is infected, it can be any of a variety of colors. It will be thick, not move to a change in pressure, and the reflection of light on it won't be clear and sharp.

Again, sounds easy. But in order to diagnose an ear infection, you need both a good view and a great deal of experience looking at the range of normal and abnormal ears. When parents ask me if they should buy their own otoscope to look at their children's ears, I caution them that it is easy to be deceived. Very often what looks like the eardrum really isn't; it's a portion of the ear canal or even a clump of wax. Moreover, if the light on the instrument is dull, the drum will look red and thick even when it's normal.

Most worrisome, however, is that a drum may look normal and still have fluid behind it or abnormal pressure, which can diminish hearing. Many pediatricians test the pressure in the middle ear after an infection, either by blowing a little whiff of air through the otoscope itself (with a special attachment) or by using tympanometry.

*Tympanograms* are often called the "drive-the-car machine" since a popular version shows a graphic of a car driving into a garage, holding the child's fascinated attention during the procedure. This is a machine that measures how well the eardrum (tympanic membrane) vibrates. It can show if there is fluid in the ear and/or if the pressure is abnormal. You can have a normal tympanogram early in an ear infection, however.

# Treatment

## TREATMENT OF AN ACUTE INFECTION

Most acute ear infections are caused by viruses; a small number, by bacteria. It would be lovely if there were an easy way to tell them apart, so that the children with viruses didn't have to get antibiotics.

But there is no easy way. To tell in any individual child, you would have to get fluid from the middle ear and examine it. The only way to do this is to plunge a needle through the eardrum and get a sample. (Indeed, this had to be done on a number of children in studies, in order to establish that most infections are, in fact, viral.)

So many pediatricians treat all, or nearly all, acute ear infections with antibiotics, relying on how the drum looks, the child's history, and the bacteria prevalent in the community to choose the right antibiotic.

There are two good reasons for treating with an antibiotic:

- Studies indicate that treatment usually shortens the pain and other symptoms of acute ear infections.
- And they also strongly suggest that antibiotics have significantly decreased serious and fatal complications of otitis, such as mastoiditis and brain abscesses.

However, studies don't show that antibiotics prevent serious otitis media: that is, fluid in the middle ear.

## TREATMENT OF CHRONIC PROBLEMS

Three kinds of ongoing middle-ear problems can drive families and physicians bonkers. These are:

- Frequent episodes of acute infections (three in six months or four in a year).
- Fluid in the middle ear that lasts for a long time.
- Fluid in the middle ear PLUS frequent episodes of acute infections.

Most pediatricians agree that the handling of these problems needs to be individualized.

Most pediatricians take into consideration the child's age, general health, family situation, and other factors in deciding how many ear infections are too many. But the most important factor is whether the child's speech and language are coming along normally, or whether the temporary frequent hearing losses are causing some delay. This can be hard to judge.

Most pediatricians are especially concerned about frequent episodes of acute otitis that repeatedly perforate the eardrum, or that don't clear completely between episodes. They worry most about persistent fluid when they know that hearing loss is interfering with speech and language development or with social and educational achievements.

Very tricky, both of these judgments.

For instance, Teddy and Ruthie both have experienced six episodes of acute otitis in the last eleven months.

Is it normal for Teddy, age two, to speak in single words rather than phrases? Perhaps yes, because boys often talk later than girls, he has a three-year-old sister who "translates" for him continually, and Teddy is a very physical young man.

Is it normal for Ruthie, age eighteen months, to have a vocabulary of fifty words? It sounds as if she's right on course, but maybe not. Ruthie's five-year-old sister was conversing in sentences at this age, and the age gap is such that Ruthie is being raised more like an only child than a second sibling.

The younger the child, the more important it is to know that hearing and language development are normal in the face of frequent infections or persistent ear fluid.

Well, suppose you and your pediatrician decide that something must be done?

## OPTIONS FOR STUBBORN CASES

### Preventive antibiotics

Sometimes giving a small dose of an antibiotic at bedtime will prevent infections from occurring so frequently. One theory has it that when children lie down for the night, fluid gets into the middle ear more readily. The antibiotic may keep infection from starting during the night hours, and in the daytime normal swallowing clears the middle ear.

### Tubes in the ears (ventilating tubes; myringostomy tubes; PETs)

These are tiny plastic or Teflon or metal tubes about half the size of your pinkie fingernail. With the child under anesthesia, the pediatric surgeon looks in through the ear canal with the aid of an operating microscope, and inserts a tube right through the eardrum. One end opens in the middle ear, the other in the ear canal. The tube ensures that air can get into the middle ear even if the eustachian tube isn't functioning well.

Ventilating tubes are so small you can't see them unless you look right in at them with the otoscope. They usually stay in place through the eardrum for six months to a year. Then they migrate back out the ear canal and come out. (In California, there is a mythic figure called the Tube Fairy who gives children presents when their tubes come out.) Usually, the drum heals over the hole without scarring.

Tubes often work very well, keeping the ears clear and hearing normal while they're in place. But they aren't perfect.

- They don't prevent the colds, sore throats, sinus infections that lead to ear infections; they just protect the eardrum. In fact, sometimes a sinus infection will cause pus to enter the middle ear and then drain out the tube into the ear canal. This can be disconcerting, but is a sign that the tube is working as it should.
- You usually have to be careful that water doesn't get into the ear canal, though some tubes are so tiny that water can't get through them into the middle ear.
- Tubes can get plugged with wax or pus. When they are plugged, they don't work at all.
- Rarely, there can be complications. The rarest and most dangerous is severe dam-

age from anesthesia or from bleeding during surgery.

- Less rarely, a tube may leave behind a hole in the eardrum that doesn't heal and has to be patched.

### Adenoids out

Adenoids are tonsil-like lumps of tissue that can squash the eustachian tube if they get too big. They also seem to be able to send germs into the middle ear if they are chronically infected. Either way, taking them out sometimes helps stop frequent ear infections. You can't see adenoids directly. You might suspect they are enlarged if your child snores loudly or breathes through his mouth all the time.

Taking out the tonsils doesn't seem to help ear infections.

### Other methods

Sometimes, pediatricians will try courses of steroids (cortisone) to get rid of stubborn ear fluid, or preventive antibiotics to keep infections from starting in the middle ear, or immunizations against common ear bacteria like pneumococcus. Sometimes, it makes sense to see if food allergies, especially to dairy products, are playing a role in frequent ear infections.

## Prevention

When you look for ways of preventing ear infections, you look to conditions that cause the eustachian tube to collapse or become blocked, or activities that send mucous, food, or fluid instead of air into the tube.

Sounds easy, doesn't it? Alas, the list of culprits includes:

*Anything that causes the eustachian tube to swell up* can block it. Colds, dirty air, allergies are all suspects. So ideally you would avoid:

- Daycare situations that predispose infants and toddlers to frequent colds. Small daycare groups with same-aged children can help, especially if all the children are regulars and not drop-ins.
- Secondhand smoke.
- Exposure to foods, animals, dust, molds, and pollens that might cause allergies.

*Activities that involve changes in the pressure of the eustachian tube* can collapse it, thus blocking it.

- Flying, especially in unpressurized planes, and especially if you don't take such precautions as making sure the child doesn't have a cold and is sucking and swallowing during ascent and descent. By the way, flying is only a hazard if there is air in the middle ear because pressure change affects only air, not fluid. The child with completely fluid-filled ears won't get worse on a flight.
- Swimming under water and diving, both regular and scuba. (I know most of the under-five set doesn't scuba, but, hey—I

live in southern California and anything's possible.)

*Enlarged adenoids can sit on the eustachian tube and obstruct it.* Signs of enlarged adenoids are: mouth-breathing, snoring, and an adenoidal voice, which basically sounds like somebody with a lot of peanut butter stuck to the roof of the mouth.

*Any enterprise that involves getting fluid back through the eustachian tube into the middle ear* can result in an infection. Here's a partial list:

- Water can go up the nose while the child is in the bathtub or swimming. It's not water going *in the ears* that causes an infection, unless there's a hole in the eardrum.
- Liquid the baby is drinking can do the same if swallowing isn't

well-coordinated with the opening of the eustachian tube. Babies who drink a bottle lying on their backs may be particularly prone to this.
- Toddlers with a runny nose who sniff or blow too hard can spurt nose-run back into the middle ear. Avoiding whatever makes the nose run in the first place (allergies, smoke and smog, colds, wind and rain, eating ice cream) might help. Teaching the child to blow gently might help, too (and there's a Nobel Prize waiting for someone who figures out how).

Many experts feel that the main hope for preventing ear infections is through developing immunizations against the most common bacterial causes. Even further into the future it might be possible to develop immunizations against the viruses that start off the ear infections.

*The Moral of the Story*
Pediatricians who believe that speech, language, and behavior can be impaired by early and frequent ear infections, or by fluid in the middle ear, tend to be very aggressive in treating them and in urging prevention.

They also monitor hearing and language skills very closely. If your child has ear infections to an extent that worries you, have an honest discussion with your pediatrician to determine whether, and why or why not, he or she is concerned.

# {TWENTY-THREE}

# The Toddler or Preschooler Who Is "Sick All the Time"

## Basics

Oh, no! Here comes your child, and there's that runny green nose again. Or, after a perfectly lovely day, out of the blue, a fever of 103°. Or those bumps you hoped were flea bites definitely, no doubt about it, look like chicken pox.

"My child is sick all the time," you wail to your pediatrician, and brace for the response: *"Don't worry. It's perfectly normal."* There is something about that phrase, uttered in an offhand and slightly impatient manner as the doctor heads out the door, that rings all kinds of alarm bells in many parents.

If this scenario sounds familiar, the best possible thing you can do is make a special relaxed visit to discuss your perceptions and worries. This should be at a time when the child is not acutely ill. When you make the appointment, make sure the people at the front desk understand you want a consultation about frequent illnesses, not a visit because your child has an acute problem. If

they say they do not know how to book such a visit, ask them to ask the pediatrician.

Gird yourself to be honest with the pediatrician. If you are scared that your child has leukemia, or cystic fibrosis, or AIDS, or some rare entity you read about in *Reader's Digest* that you just can't get out of your head, tell your pediatrician. If reassurance doesn't do the trick and you won't be able to sleep at night until your child has had the specific test, tell your pediatrician that, too. Don't be shy about it, or sneak up on it; just say, "I won't be able to sleep at night until you do the specific test for (whatever)."

## Assessment

The key to every assessment of a child with frequent illnesses is a careful history and a full physical examination, with careful attention to growth, development, and the size of tonsils and lymph nodes.

Depending on what the pattern of illnesses and the examination show, your pediatrician may order absolutely no tests. Or you may be asked to have your child undergo any combination of several tests. This doesn't mean that suspicion is high that a serious problem exists. Most often these tests result in perfectly normal values, revealing at worst allergy, enlarged adenoids, iron deficiency, or a minor and transient deficiency in immunity.

Commonly ordered tests include a complete blood count, X rays of chest, sinuses, or adenoids, and sometimes specific blood tests for immune function. Often you'll be asked to take your child for a sweat test. You wouldn't think sweat would be dignified enough to yield much medical information, but this is the definitive test for *cystic fibrosis*. C.F. is an uncommon but not rare inherited condition that gives rise to lung problems, abnormal stools, and poor growth.

If you are asked to have your child undergo a sweat test, remember that far more children are tested than turn out to have C.F. There are so many symptoms of C.F., and so many benefits from diagnosing it early, that we tend to order sweat tests pretty much at the drop of a hat.

*Make some notes to take along.*

*What is the pattern of the child's life, well and sick?*

- Are there significant periods when the child is perfectly well, playing and learning and eating with vigor? (Your child's chart may not reflect all the times she was sick but you didn't need an office visit. And it won't reflect at all how she is doing between illnesses, or if there even is a "between.")
- Is the child growing and developing normally?
- Are the illnesses common ones, such as colds, ear infections, sore throats, intestinal upsets, skin infections?

If all these questions are answered Yes, and your pediatrician has assured you that your child has a normal physical examination, there is very little if any cause for concern that some underlying problem exists. Most pediatricians will do specific tests in this situation only if parents confide that they have specific worries. (But don't feel abandoned if this is the outcome. There are still ways you can help your child handle the inevitable illnesses better. See "Treatment," below.)

*Are there any disturbing factors?*

- Is the child almost never well? So that there is no period between illnesses?
- Are there problems with growth or development?
- Has the child had more than one serious infection, one that required lab tests and either injections of medication and careful watching or hospitalization? Such infections would include meningitis, sepsis (overwhelming infection in the blood stream), pneumonia other than "walking pneumonia," or bone and joint infection.
- Has the child a record of very severe symptoms with common infections such

as chicken pox, roseola, oral thrush, or hand-foot-and-mouth?

- Does the child have chronic symptoms of cough or trouble breathing, or of chronic, long-term loose or watery stools?

If the answer is Yes to any of these questions, your child will need a thorough exam, and probably additional testing.

*Does the child have a pattern of recurrent ear infections or signs of allergy?*

- Do you think the child's hearing and/or speech may be impaired?
- Does the child have signs of underlying allergy, such as dark circles under the eyes, eczema, wheezing or chronic cough, runny nose? Is there a family history of allergy?
- Are there signs of upper airway obstruction: constant mouth breathing, loud snoring with restless sleep or holding of the breath?

If the answer is Yes to any of these questions, your pediatrician may request X rays of chest, sinuses, or adenoids; a sweat test; hearing or speech evaluation; or a specific allergy evaluation, with blood or skin tests.

## Treatment

The most likely result of such a consultation is the diagnosis of normal, and normally frequent, childhood illnesses. (In the unusual event that an underlying problem is found, your pediatrician will prescribe further testing, treatment, or referral to a specialist.)

So here you are, back with the owner of the runny nose, fever, and spots. At least you're no longer haunted by dread. Now all you have to deal with is annoyance: missed daycare, catching everything yourself, and keeping your child happy even though goopy and hot.

First, be assured; your perceptions are correct. Your child probably is sick a lot. From about Eighteen months or Two until kindergarten, children are supposed to catch about ten upper respiratory infections a year. They may also have a bout with another illness, such as gastroenteritis or a named viral disease like hand-foot-and-mouth.

However, it is also true that many children who "catch" such viruses don't actually get sick. They may stay perfectly healthy, even as their bodies generate excellent immunity to the virus, and even as they guilelessly spread that virus everywhere they go.

How can you transform your child from one who always gets sick when exposed to viruses to one who can make immunity without getting sick? And how can you yourself avoid catching viruses from your child and her apparently healthy but virus-laden buddies? Unfortunately, I have no magic answers. The only suggestions I have are mundane:

- Eliminate secondhand smoke from your child's environment. Some studies show that consistent exposure to cigarette smoke can as much as double the number of ear infections in young children. Secondhand smoke can trigger wheezing and make attacks more intense.

- If you're going to have another child, try to breast-feed or give expressed milk only, without formula or foods, for the first four to six months of life. Both ear infections and allergies have been shown to be markedly diminished by this means.

- Keep everybody's immune system in good shape with plenty of exercise, sleep, the right foods, physical affection, and humor. Drink a lot of water. If you believe that prayer is helpful, pray. If you believe in herbs, take herbs.

I don't think anybody knows for sure whether vitamin C helps or not. But we do know that too much vitamin C can cause kidney problems, and that chewing vitamin C tablets is not good for tooth enamel. Worse, we know that if a child or adult is used to taking large doses of vitamin C, and then stops doing so, the immune system doesn't like it and the child is more prone to catching illnesses for a time. On the whole, I wouldn't advise artificial sources of vitamin C, but make sure you get plenty of the natural ones: citrus fruits, cabbage, potatoes, and that babies get magic elixir, breast milk. (Apple juice contains no vitamin C unless it's added artificially. Read the label.)

- See if your daycare setting encourages more than the usual number of illnesses. Is it a very large daycare situation, or one in which the "classes" are small but once or twice a day the whole "school" of forty kids gets together to share juice, crackers, and rotavirus? Or is it drop-in care, where the children are different each session? Double the friendships and double the virus exposure. Classic situations include "drop-in" daycare at church or temple or fitness emporium. This also can happen if the child attends a small daycare once a week—but each time the children she plays with are a different group.

- Resist that impulse to get a furred or feathered pet if allergy is a suspected factor. Alas, this includes hamsters, guinea pigs, and white mice. What about goldfish?

*What about parents' own health?*
- To keep from catching the virus yourself, wash your hands before you touch your own face, especially eyes and nose. If you think you never rub your eyes or pick your nose, put a little colored chalk on your fingers and look in a mirror half an hour later. Remember, it is the apparently healthy young child who is likely to be most contagious.

*More about daycare*
- Get the handbook called *Infection Control in the Child Care Center and Preschool,* edited by Leigh G. Donowitz (Williams & Wilkins, 1991), and hash things out with the people at daycare—teachers, administrators, and other parents.

Finally, a parent's greatest challenge: don't take your child's sickness as a personal affront. Children pick up very quickly the attitude that to be sick is a failing, a loss of virtue, letting down the team.

Or, contrariwise, that to be sick often is to be special, or delicate, or particularly sensitive and virtuous.

Here's what smart parents I see try to do. First, they try to make friends with innocent but offensive symptoms. "Boy, can she shoot snot," a Ph.D. mother said admiringly of her three-year-old. Root for your child's game response, both personal and immunological, to the virus challenge.

Second, every now and then, they make a big fuss over a not-very-sick child. Not too often, but when it really counts. I still remember lemonade and forbidden comic books when I was homebound with strep throat.

Don't you have such a memory of feeling cherished?

# Severe Behavior Concerns in Very Young Children

## Autistic Spectrum Disorders (Autism, PDD [Pervasive Development Delay], and Aspberger Syndrome) and Attention Deficit Disorder (ADD)

Severe behavior problems seem to me the most difficult childhood conditions parents have to face. Not only do they suffer pain for their youngsters, and inevitably contend with guilt, rage, and helplessness, but they are often isolated from, and blamed by, those around them: friends, family, colleagues. Fortunately, the more experts find out about such problems, and the more helpful the media can be in changing the perceptions of people untouched by them.

## Autism

There has been a dramatic increase in diagnosed cases of autistic spectrum disorders, an "explosion" in which the reported cases of ASDs increased by 373% between 1980 and 1994. So if you don't know a child so afflicted, I'm pretty sure you know (or are) a worried parent or parent-to-be. What are all these disorders? What causes them? How are

they diagnosed? Can they be treated or cured—or, better yet, prevented?

Autism comes from "auto," the Greek word for self, and it refers to disturbances in a child's awareness of self versus others. A severely affected child lives in a world of her own, and does not respond to other human beings unless the need to do so is very intense. A mildly affected child doesn't have the normal ability to make relationships—to read others' facial expressions and tones of voice, to act appropriately in a social context, to understand jokes, metaphors, and figures of speech.

Many children with autism have low IQs, but some—those who are mildly affected, often labeled as having Asperger syndrome—are extremely bright, with deceptively sophisticated vocabularies and skills.

The only positive side to this "explosion" in ASDs is that the research into autistic disorders has exploded, too. Already, many affected children with early intervention make astonishing improvements. Undoubtedly, such intervention will become even more successful as we understand more about these disorders.

Autism may be suspected very early in life in a baby who does not seem to be bonding to her loving adults, showing the normal kinds of responses to strangers, engaging in the normal kinds of play with toys, or developing normal receptive and expressive speech.

For that reason, pediatricians want to know if a baby consistently shows the following behaviors:

**Six Months:** Does not turn to a familiar voice; does not turn to or respond to her own name; does not play the "baby game" of back and forth coos and smiles.

**Nine Months:** Does not turn to a beloved adult for reassurance in a strange situation. Does not make a variety of vocalizations, including some consonants.

**One Year:** Does not babble. Does not make any gestures—blow a kiss, wave Bye-Bye, point to what he wants. Does not turn to look in the direction someone else is pointing.

Usually, however, a child with this affliction is first suspected of having a problem in or after the second year of life. He or she may show any one or any combination of the following:

- Difficulty relating to others—looks away, withdraws. Frustrates others' attempts to communicate. Behaves strangely: sniffing others, laughing without apparent cause. Doesn't make eye contact, but looks at others out of the corner of his or her eyes, and doesn't "register" what is seen unless it is something of strong interest. Doesn't respond to overtures to play, cuddle, explore.

- Delayed speech, with no babbling by age Twelve Months, no single word by Sixteen Months. Abnormal speech—repeating a word or phrase over and over, with no recognition of its meaning.

- No gestural communication. By a year, does not blow kisses, imitate pat-a-cake, or point at objects.

- By Eighteen Months, no pretend play with dollies, trucks, and so on. Persists in an apparently meaningless activity—spinning, turning pages, as if hypnotized. Very restricted interest, limited to a few items. Consistently uses these few objects strangely, not as toys or for imaginative games. Instead, stares at the spinning wheels, or lines up toys again and again without ever playing with them. Becomes attached to odd categories of things, such as lids or license plates.

- Doesn't seem to learn more things, but only to change focus of preoccupation now and then.

Since these disorders pervade so many areas of human behavior, in so many individual ways, with such a variety of intensity, it is probable that there are multiple causes. A subset of cases are linked to chromosome abnormalities (such as Down syndrome) or seizure disorders.

However, the underlying abnormality or combination of abnormalities causing autism has proved elusive. There may be structural brain abnormalities, or chemical ones; there may be a general predisposition that can be triggered by something in the environment if a trigger occurs during a sensitive age period.

As for such a triggering event, there have been many candidates proposed, most of which have been discredited. It is certain that parenting styles do not cause autism. The accusation against the measles-mumps-rubella vaccine has been largely refuted. Infections during fetal life may be a possibility, but none has so far been defined.

In any event, the most useful approach to autism by parents and pediatricians is: clarify the diagnosis as soon as it is suspected; exclude other explanations for the symptoms; identify any contributing factors; obtain the best intervention as promptly as possible; try to stay as informed as possible; and seek and be open to support—from institutions, from groups, and from individuals.

Many children with Autistic Spectrum disorders make remarkable progress with the aid of intensive therapy and appropriate medication. Particularly important to parents, many of these children are able to discover pleasure, even joy, in relating to others.

What I do in this chapter is:

- Describe the difference between PDD, Autistic disorder, PDDNOS, and Asperger syndrome.
- Discuss the overlap between normal atypical behaviors and ASD behaviors.
- Discuss what kind of diagnostic testing may be appropriate.
- Summarize the conventional therapies.

- Caution and advise about Complementary and Alternative Medicines that may be suggested.
- List some helpful books and organizations.

# Autistic Spectrum Disorders

Every child with an ASD has a unique profile of strengths and problems. A diagnostic category does not tell you how much improvement a child is capable of. The terminology is not a scientific one (it's not like the terminology for, say, gallstones). Rather, these definitions have been established in order to organize current ideas about autism, so that research grants can be awarded, treatments developed, articles written, and insurance claims processed.

## PDD: Pervasive Developmental Disorder

This is the "big umbrella" for all degrees and variations on the theme of social, communication, behavioral, and cognitive problems. When a child presents with signs of PDD, a further work-up is done to identify whether a problem exists, and if so how severe. PDD is divided into two categories: Autistic disorder and Pervasive Developmental disorder not otherwise specified. The difference between the two is one of severity, with the former being more severe.

Mental retardation and seizure disorders may coexist with PDD, but are not on the list of diagnostic features.

## AUTISTIC DISORDER (CLASSIC AUTISM)

A child with autism presents with the most severe symptoms of PDD. That means the child cannot talk, does not seem to wish to communicate, and seems locked into a unique, idiosyncratic, highly constricted internal universe. She makes limited eye contact, has severe sleep problems, temper tantrums, stereotyped behaviors (flapping her hands, spinning around), fascination with parts of toys, attachment to unusual objects, distress with changes in routine, and little pretend or imitative play.

## PDDNOS

This is the diagnosis given when a child presents with PDD-type problems—trouble with speech, communication, social interaction, and ability to understand and use things in her environment—but does not have the whole array of features required to diagnose autism.

## ASPERGER SYNDROME

A child diagnosed with Asperger syndrome has features of ASD coupled with an IQ high enough to allow academic success. Such a child may do beautifully in math, physics, architecture, or related fields, but have a very hard time understanding literature. A youngster with Asperger syndrome will understand plot, but not motives, characterizations, or figures of speech such as metaphors. These youngsters have a hard time socially and in all areas of personal interaction.

## NORMAL CHILDREN WITH UNUSUAL CHARACTERISTICS

Parents may worry about a child who shows some features of autistic spectrum disorders.

Chantel, in the Eighteen Months to Two Years chapter, is one of these. She is intense, intolerant of change, and seeks repetitive motions; rocking the chair, turning and tearing the pages of her books rather than looking at the pictures. She has severe tantrums and fears. She seems to live in a world of her own much of the time.

But Chantel *wants* to communicate and *tries* to communicate. She shows that she understands just about everything her parents say and has more than ten words that she can use herself. She even puts two words together: More chair, No juice.

She has a pretty normal agenda for age (TV remote, juice). She seeks out her mother for help and glares at her when she doesn't get it. She turns to her mother for comfort when overwhelmed.

Chantel needs a lot of extra designed-specially-for-her help and attention, but then so do many children.

Nearly all babies at about nine months to fifteen months display some autistic features: They like to watch wheels spin and book pages go back and forth; they have intense fears and temper tantrums; they use toys inappropriately, picking up Barbie by her legs and smacking her like a drumstick against a kettle.

Older children are often sensitive to certain noises, prone to temper tantrums, finicky about foods and the way clothing feels. There are lots of children who delight in lining up their toys (or their fruit cocktail) by category.

But all these children display attachment behavior and a desire to communicate. They can "read" the intent of other people by their words, facial expressions, and gestures. They are not *at the mercy* of their preoccupation with spinning or turning pages, unable to stop and find something more meaningful to do. They learn a little something new every day; they are not stuck in one place.

If you have a child with delayed or sparse speech, atypical behaviors, or intense fears and tantrums, by all means speak with your pediatrician. If your worries are not set to rest, ask for a referral to a pediatric neurologist or to a developmental assessment clinic—one that is run by a medical school.

## Diagnostic Testing

There is no one specific test for these disorders. Watching the child in action, playing with toys, or interacting with a parent or caretaker, is one of the most important diagnostic tests. Taking a careful developmental history and medical history may shed light on underlying problems. A thorough family history, looking for other affected individuals, is a must, as is a thorough physical and neurological exam by the pediatrician.

However, there are other tests that may be helpful.

**Many pediatricians feel that the following ought to be performed on ALL CHIL-DREN suspected of an Autistic Spectrum disorder:**

- **Hearing evaluation:** All children with speech delay, abnormal speech, or any question of abnormal hearing, must have a hearing evaluation. This could be a test that requires cooperation (audiogram; play audiology) or one that is performed with the child asleep (brainstem auditory evoked response—BAER for short).
- **Speech evaluation:** An experienced speech pathologist can be a big help figuring out what is causing delayed or abnormal speech.
- **EEG** (electroencephalogram; "brain wave" recording): This painless, noninvasive test helps to rule out subclinical seizures—seizures that can interrupt thinking without causing any external physical signs. There is a very rare form of seizure disorder called epileptic aphasia, or Landau-Kleffner, that effects speech; treating the seizures often helps speech development.

  If a short EEG (two hours or less) doesn't show an abnormal pattern, but the neurologist is still suspicious, it may be advisable to perform an overnight EEG or even a twenty-four hour EEG.
- **Blood test for specific chromosome abnormalities:** "Fragile X" is a syndrome in which the X chromosome is easily broken. This can cause autistic features and mental retardation both in males, who have only one X chromo-

some, and in females, who have two. The intensity of the symptoms caused by fragile X varies in a very complicated way; there may well be no family history of mental retardation or autism.

While children with Down syndrome may have autistic features, this diagnosis is usually suspected and confirmed at birth.

Other more rare chromosome problems may also be searched for, if physical examination suggests their presence.

- **Blood Lead Level:** Lead poisoning can cause mental retardation and seizures. Moreover, an autistic child may be more inclined to mouth nonfood materials, and thus acquire lead poisoning on top of any underlying problem.

In children whose history or examination indicates a possible problem in these areas, the pediatrician may also order:

- **A consultation with a pediatric neurologist:** This is partly in order to assess the possibility of other diagnoses, partly to obtain assistance with prescription medication (see below), and partly because many managed care plans require it before ordering other tests, such as an EEG or chromosome study.
- **A consultation with a geneticist or dysmorphologist (specialist in human malformations):** If a child's facial or bodily features are atypical, there may be an underlying syndrome causing autism.

If so, further testing may be indicated. If there is a family history of autism or very eccentric behavior, mental retardation, or genetic or inherited problems, it's crucial to have an expert opinion on underlying causes, and on their chance of recurrence.

- **Thyroid studies:** A decreased level of thyroid hormone can cause mental retardation. However, without any physical findings to suggest this diagnosis, it is unlikely.
- **Head MRI or CT:** If the child's head is much smaller or much larger than expected, or of an abnormal shape, a brain scan may identify the reason why. If the child has a seizure disorder that could be originating in one particular area of the brain, it may be appropriate to make sure there is no tumor, cyst, or blood vessel abnormality in that area.
- **Blood tests for inborn metabolic problems:** If a child is born with a biochemical abnormality, substances can build up in his system that cause mental retardation and/or autistic behavior. There are usually other signs that this is the case: unexplained loss of consciousness, for instance, or poor growth, or findings on physical exam.
- **Blood tests for infectious diseases during fetal life:** Certain diseases of the fetus, such as cytomegalovirus or rubella, can cause mental retardation. Almost always there are signs on physical examination that this might be the case.

## Conventional Treatments and Therapies

1. **Treatment of any underlying or accompanying disorder:**
   - Hearing aids for hearing impairment.
   - Anticonvulsant medication for seizures.
   - Medication (and prevention of recurrence) for lead poisoning.
   - Thyroid hormone supplement, if deficient.
   - Diet and/or medication for inherited metabolic problems.
   - Counseling if chromosomal abnormalities discovered.

2. **Therapy in specific areas:**
   Authorities agree that these should be started at soon as a diagnosis is made, without standing back to "wait and see" if Cherub spontaneously improves. It is important to find the very best programs available. To do so, talk with your pediatrician, pediatric neurologist, developmental experts, and other parents. Read a lot, and go online: see the *Resources* section below.
   - Behavior therapy: designed to foster two-way communication, so the child begins to find pleasure, even joy, in interaction.
   - Speech therapy: again, focused on pleasure in communicating.
   - Sensory integration therapy: helping the child tolerate a variety of normal sensations, stop "perseverating" (repeating the same stereotyped motion over and over), and begin to develop more normal play and interaction with the physical world.
   - Occupational Therapy: along with sensory integration therapy, helps children to be able to accomplish the tasks of normal life.

3. **Medications for specific problems:**
   These may include medication to help alleviate obsessive-compulsive features, anxiety, hyperactivity, perseverative behaviors. They are not a replacement for, but an adjunct to, the therapies above.

## Alternative Medications, Herbs, Special Diets, and So On

The medical acronym for this category is CAM, standing for Complementary and Alternative Medicine.

For problems such as autism, where we are desperate for an explanation and cure, it's natural that there be many theories, many trials of many remedies, and many anecdotes stemming from those trials. Buried in there may be some helpful advice, but you've got to be very careful.

I strongly recommend that you call your nearest medical school library and order a copy of a fine article called "Autistic spectrum disorders: When traditional medicine is not enough," from *Contemporary Pediatrics,* Vol. 17, No. 10, October 2000 (pp. 101–116) by Susan L. Hyman, M.D., and Susan E. Levy, M.D.

This article explains very clearly how to judge any study or report about CAM. It also gives a list of sources for further information.

## Helpful Resources

Zero to Three: National Center for Clinical Infant Programs

2000 14th St. North, Suite 380
Arlington, VA 22201–2500
The Asperger Syndrome Coalition of the United States (*www.asperger.org*)
Phone: 904–745–6741
OASIS, Online Asperger Syndrome Information and Support (*www.udel.edu/bkirby/asperger*)
Phone: 703–528–4300
Fax: 703–528–6848
TDD: 703–528–0419
Autism Society of America (*www.autism-society.org*)
Autism Resources (*www.autism-resources.com*)
Autism Network for Dietary Intervention (*www.AutismNDI.com*)

For a clear-eyed look at a popular but discredited therapy, read the American Academy of Pediatrics Position Paper on "Auditory integration training and facilitated communication for autism, RE9752, *PEDIATRICS,* 1998, Vol. 102:431. You should be able to get a copy at any medical school.

## Be Optimistic But Careful

Read with your eyes open, as it were. Approach anecdotal reports with suspicion. When a study is referred to, use your common sense *and* the "Criteria for a good study," included in the reference from *Contemporary Pediatrics*, above, before you decide to embark on any new treatment. Read as much as you can, as widely as you can; this will help you to spot inconsistencies, false claims, unfounded accusations and assertions. If something sounds fishy, dangerous, or too good to be true, it probably is.

And let your friends—that includes your pediatrician!—help out in any way you can think of.

## ADD: Attention Deficit Disorder

When you come across an Under-Five who is attentive, self-controlled, who concentrates well and sits calmly, that child is probably acting in a perfectly normal fashion for age. And you probably caught him at a good moment, too: rested, well-exercised, happy with life, in a friendly and interesting environment.

When you come across an Under-Five child who doesn't pay attention, acts on impulse, can't seem to focus on something for more than a minute, and is in constant motion, that child is probably acting in a perfectly normal fashion for age, too. Maybe you caught him at a bad moment: sleepless, hungry, anxious, stressed, worried about kindergarten shots.

But what if he is always, in every situation, "at the mercy of his environment?" What if he can't sit still for Circle Time? What if he doesn't track well enough to understand Daddy's read-aloud story? What if his siblings won't play with him because he's "no fun" and "he's like a jumping bean?" What if he is always getting into trouble with the teacher for talking out of turn, making airplane noises, touching everybody around him, going *biff biff biff* with his fists into thin air, breaking things, teasing the hamster?

Poor focus, inability to concentrate, jittery, distractable, and impulsive, in *every* situation: he might have attention deficit disorder.

Most experts believe that ADD is a "neurodevelopmental" disorder in which the ability of the brain to focus its attention does not mature normally. Studies have shown very convincingly that when a child or adult has this problem, it improves only with medication. Other modalities, such as behavior modification or psychotherapy, do not change the outcome one way or the other.

However, the use of medication, especially in young children, is repugnant to many parents and pediatricians. To justify it, you need to be sure of three things: first, that the problem is severe enough that it causes trouble in everyday life: second, that the behavior is the same in every situation, not *just* at school or *just* at home, for instance; and third, that it is not due to some other underlying problem.

## DIAGNOSIS

There is no specific physical, neurological, or laboratory test that diagnoses ADD. A child with ADD has a normal physical exam, EEG (brain wave test), brain scan, blood and spinal fluid.

Instead, the diagnosis is based on excluding a long panel of other possible causes for the behavior. It is supported by obtaining the input of several adults such as parents, preschool teachers, daycare adults, and so on; and by obtaining a family history (often ADD runs in families).

The other diagnostic possibilities to be excluded include medical causes, neurodevelopmental and learning causes, and psychosocial or psychiatric causes. Here are some possible factors that may cause behaviors that seem like but are *not* ADD:

**Medical:** Side effects of daily medications, such as antihistamines; chronic sleep disturbance, such as sleep apnea; malnutrition; anemia; thyroid disturbances; and underlying medical problems causing frequent school absence.

**Neurodevelopmental:** Hearing or vision impairment, mental retardation, Tourette syndrome; uncontrolled seizures (which may be unsuspected because they are the "silent" type); PDD and autism.

**Psychosocial or psychiatric:** Severe behavior disorder, anxiety disorder, depression, psychosis; chaotic social environment; poor limit-setting at home or school; child abuse or neglect; inappropriate school placement (bored or over-challenged).

So the young child with suspected ADD

needs a complete physical examination, laboratory or neurological testing if indicated; multiple assessments from adults who know him well; and, often, a psychiatric examination.

## TREATMENT

If another cause of the behavior is found, that must be treated. If the diagnosis of ADD is strongly suspected, treatment is with stimulant medications such as Ritalin or, more often in this age group, Dexedrine. The medication related to Ritalin, called Adderall, is also a possibility.

There needs to be close monitoring of the child to make sure that the medication is, indeed, helping the situation, that the child is eating and sleeping normally, and that no side effects (nervous tics, mood changes) are noted.

At the same time, adults need to make sure Cherub is receiving lots of help establishing friendships, feeling competent at learning tasks, and getting plenty of exercise. Your pediatrician can help you figure out if your own particular Cherub might benefit from an assessment for ADD.

# Oppositional Behavior in Two and Older

Happy families, Tolstoy famously observed, are all alike.

Tolstoy was interested in the Unhappy ones. Pediatricians are very interested in the Happy ones. How they get that way and stay that way.

Happy families are *not* all alike, of course. Some travel the world as conquerors or tourists or missionaries. Others stay home and raise ducks. Some are composed of two parents and two children. Others, of a single grandparent and six adopted step-grandchildren. Some celebrate loads of religious and family rituals, throwing parties, creating decorations, feasts, and aromas to fit the occasion. Others are not so minded. "In Our Family," David announces proudly to the third grade, "our Most Important Ritual is, we don't start to eat our popcorn until the Main Feature comes on."

But the families that are happy do share one thing. They adhere to a Family Contract. Not a written one, of course; in fact, the rules of the contract are usually not even stated as rules—they're just understood and acted upon. Here are those rules:

- When a parent says something with a certain look and tone of voice, the children listen up and do what they are told, right away, without more than a sigh, compressed lips, rolling eyeballs, or soft little moan.
- Nobody in the family says really hurtful things, or commits really hurtful actions, against another family member. If that happens, the rest of the family is outraged and shocked; the guilty party is made to feel awful, and must offer atonement.

Pediatricians tend to focus on these two rules, because we are asked constantly about how to make them work. The obstacles are twofold: *Oppositional Behavior* and *Toxic Sibling Battles/Sibling Rivalry*.

Oppositional Behavior occurs when a child gets in the habit of defiance, arguing, or forcing parents to negotiate. When a child "mouths off," or physically attacks a parent, or refuses to use the potty after age Three, that's Oppositional Behavior. Once a child starts into a pattern of Opposition, it's easy for the entire parent-child relationship to become adversarial.

Toxic Sibling battles or rivalry occur when siblings "fight all the time." They bicker and whine and are constantly at each other.

Both of these violations of the family contract can be prevented and fixed (for sibling battles, please see Chapter 27). The problem is, the normal, human, instinctive parental responses to the situation are not merely unhelpful: they actually feed the flames. Here's what to do instead.

## Oppositional Behavior

### PREVENTION

As usual, the best strategy is preventing the development of oppositional behavior.

Pretty much everybody knows that there are three parenting styles: Permissive (anything goes!), Authoritarian (You do what I say, now, because I'm the Boss), and Authoritative (Here are the rules, Here is why we must obey them).

The first key to avoiding Oppositional Behavior is to realize that as a parent you get to be all three, but in sequence: you are Permissive with a young baby, gradually segue into Authoritarian as Cherub turns Five to Nine months, peak with Authoritarian around Fifteen to Twenty-four months, and then gradually segue into Authoritative by around age Three to Four.

If the Permissive Stage lasts too long, parents are likely to be shocked and angry by the normal behaviors of Cherub at Fifteen months.

If the Authoritarian Stage is started too late (say, at age Two), habits of opposition may be so engrained that parents are at their wits' end. They start to spank. It doesn't help so they spank more often.

If all does go well, by age Three, Cherub has learned what she needs to know to make it possible to segue into Authoritative Parenting. If Cherub is oppositional at Three—mouthing off, defying parental orders, Potty Resistant—then parents need to regroup and take prompt action.

By Three, Cherub should be absolutely certain of several important facts of life:

- Her parents adore her, and she loves having their attention and pleasing them. Their attention rewards her nondefiant, cooperative behavior.
- There are certain things that simply Are Not Done. These include tantrums, physical aggression against the parent or a sibling, mean teasing, saying bad words in anger, and refusing to do (or to stop doing) something that Parent has commanded. (True accidents are in a different category from Disobedience.)
- Many activities are negotiable, such as Making Mess, Making Noise, Being Silly, and Doing Something Not Already Forbidden.

She will need to be reminded about all of these Facts of Life on occasion, but the foundation has been put firmly in place.

## SQUELCHING OPPOSITIONAL BEHAVIOR

"She doesn't listen to me!" "He won't do what I say!" "He doesn't respect me!"

The preschool years are supposed to be fun for children and parents. Nobody wants to have to contend with oppositional behavior in their little ones. Your goal, then, is to eliminate, as far as possible, "back and forth" fights.

These are fights in which the parent tells the child to do, or to stop doing, some behavior; the child refuses to obey; the parent escalates the demand by yelling or threatening, the child digs in his or her heels, and the bad interchange continues until one or both parties explode or until one gives in to the other—often the parent being the one who knuckles under. Once such conflicts begin, they tend to spiral out of control, becoming more and more frequent and upsetting. At some point, the child's behavior may go truly out of control: Cherub spits in your face, or calls Mommy a "bitch" in public, or something really outrageous.

Very often, the child who is most oppositional at home behaves just fine elsewhere, and may even be well-behaved with the other parent.

There are many techniques that DON'T work for such behavior:

- Spanking may stop the immediate behavior, but not the pattern.

- Giving bribes and rewards for ordinary good behavior escalates until you're spending hundreds of dollars on toys and snacks.
- Time out backfires: the child escapes and defies you even more vigorously.
- Sympathizing with what you believe to be the child's feelings doesn't work. ("*I can see that you're angry. That's why you're hiding under the table. It's OK to be angry but it's time to come out now. Come on now, Steven. It's really time to come out. I know you don't want to but sometimes we have to do things we don't want . . ." and on and on.*)
- Having a long talk with the child and asking for cooperation is a waste of time.
- Telling Cherub firmly to listen to, or obey, or respect you brings you even more defiant behavior.

# Three Strategies for Handling Oppositional Behavior

Here are three strategies that, if used together, will prevent or very much reduce oppositional behavior in this age group. They are:

1. Make sure your child feels liked and appreciated. Not easy when you're furious most of the time.
2. Choose your battles wisely, and act with authority and firmness. This takes thought and planning. It does not come naturally.

3. Recognize the times when giving an order is asking for trouble, and have other options available. This requires even more thought and planning.

Your goal is for Cherub to understand that YOU are in charge; that you are comfortable about being in charge—that is, you have no ambivalence about making her behave; and that defiance is not an option.

## MAKE A SPECIAL EFFORT TO RESTORE PARENT-CHILD GOOD FEELING

An oppositional child knows that her parent is angry at her and that the parent spends much of the time not liking him or her. It is very hard for Cherub to behave better when she's eternally in the dog house.

*Strategy*
- Make a ten to fifteen minute appointment each day just to be with your preschooler one-on-one, without any other child or adult in the room or area. The best play is imaginative play with dolls, stuffed animals, or action figures; arts or crafts such as coloring or Play-Doh; taking a walk and talking about what you see and hear. Interactive reading, in which you talk about the story as you read it, is fine.
    *Some caveats*:
    *It has to be one on one, uninterrupted.*
    *Nobody else in the room, no phone calls.*
    *More than ten to fifteen minutes per day is* *likely to backfire, as Cherub will start to regard your undivided attention as her inalienable right.*
    *Bedtime One-on-One time doesn't count. That's a Separation Ritual.*
    *Errands, chores, classes—even Mommy and Me classes, and going to the park or library do not count. The focus and intent of the One-on-One time has to be: Parent Focuses on Interacting with Cherub.*
- Touch your child briefly, frequently, and affectionately without *saying anything*. A pat on the shoulder, a ruffle of the hair, etc.
- Take as many opportunities as possible to give positive attention. Generalized praise, however, may backfire: an oppositional child who feels unliked may respond to praise with more negativity— "It's *not* a good picture!" "I'm *not* pretty, I hate this dress!"

    Instead, try merely stating what he or she has achieved, in a happy, positive voice: "You made a big yellow circle there!" "You sang all the words of that song!" "You put on your red shirt!"
- Let your child overhear you bragging about him to others, relating specific anecdotes about how cute or funny or smart or responsible he or she is.
- Try to make most of your talk with Cherub neither blame nor praise, but conversation about what is going on in his or her life. "Your tricycle looks so shiny in the sunshine." "I wonder how many worms live under that big rock?" "I like that Frog and Toad book."

## ONLY GIVE ORDERS THAT YOU CAN ENFORCE PROMPTLY

All children need to learn that when a parent decides something is important enough to give a direct command, that command **must** be obeyed.

If you give a command that can't be enforced, this is a hard lesson to teach.

*Strategy*

1. ***Don't give a warning, as in, "Remember we don't play with scissors."***

    To a child in Oppositional Mode, that's the equivalent of saying "Go get those scissors right now and cut off your bangs."

2. ***Give a direct order or command ONLY if BOTH of the following two conditions are met:***

    • ***It's a clear-cut forbidden behavior.***

    That means that you'll spring into action if your child refuses. Putting down the sharp scissors is important; leaving the dog in peace while he's eating is important; not screaming into the telephone and hurting someone's ear is important. But picking her nose is not an important transgression, nor is singing "Jingle Bells" forty-seven times in a row.

    • **You can enforce the command immediately, on the spot.**

    You can take the scissors away, pick her up and carry her away from the dog, grab the phone out of her hand. But you *can't* stop her from picking her nose, singing "Jingle Bells," or wiggling. You *can't* physically make her keep her

shoes on, or put away her toys, or stop whining.

3. ***Don't appear to offer a choice when you are in fact giving a command.***

    Don't give an order and then say "OK?" As in, "We need to give you a blood test now, OK?" or "Time to come in now, OK?"

    Don't plead for cooperation: "Instead of playing with the scissors, how about coming down here and looking at this book?"

4. ***When you do give an order, show that you mean what you say.***

    • Make eye contact with your child. Don't smile. Give the order in a firm, louder-than-usual voice.

    • Use as few words as possible. "No scissors." "No bothering Fido."

    • Don't explain, describe, or go into detail.

    • Make the order very specific. Don't say "Be good, now." Say "No running. No shouting. Sit down."

5. ***Act Promptly.***

    • Take away the scissors as soon as the order is out of your mouth. Move Cherub away from the dog. Do so silently, strongly, and without any indecision.

6. ***Follow Through.***

    • After you have acted, remove your attention—your voice, your eye contact, your interest—for a minute or so.

    • Put on a disappointed or displeased expression, but not an angry one. Practice in front of a mirror.

- Don't draw a moral or point out a lesson. That's just letting her know you are worried that you are not in charge.
- Don't tell, or threaten to tell, anybody else about the incident. This will undermine Cherub's conviction that YOU are in charge, that YOU are unambivalent, and that YOU can make her behave.

## WHEN YOU CAN'T ENFORCE AN ORDER, DON'T GIVE ONE—HAVE OTHER ALTERNATIVES AT HAND

If you *can't* enforce your order, the back-and-forth challenge will begin. Besides, when you're not laying down the law, your child needs to learn other things from you: how to cope with the real world, how to take unimportant things lightly, how to tolerate frustration, how to be responsible for his or her actions.

### Strategy

Be sure your expectations are appropriate. For instance, most children this age eat hardly any dinner (literally, maybe only one little bite!)—not because they're "bad eaters," but because they've already had all the calories they need for the day.

Make sure he or she really can change the behavior. For instance, bedwetting is normal and frequent until age Six. At any age, the child has little control over that behavior. Nervous habits, like blinking or clearing the throat, are involuntary and don't respond to commands.

Here are some possible alternatives to giving an order you can't enforce. Think ahead of time about likely scenarios and have a game plan.

- **Let the outcome of the behavior teach the lesson.** This means you let Cherub own the behavior, instead of taking responsibility yourself. This is an appropriate response to procrastination, making a mess, or purposefully ruining toys or materials. State the unpleasant outcome of the behavior as a fact of life, not as a punishment. "No time for a story tonight. We took too much time with teethbrushing." "Not enough time to get dressed. We'll finish at school."
- **Try to make him or her think that it's his or her idea.** Only try this if Cherub and you both are in good frames of mind. If you sense that Cherub is in an adversarial mood, this will backfire. "I'll bet if you put all your dollies in one little corner of the toy chest, they'd like that a lot. They could cuddle up and tell secrets." If Cherub won't play, or responds rudely, invoke a consequence.

If the behavior is one that is annoying or distracting, but not actually offensive:

- **Lighten up.** "I'll bet that you can't twirl in the other direction for as long as you twirled in that one."
- **Consider dropping the issue.** Maybe it's not really important; maybe you're just in the habit of "getting on his or her case." Is it really important that he doesn't use the paper towel tube as a

horn? Is it really so terrible that she puts her pants on backwards? Or that he stands in the middle of the room and spins around and around?

If these suggestions don't work over a two week time, and the problem shows no signs of resolving, talk with your pediatrician. Make sure your child is physically healthy and that his or her development is on target. If no other reason for the oppositional behavior can be found, it is a very good idea to have a consultation with an expert in child guidance.

# { TWENTY-SIX }

# It's My Potty, and I'll Try If I Want To

## Potty Refusal in Three and Older

*When normally developing children Three and older refuse to use the potty, it's not because they don't know how or because they're "not ready." It's part of Oppositional Behavior—an important developmental phase, in which little children experiment with Power.*

*Potty is a great battleground for this age group, because Potty is a battle they can always win. Winning a battle is so much more satisfying than sitting on that little chair and doing what the Big People want!*

Reasoning, coaxing, demanding, rewards, and bribes do not get you anywhere. Worse: the more battles you engage in, the more stubborn Cherub becomes. Often, it becomes a State of War.

Three factors that can feed into this are:

- An attention-getting sibling.
- An absence of daily structured activity, such as preschool.
- A deficiency of daily, focused one-on-

one time with parents—especially with Mom.

All of these increase Cherub's need to feel powerful and important.

So before you go full throttle into following the Potty Program, consider Cherub's lifestyle:

- If sibling squabbling and battles go on a lot, see the section on *Siblings*, below.
- Make sure that Cherub gets about ten minutes of one-on-one time just about every day with Mom (and ideally with Dad, too). See the section in *Oppositional Behavior*, above. *Extended* one-on-one times—all day, or all afternoon—aren't necessary and may make things worse. It's the daily focusing that counts.
- Make sure that Cherub has daily age-appropriate challenges: outdoor exercise, supervised play with age-mates, crafts, puzzles and games. A good preschool is

usually ideal—but often requires that Cherub be "Potty Trained." (My suggestion: if Cherub has ever been "Potty Trained," or is clearly able to do so when he feels like it, tell the preschool that he is indeed potty competent and just needs motivation.)

*Strategy*

1. Put Cherub back in diapers (not pull-ups!) for a week or two. Don't present this as a punishment or as a "step backwards." Don't give any explanation at all—explaining will only get you an argument. Don't ask permission—don't tack on a questioning "OK?" to your statement. Present it as a decree. "It's time to take a break from the potty." Don't say anything else, no matter how often Cherub asks or whines. Stand firm.

   If Cherub takes the diaper off, put it back on, without comment, under an outfit with straps, buttons or zippers so that Cherub *can't* take it off. Don't make this yet another action Cherub can contest, another battle he or she can win! Don't start "The Program" until Cherub has stopped trying to remove the diaper. This may take several days of straps, buttons and zippers.

2. After a week or two of no battles (and also no conversation) about diapers or potty, start the following program. It will take determination and willpower, so think ahead to how you will make these strategies work. Ear plugs, a phone buddy to call in order to "be too busy"

to change the diaper, and inner resolve will help.

Remember, you are not removing love from Cherub; you are not showing anger or even annoyance. What you are doing is reversing the situation so that his or her sense of power will come from using the potty, not from ruling Mommy and Daddy.

*The program*

- Don't ask Cherub if the diaper needs changing or if he has to go potty. Instead, check once an hour by looking or feeling inside Cherub's clothing.
- *Pointedly* interrupt Cherub at play when you go to check the diaper. Don't say a word. Don't be angry or annoyed. Don't ask permission, as in "Time to check, OK?" If the diaper needs changing, do it then, but without any further comment.
- Make the act of diaper changing *pointedly* inconvenient for Cherub.
- If Cherub calls your attention to the need for a fresh diaper, do NOT change the diaper at that time. Instead, pretend (if necessary) that you are occupied with other tasks.
- Let the soiled diaper go unchanged for at least five minutes. If Cherub nags and whines for you to change the diaper, remain impervious. At some point—it may take twenty minutes or longer—he or she'll become tired of that and go off and do something else.
- Then, when you see that Cherub is enthralled in his or her own activities, *interrupt* Cherub to change the diaper.

- Don't apologize or ask permission (Don't say "OK?") but simply state, "Time to change the diaper." No other words.
- Pick up Cherub and go to a different, boring room, where you have diaper changing supplies down on the floor.
- Show Cherub by your behavior that you are completely in charge and unmoved by any of Cherub's reactions.
- Do not chase Cherub about the house. Do not recognize that Cherub is yelling, kicking, flirting, or any other behavior at all. Just tackle Cherub and get the business done. Do not say a word or look angry.
- As you change the diaper, behave as though you were a million miles away. No talk, eye contact, smiles; you are all brisk business. Not angry, not disappointed, just preoccupied—concentrating on an adult issue: tax reform, say, or genetic engineering.
- Continue to behave in this distant fashion for a couple of minutes afterward, showing no interest in what Cherub is doing.

- If Cherub runs away, don't go off in pursuit. Go back to what you were doing. Wait until Cherub settles down and starts something enthralling. Then go and proceed with the above steps, but don't let Cherub run away again.
- Once success is in sight, don't let Cherub turn it back into a power struggle.
- After a week or so of this regimen, you will notice that Cherub is showing an interest in the potty. When this happens, remain only mildly interested and helpful. Don't get too excited or triumphant; simply observe that Mr. or Mrs. Peepee or Poopoo likes to go in the potty. Don't let Cherub overhear you announcing the breakthrough to family or friends. Casually offer a choice between two sets of underpants; don't offer this as a reward, but as if you'd expected success all along.

If this doesn't work, your pediatrician and perhaps a counselor needs to get involved.

# Sibling Battles

Siblings are supposed to fight. Aren't they? Isn't that why Nature gave us backseats in cars?

That's how it seems in the media. It's a running joke. In real life, many parents really want their children to get along together, but are quick to add that "Of course I expect them to fight." When asked about that expectation, they often say that sibling fights teach children how to negotiate; give them skills valuable in the real world. I doubt it. Once a real fight starts, negotiating usually doesn't stand a chance. Look at the Mid-East.

When siblings are allowed to "work it out on their own," here's what they learn:

- The bigger and stronger child wins.
- The louder, more aggressive child wins.
- When you fight frequently, you can forget—or never learn—how to play without fighting.
- When you are always fighting, there's no reason to try to see things from your sibling's point of view—so mean teasing and sneaky nasty tricks are all just fine.

Of course, more often than not, parents simply can't stand the noise and unpleasantness, and go and break up the fight. Here's what siblings learn from *that* experience:

- I can sneak up and grab my sibling's toy. When my sibling punches me I can yell that I got punched, and Mommy or Daddy will come in and I'll say I got punched and they'll punish my sibling.
- If my sibling and I start to scream and fight, after a while Mommy or Daddy will come in, and listen to both sides of the story, and if my side is better, I can get my sibling in deep trouble.
- I can get Mommy or Daddy's attention consistently by picking a fight with my sibling.
- If we fight often enough, one of us gets the reputation of the Good Child and the other one gets the reputation of the Monster.
- The one who gets the reputation of being

The Monster can act like a Monster all the time because that's what they expect.

Parents who want to establish a Family Contract (See essay on *Oppositional Behavior*) need to nip sibling battles in the bud. That doesn't just create a more satisfying family life; it actually teaches children something important: true negotiating skills: how NOT to escalate a disagreement into a screaming or hitting fight.

To nip sibling wars in the bud, adjust your Parental Response to the age of the children.

## Siblings Who Are Both Two and Under

Children this young usually can't play together at all. A Two is just learning how to play imaginatively and socially. Two can just barely play with other Twos, who after all do have rudimentary social skills: they can talk a bit and imitate each other's games. They can get help from, and tattle to, an adult.

But asking Two to play with a younger child, without close supervision, is like asking a chimpanzee to babysit.

Not advised.

## Siblings Who Are Both Three and Under

Two children will be unable to really play together until *both* are old enough to really play. They will be unable to share toys until *both* are old enough to share. They will be unable to negotiate themselves out of a battle until *both* are able to talk.

A Three who is expected to play with, or even to entertain, a younger child doesn't understand why Fifteen Months can't join him in a pretend game, for instance, or sing the "Itsy Bitsy Spider" song. Three just thinks that the younger child is being "mean."

If Three is asked to "share" a toy with the younger child, and is rebuked for not "sharing," Three's face crumples in bewilderment and outrage. He is not being asked to "share," he is being asked to find it OK when somebody grabs his toy. This violates every rule Three is trying to learn, so laboriously, about sharing. My heart always goes out to Three when I see this happening.

**The Moral of the Story:** If you're right there, suggesting play strategies, narrating what Three is doing, praising Three for being so grown-up, intervening when Younger Sibling grabs the precious fire truck—fine.

But keep your eyes sharp and your expectations low.

When outrage and fighting occur with this age spacing, it's not the children's fault, if you get my drift.

## When Both Siblings Are Over Two, and One Is Four or Over

Finally, they are of ages that can play together. What Siblings this age need to learn is the following:

- It is up to US to play together without fighting. We get better at that as we get more experience.
- If we do start to fight, nobody is going to

come over and listen to My Side of the Story and Administer Blame and Mete Out Justice.

- If we do start to fight, somebody will come in and separate us. Which is the most frustrating and exasperating response in the world.
- Therefore, it is pointless and self-defeating to start a fight or to try to get the other sibling in trouble. Too bad. It is so satisfying to feel as if I am the favored one. Or to go into a grand sulk because I am the Martyr and It's Not Fair.
- But that's OK, because I get a satisfying amount of time alone with Mommy or Daddy regularly. I don't need to compete for it, or whine for it. It just happens.

So here is the technique that works for this more mature duo of siblings:

- Put the responsibility on the children to learn to get along. This does NOT mean letting them fight it out! On the contrary, it means separating them as soon as their interaction escalates to shouting or name-calling. Don't wait for physical aggression. As soon as you hear the signal, go in and say, "NO FIGHTING!" Just those words. Don't try to figure out who's at fault, or who did what to whom. Even if you witnessed what happened, don't judge or administer blame. Instead, separate them. Put them in two different rooms for three minutes.
- If they are squabbling about a toy or

object, go in when you can hear that they are not working it out on their own: when one of them gets that shrill note in his voice, or yells a threat. Put the TOY in "Time Out" for an hour or a day. Don't try to figure out who had it or who grabbed it.

- Make sure that each child gets about ten minutes of one-on-one time every day with each parent, especially Mommy. Extended one-on-one times—all day, or all afternoon—aren't necessary and may make things worse. It's the daily focusing that counts. See the Oppositional Behavior sheet for making One-on-One time really work.

*Addendum:*

Fighting in the backseat is a temptation because both children know that you have to pay attention to driving. Because this is a dangerous pastime, it needs to be nipped in the bud as soon as it starts. I suggest you take some practice trips to do so, while both children are still young and gullible.

That is, you tell them that you are on your way to some desirable destination. Drive there slowly, on side streets. As soon as fighting begins, pull over, stop, turn around, say "No fighting in the car!" Then turn around, drive home in silence, and put each in his or her own room for five minutes.

Let them overhear you call someone at the desirable destination and explain why you are not going to show up.

It rarely takes more than two practice trips to make the point.

# Part Four

# Glossary of Medical Terms with Pronunciations

# Glossary of

# Medical Terms with

# Pronunciations

Pronunciations below are given with a Californian accent. The emphasis on syllables is as follows: greatest emphasis on the syllable in capital letters; next greatest on the syllable in italics. As in:

TELL-uh-*vizh*-un.

**Acetaminophen** (a-*see*-ta-MEEN-ah-fen): The active ingredient in many medications (Tylenol, Tempra) that lowers fever and dampens pain. (It won't do a thing for a stuffy nose or cough.)

**Acyclovir** (ay-SIGH-klo-veer): An antivirus drug, used most often against serious or threatening herpes (see below) infections. Brand name is Zovirax (ZOE-veer-ax).

**Adenoids** (AD-in-oyds): Lumps of lymph tissue, similar to tonsils, that sit up behind the nose. When enlarged, they block breathing through the nose and contribute to middle ear infections. A child with enlarged adenoids sounds as if she is talking with a great wad of peanut butter stuck to the roof of her mouth.

**AIDS** (Acquired Immune Deficiency Syndrome): The disease caused by infection with the HIV virus. Children usually acquire this by transmission from infected mother to fetus, or by transfusion of contaminated blood, or rarely by sexual abuse. Children with AIDS cannot pass it on through casual contact or by hugging and kissing. They cannot pass it on from urine or stools. There has never been a report of a caretaker catching AIDS, or a household member catching AIDS, from an infected child. However, blood must be handled with great care—gloves and bleach.

**Allergen** (AL-er-jen): Something that can give rise to symptoms of allergy. Dust, molds, dander, pollen, bee stings, certain foods, are all potential allergens.

**Allergy** (AL-er-jee): The outcome when the immune system decides that something that is normal in the environment really is an enemy, and attacks it. Symptoms range from hay fever to asthma, from hives to anaphylaxis (see below).

**Alopecia** (*al*-oh-PEE-shyah): Absence of

hair. In young children this can happen because they twiddle and pull it out, or because they have a fungus, or rarely for other reasons.

**Alte** (AHL-tuh): Apparent Life-Threatening Event. The term for what many people call, incorrectly, a "near SIDS" or a "Near crib death." This is when caretakers find a baby who is apparently not breathing, but who is resuscitated and appears fine afterwards. An ALTE needs to be fully evaluated: often it turns out that the baby was fine all along, and had merely undergone a normal pause in breathing. But if that's not the case, the baby needs a full work-up for Apnea (see below).

**Amblyopia** (*Am*-blee-OH-pee-ah): Blindness in one eye, often preventable. It occurs when one eye has weaker vision than the other, and lets the other eye do all the work. Since the weak eye never sends impulses to the brain, the brain forgets how to make sense of images seen through that eye. If the weak eye can be made to work hard before the brain forgets, vision can be preserved. A clue to the threat of amblyopia is a "lazy" or "wandering" eye (see Strabismus, below).

**Amniotic fluid** (*am*-nee-OT-ick): The liquid in the womb that surrounds the fetus. Usually it is clear like water, unless the baby stools into it or unless there is an infection of the uterus.

**Amoxicillin** (a-MOX-i-*sill*-in): An antibiotic frequently used for childhood infections. It is related to penicillin, but is effective against a wider range of bacteria. An augmented or boosted form of amoxicillin is Augmentin.

**Anaphylaxis** (*An*-a-fil-AX-is): A life-threatening allergic reaction culminating in shock (loss of blood pressure). Fortunately, it's very rare. When it does occur, it is most often after exposure to a medication, stinging insect, or food you already know you're sensitive to. Anyone with such a sensitivity should carry a "bee sting" kit containing adrenalin to give by injection.

**Anemia** (ah-NEE-mee-ah): Lower than normal red blood cell mass. This can mean either that there are too few red blood cells, or that there is a normal number but the red blood cells are too small. It can often mean "low iron," but not necessarily. A child can be anemic with normal iron (for instance, with sickle cell anemia) or have normal red cells but still have too little iron.

**Angel kisses:** Pink patches on eyelids and forehead at birth that generally fade and disappear over the first months and years of life.

**Antibiotic** (*An*-tih-bye-OT-ic): Medication that either kills or prevents reproduction of bacteria.

**Antiviral drug:** A medication that treats viral infections. These are not called antibiotics because antibiotic means attacking a *living* enemy, and viruses aren't really living. They are bits of genetic material that have to harness the machinery of *our* cells to reproduce. There are only a few antiviral drugs—because it is tricky to get rid of the virus without harming the virus-infected cells. The only one parents are likely to encounter is acyclovir (see above).

**Anus** (AY-nus): The external opening through which poop exits from the rectum.

**Apgar score** (AP-gahr): A point system

developed by Dr. Virginia Apgar that tells how well a just-born baby is adapting to life outside the womb. It judges breathing, circulation, and central nervous system responses. The top score is 10, which very few babies receive. Above 7, most babies need no assistance; below, they may need suctioning, oxygen, or other help.

**Apnea** (AP-nee-ah): The term used when a baby, usually a newborn or young infant, pauses too long between breaths. All normal young babies have irregular breathing, especially when they sleep. A baby with apnea stops between periods of breathing long enough to cause worry. In full-term babies, this is usually about twenty seconds. That's longer than worried new parents think: time it and see.

*Sleep apnea* occurs in children whose upper airways become blocked during sleep, usually by tonsils and/or adenoids. These children are noisy, snoring, restless sleepers. During the daytime, they go around with their mouths gaping open, partly because they can't breathe through their noses and partly because they yawn a lot. This isn't a comical condition, though; it's handicapping. Tired children can't learn, behave, or have fun very easily. Sleep apnea can even be medically dangerous because it puts a strain on the heart. Upper-airway obstruction also is a cosmetic problem, causing malformation of teeth, jaw, and face structure.

**Aspiration** (As-pur-AY-shun): 1. The inhaling of something foreign into the lungs. This often causes inflammation of lung tissue called aspiration pneumonia. 2. The suctioning or removal of fluid from a body orifice or even from an internal part of the body.

**Asthma** (AZ-mah): The tendency to have wheezing episodes triggered by something in the environment or by activity. This could be due to allergy, an irritant like secondhand smoke, a change in the temperature or dryness of the air, emotional stress, or exercise (as opposed to wheezing because of a specific illness like bronchiolitis, or because of an anatomic problem, such as a mass pressing on the breathing tubes).

**Atopy** (AY-toe-pee): A condition in which the tendency towards allergy is inherited. The term is often used about children with a family tendency towards eczema, asthma, hay fever. Such children are called atopic (ay-TOP-ic).

**Augmentin** (og-MENT-in; *not* pronounced Ag-You-ment-in): An augmented, or boosted, form of the antibiotic amoxicillin. It's used when bacteria are suspected of being resistant to amoxicillin. It must be given with food, which is a good thing because children often don't like the taste of the "boosting" agent.

**Avulsion** (ay-VUL-zhun): An injury in which a piece of tissue is torn off. Not an uncommon injury to youthful fingers and toes. Don't forget to check on tetanus shot status.

**Bacteria:** The plural form of bacterium. Bacteria are one-celled animals that can cause infectious disease. Because they are independent and multiply on their own, they are more accessible to being killed or foiled by medications than viruses are. These medications are called antibiotics (see above).

**Balanitis:** (bahl-in-EYE-tis): A red, swollen penis, usually in a boy who has not been circumcised. The usual cause is infection from not cleaning under the foreskin. But yeast infections, chronic irritation from wet diapers or other sources, and even trauma (caught in the zipper!) can also play a role. Usually treated with topical or oral antibiotics and improved hygiene. If the infection keeps recurring, the boy may need to be circumcised

**Bilirubin** (BILL-ee-roo-bin): The yellow chemical that causes jaundice, or yellowing of the whites of the eyes and skin. Many people mispronounce the word as "belly-rubin." This is understandable since pediatricians judge the depth of jaundice in very young babies by pressing on the skin in different parts of the body, which certainly does look like belly rubbin'. This is because jaundice starts at the top of the body and works its way down. Bilirubin is the "breakdown" product made when red blood cells get old and die, as they are always doing. Normally, the liver puts out the bilirubin into the intestine, where it exits in poop. Too many dying red blood cells, or a liver problem, or an obstruction preventing the bilirubin from exiting into the intestine, all can cause jaundice (see below).

**Binkie:** Parentese for pacifier.

**Bronchiolitis** (*bron*-kee-o-LIE-tus): A viral illness of young babies (usually) that causes squeaky, wheezy, effortful breathing. The most common virus that causes it is RSV (respiratory syncytial virus).

**Bronchodilator** (*bron*-co-DIE-later): A medication that opens up the tubes of the lungs to make breathing easier. Used for asthma and often for bronchiolitis.

**Cat Scratch Disease (CSD):** An illness featuring swollen lymph nodes, that is caused by a bacteria (B. henselae) carried by cats. It can be passed on by a cat bite or scratch. Most often it is kittens under a year of age that carry the bacteria. While it usually features a low-grade fever and scary-looking, big swollen lymph node (in the neck or armpit), it can also give many other symptoms. A good reason to teach children not to play rough with animals.

**Carotenemia** (*kare*-o-tin-EEM-ee-yuh): An orange-yellow coloring to the skin from, believe it or not, eating too many yellow foods, and other foods that contain carotene. Like carrots. And squash, sweet potatoes, apricots. Also, spinach. You can tell it's not jaundice because in jaundice the whites of the eyes are yellow, and in carotenemia they stay white. It is often an innocent condition, but rarely can mean liver or thyroid disease. Or it can be a sign that the child has bullied caretakers into allowing him to eat ONLY those delicious yellow foods.

**Cataract** (CAT-uh-rakt): A clouding of the lens of the eye. Not the cornea: that's the clear covering over the iris. Rarely, a child can be born with a cataract, or can acquire one after eye trauma or long exposure to steroid medication.

**Catheter** (KATH-e-ter): A tube placed into any body opening or into a blood vessel, both for withdrawing fluids and delivering fluids and medication.

***C. difficile*** (SEE dif-fi-SEEL): This bacteria can be found in the intestines of perfectly healthy babies and infants, but can cause diarrhea if it makes a toxin. This can happen in children taking antibiotics, or in very young babies who are not taking them.

**Cerumen** (ser-OO-min): Euphonious and dignified name for ear wax.

**Cervical** (SUR-vick-ul): Having to do with the neck, as in cervical spine or cervical collar. (Also, the term used for the "neck" of the womb—but rarely used in this way with the pediatric age group.)

**Chicken pox** (varicella): A childhood disease with rash caused by the virus herpes zoster (see Varicella, below).

**Clitoris** (KLIT-oh-ris): The sexually sensitive, little hooded organ formed by the meeting of the labia minora, in front of the vagina and urethra.

**CMV:** See Cytomegalovirus, below.

**Congenital** (con-JEN-i-tal): Describing a condition one is born with. This doesn't necessarily mean the condition is inherited. Nor does it necessarily mean that the condition was discovered at birth.

**Conjunctivitis** (kun-*junk*-tiv-EYE-tis): Infection or inflammation of the white of the eye. Since it can be caused by viruses OR bacteria OR trauma OR a foreign body OR allergy, conjunctivitis as a term doesn't tell you much.

**Constipation:** Hard stools, or *abnormally* infrequent stools. Some children have infrequent stools as a normal condition. For instance, most breastfeeding babies more than a few weeks old don't stool more than every two to five days, but they thrive and their stools are soft and easy to pass. Beware confusing the infrequent stools of constipation with the scant, rare stools of starvation. Most often, constipation is the result of a "vicious cycle" of painful stools and withholding pooping. Such constipation may be considered a minor problem, but left untreated can turn into a chronic and major unpleasantness. Called Encopresis (see below). Rarely, constipation may result from a low thyroid function or intestinal blockage or other problem.

**Contagious:** Describing an illness that can pass from one person or animal to another. Many infections are contagious; some are not. Those that are can be passed on by several different routes. Moods and behavior are also somewhat contagious.

**Coxsackie** (cock-SACK-ee): A family of viruses that can cause a number of disease syndromes, such as hand foot-and-mouth disease and herpangina.

**Crib death:** See SIDS (Sudden Infant Death Syndrome).

**Cromolyn sodium** (KRO-mo-lyn): A unique allergy medication. It stabilizes the membrane of the mast cell, so that it can't release its "allergy toxins" during an allergy attack. Thus, it is preventive only, and must be taken regularly—as nasal spray or by inhalation.

**Croup** (CROOP): This common illness features a cough that sounds like its name—like a seal barking or goose honking. It also causes a hoarse voice and stridor—that is, trouble and noise inhaling. This is because the lining of the whole upper airway swells, and the muscles encircling the trachea go into a spasm. Croup can be caused by any of several viruses, or can be triggered by dry or dusty air.

**Cryptorchidism** (kript-OR-kid-ism): Medical term for undescended testicles—those which haven't come down into the scrotum during fetal development.

**Cystic fibrosis** (SIS-tick fie-BROH-sis): An inherited disease that causes a spectrum of lung, intestinal, and growth problems. Most often, the disease is suspected and diagnosed in infancy or toddlerhood in a child with chronic lung (and ear and sinus) infections and poor growth. In milder forms, the diagnosis may be suspected later—rarely, as late as adolescence. There are treatments, but as yet no cure, for the disease. There are treatments, but as yet no cure, for the disease. The diagnosis of CF is confirmed by a "Sweat Test" (see Glossary, below).

**Cystitis** (sis-TIE-tis): *Cyst* means bladder, and *itis* means infection or inflammation of. So cystitis usually means a bladder infection. It's not quite the same thing as a UTI (or urinary tract infection) because the term UTI could include the kidneys as well. Also, sometimes cystitis is due not to infection but to irritation.

**Cytomegalovirus,** or **CMV** (Just say "See Em Vee"): A virus that can cause mental retardation, deafness, and other problems in a fetus whose mother catches the disease. Older babies, children, and adults usually only have a mild illness, so a pregnant woman can become infected and not know it. Fortunately, many women have already had CMV in childhood, and this helps greatly to protect the fetus. Who's most at risk? The fetuses of pregnant women who aren't already immune to CMV and who might catch it. This means anyone around small children, because CMV is spread by apparently healthy children, and via all the bodily fluids that children so merrily and copiously shed. Before conceiving, it's a good idea for a woman to be tested for immunity to CMV.

**Dextromethorphan** (dex-tro-meth-OR-fan): An ingredient of many cough and cold medicines, which suppresses the cough reflex. Often abbreviated DM on labels.

**Diabetes** (dye-ah-BEE-tees): "Sugar" diabetes—a condition in which the pancreas does not make enough insulin, and the level of sugar in the blood stays much too high—can be a life-threatening disorder and requires urgent treatment. Symptoms include: increased thirst and urination, lethargy, stomach ache, vomiting, and "not acting right." Lifelong treatment with insulin (given at present as injections) and careful diet allows most children to live nearly normal lives. The condition is inherited, but in a highly complex fashion.

**Diphtheria** (Dif-THEER-ee-ah): Infection with the bacterium diphtheria—the D of the DTP vaccine, which prevents such infection. The infection causes a membrane over the throat, or, if it is contracted from dirt, over an open wound. The main danger is a toxin, or poison, created by the bacteria, which can damage organs, in particular the heart. Before the vaccine, diphtheria wiped out whole families.

**Eczema** (ECK-zima): An allergic skin disease. Babies get crusty, oozy spots; children get dry, scaly skin in creases. Scratching and frequent baths make it worse.

**EEG** (eee-eee-gee): Short for electroencephalogram, (ee-lec-troh-en-SEPH-a-low-gram) sometimes referred to as a "brainwave test." The EEG records electrical impulses in the brain, helping to determine if a child has a seizure disorder, and if so what kind—it's a guide to medication and prognosis. It doesn't hurt and doesn't deliver electricity or any radiation to the child; it just records, through leads that are pasted on to the child's scalp. Wires connect the leads to the EEG machine. The child needs to sleep through the procedure, so is usually kept up late the night before; and may require sedating medication. After the leads come off, there is often some glue on the scalp: rub it with peanut butter before shampooing, and it will come out.

**EKG** (eee-kay-gee): Short for electrocardiogram (ee-lec-troh-CAR-dee-oh-gram), a gadget for recording the electrical output of the heart. It is particularly helpful for diagnosing rhythm disturbances, enlargement of various parts of the heart, and strain on the heart muscle. It doesn't hurt, doesn't deliver any electricity to the heart, and does not require sedation (usually).

**Encephalitis** (en-*sef*-ul-EYE-tiss): An infection of the brain itself, not its surrounding membranes (that's meningitis). Encephalitis is most often caused by one of several viruses. The child with encephalitis has a headache and behavioral changes, and sometimes but not always fever or vomiting. Encephalitis can be fairly mild or it can be deadly. The three viral diseases covered by the MMR vaccine, measles, mumps, and rubella, used to be known for causing encephalitis. Chicken pox also can cause encephalitis, and perhaps that vaccine will be released by the time you read this. Herpes can cause a very serious form of encephalitis, for which acyclovir may be used as treatment.

**Encopresis** (en-co-PREE-sis): Pooping in the pants or other nontoilet places after bowel control should have been established. Usually it is a consequence of chronic constipation, and results when little bits of poop finally escape the rectum. Children with encopresis always are emotionally troubled. Usually, the emotions are due to, rather than the cause of, the encopresis. The Three who refuses to poop except in a diaper, in private, does NOT have encopresis but rather a normal developmental preoccupation with control and power.

**Endoscopy** (en-DAH-scop-ee): Examination inside a body orifice, such as the throat, airway, rectum, with a special instrument, usually a fiber optic light with a video camera attached.

**Enterobiasis** (en-ter-oh-BYE-uh-sis): Pinworm infestation. If you use the medical term, an eavesdropping Cherub won't know you're discussing scary, fascinating, gross, yucky Worms. You will be spared some histrionics.

**Enuresis** (en-yur-EE-siss): Bed-wetting long after daytime toilet skills are well established. By age five, 85 percent of children will be dry most nights; 15 percent still wet. So most pediatricians don't call bed-wetting under age five enuresis, but just normal bed-wetting.

**Epiglottitis** (ep-ee-glot-EYE-tis): A life-threatening swelling of the epiglottis, the little lid that covers the airway. Usually, the swelling is due to a bacterial infection called H. flu (hemophilus influenza). The "HIB" vaccine given in infancy seems to confer very good protection against this dreadful illness.

**Epilepsy** (EP-i-lep-see): A person with epilepsy has a tendency to have repeated convulsions or seizures. Many children will have one or two fever convulsions; this is not ever called epilepsy. Those who do have repeated seizures often can have them controlled by medications. Epilepsy is not the same thing as mental retardation. Some children with epilepsy are retarded and some children who are retarded have seizures. But most people with epilepsy have perfectly normal intelligence and behavior.

**Erythema** (*air*-i-THEEM-ah): A reddening of the skin. Usually, there is another word tacked on here:

*Erythema toxicum* (TOX-i-cum): An innocent rash of newborn babies, consisting of a white pimple on a red splotch.

*Erythema infectiosum* (in-fex-ee-OH-sum): Better known as fifth disease, or slaps; a common viral (pervovirus B19) illness of childhood, featuring bright red cheeks and a lacy rash elsewhere.

*Erythema multiforme* (mul-ti-FOR-me): A particular rash featuring spots that look like bull's-eyes or donuts or bagels. (They are called target lesions.) It can be caused by infections or by allergy to drugs or, rarely, foods. Most often the cause is not found. A rare severe form called Steven Johnson syndrome afflicts the eyes and mouth as well, and is often caused by allergy to medications.

*Erythema subitum* (SOOB-i-tum): Better known as roseola, a common viral illness, with rash following the fall in fever.

**Eustachian tube** (you-*STAY*-chee-un): The tube that connects the back of the nose and throat to the middle ear, so that air can get into the middle ear, allowing the eardrum to vibrate. Not to be confused with the ear canal, where the wax is.

**Excipient** (ex-*SIP*-ee-ent): Fancy name for the inactive ingredients in medications, such as coloring, flavoring, sweeteners, and preservatives. Allergy to these may masquerade as allergy to the medication itself.

**Expiratory** (*EX*-pri-toe-ree): Having to do with breathing out (opposite of inspiratory). The high whistling noise made breathing out with effort is called wheezing.

**External otitis** (oh-TIE-tis): Infection of the ear canal and outside of the ear, often known as swimmer's ear.

**Febrile seizure** (*feb*-rile SEE-zhur), or **fever convulsion:** An innocent but frightening convulsion or seizure caused by a sudden rise of temperature, either from reaction to vaccines or from infection. Five to 7 percent of children will have one between six months and six years of age.

**Flu:** Short for the word influenza, but a pretty meaningless term except when describing a particular epidemic caused by said virus. It is not useful to describe a set of symptoms as "flu." To some, "flu" means high fever and aches; to others, vomiting and diarrhea; and to others, cold symptoms.

**Fontanel (soft spot)** (fon-tuh-NELL): The soft place at the top of the baby's head that the skull doesn't cover. This is the place where several plates of the skull meet. These plates of bone must be "floating" and unattached to each other, both so that the skull can be molded during delivery, and so it can accommodate the rapidly growing brain. The soft spot is covered by a durable membrane; it's tough as well as vulnerable. You will often see it pulse gently with the heartbeat. It will always look slightly sunken, of course, because there isn't any skull over it. If it is *very* sunken, this is a sign of severe dehydration. It may bulge briefly during intense crying, but a bulging fontanel in a sick baby may mean meningitis. The fontanel closes by about eighteen months to two years of age.

**Functional:** A term used by medical personnel, meaning that whatever sign or symptom being discussed is real but normal, not a problem, such as a "functional" heart murmur. Or that it is a problem caused by behavior or emotion rather than an abnormality of the body per se: such as "functional" constipation.

**Gastroenteritis** (*gas*-tro-en-ter-EYE-tis): Diarrhea and vomiting, usually due to a viral infection.

**German measles:** See Rubella.

**Giardia lamblia** (JHAR-dee-ah LAM-bee-ah): A one-celled intestinal parasite, passed on through food, water, or dirty hands, that causes intestinal problems. Usually, this means diarrhea and cramping, alternating with constipation, and pale-colored, floating poops.

**Gingivostomatitis** (JIN-ji-voe-sto-ma-TIE-tis): *Gingivo* means of the gums; *stoma*, of the mouth; *itis*, infection. This infection of the mouth and gums is usually due to childhood herpes virus.

**Glands:** When used in the term "swollen glands," these aren't glands at all, really, but lymph nodes. True glands are organs that secrete a substance necessary for the body. The pancreas, thyroid, and adrenals are all glands.

**Glaucoma** (glow-COH-muh): Increased pressure in the eye. In children, it's rare and almost always congenital, and shows up as a teary eye that looks enlarged. It seems to hurt when the baby looks at a bright light, and the baby is usually fussy and fails to thrive. A teary eye with these features needs to be checked carefully; don't assume it's just a blocked tear duct. Rarely, glaucoma can occur after an eye injury.

**Hand-foot-and-mouth disease:** A disease caused by the virus coxsackie, which includes sores in the mouth and on the palms and soles. Not related to hoof and mouth (nor to putting one's foot in one's mouth and getting out of hand).

**Heart murmur:** A noise made by turbulence of the blood as it whooshes from one place to another—through the chambers of the heart, from the heart into blood vessels. Most often, a heart murmur heard in childhood is perfectly normal, just due to enthusiastic whooshing. Next often, it is due to a small opening between two heart chambers, or to an irregularity of a valve. Many of these causes require no special care at all. If care is needed, it often consists only of preventive antibiotics before dentist appointments. This is because any dental work (even cleaning and braces adjustments) gets bacteria into the bloodstream, and they do like to cling to valve irregularities and cause infection.

**Hematoma** (*heem*-a-TOE-muh): Bleeding so as to cause a swelling. Sometimes used as a fancy name for an ordinary bruise, but also for more serious bleeding, such as between the brain and the skull.

**Hemophilus influenza** (**H. flu**) (Just say "Aitch Floo"): A family of bacteria that has nothing, really, to do with the virus influenza. The name arises because this was the infection that caused fatal pneumonia after influenza in the great epidemic early in this century. H. Flu causes minor infections, like otitis media, and major ones, like meningitis and epiglottitis. The HIB vaccine gives excellent protection against the serious H. flu infections, but not against the minor ones.

**Henoch-Schönlein purpura** (**HSP**) (Just say "Aitch Ess Pee"): An uncommon but not rare disorder in which tiny blood vessels become inflamed, causing a bruiselike rash on the legs. There can be complications in the kidneys, intestine, and other organs. The cause is not known.

**Hepatitis** (*hep*-uh-TIE-tis): An infection or inflammation of the liver. Many viruses can cause this as a kind of side effect; "mono" is famous for doing so. The ones that primarily cause liver disease are called hepatitis, and named by capital letters:

*Hepatitis A* is the most common. It is carried by contaminated food, is usually innocent, and often goes undiagnosed in childhood. A vaccination is available, but not given routinely. Exposed people can get a gamma globulin shot to protect against disease.

*Hepatitis B* is also usually undiagnosed in childhood, but it is much more serious. The virus can linger lifelong and in adulthood can spring to life again and cause serious, even fatal liver disease. The earlier in childhood the disease is acquired, the greater the chance of this happening. Immunization is recommended for all children, especially those under age five. All children of mothers who are carriers of Hepatitis B should be immunized at birth, and given a gamma globulin shot to try to protect them against this disease.

*Hepatitis C* is also a dangerous hepatitis, often passed by blood transfusion. A vaccine for C should be available shortly.

**Hernia** (HER-nee-ah): An opening in a body wall that shouldn't be there, allowing a bit of the body contents to protrude through, either into another body compartment or into the outside world as a bulge. There are several varieties:

*Diaphragmatic hernia* (DIE-uh-frag-*mat*-ic): An opening in the diaphragm, the muscle that moves the lungs when we breathe, fails to close during fetal life, and a bit of intestine (rarely, a LOT of intestine) protrudes through into the chest. The rare, severe form can be life-threatening because it may keep the lungs from developing or from expanding well.

*Inguinal hernia* (IN-gwin-al): The most common hernia, in which the passage through which the testicles descend during fetal life stays open, allowing a bit of intestine to protrude into the scrotum.

*Umbilical hernia* (um-BILL-i-cul): A bit of intestine and inner abdominal lining pop out through the opening that the umbilical cord came through. Usually this is a cute normal variant, but rarely can indicate an underlying disorder such as a low thyroid function. As an innocent finding, it's most common in premature babies and African-American babies.

**Herpangina** (*herp*-an-JINE-uh): A viral syndrome often caused by coxsackie, in which there are painful sores in the back of the throat.

**Herpes** (HER-peez): A virus that causes both common, innocent, and rare, serious disease in childhood. Fortunately, most children who come into contact with herpes acquire immunity to it without ever getting sick. There are two big categories of herpes, both of which can, if necessary, be treated with the antiviral drug acyclovir (Zovirax).

1. **Herpes simplex** (SIM-plex) causes, among other things:

   *Herpes stomatitis* (sto-ma-TIE-tis): This common and usually innocent infection is nonetheless very painful and upsetting.

   *Neonatal herpes* (nee-oh-NAY-tal): A very dangerous and overwhelming infection usually transmitted from mother to baby during labor and delivery.

   *Herpetic whitlow* (her-PET-ick WIT-loe): A painful herpes sore on fingers or hands, usually where a child sucks.

2. **Herpes zoster** causes:

   *Chicken pox, or varicella* (vare-ih-SELL-uh): A usually innocent childhood disease with rash.

   *Shingles:* A painful skin rash, caused when the chicken pox virus, which stays dormant in the body lifelong, becomes aroused by stress, sun, or renewed exposure to chicken pox.

Both cause uncommon but dangerous eye infections called ocular herpes (OCK-you-ler) or keratoconjunctivitis (CARE-at-oh-con-junk-tiv-EYE-tis), which need prompt diagnosis and treatment by an ophthalmologist.

**Hives (urticaria;** *oor*-tee-KARE-ee-uh): Itchy "welts" or blotches on the skin caused by an allergic reaction to any of a number of allergens: foods, medications, insect stings, or even a viral or bacterial infection. Much of the time, the underlying cause isn't ever determined. Adrenalin by injection and antihistamines by mouth help. Sometimes steroids are needed.

**Hydrocephalus** (*high*-dro-SEF-uh-lus): Increased pressure of the fluid that bathes the inside of the brain, often caused by an obstruction to its drainage. It is rare, and found predominantly in very premature babies, children with spina bifida, brain tumors, and after bacterial meningitis. But once in a while a child with no underlying problem will have hydrocephalus, showing up as a rapidly enlarging head, often with a very protruding forehead. Treatment can often prevent or limit damage to the brain.

**Hyper:** Too much of something. Too

much of a hormone (as in hyperthyroid) or too much of a symptom (as in hyperemesis, or too much vomiting) or too much of a good thing (hyperimmune from getting a tetanus shot with every booboo). Or just too much altogether, as in "She was so hyper I wanted to put her in a padded cell!"

**Hyphema** (high-FEE-muh): Bleeding into the little chamber at the very front of the eye (between the iris—the colored ring that makes us Brown or Blue or Whatever-eyed— and the clear covering of that same ring, which is called the cornea). Such bleeding can happen after a direct blow or a piercing injury to the eye. If it's not diagnosed and treated, vision can be impaired or destroyed, which is one of several reasons why blows to the eye need to be looked at by the pediatrician, emergency doctor, or ophthalmologist.

**Hypo:** Too little of something. Hypothyroid, or hypocalcemia, or hypotension (low blood pressure). (Isn't it interesting children's behavior is never called "hypo"?)

**Ibuprofen** (eye-byu-PRO-fen): The active ingredient in such medications as Advil and Motrin, which reduces fever, pain, and swelling.

**Idiopathic** (*id*-ee-oh-PATH-ick): A word that describes a baffling symptom, disease, or condition. Something is causing it, but we don't know what.

**Immunoglobulins** (i-*myun*-oh-GLOB-you-lins): Protein molecules made by the immune system as a response to "attackers" like viruses, bacteria, and allergens.

**Impetigo** (*im*-puh-TIE-go): Not "infantigo," though it is often young children who get this skin disease with its blis-ters, crusting, and oozing. Caused by strep bacteria, sometimes with staph helping out.

**Incubation period** (*in*-kew-BAY-shun): The amount of time between someone's exposure to a germ that causes disease and the appearance of symptoms.

**Infantile spasms:** A rare form of seizures that doesn't occur before two months or after nine months of age. Babies with this rare disease often look as if they are "salaaming" with arms outstretched and head bowed with chin to chest. They do this repeatedly, several times in a row. They also show behavior changes: increased sleepiness, less playfulness, not learning new things. Early treatment may make a big difference in prognosis.

**Infectious:** Pertaining to a disease caused by an infection rather than by a metabolic disturbance or ingestion or some other cause. Infections can be caused by viruses, bacteria, fungi, algae, parasites, and yeast. Some infections are contagious and some are not.

**Influenza** (*in*-floo-EN-za): A family of viruses that cause "flu" epidemics just about every year. Symptoms generally include aches, fever, cough, and sore throat. A new vaccine is needed every year, and may even then be ineffectual, because these viruses are so clever at changing their protection against immune defenses. But all children with chronic or serious diseases or conditions need yearly flu shots.

**Inguinal:** See hernia, above.

**Inhalation:** The act of breathing in something. This could be medication in an aerosol, or something toxic like secondhand smoke.

**Inspiratory** (IN-spri-toe-ree): Occurring while breathing in rather than out. The opposite of expiratory. The noise made by effortful breathing in is called stridor.

**Intubate** (IN-toob-ate): To place a tube into the airway, to suction, to deliver oxygen, or to enable the physician to take over the whole act of breathing.

**Intussusception** (IN-tuh-suh-SEP-shun): An uncommon but not rare severe intestinal problem of infants, three months to three years old, very rare in older children. The intestine telescopes on itself, as if you wrinkled a sleeve. Most babies will scream with pain, vomit, and have stools that look like currant jelly. But others will simply become limp and lethargic and pale. URGENT care is needed. Often the "wrinkling" can be straightened out by a special enema, but sometimes surgery is required.

**Jaundice** (JAWN-diss): The yellow color to skin and whites of eyes caused by too much of the substance *bilirubin* (see above). **Jaundice is NOT normal in babies who act sick, start to become jaundiced after five days of age, in whom jaundice persists after two weeks of age, or who look REALLY yellow.** When in doubt, ask your pediatrician. See the discussions in the *Birth to Two Weeks* chapter.

**Kawasaki disease or syndrome** (cow-a-SOCK-ee): A serious disease with a blotchy rash, red, cracked lips, redness of the whites of the eyes, and fever. The serious aspect is that the coronary arteries (the ones that supply the heart with blood) can become damaged afterwards. Prompt diagnosis, treatment, and follow-up is vital. The cause isn't yet known.

(It used to be called mucocutaneous lymph node syndrome.)

**Ketosis** (kee-TOE-siss): What happens to the body chemistry of a child who has had vomiting and diarrhea, prolonged fever, or other metabolic stress. Acid substances build up that make the child feel worse and less inclined to take fluids. And that makes the ketosis worse. This vicious cycle is addressed in the discussions on Vomiting, Diarrhea, and Dehydration in Part II. Children with diabetes can also get ketosis, but they will need medical help (and insulin) to get out of the problem.

**Kidneys:** The paired organs back under the ribs that sift the blood, returning the good stuff and sending the rest out as urine.

**Labia** (LAY-bee-ya): The folds surrounding the vagina. The inner folds are the labia minora (min-OR-ah) and the outer ones the labia majora (ma-JOR-ah).

**Labial adhesions** (LAY-bee-al ad-HEE-zhuns): When they get a bit irritated, the labia minora sometimes stick together and actually fuse, so that when you look at the child's perineum (see below), there seems to be no opening. This can be very alarming: it looks as if something is terribly wrong, but it's not. There is an opening through which urine can exit; you just can't see it. Sometimes treatment is required, but the adhesions usually go away eventually on their own.

**Lactose:** Milk sugar, present in human milk, cow's milk, goat's milk, and most milk-based formulas. It is really a complicated sugar in which each molecule is composed of two little sugars (glucose and galactose). Lactose can't be absorbed from the intestine until the enzyme lactase splits it apart. No lactase, no absorption: instead, diarrhea and gas.

**Lanugo** (lan-OOG-oh): The downy body hair all fetuses grow. Most babies still have some at birth.

**Laryngitis** (LARE-in-JYE-tis): Hoarse voice. It can be caused by infection, usually viruses but sometimes bacteria. It can also be caused by little benign growths on the vocal cords, or rarely by other structures pressing on them.

**Larynx** (LARE-inks): Voice box. Not pronounced "Lare-Nix" or "Lare-ninx."

**Lazy eye:** Seems to have three meanings, so be careful. 1. An eye that wanders and lets the other eye do all the focusing and seeing. 2. A droopy eyelid (see Ptosis, below). 3. An eye that had such poor vision early in life that amblyopia develops (see Glossary entry).

**Lethargic** (leh-THAR-jick): An adjective that means different things to different people. To medical personnel, it is a red alert for behavior that is very abnormal: lying limply, not paying attention to anything. To many parents, it means "not as active as usual." Both parties are better off describing specific behavior than using this term.

**Leukemia** (lew-KEE-mee-ah): A rare cancer of the bone marrow, attacking only 2.4 out of 100,000 nonwhite children and 4.2 out of 100,000 white children worldwide; the nightmare of parents of young children. It is true that the peak age for the most common, and treatable, or I should say curable, form of leukemia is from three years to five years. However, most children with

signs and symptoms that hint of leukemia do not have it. Moreover, the two symptoms that parents worry about most—dark circles under the eyes and frequent upper respiratory illnesses—are not in the running as heralding leukemia. These are much more likely to result from allergy, daycare exposure, secondhand smoke, and fatigue.

**Liver:** The organ underneath the right ribs. It gets rid of toxic products our own bodies make or that we imbibe or ingest, and produces helpful chemicals like cholesterol. (Cholesterol is not an enemy unless it's in excess. It's the basis for many delightful substances, like sex hormones.)

**Lumbar puncture or LP** (Just say El Pee): The procedure that allows a small amount of spinal fluid to be removed, examined, and cultured, to diagnose such illnesses as meningitis. It is a test that is safely and frequently performed in young children, because early diagnosis of meningitis is critical for a good outcome. To perform the test, the child is held in a curled-up position (usually lying on the side; rarely, sitting up) and a needle is carefully inserted between the vertebrae of the lower spine, into the fluid-containing portion. It is stressful for the child (to say nothing of the parents) but no more painful than a blood draw, usually: it's the being held that they really resent.

**Lyme disease:** A difficult-to-diagnose (at this time) disease caused by bacteria and spread by certain ticks in certain parts of the country. It often starts on the skin, with a spot that spreads to become a big wobbly circle with a clear center; then there are fever and aches. Later on, joint swelling and even signs of meningitis and other neurological disorders can appear. It's a tricky disorder. *First*, it's hard to find the ticks, and a bite may go overlooked. They are tiny, about the size of a pencil dot. You need to look for "moving freckles." Happily, the tick has to be attached for twenty-four hours before it can pass the disease, so a careful, hilarious tick search with magnifying glass and flashlight every night should prevent problems. So might a thorough, daily shower and shampoo. Do this if you're hiking in Lyme country. (The borders change all the time. Ask.) *Second*, it's hard to diagnose: the skin rash may not be present, the symptoms nonspecific, and blood tests confusing. Fortunately, if suspected and treated early, the disease tends to be shorter and milder. But treatment too soon can distort the blood tests! So it's tricky, tricky, tricky.

**Lymph nodes:** These are the "glands" that get swollen when a child has "swollen glands." They're not really glands at all. Glands are properly organs like your thyroid or pancreas, organs that work all the time, making specific substances needed for the body to thrive. Lymph nodes just sit there until an infection occurs and then jump in to arrest it. Many times, parents worry that enlarged nodes may mean something ominous, such as a malignancy. Such a diagnosis is most unlikely because lymph nodes swell up with many innocent infections. However, since it's crucial to diagnose and treat serious diseases as early as possible, sometimes a bit of worry is warranted. Lymph nodes that persist without shrinking for a week after the child is well, or that feel rub-

bery or matted, or that are not associated with a clear-cut illness, or that are present in the area just above the collarbone or around the elbow, should be checked by the pediatrician.

**Mastitis** (mas-TIE-tis): A breast infection that occurs when a milk duct becomes blocked, creating a very tender hard lump, usually with a shiny red area on the skin. In nursing mothers, usually treated with heat, increased nursing, massage, and often with an antibiotic. In new babies of either sex, may require injections of medication or a stay in the hospital.

**Mastoiditis** (MAS-toyd-EYE-tis): Infection of the bone behind the ear. It used to be a feared complication of chronic middle ear infections, but has become much less common (but not unheard of) with the development of antibiotics.

**Measles:** See Rubeola.

**Meconium** (meh-CONE-ee-yum): The dark green-black stool of newborns, which has built up in the fetal intestine during the whole gestation period. It's that color because it contains concentrated bilirubin (see above).

**Meningitis** (*men*-in-JYE-tis): An infection of the membranes surrounding the spinal cord and brain. "Spinal meningitis" is a redundant term that doesn't convey any more information than the term "meningitis" by itself. The important distinction, usually, is what causes the meningitis. Viral meningitis very often is mild and doesn't usually produce problems like mental retardation, seizures, deafness, etc. Bacterial meningitis can be very serious, even fatal, even if treated promptly. Early signs **don't**

usually include a stiff neck in young children. In babies, a bulging soft spot or fontanel may be noted, but that may be subtle or hard to judge in a baby who's crying. The younger the child, the more important it is to perform a lumbar puncture (spinal tap) to rule out meningitis if the child is acting "really sick" or has a worrisome fever.

**Metabolic** (*met*-a-BALL-ic): Having to do with the body chemistry that allows us to turn food into energy, get rid of toxins, maintain a normal temperature, and grow. An inborn error of metabolism occurs when a child is born without the ability to perform one particular part of a metabolic task. When this happens, the whole "chain" of metabolism, from the missing link on, becomes distorted.

**Milia** (MILL-ee-yuh): The sprinkle of white dots on a newborn's face, especially the nose. These are tiny little skin cysts that will go away in time.

**Molluscum contagiosum** (moll-USS-cum con-*tage*-ee-OH-sum): For all that, you'd think it would be more than warts, but that's what it is, tiny viral warts, a bit shiny. Look closely and you'll see a tiny dimple in each. Easy to treat when they're new and few, harder when they're older and bolder.

**Mongolian spots:** Bluish, blackish, or grayish flat spots present in many newborns (about 80 percent of Asian, Native American, and African-American babies; about 10 percent of others). They can be anywhere on the body. They aren't bruises, but can be mistaken for them; a newborn may be thought to have been battered during delivery. Most go away by late childhood.

**MRI:** An imaging technique that doesn't use X rays. (Its true name is magnetic resonance imaging, and it does indeed involve magnets.) It is sometimes used instead of or with a CT scan to examine both soft tissues, like the brain and abdomen, and joints; any part of the body can be examined by MRI. Young children usually need anesthesia because they must hold very still and be encased in a kind of tunnel for the procedure.

**Mycobacteria:** A group of bacteria that includes tuberculosis and leprosy. Often hard to diagnose and/or treat.

**Mycoplasma** (MIKE-oh-PLAZ-muh): A tiny bacterium that often causes "walking pneumonia," mycoplasma has some viruslike characteristics. It usually responds to erythromycin or tetracycline (which is not given to children) but not to other antibiotics. It is not related to mycobacteria.

**Nasolacrimal duct** (NAZE-oh-LACK-rimal): The tube or duct that carries tears (which are made in a gland under the upper eyelid) from the eye into the nose, where the lining absorbs the moisture.

**Nauseated vs. Nauseous:** Nauseated means feeling as if you're going to vomit. Nauseous, believe it or not, means *causing* nauseated feeling in somebody else. Most people use the word "nauseous" to mean "nauseated," and usually that's fine. But it is very disturbing to those who are fastidious, or overfastidious, about words. "I don't know any nauseous children, myself," she said fastidiously.

**Neonatal** (NEE-oh-NAY-tal): Having to do with the newborn, up to two months of age.

**Nephritis** (nef-RY-tis): An inflammation, swelling, and general irritation of the kidney NOT produced by infection (see Pyelonephritis).

**Nephrotic syndrome** (nef-RAH-tic): A kidney condition in young children in which the kidneys put out too much of the blood's protein into the urine. This causes the blood to become thin, so watery fluid leaks out into the tissues. Usually parents note puffy eyes in the morning, which gradually clear as the day goes on; then a distended belly and puffy feet. The child tends to be lethargic and urinates small, infrequent amounts. There are many causes for the kidneys to leak protein, and each case must be handled individually.

**Neuroblastoma** (NOO-roh-blast-OH-muh): This tumor is rare, attacking only about sixteen children per million, but it is the most common abdominal malignancy of infants. It usually is found in children under the age of two years. It arises from nerve cells of the fetus. These aren't "thinking" nerve cells, but nerve cells that run the automatic functions of the body—digestion, breathing, dilating, and constricting the iris of the eye, and so on. So the symptoms of this tumor can be myriad. Most often, however, it appears as a mass in the belly.

**Newborn Screening Test:** A test performed on a newborn from a few drops of blood from the heel. This test screens for several diseases and conditions. The best known is PKU (see below), and the test is often called the PKU test. The specific conditions screened for vary from state to state and from nation to nation, but all of the conditions cause either severe mental retardation or severe illness if they go untreated. The screening test detects the condition before any symptoms appear. If treatment can be given before the child has symptoms, the outlook for normal development and growth is much better. Some conditions require a special diet (PKU, galactosemia). Low thyroid function requires hormone supplements. The inherited anemias (sickle cell, thalassemia) require careful prevention of infection and low blood levels.

**Nursemaids' elbow:** Sexist name for subluxed elbow (see below).

**Nursing bottle caries:** Cavities, often devastating, in the teeth of babies who take a bottle to bed with them. The sugar (albeit natural) in milk and juice coats the teeth all night. Some children have to have all the teeth pulled (under general anesthesia) and metal caps put in. Children who snack frequently, or constantly sip juice or soft drinks, are at the same risk.

**Organic** (or-GAN-ic): A term used by medical personnel about a disease, meaning that it is caused by an identifiable physical agent. Opposite of "functional."

**Otitis externa** (oh-TIE-tis ex-TER-na): Infection in the ear canal, the outer side of the eardrum, "where the wax is."

**Otitis media** (oh-TIE-tis MEE-dee-yuh): Infection in the middle ear, on the "inside" aspect of the eardrum.

**Pancreas** (PAN-kree-us): The organ tucked away near the stomach that produces digestive enzymes and insulin.

**Parasite:** An animal that lives in and off of another animal—e.g., your child. Some parasites are microscopic and one-celled, like giardia lambia (see above). Others are small but quite visible, like pinworms; and others are repellantly obvious, like roundworms.

**Parenteral** (par-EN-ter-al): Given by needle into a muscle or by vein.

**PAT,** or **paroxysmal atrial tachycardia** (Just say "Pee Ay Tee"): The most common abnormal heart rhythm in childhood, but this doesn't mean it's a very frequent event. Abnormal rhythms happen when the heartbeat loses its "feedback" regulation from the body, so that it doesn't adjust its rate to what the body needs. In P.A.T., the heart rate goes to very fast very suddenly—so fast you won't be able to count the pulse beats. This is despite the fact that the child is very still. The smaller the child, the more exhausting this is, and the more dangerous. A baby with P.A.T. will seem ill, with poor color. If you go to feel the pulse, you won't be able to count it. Urgent treatment is needed. (Most of the time, there's nothing wrong with the heart itself. Just with how fast it's beating.)

**Pathologic:** A term used by medical personnel, meaning that something is abnormal or caused by a disease process. A bit stronger than "organic." Opposite of "functional."

**PDA,** or **patent ductus arteriosus** (Just say "Pee Dee Aye"): One of the most common causes of a heart murmur in babies. It occurs when a blood vessel that is crucial in fetal life (because it directs blood away from the nonbreathing lungs) fails to close soon after birth. In premature babies, medicine can often close it off; in full-term babies, surgery is usually needed.

**Pectus excavatum** (PEK-tus ex-ca-VAH-tum): A scooped-out breastbone (or sternum) present from birth, probably caused by the baby's position in the womb. On occasion, may be severe enough to consider surgical reworking of the sternum: a major undertaking.

**Pediculosis** (ped-ICK-you-LOW-sis): Head or body lice.

**Perinatal** (PAIR-ee-NAY-tal): Having to do with the well-being of the fetus and very young newborn.

**Perineum** (PAIR-in-EE-yum): The region known as "down there" by bashful adults everywhere. It includes the penis, scrotum, and anus in boys, and the clitoris, labia minora, labia majora, urethra, vagina, hymen, and anus in girls.

**Pertussis** (per-TUSS-iss): The bacterial infection that causes whooping cough, against which all children should be immunized. The P in the DPT vaccine.

**Petechiae** (pet-EEK-ee-yi): Tiny red spots that look like bleeding into the skin and don't blanche to skin color when you press on them. They can be innocent, but may be a sign of a serious infection or problem with the blood-forming system (bone marrow), so always must be checked out.

**PFAPA** (Just say the letters): A syndrome in which a child regularly, every four weeks—like clockwork, comes down with *P*eriodic *F*ever, *A*phthous ulcers (sores in mouth/throat), *P*haryngitis (sore throat), and *A*denitis (swollen neck glands). May be treated, once diagnosed by lab tests, with oral cortisone or tonsillectomy.

**PKU,** or **phenylketonuria** (Just say "Pee Kay You"). An inherited inability to properly metabolize a particular amino acid—that is, a component of protein. Since it can't be metabolized, this amino acid builds up and turns into a toxin that poisons the brain. Babies with PKU should not nurse or take regular formulas. They need a SPECIAL formula that doesn't contain that amino acid. This is a lifelong condition, and the special diet can prevent mental retardation if given early enough in life. PKU screening is the name often given to the Newborn Screening Test—a blood test done shortly after birth that screens for this and several other inherited conditions.

**Placenta** (plah-SEN-tuh): The magical organ attached on one side to the uterus and on the other to the umbilical cord, through which the fetus gets nutrition and oxygen and disposes of wastes. It is part fetal and part maternal, and why the mother doesn't reject it as foreign tissue isn't known.

**Pneumocardiogram** (NU-MOH-CAR-dee-oh-gram): A study to monitor a child's breathing and heart rate over a long period, including while asleep, to detect apnea or abnormal heart rate.

**Pneumonia** (noo-MONE-yuh): An infection or inflammation of the lung tissue itself—the microscopic little air sacs, not the tubes of the lung. It can be caused by viruses, bacteria, chemicals, or irritation (for example, from aspirated food or foreign body).

**Port-wine stain:** A birthmark, the color of dark wine, different from other birthmarks because: 1. It is present right at birth, and many "birthmarks" aren't. 2. It can signify

problems beneath the skin, such as eye and brain abnormalities, so it is taken seriously. 3. It is permanent if left untreated, but can usually be nearly erased by laser therapy.

**Proctitis** (prock-TIE-tis): Infection of the rectal area, often caused by such innocent nuisances as strep and yeast. Pediatricians always consider the possibility of sexual abuse, but most cases are simple problems of hygiene and childhood.

**Ptosis** (TOE-siss): A droopy eyelid. There are many causes, from injury to nerve disease, so such a droop needs to be checked.

**Purpura** (PUR-pur-a): Bruiselike spots that don't blanche when pressed; really, tiny bleeds into the skin. These are *always* a sign of a problem, but some causes are highly urgent and others a bit less so. It is safest to treat purpura as an urgent problem, as it may herald an overwhelming and potentially fatal infection.

**Put down:** Parentese to place a lulled baby in a cradle or crib or bed for nap or sleep.

**Pyelonephritis** (PIE-lo-nef-RY-tis): Inflammation of the kidneys due to infection.

**Reduce:** A medical verb that can mean "to fix or correct." You can reduce a hernia by poking the protruding part back into the abdomen; you can reduce a fracture by aligning the bones straightly; you can reduce an intussusception by unwrinkling the intestine. I guess you just "reduce" the problem.

**Reflux** (REE-flux): When a fluid goes in the wrong direction in the body. Gastroesophageal (GAS-troh-ee-*soph*-uh-JEEL) reflux is when stomach contents go back up the esophagus. Urinary reflux is when urine goes back up the ureters to the kidneys rather than out the urethra into the outside world.

**Renal** (REE-nal): Having to do with the kidneys.

**Resuscitation** (ree-*suss*-i-TAY-shun): Using techniques to get air into the lungs and make the blood circulate to restore breathing and heartbeat when they have stopped (or, in the case of the newborn, breathing never started). CPR stands for cardiopulmonary resuscitation.

**Retinoblastoma** (RET-in-oh-blast-OH-ma): A tumor of the eye, which, though very uncommon, is important to diagnose early. It attacks one in about eighteen thousand children. While it is present at birth, it usually isn't diagnosed until the baby is months old; the average age at diagnosis is two years of age. Parents who think that the baby's eye "doesn't look right" or that the pupil reflects white, not red, with a camera flash should have the baby examined. With early treatment, the prognosis for a normal life (but not normal vision in the affected eye) is excellent.

**Retractions:** Sucking in of extra muscles to help breathing when it's difficult. These muscles include abdominal muscles, muscles between the ribs, and muscles over the collar bone. Always a sign of trouble breathing.

**Reye syndrome** (rye, not ray): A rare disease of the liver and brain that, when it does occur, tends to happen after influenza or chicken pox, especially if a child has taken real aspirin rather than acetaminophen for fever and pain. Sleepiness, vomiting, headache, and bizarre behavior are key symptoms. Urgent care is needed.

**Rheumatic fever** (rew-MAT-ick): An uncommon disease that causes swollen painful joints and often disease of a heart valve. It occurs weeks after a strep throat, probably as a complication of the way the immune system handles the strep bacteria. The best way to prevent rheumatic fever is to get rid of strep in the body fairly soon after infection. It's not the same thing as scarlet fever at all.

**RhoGAM** (ROH-gam): A special kind of immune globulin given to pregnant or just-delivered women who are Rh negative. It prevents the mother's blood from making antibodies to red blood cells from an Rh positive fetus. Such red blood cells can leak into the mother's circulation during late pregnancy and labor. If she makes antibodies, those can cross through the placenta and destroy the red cells of the NEXT fetus, giving jaundice and anemia. The shot is important, and, like all gamma globulin shots, it does hurt.

**Roseola** (*rose*-ee-OH-la): A virus-caused illness rash that just about every child gets sooner or later, but most often between six months and three years of age.

**Rotavirus** (RO-ta-vi-rus): A virus that causes diarrhea in infants, especially in the winter, especially at daycares.

**RSV,** or **Respiratory syncytial virus** (Just say "Ar Ess Vee"): A virus that causes most cases of bronchiolitis.

**Rubella** (roo-BELL-a): Not rubeola. Rubella is three-day, or German, measles, a virus mild in most people but potentially devastating to the unborn whose mother gets a case. Immunization via the MMR vaccine has kept this tragedy under control over the last twenty-five years.

**Rubeola** (*roo*-bee-OH-la): True measles. A serious disease, blotchy rash, red watery eyes, runny nose, and cough; bad enough, but complications such as pneumonia and encephalitis can occur as well. There's not a cure, but there is a prevention: the MMR vaccine.

**Scabies** (SKAY-beez): A common rash, very itchy—especially at night—caused by tiny mites that burrow under the skin, often hard to diagnose because it can masquerade as nearly a dozen other conditions.

**Scope:** A catch-all verb ("to scope him") used to mean looking into any body orifice, usually with a fiber-optic instrument.

**Scrotum** (SKRO-tum): The sac that holds the testicles.

**Seborrhea** (*seb*-or-REE-ah): Too much oil on scalp and face. Seborrheic dermatitis is the red scaly rash that seborrhea helps cause. It's most common in very young babies.

**Seizure** (SEE-zhur): The medically preferred name for a convulsion or fit. A seizure is the behavior, often scary, that occurs when there is an unregulated electrical discharge from the brain. Seizures in young children most often occur when a rapidly climbing fever triggers one, due to a temporary, innocent immaturity of the brain. A seizure in and of itself is very scary but not dangerous unless it puts the child in a dangerous situation (e.g., while swimming) or it goes on for a very, very, very long time. However, the underlying cause of the seizure can be dangerous: for instance, seizures from poisoning, brain swelling, or central nervous system infection.

**Sepsis** (SEP-sis): An overwhelming and serious infection, usually bacterial, in the blood.

**Septal defect** (SEP-tal DEE-fekt): A hole in the wall, or septum, that divides the right side of the heart from the left. A hole in the septum between the two lower chambers, the ventricles, is the most common congenital heart problem. (See VSD, below.)

**Sickle cell anemia, disease, and trait:** A hereditary disorder of red blood cells. This is a complex of conditions affecting many African-American people, though most go through life with no symptoms because they carry the trait but do not have the disease. If two people with the trait have a child, however, there is a one in four chance that child will have the disease, which is a severe, chronic, and life-threatening one. One in 650 African-Americans have this disease. In sickle cell disease or anemia, the hemoglobin in the red blood cells is abnormal, so that when they are deprived of oxygen the cells curl up into a "sickle" shape. When they do so, they clog blood vessels, causing severe pain in belly, limbs, back, and elsewhere. Moreover, the spleen (which gets rid of old red blood cells) is damaged by all these abnormal cells, and can't produce normal immunity, so children with Sickle Cell Disease can get overwhelming infections. Treatment is very helpful, but there is no cure as yet; perhaps gene therapy will provide one eventually. In sickle cell trait, which affects about 8 percent of African-Americans, there is enough normal hemoglobin that the sickling cells don't cause problems most of the time, except in conditions when oxygen pressure is low, at very high altitudes, for instance. Routine newborn screening for sickle cell disease is done in several states. It allows treatment for disease to start early; in fact, babies with the disease are started on daily penicillin at only three months of age. Screening also allows carriers of the trait to know about their status before planning a second child (or a trip to high altitudes). While sickle cell is found mostly in black-skinned people, it is also found in people of Mediterranean, Middle Eastern, and Indian descent, and, indeed, all racial groups.

**SIDS** (sidz), **an acronym for Sudden Infant Death Syndrome, or Crib Death:** The sudden and unexplained death of a previously healthy baby under a year of age. This is uncommon, between two and eight per one thousand live births. There are some risk factors that can be avoided. Some of these are pregnancy risk factors, such as smoking and drug use, and inadequate prenatal care. Other risk factors can be prevented after the baby is born: secondhand smoke is the biggest. Bundling the baby too warmly is another. Recent evidence now also points to sleeping position as a risk factor. Contrary to common sense, the position of highest risk is having the baby sleep on her belly. Research shows this position can actually block the airway. Babies should be put to sleep on their sides or their backs until they can move well enough to assume their own position of choice. This does not apply to some premature babies and babies with reflux of stomach contents, who may need to sleep on the belly. Well-controlled studies have shown that SIDS is *not* related to the DPT vaccine. **Near-SIDS** refers to a baby who has been found in the crib lifeless but who has been resuscitated. A baby with such an experience should be hospitalized briefly to try to determine what occurred and whether precautions need to be taken to prevent a recurrence. Such an event is sometimes called an **ALTE** (ALL-tuh), or apparent life-threatening event.

**Sphincter** (SFINK-ter): Any circular muscle that squeezes shut, as in anal sphincter.

**Spleen:** The organ under the left ribs, which gets rid of old red blood cells and produces important immunity. Since it is blood-filled, it can rupture when injured. And it tends to become swollen with some viral diseases, like mono.

**Staph** (staff): A particular family of bacteria that tends to cause and complicate skin infections, though any site in the body can be infected.

**Steroids** (STARE-oids): Hormones put out by the adrenal gland, or medications that simulate them, such as prednisone and Decadron. They reduce inflammation and swelling. These are strong medications and can have important side effects.

**Stomatitis** (*sto*-ma-TIE-tis): Mouth infection. (See Herpes, above.)

**Strabismus** (stra-BIZ-mus): A wandering or lazy eye (not eyelid) that doesn't focus with the other. Usually, the vision in this eye is not as good as the one that does focus, so it lets the "good eye" do all the work. (See Amblyopia, above.)

**Streptococcus** (say "Strep"): A family of bacteria that causes many childhood infections, most notably strep throat and strep vaginitis.

**Subluxed elbow** (nursemaids' elbow; subluxed radial head): A very common incident in young children in which a yank on the hand or arm pulls the elbow out of whack. The elbow bone gets pulled out of its cuff and traps a ligament, or tissue band. See the section on *Arms* in *First Aid*, Chapter 13 of Part II.

**Sulfa drugs:** Medicines used for urinary tract and middle ear infections, mostly, but also crucial for the "AIDS pneumonia," pneumocystis carinii. Includes such familiar names as gantrisin, bactrim, septra. Allergic reactions, and reactions from sun sensitivity are not uncommon, and parents need to watch out for rashes.

**Suppository** (suh-PAHZ-i-*tor*-y): Medication placed into the rectum (sometimes the vagina, but not in our age group). These are given either to produce a poop from the constipated, or to administer medication when it can't be given (or won't be taken) by mouth. Absorption is tricky, which is why all pediatric medicines can't come in suppository form.

**Sweat test:** A laboratory test for the disease cystic fibrosis. It involves stimulating and then collecting and analyzing a little sweat, looking for too much chloride. Sweat tests are done routinely on babies with wheezing, problems growing and gaining weight, diarrhea and foul stools, and a number of other symptoms. Most children who have sweat tests performed do not have cystic fibrosis. It's important to do the test because it's the only way to diagnose this condition, which has many symptoms and signs and a range of severity.

**Swimmer's ear:** See Otitis, External, above.

**Symptom:** An expression of a disease, such as pain, that the patient complains about. Subtly different from "sign" of a disease. A sign of a disease is something somebody else can assess: elevated temperature, increased rate of breathing, blue or pale skin.

**Syncope** (SIN-co-pee): Fainting or passing out, losing consciousness without an injury or after a seizure.

**Syndrome:** A collection of signs and symptoms that all go together under one label, even though there may be several different underlying causes.

**Tachycardia** (*tak*-ee-CAR-dee-ah): Rapid heartbeat from any cause.

**Testicles:** The two "balls" in the scrotum that manufacture sperm.

**Thalassemia** (*thal*-uh-SEEM-ee-yuh): An inherited group of conditions that can cause anemia. The anemia is due to abnormal hemoglobin, the molecule that makes a red blood cell a red blood cell. Many Americans carry a gene for one of the thalassemias, and if two people with such a gene have children, each child has a one in four chance of having a thalassemia disease. Such children are anemic and grow poorly. They need blood transfusions frequently, and care from a pediatric hematologist. Most people with the trait, however, have no symptoms or only mild anemia, which often gets treated, inappropriately, with iron. A child with ancestors from Africa, the Mediterranean, the Mideast, or Southeast Asia who has anemia (a "low hemoglobin" or "low hematocrit") may well be carrying the thalassemia trait. Some states screen for thalassemia in the Newborn Screening Test.

**Thyroid:** The gland in the neck that produces thyroid hormone, crucial for growth, intelligence, and sexual maturation.

**Tic:** A nervous habit, like blinking or throat clearing.

**Tick:** An insect that bites animals and people and sometimes carries disease (Lyme disease, Rocky Mountain spotted fever, tick paralysis).

**Tonsillitis:** Any infection or swelling of the tonsils, whether due to bacterial or viral causes.

**Tonsils:** Lumps of lymph tissue found on each side of the throat in young children. They are usually smooth and round and don't get in the way of swallowing or breathing, but can become enlarged.

**Torticollis** (*tor*-ti-COLL-is): Abnormal position of the head and neck, with the head tilted and/or turned to one side. In a young baby, it may indicate a snug fit in the womb, so other problems—especially hip dislocations—should be looked for. If it comes on in older children, there are so many causes that pediatric advice is necessary.

**Toxic:** Describing the appearance of a child who is very sick, possibly with sepsis (infection of the bloodstream). Such a child typically is whimpering or very irritable or limp, with poor color, unable to pay attention or make much eye contact—usually with, but regardless of, fever.

**Tympanic membrane** (tim-PAN-ick): The eardrum.

**Tympanogram:** A reading of the eardrum's ability to vibrate, produced by a machine called a tympanometer. This doesn't hurt, and is fast. It works by reflected sound, and doesn't deliver any electricity or rays into the head or ear.

**Ultrasound:** An imaging technique that can look at soft tissues of the body without delivering radiation. Most of the time a child doesn't even need sedation: it doesn't hurt, and it doesn't matter if the child wiggles a little. Many parts of the body can be viewed by ultrasound: brains and spinal cords in very young infants, and for everybody, kidneys and abdomens. It's often abbreviated UTZ (say "You Tee Zee").

**Umbilical** (um-BILL-i-cal): Having to do with the umbilical cord or belly button.

**Umbilical granuloma** (*gran*-u-LOME-ah): Grayish white lump of tissue where the cord came off, reflecting delayed healing of the cord. Usually needs to be treated with the chemical silver nitrate by the pediatrician.

**Ureters** (YOUR-eh-tours): The pair of tubes, each connecting a kidney to the bladder, that transport urine.

**Urethra** (you-REETH-rah): The tube through which urine exits from the bladder to the outside world.

**Urinary tract infection,** or **UTI** (say "You Tee Eye"): A term referring to infection of the bladder and/or kidneys, often used when the physician isn't sure if the kidneys are involved as well as the bladder.

**Uterus** (YOU-ter-us): The hollow muscular organ also known as the womb—a very small organ in girls under five.

**Uvula** (YOUVE-you-la): The little thing that dangles down from the rear of the roof of the mouth. Usually it's single; sometimes it's split like an upside-down Y. It assists with swallowing, and sometimes gets infected, often with strep.

**Vagina** (vuh-JINE-ah): The pocketlike structure leading to the uterus and opening at the hymen.

**Vaginitis** (*vaj*-in-EYE-tis): Irritation of the vagina with or without infection.

**Varicella** (*vare*-ih-SELL-uh): Chicken pox, the disease caused by herpes zoster.

**Vascular** (VAS-kew-lur): Anything having to do with the blood vessels.

**Vasculitis** (*vas*-kew-LIE-tis): An inflammation of the blood vessels.

**VCUG,** or **voiding cystourethrogram** (Just say "Vee See You Jee"): A study often performed on a baby or young child who has had a urinary tract infection, or whose urinary stream seems weak or forced. The study shows if reflux is taking place. Reflux occurs when urine goes back up the ureters into the kidneys rather than out the urethra into the outside world. A catheter is placed through the urethra into the bladder, dye is injected, and then X rays are taken as the child urinates. Sometimes called RUG (say "Are You Jee") for "retrograde ureterogram."

**Ventricles** (VEN-tri-calz): The two lower pumping chambers of the heart. The left one pumps red blood to the body. The right one pumps blue blood to the lungs.

**Ventricular septal defect,** or **VSD** (Just say "Vee Ess Dee"): A hole in the wall between the two ventricles. This is the most common congenital heart abnormality, a fre-quent cause of a heart murmur in the first few weeks or months. It is never heard at birth because the heart is still adjusting to pumping blood to the lungs. Many holes close spontaneously; some never do, but are so small they don't cause problems; some must be closed surgically.

**Vernix** (VER-nix): The creamy-cheesy material that coats fetuses in the womb. Full-term and postterm babies don't have much; preemies can be coated.

**Virus:** An exceedingly tiny cause of infectious disease, from colds to measles and mumps to AIDS. A bit of DNA (genetic material) wrapped in a coat of protein. Viruses take over the body's cells in order to replicate, so it is very hard to get rid of them without damaging the body itself. Antibiotics don't work on viruses. See the essay in Part III.

**Vital signs:** Temperature, pulse, rate of breathing, and blood pressure.

**Vitiligo** (vi-tuh-LYE-goh): Pale spots on the skin from lack of pigment. Vitiligo can occur after a skin infection, or from hormone or other problems, but usually is mysterious and innocent, though a cosmetic bother.

**Wheeze:** The sound made when the tubes of the lung clamp down. It's a high-pitched whistle made while breathing *out*.

**Wheezing** (WEEZ-ing, not Weezelling): Repeated wheezes while breathing, usually with effort getting the air out. If the cause is not found to be a specific infection or anatomic problem, and the wheezing occurs more than once, the child is said to have asthma.

**Whipperdill:** Parentese for a flamboyant, spectacular, unusual, but nonserious traumatic event, usually a fall.

**Whitlow:** See Herpetic whitlow under Herpes.

**Whooping cough:** A syndrome in which a baby or child *whoops*. This means the baby "loses his breath" through forced coughing, and to catch it inhales air with a mighty, effortful "whoop." It is often caused by pertussis (see above).

**Wilms' tumor:** The second most common malignant abdominal tumor of childhood, after neuroblastoma (see above), found in children between the ages of two and five years. This is a tumor of the kidney, and usually is found as a mass in the abdomen, though it sometimes presents as a severe stomachache. Even though it is the second most common tumor, it is still rare, affecting only about eight children out of a million. If found and treated relatively early, the prognosis is usually good.

**Yeast:** A microscopic organism that is ubiquitous. In children under the age of five, it causes thrush in the mouth (see Common Illnesses with [More or Less] Unfamiliar Names in Part II of this book) and diaper rash. It doesn't cause an infection in the vagina until the hormones of puberty start to be secreted. Children on antibiotics, or who have consistently wet bottoms (wet bathing suits, inadequate diapering) or irritated skin that stays moist, all tend to get more yeast. Children with an impaired immune system may get very severe, stubborn yeast. There is no evidence that normal children, without an immune deficiency and without a catheter into a blood vessel, ever get "systemic" yeast.

**Zovirax:** (ZOH-veer-ax): See acyclovir.

# Index

Kerion, 513
Ketosis, 516, 520
Kidnapping, 340, 351–52, 393, 401
Kidneys, 57, 128, 430, 506–8
Kindergarten, 364, 395–97, 403–4
Kling bandages, 198
KY jelly, 51

Labia, 502
    adhesions, 175, 202, 503
Lactation counselors, 9, 47, 48
Lactose intolerance, 11
La Leche league, 9
Language skills
    4 to 6 months, 131, 133, 136, 150
    6 to 9 months, 160–61
    9 months to 1 year, 99, 183, 185, 190, 200, 208–9
    1 year to 18 months, 126, 127, 225, 233, 241, 248
    18 months to 2 years, 181, 213, 256–57, 260–61
    2 to 3 years, 279–80, 289–90, 302
    3 to 4 years, 318–20, 330–31, 342
    4 to 5 years, 364, 377–78
    cooing and, 76, 595
    ear infection and, 200, 241, 469, 580, 584–85, 588
    evaluation of, 598
    newborn hearing problem and, 18
    pacifier use and, 32, 193, 209
    stuttering and, 279, 330–32, 378
    teachers and, 302
    vocalization, 107, 133, 595

Laryngitis, 388, 479
Laryngotracheobronchitis. *See* Croup
Lazy eye, 232, 329, 464, 466
Lead, 144, 183, 246
    in water, 49, 84, 98
Lead poisoning, 177, 204, 243, 501, 599
Leashes, baby, 198, 217, 253
Legs, 514
    aches, 386
    bowed, 39, 92, 514, 546
    growing pains, 386, 516
    injuries to, 449–50, 515
    of newborns, 39, 92
    rickets, 43, 515
    toeing in, 233, 514
    *See also* Feet; Limping
Leukemia, 54, 243, 299–300, 389
Lice, 341, 466, 513
Lidocaine, 480
Limit-setting
    2 weeks to 2 months, 71
    2 to 4 months, 106–7, 114
    4 to 6 months, 134–35
    6 to 9 months, 155, 159–60, 179
    9 months to 1 year, 183, 186, 188–89, 209
    1 year to 18 months, 221, 227–32, 248
    18 months to 2 years, 256, 259–60
    2 to 3 years, 287–89, 311
    3 to 4 years, 323–29
    4 to 5 years, 369–77
    for adults, 31–32
    for toddler siblings, 126, 127
Limping, 300, 341, 449, 515
    toxic synovitis, 242, 299, 341, 515

Lipids, 550
Lips
    biting through, 243, 446
    blisters on, 36
    bluish, 61
    cuts to, 446
Lisping, 330
Listening skills, 375
Liver, 56, 58, 507, 508
Lockjaw, 454
Lomitil, 495
Lost children, 384–85
Lotrimin, 89
Lozenges, 297, 480, 490
Lungs, 430, 481, 576
Lying, 373
Lyme disease, 512
Lymph nodes, 522, 523

Mad cow disease, 11
Magical thinking, 318, 339, 368
    about sickness, 361, 386, 387
    toilet skills and, 304, 338
Managed care, 26–27
Manipulation, 327–28, 401
Manners, 275, 290, 355–56, 366–67, 374–77, 404
Mastitis, 46–47
Mastoiditis, 469, 470, 566, 579, 584
Masturbation, 502
    6 to 9 months, 160
    9 months to 1 year, 209
    1 year to 18 months, 238–39
    18 months to 2 years, 265
    2 to 3 years, 294
    3 to 4 years, 349
    4 to 5 years, 391, 394, 404
    parent's attitude and, 346
    sexual abuse/molestation and, 351

Newman, Jack, 3, 43
Nicotine, 72
Nightmares. *See* Bad dreams
Night terrors, 432–34
    6 to 9 months, 163–64
    1 year to 18 months, 242
    2 to 3 years, 285–86, 299
    3 to 4 years, 342
    4 to 5 years, 388
Nitrates, 49, 84, 141
"No" saying, 159–60
Nose, 474–76, 481
    foreign bodies in, 341,
        344–45, 442–43, 516
    head bonks and, 206, 438, 444
    injuries, 444–45
    of newborns, 36, 62, 91
    picking of, 343, 443, 475
    runny, 91, 475–76
    stuffy, 117, 474–75
Nosebleeds, 343, 388, 443–44,
    475
Nose drops/spray, 387, 474
    medicated, 117, 575
    saline, 54, 117, 126, 475
Nursemaid's elbow, 242, 296,
    299, 342, 388, 447–49
Nursery monitors, 54
Nursing. *See* Breastfeeding
Nursing-bottle mouth, 478–79
Nursoy, 42
Nutramigen, 11, 12
Nuts, 168, 193, 340, 424
Nystatin, 91

Oatmeal baths, 531
Obesity, xiv-xv, 540–60
    family history, 109, 164, 167,
        544
    medical complications, xv, 546
    prevention, 547–48, 549

"Oedipal" stage, 312
Oppositional behavior, xiii-xiv,
    198, 604–10
    1 year to 18 months, 221
    18 months to 2 years, 259–60
    2 to 3 years, 275, 277–78,
        287–89
    3 to 4 years, 317–18, 324–28,
        356
    4 to 5 years, 366–67
    handling strategies, 606–10
    prevention, 605–6
Oral medicine, 387
Orchitis, 568

Pacifiers, 32, 115, 131, 150
    giving up, 193, 209
Pads, safety, 383
Pain
    during circumcision, 20, 21
    as illness sign, 116
    stomachaches and, 492
    vomiting and, 122
Parathyroid glands, 501
Paregoric, 495
Parents
    authoritarian, 228, 605
    authoritative, xiv, 369–77,
        605
    bonding with baby, 30–31,
        134
    of colicky babies, 75
    extended separation from,
        96–97
    illness responses, 409–13
    kidnapping by, 352, 393, 401
    mistake-making by, 373–74
    permissive, 605
    sleeping with, 54, 93, 153,
        162
    time alone together, 125

of 2–year-olds, 278–83
    *See also* Divorce effects
Pavlik harness, 39, 515
PDD, 596, 597
PDDNOS, 597
Peanut butter, 168
Peanuts, 168, 193, 424
Pedialyte, 54, 64, 115, 118, 495,
    497
Pediatricians
    calls to, 411–13
    communicating with, 87, 417,
        496
    divorced parents and, 214
    postpartum follow-up visit, 28
    prenatal visits, 3–4, 24–26
    selecting after move, 100
    *See also* Well-child visits
Peeing, 505–9
    in bath water, 112
    bed-wetting, 338, 506, 609
    birth to 2 weeks, 28, 40, 46,
        51, 52
    dehydration and, 427–28, 507
    diarrhea and, 118
    frequent, 506
    infrequent, 506–7
    names for, 308
    painful, 506
    pants-wetting, 509
    stomachaches and, 492
    in various receptacles, 306
    *See also* Toilet skills
Penicillin, 522
Penis, 502–5
    cancer of, 20, 21
    circumcised, 4, 20–22, 38,
        51–52, 54, 64, 85
    curved (Chordee), 37
    foreskin care, 51, 294, 334,
        381
    hypospadias, 38, 503–4

Prenatal concerns. *See* Pregnancy
     concerns
Preschool
     2 to 3 years, 283–84, 301
     3 to 4 years, 315, 334–35
     4 to 5 years, 397
Prescription drugs, 27
Pretend play. *See* Fantasy play
Prevnar vaccine, 101, 488
Primary care providers, 26
Procrastination, 318, 325–26,
     366, 609
Prolactin, 9
Prosobee, 42, 118
Protein, 168
     in breast milk, 8
     in cow's milk, 11, 12, 166, 236
Prune juice, 119, 195, 499
Ptosis, 35
Puberty, 291, 537
     obesity and, xv, 292, 542,
        546, 549
Pubic hair, 504
Puncture wounds, 449, 450
Punishment. *See* Discipline;
        Limit-setting; Spanking
Purpura, 510
Pyelonephritis, 57
Pyloric stenosis, 63, 92–93


Rabies, 453–54, 455, 567
Raisins, tooth decay from,
     478–79
Rashes, 509–11, 524–25
     antibiotic use and, 565
     baby acne, 85, 88–89
     blood clotting and, 443
     diaper, 89, 139, 142, 498, 510
     facial, 139, 142
     Fifth disease, 201, 341, 513,
        524–25

as food reaction, 139, 142
     impetigo, 343
     metal-contact, 112
     newborns and, 34, 61
     roseola, 174, 202, 513, 525
     scarlet fever, 513, 521–22,
        529
     with streaks, 511
     urgent problems, 510–11
RAST (blood test), 574, 578
Reactive airway disorder. *See*
        Asthma
Reading, 336–37, 355, 364
     at three years, 337
     *See also* Books
Reciprocity, 94, 123
Rectal thermometer, 54, 115,
        126
Reglan, 48
Relatives, death of, 362, 403
Reminder bottles. *See*
        Breastfeeding
Resentment, 369
Respiratory illness, 54
     2 to 3 years, 298
     3 to 4 years, 342
     4 to 5 years, 385
     headaches and, 460
     obesity and, 546
     stiff neck and, 439
     *See also* Breathing difficulty
Respiratory syncytial virus, 98,
        170
     bronchiolitis and, 98, 121,
        174, 485–87, 526
     prevention of, 487
Retina, 466
     detached, 390, 440
Retinoblastoma, 243
Retrograde urethrogram
        (VCUG), 508
Reye syndrome, 431, 495, 532

Rh blood factor, 17, 58
Rheumatic fever, 480, 521, 522,
        566
Rhino kicker. *See* Pavlik harness
Rhinoprobe, 573
RhoGAM injections, 17, 58
Ribavirin, 486
Ricelyte, 118, 497
Rickets, 43, 515
Ringworm, 341, 513
Ritalin, 603
Role-playing, 310
Rolling over
     2 weeks to 2 months, 76, 77
     2 to 4 months, 108
     4 to 6 months, 131, 136, 137
     6 to 9 months, 161
Roseola, 173–74, 202, 387, 513,
        525
Rotavirus diarrhea, 98, 121, 170,
        172, 174, 387, 526–27
RSV-IGIV, 487
Rubella. *See* German measles
Rubeola. *See* Measles
Rudeness, 371


Safety
     birth to 2 weeks, 54
     2 weeks to 2 months, 86–87
     2 to 4 months, 114, 115
     4 to 6 months, 131, 143–45
     6 to 9 months, 155, 159, 172,
        180
     9 months to 1 year, 189, 197,
        198
     1 year to 18 months, 241
     18 months to 2 years, 268–69
     2 to 3 years, 296–97
     3 to 4 years, 339–40
     4 to 5 years, 383–85, 390
     Baby-proofing list, 115

Trembling, 40
Triceram, 577
Tricycles, 295, 296, 340
Trust development, 31, 94, 123
Tuberculosis, 215, 489
Tweezers, 199
Twitching/jerkiness, 120, 423, 466–67. *See also* Fits (seizures, convulsions)
Tylenol. *See* Acetaminophen
Tympanograms, 583

*Ultimate Breastfeeding Book of Answers, The* (Newman and Pitman), 3, 43
Umbilical cord, 36–37, 51, 68
  care of, 20, 37, 54
  infection of, 37, 57, 61, 63
  sleep positions and, 28, 31
Umbilical granuloma, 37
Umbilical hernias, 37, 92
Unconsciousness, 206
Undescended testicles, 38, 176, 203, 504
Urinalysis, 20–21, 176, 507–8
Urinary tract infections, 176, 203, 381, 389
  antibiotics for, 508, 562, 565
  delayed urination and, 338
  fever and, 147, 176, 202
  frequent urination and, 506
  labial adhesions and, 175, 503
  prevention, 508–9
  sexual abuse/molestation and, 351, 381
  smelly urine and, 507
  testing for, 508–9
  uncircumcised boys and, 20–21, 176n

Urine
  abnormal, 507–9
  infection of, 109, 506, 508
  *See also* Peeing
Utensil use
  chopsticks, 290, 332, 355, 375, 378
  feeding self, 155, 193, 209, 217, 265
  forks, 217, 265, 290, 332, 355, 375
  spoons, 110, 141–42
UTIs. *See* Urinary tract infections
Uvula, 477, 479, 522

Vaccination. *See* Immunization
Vagina, 502–5
  bleeding from, 38
  foreign bodies in, 341, 344, 505, 516
  inflamed, 343, 351, 388
  irritation of, 195, 294, 503, 504–5, 507, 509
  labial adhesions, 175, 202, 503
  names for, 307
  yeast infections, 505
Vaginitis, 294, 334, 341
Varicella. *See* Chicken pox
VCUG, 508
Vegetables, 167–68, 237, 293, 334, 380
  baby foods, 141, 142
Vegetarian diets, 294
Video games, 383–84
Videos, 303, 315, 354, 369, 400
Viruses, 32, 124, 145–46, 170, 172
  2 to 3 years and, 298
  antibiotics and, 561–63, 566

breastfeeding transmission of, 9, 10, 81
common, 174, 201–2, 242, 299
herpes. *See* Herpes
rabies, 453–54, 455, 567
rarer, 174, 202, 242, 299
roseola, 173–74, 202, 525
stomachaches and, 491, 492–93
toxic synovitis, 242, 299, 341, 515
*See also* Colds; Infectious diseases
Vision
  birth to 2 weeks, 34–35
  2 weeks to 2 months, 76, 91
  2 to 4 months, 107, 129
  4 to 6 months, 135
  6 to 9 months, 160
  9 months to 1 year, 189–90, 191
  1 year to 18 months, 232
  18 months to 2 years, 261
  2 to 3 years, 289
  3 to 4 years, 329
  4 to 5 years, 377
  color blindness, 377
  mononucleosis and, 523
  nearsightedness, 329
  prekindergarten testing, 403–4
Visitors, 32, 45
Vitamin A, 48, 237, 529
Vitamin B complex, 168, 237, 294
Vitamin C, 48, 237, 592
Vitamin D, 103, 236, 237, 294
  hyperparathyroid and, 501
  rickets and, 43, 515
  supplements, 43
Vitamin E, 237